Springer Series in Biophysics 16

Series editor: Boris Martinac

For further volumes:
http://www.springer.com/series/835

Preface

During the last decade, life sciences have experienced a major shift from *analytical* to *integrative* approaches that can be globally defined as *Systems Biology*. In this overall new landscape, the importance of complex systems and whole-system approaches has become paramount.

The volume that you have in your hands represents the collective effort of a group of dedicated and accomplished researchers in the nascent field of systems biology. Although a rather recent *renaissant* research endeavor, systems biology has a long ancestry that goes back as far as Newton, Leibniz, Mendel, Poincaré, Bernard, Wiener, and von Bertalanffy, amongst many others. The roots are not only strong but diverse; they encompass mathematics, computer science, physiology, genetics, engineering, and biology.

Cells, organisms, and ecosystems consist of a large number of usually nonlinearly interacting parts that exhibit complex behavior while exchanging matter and energy with their environment. *Systems biology* represents a holistic approach for analyses of structural and functional interactions between components rather than individual elements. Vast data gathering from -omics technologies (i.e., gen-, transcript-, prote-, and metabol-omics), together with the growing capability of generating computational models, have allowed for a massive integration and interpretation of new information. Noninvasive imaging technologies used together with intracellular probes are increasing our ability to monitor the spatiotemporal dynamics of cellular, metabolic, and signaling processes in living systems. As such, systems biology can integrate multiple spatial and temporal scales and has the potential to allow new insights into fundamental mechanisms involved in, e.g., human health and disease.

Cellular mass–energy transformations comprise networks of metabolic and transport processes represented by the metabolome and fluxome, which account for the complete set of metabolites and fluxes in a cell. The information-carrying networks include the genome, transcriptome, and proteome that represent the whole set of genes, transcripts, and proteins, respectively, present in a cell. Signaling networks mediate between the genome–transcriptome–proteome and metabolome–fluxome and, as such, play the crucial role of influencing the unfolding of cell function in space and time.

Network is a central concept to systems biology. The study of network properties, and how these control the behavior of cells and organisms, constitutes a main focus of systems biology. A major unsolved biological problem is to understand how a cell works and what goes wrong in pathology. However, in order to achieve this goal we need to unravel how the mass–energy and information networks of the cell interact with each other while being modulated (activated or repressed) by signaling networks to produce a certain phenotype or (patho)physiological response. This novel perspective constitutes a distinctive feature of this volume, thus allowing it to differ from previously published books on systems biology.

If *information* is organized data (and we have a plethora), *knowledge* organized information, and *wisdom* organized knowledge, then systems biology is at the interphase between information and knowledge. We are learning to think and act systemically, to organize catalogs of data into meaningful information, and to distil knowledge from that learning process. To what kind of new wisdom is this emerging knowledge leading us? Although we are far from being there yet, a few lessons have been learned along the way.

Certainly, life is more complex and far-reaching than our genes, at least by the numbers. This is one of the first lessons gleaned from sequencing the genome of species with diverse lineage and evolutionary paths: the number of genes and core proteomes does not correlate with their apparent complexity. For example, the basic proteome of the human genome is not much larger than that of the fly and the worm, but human complexity is. Therefore, where does complexity lie? If diversity and number of functions cannot be directly connected to genes, then we have at least two possibilities. One is that genes are subjected to some combinatorial process that elevates exponentially their numbers (e.g., by alternative splicing), coding diversity, and functional outcomes. Another is the spatiotemporal unfolding of gene expression that, in interaction with the environment, modifies and is modified in a combinatorial manner to give rise to multiple functions. The unfolding in space and time of gene expression would proceed as presciently suggested by the philosopher-scientist Evelyn Fox Keller who wrote, right at the turn of this century, these words referring to developmental genetics:

> . . .we could describe the fertilized egg as a massively parallel and multilayered processor in which both programs (or networks) and data are distributed throughout the cell. The roles of data and program here are relative, for what counts as data for one program is often the output of a second program, and the output of the first is data for yet another program, or even for the very program that provided its own initial data. For some developmental stages, the DNA might be seen as encoding programs or switches that process the data provided by gradients of transcription activators. Or, alternatively, one might say that DNA sequences provide data for the machinery of transcription activation (some of which is acquired directly from the cytoplasm of the egg). In later developmental stages, the products of transcription serve as data for splicing machines, translation machines, and so on. In turn, the output from these processes make up the very machinery or programs needed to process the data in the first place.

More than a decade later we could translate these ideas into the more precise concept about iteratively interacting networks of mass–energy, information, and signaling, which is precisely the subject of this book. A basic principle of living systems is worth noting at this point: unicellular or multicellular organisms make themselves. This essential defining property of living systems, in general, demands a circular causality in which these different networks are both input and output data, i.e., they provide metabolite precursors, second messengers, and transcriptional factors, and they are supplied with substrates, effectors, and signals—as suggested by Fox Keller's quotation. In these circular loops lies the self-determination of the living, and from their nonlinear dynamics involving feed-back and feed-forward autocatalysis and other interactions, with their potential for self-organization and emergent novelties, results the diversity and distinctiveness of life. According to this perspective then, we should probably look much more into the dynamics of how these different networks evolve and interact in time and space in order to find the unique complexity of yeast, mice, flies, worm, or humans.

The book comprises 13 chapters: the first two introductory and the remaining ones organized in four blocks devoted to the systems biology of signaling networks, cellular structures and fluxes, organ function, and microorganisms.

Chapter 1 explores the historical roots of the twenty-first century approach to systems biology tracing from its origins in dynamics and the invention of differential calculus, physiology, self-organized systems, biochemistry, bioenergetics, and molecular biology to the currently accepted networks approach. Chapter 2 gives an overview of the three types of networks involved in the interactive unfolding of the spatiotemporal organization of living systems: mass–energy, information, and signaling. Chapter 3 describes a quantitative approach to signaling from the perspective of metabolic control analysis. Chapter 4 addresses the novel regulatory features bestowed by microRNAs to the mass–energy transducing networks. Chapter 5 analyzes (from a combined experimental–computational approach) the energetic and redox behavior of mitochondrial networks, along with the signaling role of reactive oxygen species. Chapter 6 highlights the role of adenylate kinase in metabolic AMP-dependent signaling involved in cellular sensing of energetic status and the response to stress. Chapters 7–9 address from different viewpoints the systems biology of the organization in space and time of cellular macromolecular structures and its impact on fluxes through mass–energy networks. Chapters 10 and 11 describe systems organization across and between different temporal and spatial scales from the molecular to the organ levels, namely, as applied to the heart. Chapters 12 and 13 approach the systems biology of network organization from two different angles; in the case of yeast the overall temporal organization of mass–energy, information, and signaling networks exhibited by this unicellular eukaryote in self-synchronized chemostat cultures is presented and analyzed, as the sole model example of in vivo deconvolution of the time structure of a living system at present available, whereas Chap. 13 reviews systems biology approaches as applied to the engineering of mass–energy transforming networks.

Miguel A. Aon
Valdur Saks
Uwe Schlattner

Contents

Part I
Introduction

Chapter 1
From Physiology, Genomes, Systems, and Self-Organization to Systems Biology: The Historical Roots of a Twenty-First Century Approach to Complexity

M.A. Aon, D. Lloyd, and V. Saks

Abstract Systems Biology represents a new paradigm aiming at a whole organism-level understanding of biological phenomena, emphasizing interconnections and functional interrelationships rather than component parts. Historically, the roots of Systems Biology are multiple and of a diverse nature, comprising theoretical and conceptual developments, mathematical and modeling tools, and comprehensive analytical methodologies aimed at listing molecular components.

As a systemic approach, modern Systems Biology is deeply rooted in Integrative Physiology from which it inherits two big foundational principles: (1) a non-reductionist, integrative, view and (2) the capability of defining the context within which genes and their mutations will find meaning.

1.1 From Integrative Physiology to Systems Biology

> Yet, biological questions do not end in the gene at all: they start there. (Ball 2004)

> (. . .) physiology, or whatever we wish to call that part of the science of the logic of life that deals with bodily function and mechanism, will not only continue to exist as an identifiable body of knowledge: it will be indispensable to the proper interpretation of molecular biology itself. (Noble and Boyd 1993)

M.A. Aon (✉)
School of Medicine, Johns Hopkins University, Baltimore, MD 212051, USA
e-mail: maon1@jhmi.edu

D. Lloyd
School of Biosciences, Cardiff University, Main Building, Museum Avenue, Cathays Park, Cardiff CF 10 3AT, Wales, UK

V. Saks
INSERM U1055 Laboratory of Fundamental and Applied Bioenergetics, Joseph Fourier University, Grenoble, France

Laboratory of Bioenergetics, National Institute of Chemical Physics and Biophysics, Tallinn, Estonia

M.A. Aon et al. (eds.), *Systems Biology of Metabolic and Signaling Networks*, Springer Series in Biophysics 16, DOI 10.1007/978-3-642-38505-6_1,

The French physiologist Claude Bernard in his classical: "L'introduction a l'étude de la médicine expérimentale" (1865) stated that the control of the environment in which molecules function is at least as important as the identification of the organic molecules themselves, if not more. In an incisive essay, published almost 20 years ago, Noble and Boyd (1993) put forward the following three aims with which Physiology should be concerned, beyond merely determining the mechanisms of living systems: (1) integrative questions of order and control; (2) self-organization, in order to link how such order may have emerged; and (3) make this challenge exciting and possible. This is precisely what has happened, and in the meantime the emergence of System Biology emphasizes interconnections and relationships rather than component parts.

To describe a biological system we need to know the structure, the pattern of organization, and the function (Capra 1996; Kitano 2002a). The first refers to a catalog of individual components (e.g., proteins, genes, transcriptional factors), the second insert as to how the components are wired or linked between them (e.g., topological relationships, feedbacks), and third how the ensemble works (e.g., functional interrelationships, fluxes, response to stimuli, growth, division).

The analytical phase of biology has led to a detailed picture of the biochemistry of living systems and produced wiring diagrams connecting chemical components and processes such as metabolic, signaling, and genetic regulatory pathways. Integration of those processes to understand properties arising from their interaction (e.g., robustness, resilience, adaptation) and ensuing dynamics from collective behavior have become a main focus of Systems Biology(Noble 2006; Saks et al. 2009).

Systems Biology started to emerge as a distinct field with the advent of high throughput, -omics technologies, i.e., gen-, transcript-, prote-, and metabol-omics. Massive data gathering from -omics technologies, together with the capability for generating computational models, have made possible the massive integration and interpretation of information. High throughput technologies combined with the growing facility for constructing mathematical models of complicated systems constitute the core of Systems Biology. As such, Systems Biology has the potential to allow us gaining insights into not only the fundamental nature of health and disease but also with their control and regulation.

1.2 Dynamics, the Invention of Calculus, and the Impossibility of Prediction

Newton is considered the inventor of the science of dynamics, and he shares with Leibniz the invention of differential calculus (Gleick 2003; Mitchell 2009). Born the year after Galileo—who had launched the scientific revolution—died, Newton introduced the laws of motion that laid out the foundations of dynamics. These laws—constant motion, inertial mass, and equal and opposite forces—that apply to objects on earth and in heavens as well, gave rise to the notion of a "clockwork universe." This led Laplace to assert that given Newton's laws and the current position and velocity of every particle in the universe, it would be possible to predict everything for all time.

Poincaré, one of the most influential figures in the development of the modern field of dynamical systems theory, described a sensitive dependence to initial conditions in dealing with the "three body problem"—the motion of a third planet orbiting in the gravitational field of two massive planets (Poincare 1892). This finding rendered prediction impossible from knowledge of the situation at an initial moment to determine the situation at a succeeding moment. Otherwise stated: "...even if we knew the laws of motion perfectly, two different sets of initial conditions...even if they differ in a minuscule way, can sometimes produce greatly different results in the subsequent motion of the system" (Mitchell 2009). The discovery of chaos in two metaphorical models applied to meteorology (Lorenz 1963) and population dynamics (May 1974), and two different mathematical approaches—differential continuous and difference discrete equations, respectively—[see (Gleick 1988) and (May 2001) for historical accounts] introduced the notion that irregular dynamic behavior can be produced from purely deterministic equations. Thus, the intrinsic dynamics of a system can produce chaotic behavior independently of external noise. The existence of chaotic behavior with its extreme sensitivity to initial conditions limits long-term predictability in the real world.

The power of using mathematical modeling based on differential calculus was shown in two papers published in 1952 by Turing and Hodgkin & Huxley (Hodgkin and Huxley 1952; Turing 1952). Turing employed a theoretical system of nonlinear differential equations representing reaction-diffusion of chemical species ("morphogens" because he was trying to simulate morphogenesis). With this system Turing attempted to simulate symmetry breaking, or the appearance of spatial structures, from an initially homogeneous situation. This class of "conceptual modeling" contrasts with the approach adopted by Hodgkin and Huxley (1952) in which they modeled electrical propagation from their own experimental data obtained in a giant nerve fiber to account for conduction and excitation in quantitative terms. The "mechanistic modeling" approach of Hodgkin and Huxley is an earlier predecessor of the experimental–computational synergy described in the present book. These two works had a long-lasting influence in the field of mathematical modeling applied to biological systems.

1.3 From Multiple Interacting Elements to Self-Organization

Systems Biology aims at "system-level understanding of biological systems" (Kitano 2002b). It represents an approach to unravel interrelations between components in "multi-scale dynamic complex systems formed by interacting macromolecules and metabolites, cells, organs, and organisms" (Vidal 2009).

One of the fundamental problems addressed by Systems Biology is about the relation between the whole and its component parts in a system. This problem, that pervades the history of "systems thinking" in biology (Haken 1978; Nicolis and Prigogine 1977; Von Bertalanffy 1950), begs the central question of how macroscopic behavior arises from the interaction between the elementary components of a system. It represents a connecting thread that different generations of scientists have formulated and attempted to solve in their own conceptual and methodological ways with the technologies available at the time (Junker 2008; Skyttner 2007; Yates 1987).

The notion that variation in any element affects all the others bringing about changes in the whole system is one of the foundations of systemic thinking. However, interactions in a biological system are directed and selective: this result in organization obeying certain spatial and temporal constraints. For example, in cellular systems molecular components exist either individually or as macromolecular associations or entire structures such as cytoskeleton, membranes, or organelles. Interactions are at different degrees of organization and can be visualized at *structural* or *morphological* levels assessed on molecular or macroscopic scales. However, biological interactions are not random and in organized systems they follow certain *topological* properties, i.e., more or less and preferentially connected to each other. For example, molecular–macromolecular functional interactions are ruled by thermodynamics and stereo-specificity. Finally, function in biological systems is a *dynamic process* resulting from interactions between structurally arranged components under defined topological configurations.

Dynamic organization is the realm where complexity manifests as a key trait of biological systems. As a matter of fact, biological systems are complex because they exhibit nontrivial emergent and self-organizing behaviors (Mitchell 2009). Emergent, self-organized behavior results in macroscopic structures that can be either permanent (e.g., cytoskeleton) or transient (e.g., Ca^{2+} waves), and have functional consequences. Indeed, macroscopically self-organized structures are dissipative ["dissipative structures": (Nicolis and Prigogine 1977)], i.e., they are maintained by a continuous flow of matter and energy. Dissipative structures emerge as complexity increases from cells to organisms and ecosystems that are thermodynamically open, thus subjected to a constant flux of exchange of matter (e.g., substrates in cells) and energy (e.g., sunlight in ecosystems such as forests) far from thermodynamic equilibrium. Therefore, emergent macroscopic properties do not result merely from static structures, but rather from dynamic interactions occurring both within the system and between the system and its environment (Jantsch 1980).

A remarkable example of the latter is given by the adaptation of an organism's behavior to its environment that depends upon biological rhythm generation. The role of biological clocks in adapting cyclic physiology to geophysical time was highlighted by Sweeney and Hastings (1960). Timing exerted by oscillatory mechanisms is foundational of autonomous periodicity, playing a pervasive role in the timekeeping and coordination of biological rhythms (Glass 2001; Lloyd 1992). Winfree (1967) pioneered the analysis of synchronization among coupled oscillators in a network, later refined by Kuramoto (1984) [reviewed in (Strogatz 2003)].Considering idealized systems of nearly identical weakly coupled sinusoidal oscillators, Winfree found that below a certain threshold of coupling, each oscillator runs at its own frequency, thus behaving incoherently until a further increase in coupling overcomes the threshold for synchronization (Winfree 1967, 2002). This synchronization event was characterized as the analog of a phase transition, revealing an insightful connection between nonlinear dynamics and statistical physics (Strogatz 2003).

1.4 Dynamics in Developing Systems

Feedback is a prominent source of nonlinear behavior, and biological systems exhibit both negative and positive types of feedback. The central importance of negative feedback as a control device in biological systems was formulated by Wiener (1948). The discovery of negative feedback devices in a variety of biological systems revealed the universality and simplicity of this control mechanism, whereby a process generates conditions which discourage the continuation of that process. End-product inhibition [later renamed "allosteric" inhibition by Monod and Jacob (1961)] is a prominent example of the latter; Umbarger (1956) and Pardee and Yates (1956) showed that the end product in the biosynthesis of isoleucine or pyrimidine inhibited the pathway. Feedback control was highlighted as a mechanism of avoiding behavioral extremes, echoing the concept of constancy of the *milieu intérieur* by Claude Bernard [1865; 1927 translation by Green (Bernard 1927)] and Walter Cannon's (1932) notion of homeostasis. Long before the discovery of feedback inhibition, Max Delbruck had introduced a mathematical model of mutually inhibiting chemical reactions (Delbruck 1949). By such a system of cross-feedback, two independent metabolic pathways can switch between stable steady states under unaltered environmental conditions or as a response to the stimulus of transient perturbations.

Positive feedbacks like autocatalysis are also ubiquitous in biology; their importance as a source of instability giving rise to bifurcation and nonlinear behavior was put forward by Turing (1952) in the context of morphogenesis. Turing's pioneer work demonstrated that an autocatalytic reaction occurring in an initially uniform or isotropic field, when coupled to the transport of matter through diffusion, can produce symmetry breaking visualized as spatial patterns. This work was ground breaking because it explained that stable spatial structures could arise—without assuming a preexistent pattern—through self-organization arising from bifurcations in the dynamics. This work opened the way to a reaction-diffusion theory of pattern formation (Meinhardt 1982) rather than its original goal "to account for the main phenomena of morphogenesis" (Turing 1952). Later on, Wolpert (1969) proposed "positional information" to account for the mechanisms by which cells seem to know where they are. In order to differentiate, cells interpret their position according to the concentration of a morphogen in a gradient. Wolpert's concept of positional information became important thanks to studies of Nusslein-Volhard and Wieshaus when they first identified the genes required for the formation of the body plan of the *Drosophila* embryo, and then showed how these genes were involved in morphogenesis. A main finding was that even before fertilization a pattern, formed by the differential distribution of specific proteins and mRNA molecules, is already established. As a consequence of differential rates of transcription, gradients in the concentrations of the new mRNA molecules and proteins are generated. Position along the antero-posterior axis of the Drosophila body plan is determined by a cascade of events that is initiated by the initial localization of *bicoid* mRNA or with the gradient of *bicoid* protein to which that localization gives rise (Driever and Nusslein-Volhard 1988a). These authors stated that the *bicoid* protein has the properties of a morphogen that autonomously determines position in the anterior half of the embryo (Driever and Nusslein-Volhard 1988b). This discovery enabled the combination of three keywords: diffusion, gradient, and morphogen since the distribution of the *bicoid* protein—a morphogen—is made at a source and both diffuses and is broken down, thereby generating a gradient.

Another big achievement in the field of developmental dynamics was given by the realization that the mammalian genome does not undergo irreversible change in the course of development. Initially, Gurdon (Gurdon et al. 1979) showed that at least in frogs the nucleus of a fully differentiated cell can be reprogrammed when transferred into an enucleated zygote. The sheep, Dolly, was the first mammal to be cloned by transferring the nucleus of an adult cell into an enucleated oocyte of another of the same species (Wilmut et al. 1997).

1.5 Metabolism—Epigenetics—Genetics: Waddington and the Epigenetic Landscape

Life requires both nucleic acid and a metabolic system for self-maintenance. The emergence of living systems as we know them could have come about as a result of a symbiotic fusion between a rapidly changing set of self-reproducing but error prone nucleic acid molecules and a more conservative autocatalytic metabolic system specializing in self-maintenance [Dyson (1985), quoted by Fox Keller (2000)]. Along this line of reasoning, early on Waddington (1957) had already introduced the concept of "epigenetic landscape" to describe cellular differentiation beyond genetic inheritance. Metaphorically, the "epigenetic landscape" can be visualized as "a mountainous terrain whose shape is determined by . . . the influence of genes. . .The valleys represent possible pathways along which the development of an organism could in principle take place. A ball rolls down the landscape, and the path it follows indicates the actual developmental process in a particular embryo" (Saunders and Kubal 1989). The epigenetic landscape illustrates the fact that isolated genes will have little effect on the shape of the landscape, which will depend more on the underlying dense gene interactive network. The outcome of a developmental process (represented by the rolling ball) depends on its dynamic trajectory, and its path through the different valleys (i.e., differentiation states) can arise from bifurcations. Morgan (1934) had previously noted that different groups of genes will come into action as development proceeds.

Waddington (1957) stressed the important implications of time in biology, distinguishing the biochemical (metabolic), developmental (epigenetic), and evolutionary, as the three realms in which time plays a central role in biology. Later, Goodwin (1963) adopts the metabolic, epigenetic, and genetic systems as basic categories for defining a system (e.g., cell) with respect to its environment.

The concept of epigenesis is a precursor of what is now known as "epigenetics," a whole new research field. Historically, the term "epigenetics" was used to describe events that could not be explained by genetic principles. Originally, Waddington defined epigenetics as "the branch of biology which studies the causal interactions between genes and their products, which bring the phenotype into being" [Waddington (1942), quoted in Goldberg et al. (2007)]. Consequently, a phenotypic effect or an organism following a developmental path is not only brought about by genetic variation but also by the environment.

Today, we know that in addition to primary DNA sequence information, much of the information regarding when and where to initiate transcription is stored in covalent modifications of DNA and its associated proteins. Modifications along the chromatin involve DNA cytosine methylation and hydroxymethylation, and acetylation, methylation, phosphorylation, ubiquitination, and SUMOylation of the lysine and/or arginine residues of histones are thought to determine the genome accessibility to transcriptional machinery (Lu and Thompson 2012). Recent data indicate that information about a cell's metabolic state is also integrated into the regulation of epigenetics and transcription; cells constantly adjust their metabolic state in response to extracellular signaling and/or nutrient availability. One of the challenges is to visualize how levels of metabolites that control chromatin modifiers in space and time, translate a dynamic metabolic state into a histone map (Katada et al. 2012).

1.6 The Core of the Living: Biochemistry and Genomes

The elucidation of the basic biochemistry of living systems and the recognition of its similarity across kingdoms and phyla represent major achievements of the twentieth century research in biology. The description of metabolic pathways, mechanisms of energy transduction and of genetic transmission, replication, regulation, and expression stand out as main ones.

The pathways utilized by cells to break down carbohydrates and other substrates like lipids, roughly divided into glycolysis, respiration, and β-oxidation, were already known to biochemists by the 1950s. Respiration includes the complete breakdown of the two carbon unit acetyl-CoA into carbon dioxide—discovered by

Krebs (1953; Krebs and Johnson 1937)—and the transfer of electron from NADH to molecular O_2 through the respiratory chain to produce water—pioneered by the discovery of cytochromes that changed their spectroscopic properties in the presence of O_2 (Keilin 1929).

Lipmann (1941) proposed that ATP is the universal carrier of biological energy when the phosphate bond energy released from its hydrolysis is used to drive most biochemical reactions that require energy. However, missing from this picture was the regeneration of ATP that involves the phosphorylation of ADP with energy provided by the oxidative breakdown of foodstuffs, hence "oxidative phosphorylation." This riddle was solved by the "chemiosmotic hypothesis." As postulated by Mitchell (1961), the "chemiosmotic hypothesis" proposed that the energy released by respiration is used by the respiratory enzymes to transport protons across the mitochondrial membranes building up a proton motive force (pmf) composed of an electric potential and an osmotic component (Mitchell 1961). This pmf is used by the ATP synthase to phosphorylate ADP [see (Weber 2005), for a useful historical and epistemological account]. The system is self-regulated by the availability of ADP (Chance and Williams 1956).

The fact that DNA is the carrier of biological specificity in bacteria was demonstrated directly by Avery et al. (1944) and Hershey and Chase (1952). Watson and Crick (1953) introduced the double helix model of the DNA thus providing a mechanism for self-replication and fidelity; complementary base-pairing ensured both replication and conservation. However, "indications that the cell was involved in the maintenance of genetic stability had begun to emerge from studies of radiation damage in bacteria and bacterial viruses (phages), especially from the discovery that certain kinds of damage could be spontaneously reversed" (Fox Keller 2000).

The association of the sequence of bases in the DNA and a protein came after the direct demonstration of the synthesis of a polymer string of the amino acid phenylalanine from a uniform stretch of nucleic acid consisting of a single nucleotide (uridine) (Nirenberg and Matthaei 1961). The central dogma was born; "DNA makes RNA, RNA makes protein, and proteins make us" (Crick 1957).

In 1961, Jacob and Monod introduced the concept of *genetic program*, extending their success, and in analyzing the operon as a mechanism of regulation of enzyme synthesis in *Escherichia coli* (Jacob and Monod 1961). This provided a more general description of the role of genes in embryonic development (Fox Keller 2002). These investigations led to the proposal of "structural" and "regulatory" genes thereby locating in the genome the program as a means of controlling its own execution, i.e., structural genes and regulatory elements are coordinated by the product of a regulatory gene. At present, the genome sequencing of more than 350 species, including *Homo sapiens*, and the informational content of genes and proteins systematized in databases constitute a fertile field for data mining and the ground work for exploring genetic interrelationships within and between species and their evolutionary meaning.

1.7 Scaling: A Fundamental Concept in Systems Biology

Size is a crucial biological property. As the size and complexity of a biological system increases, the relationship among its different components and processes must be adjusted over a wide a range of scales so that the organism can continue to function (Brown et al. 2000). Otherwise stated, the organism must remain self-similar. Self-similarity is a main attribute of fractals—a concept introduced by Mandelbrot (1977)—therefore the relationships among variables from different processes can be described by a fractal dimension or a power function. Geometrically, fractals can be regarded as structures exhibiting *scaling* in *space:* this is because their mass as a function of size, or their density as a function of distance, behave as a power law. If a variable changes according to a *power law* when the parameter on which it depends is growing linearly, we say it *scales*, and the corresponding exponent is called *scaling exponent, b*:

$$Y = Y_o M^b \tag{1.1}$$

where Y can be a dependent variable, e.g., metabolic rate; M is some independent variable, e.g., body mass, while Y_o is a normalization constant (Brown et al. 2000). If $b = 1$, the relationship represented by (1.1) is called isometric, whereas when $b \neq 1$ is called allometric—a term coined by Julian Huxley (1932). An important allometric relationship in biology is the existing between metabolic rate and body mass, first demonstrated by Kleiber (1932). Instead of the expected $b = 2/3$ according to the surface law (i.e., surface to volume area), Kleiber showed that $b = 3/4$ (i.e., 0.75 instead of 0.67), meaning that the amount of calories dissipated by a warm-blooded animal each day scales to the ¾ of its mass (Whitfield 2006).

Scaling not only applies to spatial organization but to temporal organization as well. The dynamics of a biological system—visualized through time series of its variables (e.g., membrane potential, metabolites concentration)—exhibits fractal characteristics. In this case, short-term fluctuations are intrinsically related to the long-term trends through statistical fractals. On these bases, we can say that scaling reflects the interaction between the multiple levels of organization exhibited by cells and organisms, thus linking the spatial and temporal aspects of their organization. The discovery of chaotic dynamics by Lorenz (1963) and criticality in phase transitions by Wilson (1983) enabled the realization that *scaling* is a common fundamental and foundational concept of chaos and criticality as it is with fractals.

1.8 Networks

The concept of networks is basic for understanding biological organization. Specifically, networks enable address of the problems of collective behavior and large-scale response to stimuli and perturbations exhibited by biological systems (Alon 2007; Barabasi and Oltvai 2004). Scaling and topological and dynamical organization of networks are intimately related concepts. Networks exhibit scale-free topologies and dynamics. Topologically, networks are scale free because most of the nodes in a network will have only a few links and these will be held together by a small number of nodes exhibiting high connectivity (Barabasi 2003). Dynamically, the scale-free character of networks manifests as diverse frequencies across multiple and highly dependent temporal scales (Aon et al. 2012; Lloyd et al. 2012).

The present topological view of networks evolved from multiply and randomly interacting elements in a system (Erdos and Renyi 1960) to "small worlds" (Watts and Strogatz 1998) to "scale-free" networks (Barabasi 2003). From the classical work of Erdos and Renyi based on random graphs, in which every node is linked to other node irrespective of their nature and connectivity, the "small world" concept introduced the notion that real networks as disparate as the neural network of the worm *Caenorhabditis elegans*, or those of power grids exhibit high clustering (i.e., densely connected subgraphs) and short path lengths. Barabasi and collaborators presented the view that nodes in a network are held together by a small number of nodes exhibiting high connectivity, rather than most of the nodes having the same number of links as in "random" networks (Barabasi and Albert 1999; Barabasi and Oltvai 2004). The "scale-free" organization of networks expresses the fact that the ratio of highly connected nodes or "hubs" to weakly connected ones remains the same irrespective of the total number of links in the network (Albert and Barabasi 2002; Helms 2008). Mechanistically, it has been proposed that the scale-free topology of networks is based on growth and preferential attachment (Albert and Barabasi 2002; Barabasi 2003).

The networks approach was introduced into biochemistry as metabolic control analysis (MCA). Independently developed in the second half of the past century by Kacser and Burns (1973) and Heinrich and Rapoport (1974), MCA represents an experimental approach with mathematical bases founded on the kinetics of enzymatic and transport networks in cells and tissues. MCA deals with networks of reactions of any topology and complexity to quantifying the control exerted by each process on systemic and local levels (Fell 1997; Westerhoff et al. 2009). Metabolic flux analysis (MFA), also called flux balance analysis, represents another methodological approach to the study of reaction networks (Savinell and Palsson 1992a, b). Developed in the 1990s MFA is based on stoichiometric modeling and accounts for mass–energy relationships among metabolic network components.

1.9 Systems Biology: A Twenty First Century Approach to Complexity

Our potential to address and solve increasingly complex problems in fundamental and applied research has expanded enormously. The following developments underscore our possibilities to address increasingly complex behavior in complex systems (Aon et al. 2012; Cortassa et al. 2012):

- The ability and the computational power to mathematically model very complicated systems, and analyze their control and regulation, as well as predict changes in qualitative behavior
- An arsenal of theoretical tools (each with its own plethora of methods)
- High throughput technologies that allow simultaneous monitoring of an enormous number of variables
- Automation and accessibility of databases by newly developing methods of bioinformatics
- Powerful imaging methods and online monitoring systems that provide the means of studying living systems at high spatial and temporal resolution of several variables simultaneously
- The possibility of employing detailed enough bottom-up mathematical models that may help rationalize the use of key integrative variables, such as the membrane potential of cardiomyocytes or neurons, in top-down conceptual models with a few state variables.

A Complex Systems Approach integrating Systems Biology with nonlinear dynamic systems analysis, using the concepts and analytical tools of chaos, fractals, critical phenomena, and networks has been proposed (Aon and Cortassa 2009). This approximation is needed because the focus of the integrative physiological approach applied to biology and medicine is shifting toward studies of the properties of complex networks of reactions and processes of different nature, and how these control the behavior of cells and organisms in health and disease (Cortassa et al. 2012; Lloyd and Rossi 2008; Saks et al. 2007, 2012).

The mass–energy transformation networks, comprising metabolic and transport processes (e.g., metabolic pathways, electrochemical gradients), give rise to the metabolome and fluxome, which account for the whole set of metabolites and fluxes, respectively, sustained by the cell. The information-carrying networks include the genome, transcriptome, and proteome, which account for the whole set of genes, transcripts, and proteins, respectively, possessed by the cell. Signaling networks modulate (activating or repressing) the interactions between information and mass–energy transducing networks, thus mediating between the genome–transcriptome–proteome and metabolome–fluxome. As such, signaling networks pervade the whole cellular network playing the crucial role of influencing the unfolding of its function in space and time. The output of signaling networks consists of concentration levels of intracellular metabolites (e.g., second messengers such as cAMP, AMP, phosphoinositides, reactive oxygen, or nitrogen species), ions, proteins or small peptides, growth factors, and transcriptional factors.

The underlying difficulty of the question of how the mass–energy and information networks of the cell interact with each other to produce a certain phenotype arises from the dual role of, e.g., metabolites or transcriptional factors; *they are at the same time a result of the mass–energy or information networks while being active components of the signaling networks that will activate or repress the networks that produced them* (see Chap. 2). The presence of these loops, in which the components are both cause and effect, together with their self-organizing properties, constitute the most consistent defining trait of living systems and the source of their inherent complexity (Cortassa et al. 2012). Indeed, it is increasingly recognized that the regulatory state of a cell or tissue, as driven by transcription factors and signaling pathways, can impose itself upon the dynamics of metabolic state, but the reciprocal—the feedback of metabolic state on regulatory state—must be equally true (Katada et al. 2012; Lu and Thompson 2012; McKnight 2010). Along this vein, one of the main undertakings of this book is to understand how the components and dynamics of signaling networks affect and is affected by the other cellular mass–energy and information networks in health and disease to produce a certain phenotype or a cellular response under defined physiological conditions.

References

Albert R, Barabasi AL (2002) Statistical mechanics of complex networks. Rev Modern Phys 74:47–97

Alon U (2007) An introduction to systems biology. Design principles of biological circuits. Chapman & Hall/CRC, London

Aon MA (2013) Systems biology of networks: the riddle and the challenge. In: Aon MA, Saks V, Schlattner U (Eds) Systems biology of metabolic and signaling networks. Springer, Heidelberg

Aon MA, Cortassa S (2009) Chaotic dynamics, noise and fractal space in biochemistry. In: Meyers R (ed) Encyclopedia of complexity and systems science. Springer, New York

Aon MA, Cortassa S, Lloyd D (2012) Chaos in biochemistry and physiology. In: Meyers RA (ed) Advances in molecular biology and medicine: systems biology. Wiley-VCH Verlag GmBH & Co. KGaA, Weinheim, pp 239–275

Avery OT, Macleod CM, McCarty M (1944) Studies on the chemical nature of the substance inducing transformation of Pneumococcal types : induction of transformation by a Desoxyribonucleic acid fraction isolated from Pneumococcus type Iii. J Exp Med 79:137–58

Ball P (2004) The self-made tapestry. Pattern formation in nature. Oxford University Press, Oxford, UK

Barabasi AL (2003) Linked. Plume, New York

Barabasi AL, Albert R (1999) Emergence of scaling in random networks. Science 286:509–12

Barabasi AL, Oltvai ZN (2004) Network biology: understanding the cell's functional organization. Nat Rev Genet 5:101–13

Bernard C (1927) Introduction a l'etude de la medicine experimentale. McMillan, New York

Brown JH, West GB, Enquist BJ (2000) Scaling in biology. Oxford University Press, New York

Cannon WB (1932) The wisdom of the body. Norton, New York

Capra F (1996) The web of life. Anchor books Doubleday, New York

Chance B, Williams GR (1956) The respiratory chain and oxidative phosphorylation. Adv Enzymol Relat Sub Biochem 17:65–134

Cortassa S, Aon MA, Iglesias AA, Aon JC, Lloyd D (2012) An introduction to metabolic and cellular engineering, 2nd edn. World Scientific, Singapore

Crick FH (1957) On protein synthesis. Symp Soc Exp Biol 12:138–63

Delbruck M (1949) Discussioneditor Unites biologiques douees de continuite genetique. Editions du CNRS, Lyon, p 33

Driever W, Nusslein-Volhard C (1988a) The bicoid protein determines position in the Drosophila embryo in a concentration-dependent manner. Cell 54:95–104

Driever W, Nusslein-Volhard C (1988b) A gradient of bicoid protein in Drosophila embryos. Cell 54:83–93

Dyson F (1985) Infinite in all directions. Harper & Row, New York

Erdos P, Renyi A (1960) On the evolution of random graphs. Publ Math Inst Hung Acad Sci 5:17–61
Fell D (1997) Understanding the control of metabolism. Cambridge University Press, London
Fox Keller E (2000) The century of the gene. Harvard University Press, Cambridge, MA
Fox Keller E (2002) Making sense of life. Explaining biological development with models, metaphors, and machines. Harvard University Press, Cambridge, MA
Glass L (2001) Synchronization and rhythmic processes in physiology. Nature 410:277–84
Gleick J (1988) Chaos: making a new science. Penguin, New York
Gleick J (2003) Isaac Newton. Pantheon, New York
Goldberg AD, Allis CD, Bernstein E (2007) Epigenetics: a landscape takes shape. Cell 128:635–8
Goodwin BC (1963) Temporal organization in cells. A dynamic theory of cellular control processes. Academic, London
Gurdon JB, Laskey RA, De Robertis EM, Partington GA (1979) Reprogramming of transplanted nuclei in amphibia. Int Rev Cytol Suppl 1979:161–78
Haken H (1978) Synergetics. Springer, Berlin, Heidelberg
Heinrich R, Rapoport TA (1974) A linear steady-state treatment of enzymatic chains. General properties, control and effector strength. Eur J Biochem 42:89–95
Helms V (2008) Principles of computation cell biology. Wiley-VCH Verlag GmbH & Co. KGaA, Weinheim
Hershey AD, Chase M (1952) Independent functions of viral protein and nucleic acid in growth of bacteriophage. J Gen Physiol 36:39–56
Hodgkin AL, Huxley AF (1952) A quantitative description of membrane current and its application to conduction and excitation in nerve. J Physiol 117:500–44
Huxley JS (1932) Problems of relative growth. Methuen, London
Jacob F, Monod J (1961) Genetic regulatory mechanisms in the synthesis of proteins. J Mol Biol 3:318–56
Jantsch E (1980) The self-organizing universe. Scientific and human implications of the emerging paradigm of evolution. Pergamon, New York
Junker BH (2008) Networks in biology. In: Junker BH, Schreiber F (eds) Analysis of biological networks. Wiley-Interscience, Hoboken, NJ, pp 3–14
Kacser H, Burns JA (1973) The control of flux. Symp Soc Exp Biol 27:65–104
Katada S, Imhof A, Sassone-Corsi P (2012) Connecting threads: epigenetics and metabolism. Cell 148:24–8
Keilin D (1929) Cytochrome and respiratory enzymes. Proc Royal Soc Lond B: Biol Sci 104: 206–52
Kitano H (2002a) Computational systems biology. Nature 420:206–10
Kitano H (2002b) Systems biology: a brief overview. Science 295:1662–4
Kleiber M (1932) Body size and metabolism. Hilgardia 6:315–53
Krebs HA (1953) Nobel lecture. The citric acid cycle
Krebs HA, Johnson WA (1937) Metabolism of ketonic acids in animal tissues. Biochem J 31: 645–60
Kuramoto Y (1984) Chemical oscillations, waves, and turbulence. Springer, Berlin
Lipmann F (1941) Metabolic generation and utilization of phosphate bond energy. Adv Enzymol 18:99–162
Lloyd D (1992) Intracellular time keeping: epigenetic oscillations reveal the functions of an ultradian clock. In: Lloyd D, Rossi EL (eds) Ultradian rhythms in life processes. Springer, London, pp 5–22
Lloyd D, Rossi EL (2008) Epilogue: a new vision of life. In: Lloyd D, Rossi EL (eds) Ultradian rhythms from molecules to mind. Springer Science+Business Media BV, New York, pp 431–39
Lloyd D, Cortassa S, O'Rourke B, Aon MA (2012) What yeast and cardiomyocytes share: ultradian oscillatory redox mechanisms of cellular coherence and survival. Integr Biol (Camb) 4:65–74
Lorenz EN (1963) Deterministic nonperiodic flow. J Atmos Sci 20:130–41

Lu C, Thompson CB (2012) Metabolic regulation of epigenetics. Cell Metabol 16:9–17
Mandelbrot BB (1977) The fractal geometry of nature. WH Freeman, New York
May RM (1974) Biological populations with nonoverlapping generations: stable points, stable cycles, and chaos. Science 186:645–7
May RM (2001) Stability and complexity in model ecosystems. Princeton University Press, Princeton, NJ
McKnight SL (2010) On getting there from here. Science 330:1338–9
Meinhardt H (1982) Models of biological pattern formation. Academic, London
Mitchell P (1961) Coupling of phosphorylation to electron and hydrogen transfer by a chemi-osmotic type of mechanism. Nature 191:144–8
Mitchell M (2009) Complexity. A guided tour. Oxford University Press, New York
Monod J, Jacob F (1961) Teleonomic mechanisms in cellular metabolism, growth, and differentiation. Cold Spring Harb Symp Quant Biol 26:389–401
Morgan TH (1934) Embryology and genetics. Columbia University Press, New York
Nicolis G, Prigogine I (1977) Self-organization in nonequilibrium systems : from dissipative structures to order through fluctuations. Wiley, New York, p 491, p xii
Nirenberg MW, Matthaei JH (1961) The dependence of cell-free protein synthesis in E. coli upon naturally occurring or synthetic polyribonucleotides. Proc Natl Acad Sci USA 47:1588–602
Noble D (2006) The music of life. Biology beyond the genome. Oxford University Press, Oxford
Noble D, Boyd CAR (1993) The challenge of integrative physiology. In: Boyd CAR, Noble D (eds) The logic of life. The challenge of integrative physiology. Oxford University Press, New York, pp 1–13
Pardee AB, Yates RA (1956) Control of pyrimidine biosynthesis in Escherichia coli by a feed-back mechanism. J Biol Chem 221:757–70
Poincare JH (1892) Les methodes nouvelles de la mechanique celeste. Washington, DC: English translation NASA (1967)
Saks V, Monge C, Anmann T, Dzeja PP (2007) Integrated and organized cellular energetic systems: theories of cell energetics, compartmentation, and metabolic channeling. In: Saks V (ed) Molecular system bioenergetics: energy for life. Weinheim, Wiley-VCH Verlag GmbH & Co. KGaA, pp 59–109
Saks V, Monge C, Guzun R (2009) Philosophical basis and some historical aspects of systems biology: from Hegel to Noble—applications for bioenergetic research. Int J Mol Sci 10: 1161–92
Saks V, Kuznetsov AV, Gonzalez-Granillo M, Tepp K, Timohhina N, Karu-Varikmaa M, Kaambre T, Dos Santos P, Boucher F, Guzun R (2012) Intracellular energetic units regulate metabolism in cardiac cells. J Mol Cell Cardiol 52:419–36
Saunders PT, Kubal C (1989) Bifurcation and the epigenetic landscape. In: Goodwin BC, Saunders PT (eds) Theoretical biology. Epigenetic and evolutionary order from complex systems. Edinburgh University Press, Edinburgh, pp 16–30
Savinell JM, Palsson BO (1992a) Optimal selection of metabolic fluxes for in vivo measurement. I. Development of mathematical methods. J Theor Biol 155:201–14
Savinell JM, Palsson BO (1992b) Optimal selection of metabolic fluxes for in vivo measurement. II. Application to Escherichia coli and hybridoma cell metabolism. J Theor Biol 155:215–42
Skyttner L (2007) General systems theory. Problems, perspective, practice. World Scientific Publishing Co, Singapore
Strogatz SH (2003) Sync. The emerging science of spontaneous order. Hyperion books, New York
Sweeney BM, Hastings JW (1960) Effects of temperature upon diurnal rhythms. Cold Spring Harb Symp Quant Biol 25:87–104
Turing AM (1952) The chemical basis of morphogenesis. Phil Trans R Soc Lond B237:37–72
Umbarger HE (1956) Evidence for a negative-feedback mechanism in the biosynthesis of isoleucine. Science 123:848
Vidal M (2009) A unifying view of 21st century systems biology. FEBS Lett 583:3891–4
Von Bertalanffy L (1950) The theory of open systems in physics and biology. Science 111:23–9
Waddington CH (1942) Endeavour 1:18–20
Waddington CH (1957) Strategy of the genes. Allen & Unwin, London

Watson JD, Crick FH (1953) Genetical implications of the structure of deoxyribonucleic acid. Nature 171:964–7

Watts DJ, Strogatz SH (1998) Collective dynamics of 'small-world' networks. Nature 393:440–2

Weber M (2005) Philosophy of experimental biology. Cambridge University Press, Cambridge

Westerhoff HV, Kolodkin A, Conradie R, Wilkinson SJ, Bruggeman FJ, Krab K, van Schuppen JH, Hardin H, Bakker BM, Mone MJ, Rybakova KN, Eijken M, van Leeuwen HJ, Snoep JL (2009) Systems biology towards life in silico: mathematics of the control of living cells. J Math Biol 58:7–34

Whitfield J (2006) In the beat of a heart. Life, energy, and the unity of nature. Joseph Henry Press, Washington, DC

Wiener N (1948) Cybernetics: or control and communication in the animal and the machine. MIT Press, Boston

Wilmut I, Schnieke AE, McWhir J, Kind AJ, Campbell KH (1997) Viable offspring derived from fetal and adult mammalian cells. Nature 385:810–3

Wilson KG (1983) The renormalization group and critical phenomena. Rev Modern Phys 55: 583–600

Winfree AT (1967) Biological rhythms and the behavior of populations of coupled oscillators. J Theor Biol 16:15–42

Winfree AT (2002) Oscillating systems. On emerging coherence. Science 298:2336–7

Yates FE (1987) Self-organizing systems. General introduction. In: Yates FE (ed) Self-organizing systems. The emergence of order. Plenum, New York, pp 1–14

Chapter 2
Complex Systems Biology of Networks: The Riddle and the Challenge

Miguel A. Aon

Abstract There is no direct relationship between metabolite, mRNA, protein, and gene; the expression of a gene is not necessarily correlated with the abundance of the corresponding protein product, and the activity of a protein may depend on posttranslational modifications, e.g., phosphorylation, redox-modulation/modification, and acetylation. It is believed that the diverse nature and outcomes of networks composed of genes, transcripts, proteins, and metabolites remain an obstacle for tracing the flux from genes to proteins in order to be able to capture or explain developmental programs or the underlying mechanisms of a disease. A different approach is needed to address this problem, and accordingly an alternative view based on the dynamic integration of three different kinds of networks, mass–energy, information, and signaling, is proposed and developed in this chapter. From this perspective, the spatio-temporal expression of mass–energy transformation and information-carrying networks is modulated by signaling networks associated with fundamental cellular processes such as cell division, differentiation, and autophagy. The dynamic network of reaction fluxes (i.e., the fluxome) represents the ultimate integrative outcome of the whole process. This approach—which accounts for the basic biological fact that cells and organisms make themselves—can only be realized by networks connected by overall cyclic topologies. Thereby, the output of mass–energy/information networks, composed of proteins, transcriptional factors, metabolites, is at the same time input for signaling networks which output activates or represses those same networks that produced them.

(. . .) If the genes are "essentially the same," what then is it that makes one organism a fly and another a mouse, a chimp, or a human? The answer, it seems, is to be found in the structure of gene networks—in the way in which genes are connected to other genes by the complex regulatory mechanisms that, in their interactions, determine when and where a particular gene will be expressed. But unlike the sequence of the genome, this regulatory

M.A. Aon (✉)
Division of Cardiology, Johns Hopkins University, Baltimore, MD 21205, USA
e-mail: maon1@jhmi.edu

M.A. Aon et al. (eds.), *Systems Biology of Metabolic and Signaling Networks*,
Springer Series in Biophysics 16, DOI 10.1007/978-3-642-38505-6_2,

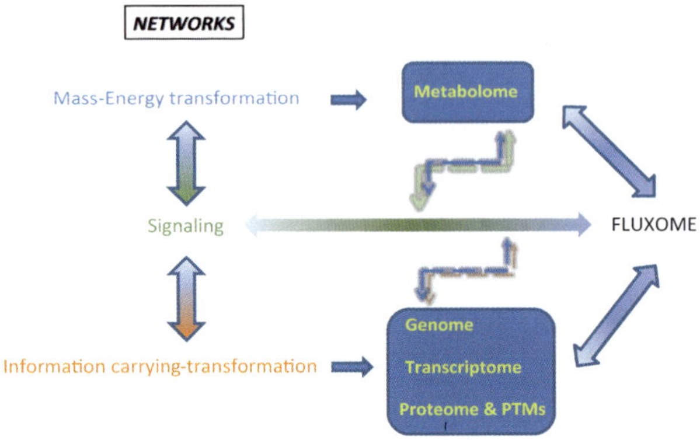

Fig. 2.1 The fluxome and the overall integration of mass–energy/information and signaling networks. Signaling networks connect and modulate the mass–energy–information networks. The fluxome represents the complete ensemble of fluxes in a cell, and as such it provides a true dynamic picture of the phenotype because it captures, in response to the environment, the metabolome (mass–energy) in its functional interactions with the information (genome, transcriptome, proteome, and posttranslational modifications, PTMs) and signaling networks (Cortassa et al. 2012). As a result of this integration between several cellular processes, the fluxome represents a unique phenotypic signature of cells (Cascante and Marin 2008).
The double sense of the arrows denote reciprocal interactions and an overall cyclic topology and connectivity that results in circular causality. Thus, an output from a network (metabolome, e.g., ROS or AMP:ATP ratio) is the input of the next network (signaling, e.g., AMPK network), which after processing will feedback on the same network that produced the initial triggers (e.g., ROS, AMP), thus modulating their levels.

> circuitry is not fixed: it is dynamic rather than static, a structure that is itself changing over the course of the developmental cycle. It is just this dynamic system that I am calling the developmental program." (Fox Keller 2000)

> If regulatory state (transcription factors, signaling pathways, etc.) is accepted to control metabolic state, is it not also unconditionally certain that metabolic state will reciprocally control the regulatorystate itself? Understanding this reciprocity, and digging to the bottom of it, is where the future lies (McKnight 2010)

Cell function can be visualized as the outcome resulting from the unfolding in space and time of three different kind of interacting networks: mass–energy, information, and signaling (Fig. 2.1). *Mass–energy transformation networks* comprise metabolic and transport processes, e.g., metabolic pathways and electrochemical gradients, that give rise to the metabolome. *Information-carrying networks* include the genome, transcriptome, and proteome, which account for the whole set of genes, transcripts, proteins, and their posttranslational modifications, respectively. *Signaling networks*, distinct in composition, dynamics, and topology, modulate by activating or repressing the function in space and time of the mass–energy/information networks to which they relate, e.g., metabolome, genome. The overall outcome of this process is the phenotype represented by the fluxome, which accounts for the whole set of fluxes sustained by a diverse range of processes associated with vital cellular functions such as division, differentiation, autophagy, apoptosis/necrosis, or the response to key environmental signals such as starvation or hypoxia. As such, the fluxome provides a true dynamic picture of the phenotype thereby constituting a unique phenotypic signature of cells (Cascante and Marin 2008) while integrating a myriad of cellular processes. In the mouse, for example, there are only ~600 metabolites (i.e., low-molecular-weight intermediates) (Griffin 2006), when as there are ~10,000 proteins, and ~22,000 protein-encoding genes (Cortassa et al. 2012). Thus, an unique advantage of fluxomics over genomics and proteomics is that the former is based on information from metabolites, which are far fewer than genes or proteins (Gherardini and Helmer-Citterich 2013; Raamsdonk et al. 2001).

The riddle is schematized in Fig. 2.1 and can be summarized as follows. Transcriptional factors, proteins, and metabolites are, at the same time, the products of mass–energy/information networks and their modulators by participating in the signaling networks that activate or repress the same networks that produced them. The presence of these control loops, in which network components are both cause and effect, together with their self-organizing properties sustained by a continuous exchange of energy and matter with the environment, is where the riddle of the unique complexity of the living state lies.

2.1 Signaling Networks: Connecting and Modulating the Mass–Energy-Information Networks

Information (e.g., gene, mRNA, and protein circuits) and signaling (e.g., AMPK, MAPK) networks can be clearly distinguished, by the following differences (Kiel et al. 2010):

- Signaling systems operate rapidly (ms to min) whereas transcriptional responses are slow, ranging from minutes (prokaryotes) to hours (eukaryotes)
- Subcellular localization plays an important role in signaling
- Protein structure and folding are involved in signaling (Mitrea and Kriwacki 2013); these processes are less predictable than DNA conformational changes present in information networks
- Genetic circuits tend to be noisy because they involve fewer molecules compared with signaling pathways, which usually involve larger number of molecular steps and thus tend to be less stochastic
- Amplification cascades occur in signaling thus spontaneous activation is avoided through negative feedback regulation or duplicated triggering signal.

Time-dependent regulation is of utmost importance for cellular responses, resulting from sudden, transient changes in environmental conditions. The earliest cellular response to an external cue usually consists in the activation of upstream signaling networks, which in turn regulate transcription factors. The modulation of gene expression therefore represents a later event (Gherardini and Helmer-Citterich 2013). The rapid relaxation provided by molecular mechanisms involved in signaling networks is crucial for fast adaptation (Aon and Cortassa 1997; Aon et al. 2004; Lloyd et al. 1982). Signaling networks exhibit their own intrinsic dynamics (Bhalla 2003; Bhalla and Iyengar 1999; Eungdamrong and Iyengar 2004) (see Chap. 4). Several different kinds of dynamic behaviors have been described, among them ultrasensitivity, bistability, hysteresis, and oscillations (Dwivedi and Kemp 2012; Kholodenko et al. 2012). Ultrasensitive behavior arises in protein modification cycles, whereas bistability stems from positive feedback loops, e.g., MAPK cascades, present in signaling cascades that may result in all- or none responses. Positive feedback loops alone or in combination with negative feedbacks can trigger oscillations. Emergent properties such as negative-feedback amplification could be demonstrated in the Raf-MEK-ERK signaling network with negative feedbacks. The "negative-feedback amplifier" confers resistance to perturbations of the amplifier resulting in resistance to inhibitors (e.g., that account for drug resistance) (Kholodenko et al. 2012).

The output of mass–energy networks is the metabolome as constituted by the ensemble of intracellular metabolites, e.g., cAMP, AMP, phosphoinositides, Reactive oxygen species (ROS), or nitrogen (RNS) species (Fig. 2.2). The outcome of information networks is represented by mRNAs, proteins or small peptides, and growth and transcriptional factors. Metabolites such as ROS, cAMP, and ADP exhibit a dual role; on the one hand, they are essential constituents from mass–energy networks that produce them, but on the other hand, they may act as intracellular sensors/messengers with allosteric effects (positive or negative) that react on enzymatic activities present in signaling networks. These dual roles of metabolites compose crucial cellular mechanisms in response to increasing environmental challenges (e.g., oxygen or substrate restriction) or cues (e.g., light, temperature). For example, the alterations of AMP:ATP ratio in response to starvation activates the AMPK signaling pathway, and at the same time AMP functions as an allosteric activator of the AMP kinase within the signaling network thus contributing to modulation of its dynamics. This results in the activation of catabolic and repression of anabolic processes thereby modulating the spatio-temporal unfolding of the mass–energy networks by, e.g., favoring organelle autophagy over biogenesis.

The spatio-temporal dynamics of the fluxome (Fig. 2.2) changes in response to signaling networks, through which cells can modulate, suppress, or activate gene expression (transcription, translation), whole metabolic pathways (e.g., respiration and gluconeogenesis during carbon catabolite repression), or specific enzymatic reactions within them.

Signaling networks are characterized by specific: (1) components and mechanisms; (2) metabolic pathway targeted; (3) conditions for signaling activation; and (4) physiological response (Cortassa et al. 2012). Each one of these characteristics can be identified in the AMP-activated protein kinase (AMPK) signaling pathway.

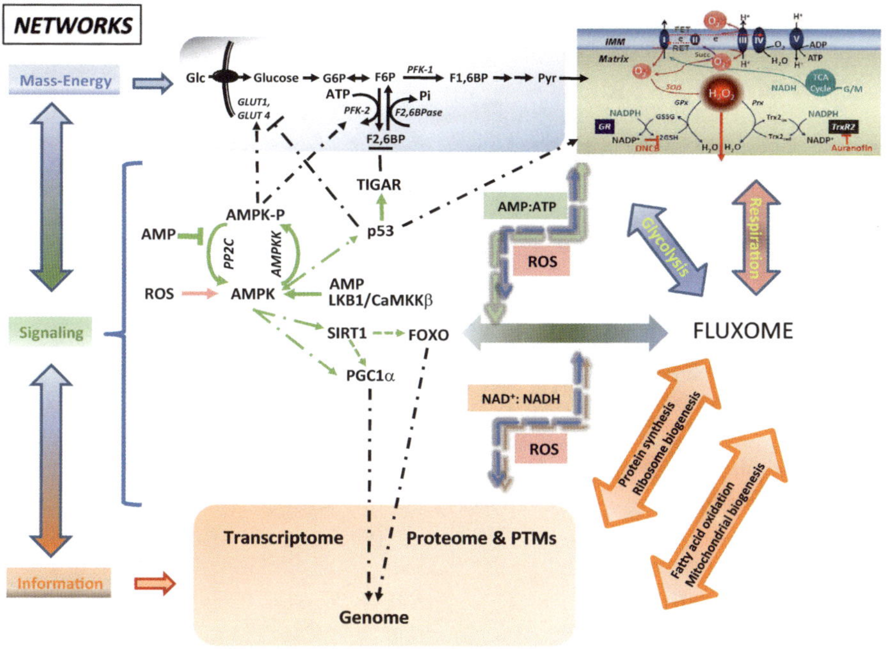

Fig. 2.2 The AMP-activated protein kinase (AMPK) signaling network and interactions with mass–energy and information networks. This figure depicts the main components of the AMPK signaling network and its interactions with the metabolome, genome-transcriptome-proteome, and other signaling paths.

AMPK: this network includes the kinase (AMPKK) and the phosphatase (PP2C), representing an ultrasensitive cellular energy sensor, as it is allosterically modulated by AMP (Hardie and Hawley 2001; Hardie et al. 1999). Environmental stressors such as starvation or hypoxia produce changes in the metabolome (e.g., rising AMP and falling ATP). An increased AMP binds to sites located on the γ subunit of AMPK, whereas high concentrations of ATP are inhibitory. Apart from being an allosteric activator, AMP also inhibits dephosphorylation of AMPK. AMPK is activated 1,000-fold by the combined effect of activation by its upstream kinases, neuronally enriched calcium/calmodulin-dependent protein kinase β (CaMKKβ) or LKB1, together with its allosteric stimulator AMP (Suter et al. 2006).

Targets of the AMPK signaling network are components of the metabolome in glycolysis (PFK-2, F2,6BP, GLUT), oxidative phosphorylation (ROS), and other pathways; the latter are not indicated but may comprise fatty acid (FA) oxidation and anabolism, e.g., triacylglycerol synthesis, glycogen, FAs, protein, and cholesterol. In the feedback from AMPK signaling to the metabolome, the dashed lines indicate activation through phosphorylation by AMPK-P of PFK-2 and glucose transport by increasing the levels of glucose transporters (GLUT1 and GLUT4). The increase in PFK-2 activity augments the level of the allosteric regulator F2,6BP that in turn activates PFK-1; the activity of the latter is also enhanced by the decrease in ATP. Thus, activation of glycolysis under ischemic conditions results in alteration of the fluxome as a result of the concerted action of AMPK signaling and the metabolome.

SIRT1: depicted is the interaction of the AMPK network with SIRT1 and their impact on the acetylation status of PGC-1α and other transcriptional regulators such as the FOXO family of transcription factors. Activation of AMPK in muscle by means of pharmacological intervention (metformin) or physiology (fasting or exercise) triggers an increase in the NAD^+/NADH ratio which activates SIRT1. AMPK induces the phosphorylation of PGC-1α and primes it for subsequent deacetylation by SIRT1(Canto et al. 2009). Deacetylation of PGC-1α is a key mechanism by which AMPK triggers PGC-1α activity in cultured myotubes and in skeletal muscle.

2.2 Reciprocal Interactions Between Signaling and Mass–Energy/Information Networks: Two Case Studies

2.2.1 *The AMPK Signaling Network*

This evolutionarily conserved signaling network functions as a cellular switch that activates catabolic pathways and turns off anabolic processes thereby restoring cellular energy levels (Poels et al. 2009). In physiological situations, AMPK senses energy deficiency (in the form of an increased AMP/ATP ratio), but it is also activated by metabolic stress such as glucose or oxygen deprivation triggering transient behavior regulation (Fig. 2.2). It has recently been shown that mitochondria-generated ROS induces autophagy mediated by the AMPK pathway under starvation conditions (Li et al. 2013). The decline in the responsiveness of AMPK signaling toward cellular stress with aging impairs metabolic regulation, increases oxidative stress, and reduces autophagic clearance (Salminen and Kaarniranta 2012).

The AMPK signaling network is paradigmatic, because the molecular components and mechanisms involved (i.e., kinetic properties of AMPK toward main effectors), demonstrating physiological impact as well as conditions in which the signaling operates, are all well understood and thus clearly identifiable (Fig. 2.2). As a specific example of the AMPK signaling network function in the context of ischemia in the heart: (1) components: AMPK allosterically modulated by AMP and phosphorylation (Hardie and Hawley 2001); (2) targets (changes in the metabolome): 6-phosphofructo-2-kinase (PFK-2) activity, fructose 2,6-bisphosphate (F2,6BP)

Fig. 2.2 (continued) SIRT1 appears to be a mediator of AMPK action on PGC-1α transcriptional activity. The acute actions of AMPK on lipid oxidation (fluxome) alter the balance between cellular NAD^+:NADH (metabolome), which acts as a messenger to activate SIRT1 (signaling), and the latter closes the circle by acting on the genome (information) which then again modifies the fluxome (mitochondrial respiration, lipid oxidation).
Tumor suppressor protein P53:a transcription factor that acts in response to cellular stress signals (e.g., DNA damage, hypoxia, oxidative, and nitrosative stress) and is redox sensitive because of the presence of conserved Cys residues that contain redox-sensitive thiol groups (see Fig. 2.3) (Vurusaner et al. 2012). P53 is also able to inhibit the nutrient-sensitive kinase target of rapamycin complex 1, mTORC1, by activation of AMPK, which is subsequently followed by induction of autophagy (Li et al. 2013; Melnik 2012; Poels et al. 2009). The interaction between p53 and TOR plays an important role in normal cell growth and proliferation (Jones et al. 2005), and it is likely that the AMPK-dependent induction of autophagy by p53 contributes to its role in tumor suppression (Poels et al. 2009).
Abbreviations: *AMPK* AMP-activated protein kinase, *CaMKKβ* calcium/calmodulin-dependent protein kinase kinase β, *p53* tumor suppression protein, *TOR* target of rapamycin, *PGC-1α* peroxisome proliferator-activated receptor-γ coactivator 1α, *PTMs* posttranslational modifications

concentration, and glucose transporters (GLUT1 and GLUT4) levels and translocation; (3) conditions for signaling activation: any stress that interferes with ATP synthesis and readily affects the AMP:ATP ratio, e.g., interruption of blood supply (ischemia); (4) physiological response (changes in the fluxome): activation of glycolysis that increases ATP availability (Marsin et al. 2000) [see also Chap. 11 in this book, Chap. 10 in Cortassa et al. (2012) and Fig. 2.2 for further explanation].

2.2.2 *ROS-Signaling Networks*

Redox signaling can be exemplified by the regulation of protein activity and the transduction of signals to downstream proteins through oxidative modification of reactive cysteine (Cys) residues by ROS (Finkel 2000; Paulsen and Carroll 2010). Cellular functions can be signaled by ROS in essentially two ways (Fig. 2.3): (1) through direct oxidation of specific Cys or (2) indirectly through changes in the activity of kinases or phosphatases that in turn modulate protein phosphorylation. The switch-like nature of the sulfenic acid (SOH) and disulfides that are formed after the initial reaction of a Cys thiolate with H_2O_2 and by reaction of SOH with neighboring Cys or reduced glutathione (GSH), explains their potential to function as reversible modifications that regulate protein function, analogous to phosphorylation (Haddad 2004). For example, myofilament activation and contractile function may be altered during oxidative stress by direct oxidative modifications of specific sites on contractile proteins or by ROS-induced changes in the activity of kinases or phosphatases that regulate sarcomeric protein phosphorylation (Santos et al. 2011; Sumandea and Steinberg 2011).

Another relevant example is given by the tumor suppressor protein p53, a transcriptional factor that in response to environmental challenge (e.g., hypoxia, oxidative stress) can sense cellular redox status. When p53 is oxidized by ROS its DNA binding capacity is decreased (Sun et al. 2003). Thus, under stressful conditions, ROS from the metabolome oxidizes p53: the latter when oxidized decreases its DNA binding capacity (Sun et al. 2003) thus inhibiting gene expression (genome) (Fig. 2.3). In turn, p53 can influence the metabolome through decreasing F2,6BP and glucose transporter levels that affect the fluxome by diminishing glycolysis and stimulating mitochondrial respiration (Fig. 2.2) (Lago et al. 2011). p53 is also able to interact with the AMPK signaling network inducing its activation after inhibition of the nutrient-sensitive kinase mTORC1: these effects are followed by induction of autophagy (Fig. 2.2).

2.2.3 *Sensing H_2O_2 through Cysteine Oxidation*

Cells can "sense" changes in redox balance through the specific reactions of H_2O_2 (D'Autreaux and Toledano 2007; Paulsen and Carroll 2010; Pourova et al. 2010; Schroder and Eaton 2009). In proteins, the thiol side chain of the amino acid Cys is

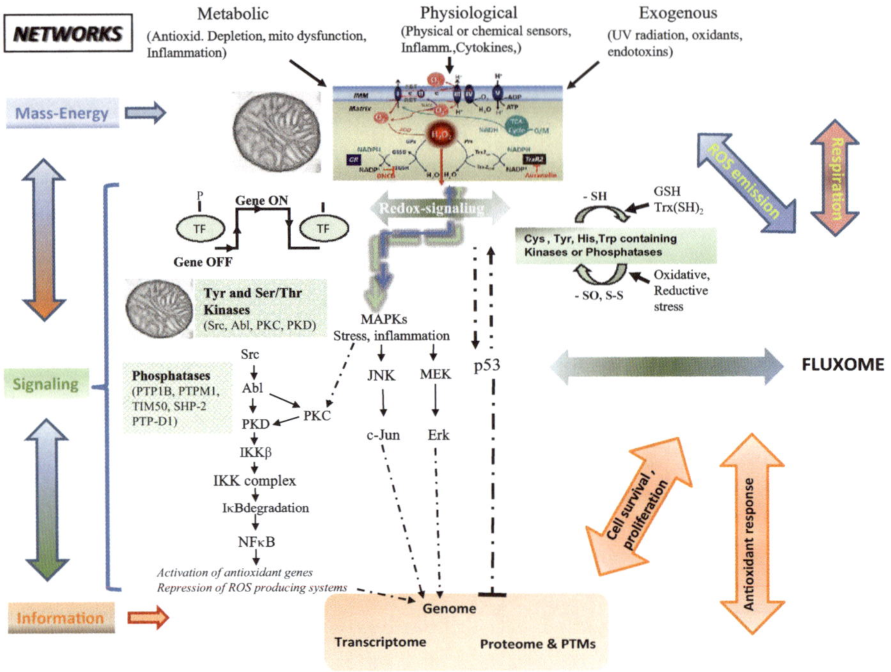

Fig. 2.3 ROS-activated signaling networks and interactions with mass–energy and information networks. Reactive oxygen species (ROS) can signal cellular functions through direct oxidation of specific Cys or indirectly through changes in the activity of kinases or phosphatases that in turn modulate protein phosphorylation. The redox-sensitive family of components from signaling networks use Cys motifs as redox-sensitive sulphydryl (-SH) switches (Paulsen and Carroll 2010). ROS-activated signaling networks comprise two distinct protein families—the Mitogen Activated Protein Kinase (MAPK) and the redox-sensitive kinases (Waris and Ahsan 2006).
MAPKs: serine (Ser)/threonine (Thr) kinases that transduce signals from the cell membrane to the nucleus in response to a wide range of stimuli, and modulate gene expression through phosphorylation of a wide array of transcription factors. MAPKs consist of three family members: the extracellular signal-regulated kinase (ERK); the c-Jun NH2-terminal kinase (JNK); and the p38 MAPK (Wada and Penninger 2004). MAPKs regulate processes important in carcinogenesis including proliferation, differentiation, and apoptosis (Waris and Ahsan 2006).
NFκB: the modification of gene expression by ROS has direct effects on cell proliferation and apoptosis through the activation of transcription factors including MAPK and NFκB pathways. The NFkB signaling network is significantly altered by dysregulated ROS that activates NFkB signaling through elimination of the IkB inhibitor. An increase in ROS levels induces the activation of the IkB kinase (IKK), which in turn phosphorylates IkB, leading to its proteasome-dependent degradation (Maryanovich and Gross 2012),while releasing NFκB for nuclear translocation and gene transcription (Chiu and Dawes 2012). NF-κB modulates the expression of several genes involved in cell transformation, proliferation, and angiogenesis, including bcl-2, bcl-x*L*, SOD (Chiu and Dawes 2012; Waris and Ahsan 2006). The NFκB link to carcinogenesis derives from of its roles in inflammation, differentiation, and cell growth (Rahman et al. 2006). Carcinogens and tumor promoters (e.g., UV radiation, phorbol esters, asbestos, alcohol) activate NF-κB.
Src and Abl: tyrosine kinases that constitute the signaling pathway leading to H_2O_2-mediated tyrosine phosphorylation of PKD that enhances NFkB activation (Storz and Toker 2003).
Protein tyrosine phosphatases (PTPs): regulated by H_2O_2 through induction of intramolecular disulfide bond formation that inactivates PTPs (Paulsen and Carroll 2010). The activity of the PTP

particularly sensitive to oxidation. Thiolate anions (RS^-) are intrinsically better nucleophiles and show enhanced reactivity with H_2O_2, compared to the thiol form (Winterbourn and Metodiewa 1999). Thus, the p*K*a value of the thiol group can modulate Cys reactivity. Other determinants of Cys reactivity toward H_2O_2 include access of the oxidant to its target and the presence of specific binding sites, e.g., low pKa catalytic Cys from peroxiredoxins or protein tyrosine phosphatases that react with $H_2O_2$with different second-order rate constants likely due to the unique conformation of their active site (Paulsen and Carroll 2010).

Thus, the initial reaction of a Cys thiolate with H_2O_2 yields a SOH, which once formed can lead to formation of additional posttranslational modifications (PTMs). The stability of a SOH is influenced, in part, by the presence of nearby Cys residues and by the accessibility of the modification site to GSH (Paulsen and Carroll 2010). The reaction of SOH with either a neighboring cysteine or GSH will generate a disulfide bond that, in the case of GSH, corresponds to *S*-glutathiolation (Mieyal et al. 2008). Both disulfide products can be reduced back to the thiol by the action of either the GSH/glutathione reductase or the thioredoxin/thioredoxin reductase systems (Aon et al. 2012a; Berndt et al. 2007; Ghezzi and Di Simplicio 2009; Stanley et al. 2011).

SOH can undergo further reaction with H_2O_2 to generate the SO_2H (sulfinic) and SO_3H (sulfonic) oxoforms, though the rates of these reactions are slower than those observed for a thiolate (Hugo et al. 2009). Both the SO_2H and SO_3H modifications are considered irreversible, and the latter is deemed a hallmark of diseases such as cancer, diabetes, cardiovascular, and neurodegenerative disorders that are associated with oxidative stress (Aggarwal and Makielski 2013; Andersen 2004; Jeong et al. 2012; Kembro et al. 2013; Klaunig and Kamendulis 2004; Leloup et al. 2011; Lowell and Shulman 2005). In a subset of eukaryotic peroxiredoxins, the SO_2H modification can be reversed by sulfiredoxin (Biteau et al. 2003). To prevent over oxidation of critical Cys residues, SOH may be converted to a disulfide or be *S*-glutathiolated or form sulfenamide and hypervalent sulfur species (Paulsen and Carroll 2010).

Fig. 2.3 (continued) family (e.g., PTP1B, PTPα) of phosphatases can be SOH-regulated, which is facilitated by the low p*K*a of catalytic Cys that can oxidize to SOH with concomitant inactivation. Protein kinases also undergo redox control.
Tumor suppressor p53: its gene is mutated in 30–50 % of human cancers, representing a checkpoint protein that elicits cell cycle arrest, DNA repair, and apoptosis in response to stressors (Sun et al. 2003). To perform its tumor suppressor activity p53 binds, as a tetramer, to DNA elements within promoters of its target genes and enhances transcription. P53 is sensitive to redox signaling: oxidation of Cys residues (some of the ten present in p53), and formation of disulfide bonds inhibits p53 tetramerization and DNA binding activity (Lago et al. 2011; Sun et al. 2003). P53 stimulates mitochondrial respiration and decreases glycolysis by affecting F2,6BP and the plasma membrane glucose transporters (Lago et al. 2011).
Abbreviations: *NFκB* nuclear factor κB, *MAPK* mitogen-activated protein kinase, *PKD* protein kinase D, *PKC* protein kinase C, *Src, Abl* tyrosine kinases, *IKK* IkB kinase, *JNK* c-Jun N-terminal kinase, *MAP kinase* mitogen-activated protein kinase, *β-MHC* b-myosin heavy chain

We should also keep in mind that, under oxidative stress, Cys thiols that are not redox sensors can also become oxidized. Thus, it is important to differentiate true Cys redox sensors that participate in redox signaling from other Cys that become oxidized but without biological consequences (Chiu and Dawes 2012).

2.3 Complex Systems Biology of Networks

The intricate networks of reactions and processes within living systems (Figs. 2.1, 2.2, and 2.3) exhibit complex dynamic behavior (Lloyd and Lloyd 1993, 1995; Lloyd and Stupfel 1991) (see Chaps. 12, 8 and 5). This complexity arises in part from the existence of multiple topological, structural, as well as functional interactions among components of these networks organized as molecular (e.g., enzymes), supra-molecular (e.g., cytoskeleton, respiratory, or enzymatic supercomplex), and organellar assemblies (e.g., in mitochondria) (see Chaps. 7, 8 and 11). Consequently, a full description of a biological system involves the *structure*, the *pattern of organization*, and the *function* (Capra 1996; Kitano 2002). *Structure* refers to the catalog of individual components (e.g., proteins, genes, enzymes, transcriptional factors); *pattern of organization* indicates how the components are wired (linked) and organized (e.g., topological relationships, morphology, feed-forward, and feed-back), and *function* implies how the ensemble works, i.e., unfolding in space and time of functional interrelationships, mass–energy-information fluxes, response to stimuli, growth, division (Figs. 2.2 and 2.3).

The collective dynamic function of networks is characterized by novel properties that cannot be anticipated from the behavior of network components in isolation. These novel properties are called *emergent*. As a fundamental trait of complexity, emergence is a manifestation of the interdependent function of processes within cells, organs, organisms (see Chap. 10). Ultimately, what we seek is to understand how function is coordinated in a cell that exhibits spatially distributed and compartmentalized subsystems, and the dynamics which unfolds simultaneously, although in sequentially consecutive temporal scales. Consequently, in the following, we attempt to dissect key organizational and functional traits of living cellular systems.

- *Function occurs in spatially distributed, compartmentalized, systems in which process dynamics unfolds in different successive but overlapping temporal scales*

  Processes of different nature occur in distinct compartments connected by transport mechanisms, temporally unfolding on different timescales from few milliseconds (electric), hundreds of milliseconds (mechanical) to few seconds (energetic) (see Chaps. 11 and 5).

- *Network organization*
  Mass–energy/information/signaling networks exhibit an overall loop topology. They comprise reaction and transport processes, and some nodes in these networks represent hubs since they exhibit multiple inputs and outputs while most nodes only possess a few of them. This feature confers these networks the trait of "scale-free." The topologically circular connectivity present in these networks bestows them with self-making and -maintaining properties that combined with continuous energy and matter exchanges allow them to self-organize in space and time (Aon and Cortassa 2009; Aon et al. 2012b). Although constructed with a high degree of redundancy that confers these networks resilience to attack (Barabasi 2009), under stress they may reach critical conditions that make them collapse, especially if hubs fail (see Chap. 5).
- *Top—down and bottom—up interrelationships (heterarchies)*
  Cells, tissues, and organs can be viewed as networks within networks. One of the most distinctive features of these networks is that all components interact one way or another, constituting a heterarchy (Aon and Cortassa 2012; Aon et al. 2012b; Lloyd and Rossi 2008; Yates 1993). In a heterarchy, but unlike in a strict hierarchy, interactions between network components and relevant functional interrelationships (including regulatory ones) flow between levels in both directions, top-down and bottom-up. This has important consequences for control and regulation of integrated metabolic and transport networks where every reaction, metabolite, ion, and process, may contribute, although to differing extents, to the overall control and regulation of the network (Aon and Cortassa 2012; Cortassa et al. 2012).
- *Control is distributed, and operates through "diffuse loops"*
  Systemic analysis of extensive networks as given by Metabolic Control Analysis shows that every process (edge, e.g., enzymatic reaction, channel) controls and is controlled by every other process in the network. However, the strength of control exerted by different processes may vary significantly, a trait that relates the fact that control is distributed (see Chaps. 3, 9 and 13). In the case of nodes (e.g., metabolites, second messengers, cofactors), every node can regulate other processes and in turn can be controlled (e.g., its concentration) by a process. The character of "distributed control" relates the fact that different processes (edges) exert control, and can be "diffuse" as well as direct. A diffuse control was first described in an integrated computational model of the cardiomyocyte and corresponds to the control exerted by seemingly indirectly related processes through shared ubiquitous cofactors such as ATP, ADP, and Ca^{2+} (Cortassa et al. 2009a, b).

  In networks involving various compartments, not all the control of the flux, e.g., in an organelle, resides within the organelle itself. In the heart, the control of mitochondrial respiration is exerted by cytoplasmic and sarcolemmal membrane-linked processes, e.g., the myofibrillar and Na/K ATPases, in addition to processes residing in the mitochondrion (Aon and Cortassa 2012). This is especially true under working conditions, when the interaction between cytoplasmic and mitochondrial processes is quantitatively more important.

- *Scaling—Fractal dynamics*

  Scaling tackles the question of functional coordination in a living cell that exhibits spatially distributed and compartmentalized subsystems with time constants in sequentially consecutive and overlapping temporal scales. Scaling involves both spatial and temporal levels of organization and reveals the interdependence between processes happening at different spatio-temporal coordinates (Aon and Cortassa 2009; Aon et al. 2012b; Lloyd et al. 2012). Genome-wide expression (transcriptome, ~5300 transcripts) during the time frame period provided by the ~40 min ultradian clock revealed the existence of two blocks of redox superclusters manifested in two phases of ~600 and ~4,700 maximally expressed genes during active respiration (oxidative) and low respiration (reductive), respectively (Klevecz et al. 2004; Lloyd and Murray 2005) (see also Chap. 12). Within the 40 min time frame of the clock, there is a 10–15 min period conducive to DNA synthesis and replication, a time window that opens during the reductive phase of the clock cycle: this suggests an evolutionary strategy to avoid oxidative damage.

  A bottom up modeling strategy provides an insight into how scaling arises, and what it reveals. For the sake of example, during a heartbeat, macroscopic and measurable properties of the cardiac cell such as action potentials, cell shortening-relaxation, and concomitant Ca^{2+} transients emerge from the integrated dynamic behavior of excitation–contraction and mitochondrial energetics (Aon and Cortassa 2012; Aon et al. 2012b). Underlying key electromechanical macroscopic functional properties, fast ionic currents operating in the few milliseconds range are revealed. These, in turn, are fueled by relatively slower (few seconds) mitochondrial energetic processes involving rapid transport processes in different subcellular compartments: sarcolemma, mitochondria, sarcoplasmic reticulum (see Chaps 5 and 10). The processes involved in the phenomenon of a heartbeat are simultaneous, and their apparent sequential nature results from the differential relaxation properties exhibited by the processes involved. Thus, the scale-free dynamic behavior exhibited by mitochondrial network energetic-redox function is based on the simultaneous operation of processes of different nature (electrical, mechanical, metabolic) in distinct compartments. Faster to slower temporal relaxation reflects the time it takes a process to return to the state previous to the stimulation that elicited the response, e.g., the initial potential depolarization triggered by the opening of Na channels in the sarcolemma.

  The inverse power law behavior of the power spectrum and the invariant relative dispersion across temporal scales obtained from the analysis of experimentally obtained time series in yeast and cardiac mitochondria support the existence of scale-free dynamics. The multi-oscillatory behavior of yeast and heart cells corresponds to statistical fractal dynamics, a behavior consistent with scale-free dynamics spanning a wide range of frequencies of at least three orders of magnitude (Aon et al. 2007, 2008; Lloyd et al. 2012). Scale-free temporal organization for organelle, cell, and organism implies timekeeping occurring across temporal scales in living systems (Aon et al. 2008; Sasidharan et al. 2012) (see also Chap. 12).

2.4 Concluding Remarks

The fundamental complexity and uniqueness of living systems resides in their capacity for self-making and -repairing (Luisi 2006; Varela et al. 1974). This distinctive trait is possible to be accomplished through closed loop topologies of nonlinearly interrelated processes operating in thermodynamically open systems thereby subjected to continuous energy dissipation and exchange of matter, e.g., substrates, and gases.

Another consequence of the self-making ability of living systems is that some network components (nodes, e.g., metabolites like AMP or TFs such as NFκB) can be both cause and effect at the same time (or output and input) for the same network, i.e., mass–energy and information, respectively, in these examples. A plethora of computational and experimental network-based methods are being developed and applied to different biological systems including complex diseases (Cho et al. 2012; Kholodenko et al. 2012; Neph et al. 2012; Przytycka and Cho 2012). It is worth remarking that the data and meaningful information that these approaches can provide are just the starting point for testable hypotheses.

The dynamic diversity arising from the interactions between spatially distributed mass–energy/information/signaling networks organized in circularly connected topologies has potentially explosive combinatorial possibilities. The modulation exerted by signaling networks on the spatio-temporal unfolding of mass–energy/information networks, together with the countless available dynamic paths emerging from these interactions, generates both the uniqueness and diversity of living creatures. Interestingly, recent findings have highlighted the marked cell type specificity of human transcriptional regulatory networks, with only ~5 % of overlap across 41 tested cell types, thereby underscoring the high regulatory diversity within humans (Neph et al. 2012).

Acknowledgments This work was performed with the financial support of R01-HL091923 from NIH.

References

Aggarwal NT, Makielski JC (2013) Redox control of cardiac excitability. Antioxid Redox Signal 18:432–68

Andersen JK (2004) Oxidative stress in neurodegeneration: cause or consequence? Nat Med 10 (Suppl):S18–25

Aon MA, Cortassa S (1997) Dynamic biological organization. Fundamentals as applied to cellular systems. Chapman & Hall, London

Aon MA, Cortassa S (2009) Chaotic dynamics, noise and fractal space in biochemistry. In: Meyers R (ed) Encyclopedia of complexity and systems science. Springer, New York

Aon MA, Cortassa S (2012) Mitochondrial network energetics in the heart. Wiley Interdiscip Rev Syst Biol Med 4:599–613

Aon MA, Cortassa S, Lloyd D (2012a) Chaos in biochemistry and physiology. In: Meyers R (ed) Encyclopedia of molecular cell biology and molecular medicine: systems biology. Weinheim, Wiley-VCH
Aon MA, Cortassa S, O'Rourke B (2007) On the network properties of mitochondria. In: Saks V (ed) Molecular system bioenergetics: energy for life. Wiley-VCH, Weinheim, pp 111–35
Aon MA, O'Rourke B, Cortassa S (2004) The fractal architecture of cytoplasmic organization: scaling, kinetics and emergence in metabolic networks. Mol Cell Biochem 256–257:169–84
Aon MA, Roussel MR, Cortassa S, O'Rourke B, Murray DB, Beckmann M, Lloyd D (2008) The scale-free dynamics of eukaryotic cells. PLoS One 3:e3624
Aon MA, Stanley BA, Sivakumaran V, Kembro JM, O'Rourke B, Paolocci N, Cortassa S (2012b) Glutathione/thioredoxin systems modulate mitochondrial H_2O_2 emission: an experimental-computational study. J Gen Physiol 139:479–91
Barabasi AL (2009) Scale-free networks: a decade and beyond. Science 325:412–3
Berndt C, Lillig CH, Holmgren A (2007) Thiol-based mechanisms of the thioredoxin and glutaredoxin systems: implications for diseases in the cardiovascular system. Am J Physiol Heart Circ Physiol 292:H1227–36
Bhalla US (2003) Understanding complex signaling networks through models and metaphors. Prog Biophys Mol Biol 81:45–65
Bhalla US, Iyengar R (1999) Emergent properties of networks of biological signaling pathways. Science 283:381–7
Biteau B, Labarre J, Toledano MB (2003) ATP-dependent reduction of cysteine-sulphinic acid by S. cerevisiae sulphiredoxin. Nature 425:980–4
Canto C, Gerhart-Hines Z, Feige JN, Lagouge M, Noriega L, Milne JC, Elliott PJ, Puigserver P, Auwerx J (2009) AMPK regulates energy expenditure by modulating NAD+ metabolism and SIRT1 activity. Nature 458:1056–60
Capra F (1996) The web of life. Anchor books Doubleday, New York
Cascante M, Marin S (2008) Metabolomics and fluxomics approaches. Essays Biochem 45:67–81
Chiu J, Dawes IW (2012) Redox control of cell proliferation. Trends Cell Biol 22:592–601
Cho DY, Kim YA, Przytycka TM (2012) Chapter 5: Network biology approach to complex diseases. PLoS Comput Biol 8:e1002820
Cortassa S, Aon MA, Iglesias AA, Aon JC, Lloyd D (2012) An introduction to Metabolic and Cellular Engineering, 2nd edn. World Scientific, Singapore
Cortassa S, O'Rourke B, Winslow RL, Aon MA (2009a) Control and regulation of mitochondrial energetics in an integrated model of cardiomyocyte function. Biophys J 96:2466–78
Cortassa S, O'Rourke B, Winslow RL, Aon MA (2009b) Control and regulation of integrated mitochondrial function in metabolic and transport networks. Int J Mol Sci 10:1500–13
D'Autreaux B, Toledano MB (2007) ROS as signalling molecules: mechanisms that generate specificity in ROS homeostasis. Nat Rev Mol Cell Biol 8:813–24
Dwivedi G, Kemp ML (2012) Systemic redox regulation of cellular information processing. Antioxid Redox Signal 16:374–80
Eungdamrong NJ, Iyengar R (2004) Computational approaches for modeling regulatory cellular networks. Trends Cell Biol 14:661–9
Finkel T (2000) Redox-dependent signal transduction. FEBS Lett 476:52–4
Fox KE (2000) The century of the gene. Harvard University Press, Cambridge, MA
Gherardini PF, Helmer-Citterich M (2013) Experimental and computational methods for the analysis and modeling of signaling networks. N Biotechnol 30:327–32
Ghezzi P, Di Simplicio P (2009) Protein glutathiolation. In: Jacob C, Winyard PG (eds) Redox signaling and regulation in biology and medicine. Weinheim, WILEY-VCH Verlag Gmbh & Co. KGaA, pp 123–41
Griffin JL (2006) The Cinderella story of metabolic profiling: does metabolomics get to go to the functional genomics ball? Philos Trans R Soc Lond B Biol Sci 361:147–61
Haddad JJ (2004) Oxygen sensing and oxidant/redox-related pathways. Biochem Biophys Res Commun 316:969–77

Hardie DG, Hawley SA (2001) AMP-activated protein kinase: the energy charge hypothesis revisited. Bioessays 23:1112–9

Hardie DG, Salt IP, Hawley SA, Davies SP (1999) AMP-activated protein kinase: an ultrasensitive system for monitoring cellular energy charge. Biochem J 338(Pt 3):717–22

Hugo M, Turell L, Manta B, Botti H, Monteiro G, Netto LE, Alvarez B, Radi R, Trujillo M (2009) Thiol and sulfenic acid oxidation of AhpE, the one-cysteine peroxiredoxin from Mycobacterium tuberculosis: kinetics, acidity constants, and conformational dynamics. Biochemistry 48:9416–26

Jeong EM, Liu M, Sturdy M, Gao G, Varghese ST, Sovari AA, Dudley SC Jr (2012) Metabolic stress, reactive oxygen species, and arrhythmia. J Mol Cell Cardiol 52:454–63

Jones RG, Plas DR, Kubek S, Buzzai M, Mu J, Xu Y, Birnbaum MJ, Thompson CB (2005) AMP-activated protein kinase induces a p53-dependent metabolic checkpoint. Mol Cell 18:283–93

Kembro JM, Cortassa S, Aon MA (2013) Mitochondrial ROS and arrhythmias. In: Laher I (ed) Systems biology of free radicals and anti-oxidants. Springer, Heidelberg

Kholodenko B, Yaffe MB, Kolch W (2012) Computational approaches for analyzing information flow in biological networks. Sci Signal 5:re1

Kiel C, Yus E, Serrano L (2010) Engineering signal transduction pathways. Cell 140:33–47

Kitano H (2002) Systems biology: a brief overview. Science 295:1662–4

Klaunig JE, Kamendulis LM (2004) The role of oxidative stress in carcinogenesis. Annu Rev Pharmacol Toxicol 44:239–67

Klevecz RR, Bolen J, Forrest G, Murray DB (2004) A genomewide oscillation in transcription gates DNA replication and cell cycle. Proc Natl Acad Sci USA 101:1200–5

Lago CU, Sung HJ, Ma W, Wang PY, Hwang PM (2011) p53, aerobic metabolism, and cancer. Antioxid Redox Signal 15:1739–48

Leloup C, Casteilla L, Carriere A, Galinier A, Benani A, Carneiro L, Penicaud L (2011) Balancing mitochondrial redox signaling: a key point in metabolic regulation. Antioxid Redox Signal 14:519–30

Li L, Chen Y, Gibson SB (2013) Starvation-induced autophagy is regulated by mitochondrial reactive oxygen species leading to AMPK activation. Cell Signal 25:50–65

Lloyd AL, Lloyd D (1993) Hypothesis: the central oscillator of the circadian clock is a controlled chaotic attractor. Biosystems 29:77–85

Lloyd AL, Lloyd D (1995) Chaos—Its significance and detection in biology. Biol Rhythm Res 26:233–52

Lloyd D, Cortassa S, O'Rourke B, Aon MA (2012) What yeast and cardiomyocytes share: ultradian oscillatory redox mechanisms of cellular coherence and survival. Integr Biol (Camb) 4:65–74

Lloyd D, Murray DB (2005) Ultradian metronome: timekeeper for orchestration of cellular coherence. Trends Biochem Sci 30:373–7

Lloyd D, Poole RK, Edwards SW (1982) The cell division cycle: temporal organisation and control of growth and reproduction. Academic, London, p 523

Lloyd D, Rossi EL (2008) Epilogue: a new vision of life. In: Lloyd D, Rossi EL (eds) Ultradian rhythms from molecules to mind. Springer Science+Business Media B.V., New York, pp 431–39

Lloyd D, Stupfel M (1991) The occurrence and functions of ultradian rhythms. Biol Rev Camb Philos Soc 66:275–99

Lowell BB, Shulman GI (2005) Mitochondrial dysfunction and type 2 diabetes. Science 307:384–7

Luisi PL (2006) The emergence of life. From chemical origins to synthetic biology. Cambridge University Press, Cambridge, UK

Marsin AS, Bertrand L, Rider MH, Deprez J, Beauloye C, Vincent MF, Van den Berghe G, Carling D, Hue L (2000) Phosphorylation and activation of heart PFK-2 by AMPK has a role in the stimulation of glycolysis during ischaemia. Curr Biol 10:1247–55

Maryanovich M, Gross A (2012) A ROS rheostat for cell fate regulation. Trends Cell Biol 23:129–34
McKnight SL (2010) On getting there from here. Science 330:1338–9
Melnik BC (2012) Leucine signaling in the pathogenesis of type 2 diabetes and obesity. World J Diabetes 3:38–53
Mieyal JJ, Gallogly MM, Qanungo S, Sabens EA, Shelton MD (2008) Molecular mechanisms and clinical implications of reversible protein S-glutathionylation. Antioxid Redox Signal 10:1941–88
Mitrea DM, Kriwacki RW (2013) Regulated unfolding of proteins in signaling. FEBS Lett 587:1081–8
Neph S, Stergachis AB, Reynolds A, Sandstrom R, Borenstein E, Stamatoyannopoulos JA (2012) Circuitry and dynamics of human transcription factor regulatory networks. Cell 150:1274–86
Paulsen CE, Carroll KS (2010) Orchestrating redox signaling networks through regulatory cysteine switches. ACS Chem Biol 5:47–62
Poels J, Spasic MR, Callaerts P, Norga KK (2009) Expanding roles for AMP-activated protein kinase in neuronal survival and autophagy. Bioessays 31:944–52
Pourova J, Kottova M, Voprsalova M, Pour M (2010) Reactive oxygen and nitrogen species in normal physiological processes. Acta Physiol (Oxf) 198:15–35
Przytycka TM, Cho DY (2012) Interactome. In: Meyers RA (ed) Systems biology: advances in molecular biology and medicine. Wiley-VCH Verlag GmbH & Co. KGaA, Weinheim, pp 85–115
Raamsdonk LM, Teusink B, Broadhurst D, Zhang N, Hayes A, Walsh MC, Berden JA, Brindle KM, Kell DB, Rowland JJ, Westerhoff HV, van Dam K, Oliver SG (2001) A functional genomics strategy that uses metabolome data to reveal the phenotype of silent mutations. Nat Biotechnol 19:45–50
Rahman I, Yang SR, Biswas SK (2006) Current concepts of redox signaling in the lungs. Antioxid Redox Signal 8:681–9
Salminen A, Kaarniranta K (2012) AMP-activated protein kinase (AMPK) controls the aging process via an integrated signaling network. Ageing Res Rev 11:230–41
Santos CX, Anilkumar N, Zhang M, Brewer AC, Shah AM (2011) Redox signaling in cardiac myocytes. Free Radic Biol Med 50:777–93
Sasidharan K, Tomita M, Aon M, Lloyd D, Murray DB (2012) Time-structure of the yeast metabolism in vivo. Adv Exp Med Biol 736:359–79
Schroder E, Eaton P (2009) Hydrogen peroxide and cysteine protein signaling pathways. In: Jacob C, Winyard PG (eds) Redox signaling and regulation in biology and medicine. Wiley-VCH Verlag Gmbh & Co. KGaA, Weinheim, pp 181–195
Stanley BA, Sivakumaran V, Shi S, McDonald I, Lloyd D, Watson WH, Aon MA, Paolocci N (2011) Thioredoxin reductase-2 is essential for keeping low levels of H(2)O(2) emission from isolated heart mitochondria. J Biol Chem 286:33669–77
Storz P, Toker A (2003) Protein kinase D mediates a stress-induced NF-kappaB activation and survival pathway. EMBO J 22:109–20
Sumandea MP, Steinberg SF (2011) Redox signaling and cardiac sarcomeres. J Biol Chem 286:9921–7
Sun XZ, Vinci C, Makmura L, Han S, Tran D, Nguyen J, Hamann M, Grazziani S, Sheppard S, Gutova M, Zhou F, Thomas J, Momand J (2003) Formation of disulfide bond in p53 correlates with inhibition of DNA binding and tetramerization. Antioxid Redox Signal 5:655–65
Suter M, Riek U, Tuerk R, Schlattner U, Wallimann T, Neumann D (2006) Dissecting the role of 5′-AMP for allosteric stimulation, activation, and deactivation of AMP-activated protein kinase. J Biol Chem 281:32207–16
Varela F, Maturana H, Uribe R (1974) Autopoiesis: the organization of living systems, its characterization and a model. Biosystems 5:187–96
Vurusaner B, Poli G, Basaga H (2012) Tumor suppressor genes and ROS: complex networks of interactions. Free Radic Biol Med 52:7–18

Wada T, Penninger JM (2004) Mitogen-activated protein kinases in apoptosis regulation. Oncogene 23:2838–49

Waris G, Ahsan H (2006) Reactive oxygen species: role in the development of cancer and various chronic conditions. J Carcinog 5:14

Winterbourn CC, Metodiewa D (1999) Reactivity of biologically important thiol compounds with superoxide and hydrogen peroxide. Free Radic Biol Med 27:322–8

Yates FE (1993) Self-organizing systems. In: Boyd CAR, Noble D (eds) The logic of life. The challenge of integrative physiology. Oxford University Press, New York, pp 189–218

Part II
Systems Biology of Signaling Networks

Chapter 3
The Control Analysis of Signal Transduction

Hans V. Westerhoff, Samrina Rehman, Fred C. Boogerd, Nilgun Yilmaz, and Malkhey Verma

Abstract This chapter discusses how metabolic control analysis (MCA) and generalisations thereof such as hierarchical control analysis (HCA) may help to understand the control of cell function through signal transduction, as well as the control of signal transduction itself. It reviews the key concepts of MCA paying attention to their applicability to signal transduction. Control analysis has already led to major insights into signal transduction such as that control of signalling tends to be distributed over multiple components and that the phosphatases are as important as, or more important than, the kinases. Examples of applications of control analysis in the medical domain are discussed.

H.V. Westerhoff (✉)
Manchester Centre for Integrative Systems Biology, Manchester Institute of Biotechnology, School for Chemical Engineering and Analytical Science, University of Manchester, Manchester M1 7DN, UK

Synthetic Systems Biology, Swammerdam Institute for Life Sciences, University of Amsterdam, Amsterdam, The Netherlands

Department of Molecular Cell Biology, Netherlands Institute for Systems Biology, VU University Amsterdam, Amsterdam, The Netherlands
e-mail: Hans.Westerhoff@manchester.ac.uk

S. Rehman • M. Verma
Manchester Centre for Integrative Systems Biology, Manchester Institute of Biotechnology, School for Chemical Engineering and Analytical Science, University of Manchester, Manchester M1 7DN, UK
e-mail: Samrina.Rehman@manchester.ac.uk; Malkhey.Verma@manchester.ac.uk

F.C. Boogerd • N. Yilmaz
Department of Molecular Cell Biology, Netherlands Institute for Systems Biology, VU University Amsterdam, Amsterdam, The Netherlands
e-mail: fred.boogerd@falw.vu.nl; nilguenyilmaz@gmail.com

M.A. Aon et al. (eds.), *Systems Biology of Metabolic and Signaling Networks*,
Springer Series in Biophysics 16, DOI 10.1007/978-3-642-38505-6_3,

3.1 From Physics to Life Through Systems Biology

For many years, in what has been referred to as the reductionist approach, it was believed that once all the parts of a living organism would have been characterised, the whole organism or any subsystem thereof would be understood without further ado. Obtaining the complete genome sequence of the human was partly driven by this motivation: knowing all the genes would imply understanding of the corresponding living system. For cell biology, the system could correspond to the functioning of a particular cell, the components being all the pathways, genome-wide, inclusive of signal transduction, gene expression, and metabolism. For biochemistry, a subsystem might be a particular metabolic pathway, and the components would be all enzymes and metabolites in that pathway. Such pathways can now be found as elementary modes or extreme pathways in the consensus metabolic map of the human (Thiele et al. 2013).

Is the concept that the whole is the sum of the parts always true, always false, or does it depend on the property one takes into consideration and the type of system at hand? Well, there are cases where the concept does apply: If the focus of interest were the mass of an organism, or the number of grams of carbon flowing out of a system, then the reductionist approach would be perfectly alright; the whole mass is the sum of the masses of the components and the carbon influx is the simple sum of the number of grams of carbon flowing through all fluxes into the system. Also, the sum of all the fluxes that consume leucine or virtually any other metabolite in any living organism must equal the sum of all the fluxes that produce it, at steady state, a property defining much of flux and flux balance analysis, which thereby is a powerful linear methodology leading to analytical solutions (Westerhoff and Palsson 2004).

In many such cases where the whole equals the sum of the parts, biology has in common with physics. Biology is also different from physics however: biology relates to function (Boogerd et al. 2005; Westerhoff et al. 2009b; Kolodkin et al. 2012), where function is defined by what promotes maintenance and amplification in a dynamic environment. This functional aspect requires improving on physics in terms of accelerating processes that also happen in physics, such as the breakdown of glucose to lactic acid. It also requires carrying out processes that are impossible in physics alone. It requires robustness of these processes vis-à-vis intrinsic noise as well as perturbations or extrinsic noise. And it requires proper adjustment of processes when conditions change.

The acceleration has been achieved by the evolution of protein-based catalysts, the concentration of which a living organism can increase by increasing gene expression. The "impossible processes" have become possible by enzymes and networks that couple "endergonic" processes, which are uphill in terms of Gibbs free energy, to other processes that are downhill (Caplan and Essig 1969; Westerhoff and Van Dam 1987). The synthesis of ATP from ADP and inorganic phosphate, for instance, is thermodynamically impossible at the intracellular concentrations of the reactants, yet proceeds because it is coupled to the oxidation of NADH through a network of processes that involves the electrochemical potential difference for protons across the inner mitochondrial membrane (Mitchell 1961). The required robustness towards fluctuations and intrinsic noise is a necessary property of the stable stationary state a system relaxes to; the balancing of control coefficients and component properties this requires is at the basis of some of the laws of metabolic control analysis (Westerhoff and Chen 1984). The robustness against sustained perturbations in parameter values is greatly enhanced by the feature that biological functions depend on the integral of multiple processes that would typically be perturbed independently (Quinton-Tulloch et al. 2013).

In all these cases it is particular interactions between processes or substances that lead to what is essential for biology. These particular interactions deviate from mainstream physics in that they do not correspond to the simplest case but rather to the most functional case. They may lead away from Onsager's precise reciprocity relations (Westerhoff and Dam 1987) that are valid near equilibrium (Cortassa et al. 1991). It is these particular cases of physics, away from mainstream physics, that "systems biology" should focus on.

With all this, is the whole still the sum of the parts, or is it different from that? Well, physiology continues to be the sum of the parts in the way the parts are active in vivo and in situ. However, when together, the parts behave in ways that are different from how they behave in isolation. One reason is that the conditions *in vivo* are different from what they are when we characterise the parts in isolation and the differences matter for the activity of the parts (van Eunen et al. 2010; García-Contreras et al. 2012). However, a second and more important reason is that these conditions themselves are influenced by the components. This may put in place a regulatory loop through which a component on itself depends on the response the other components' activities exhibit to changes in its behaviour. It is this aspect of regulatory looping that we cannot measure by assaying each component in isolation of all other components, even if we perform this measurement under otherwise in vivo conditions. It is in this second aspect where the functioning whole may differ essentially from the sum of the parts functioning in isolation.

We can only evaluate this aspect properly by measuring the components together, in situ (i.e. by observing physiology), or by measuring their properties, inclusive of their response properties, in vitro, and then reconstructing their collective behaviour in silico, i.e. in a computer model (Westerhoff et al. 2009b). The latter strategy enables understanding as well as observation, as it allows one to change parameter values in the model and evaluate the consequences. This strategy is the essence of systems biology: systems biology is about the difference between the whole and the sum of the parts in isolation (Alberghina and Westerhoff 2005). Thereby the essence of systems biology are the regulatory loops (or spirals) that enable a system's component to influence its own behaviour in ways that depend on the activity of other components, and this then for all components that are involved. Since virtually all molecules of a living cell are connected (Wagner and Fell 2001), this makes systems biology an activity that is in principle genome-wide.

Components interact in various ways. A direct mode of interaction is that of metabolism, where a substrate and a product of an enzyme in a metabolic pathway communicate with each other through their effects on the enzyme's catalytic rate, or where a product of a metabolic pathway interacts allosterically with an enzyme near the beginning of the pathway. More indirectly, a metabolite may activate a protein kinase that phosphorylates and thereby inactivates an enzyme in the pathway which affects the concentration of a metabolite, or the metabolite may activate a whole chain of signal transduction reactions and influences the expression of the gene encoding one of the enzymes in the pathway that control the concentration of the metabolite. The latter two cases are examples where signal transduction is involved. As we mentioned above, it is the looping of the interactions that causes the whole to differ from the sum of the parts. In our previous analyses we have called systems engaged in such looping "democratic" control systems. Even though in principle the DNA level encodes the other levels, the expression of the encoded information often requires the involvement of those other levels, again causing regulatory looping, at least in the "democratic" cases that are common to living systems (Westerhoff et al. 1990; Kahn and Westerhoff 1991).

More, in general, processes in living organisms engage in such regulatory looping, if not through metabolic, signal transduction, or classical gene expression networks (Westerhoff et al. 1990), then through RNAs [sense, anti-sense, or micro (Hendrickson et al. 2009)], or through dynamic ultrastructure (Westerhoff et al. 1990, 2009a; van Driel et al. 2003). In order to grasp the circular or spiralling causality that ensues from the looping (Boogerd et al. 2005; Westerhoff et al. 2009b), one needs to look at the operation and integration of several simultaneous processes, as functions of time or all parameters involved. Since the sum of a negative and a positive effect may be important, the experiments need to be precise and the analysis quantitative (Westerhoff and Palsson 2004). It is in learning to appreciate these concentrations of the various levels of cell functioning that systems biology should help. In this chapter we shall focus on how metabolic control analysis has done this for signal transduction.

3.2 Control Analysis

Biochemists have used various investigative methods for the identification of what they referred to as rate-limiting enzymes. These enzymes were envisaged mostly to reside at the beginning of a pathway and were supposed to catalyse essentially irreversible reactions (i.e. reactions with very large equilibrium constants). In an often pronounced but for linear pathways obviously erroneous concept, the rate-limiting enzyme operated at a lower velocity than the other (downstream) enzymes in the pathway and therefore "controlled" the pathway. If one wanted to increase the throughput of the pathway it was supposed then to suffice to increase the amount of that enzyme only. For many decades these ideas concerning the control of pathways were presented, and experimental observations were interpreted so as to fit these generalisations (Thomas and Fell 1998). It was not until various individuals and groups began to criticise these concepts, began to examine where they came from, and began to explore the properties of metabolic pathways composed of enzymes and transporters exhibiting the known nonlinear kinetics with respect to the amount of substrates, products, and effectors (Michaelis and Menten 1913) that a paradigm shift towards systems biology occurred. With the limited computing power then available, many still approached this through computer simulations. Jim Burns and Henrik Kacser (Kacser and Burn 1973) were one team, performing numerical simulations on an analogue computer examining what one should expect for the dependence of flux on gene dosage. Reinhart Heinrich and Tom Rapoport (Heinrich and Rapoport 1974) were another team, examining how the flux through a model of the erythrocyte's glycolytic pathway would depend on the activities of its enzymes. Their computations led to the realisation that in biochemical pathways, because of the presence of back pressure effects of substances downstream on upstream reactions, there was no *a priori* reason why all control of pathway flux should reside in one enzyme, or why one should expect one gene to be dominant in terms of controlling the pathway flux. Rather one should expect recessiveness for all enzyme genes (Kacser and Burn 1973). This led to metabolic control analysis (MCA), a pendant of biochemical systems theory (BST) (Savageau 1976).

Metabolic control analysis has played at least five important roles in biology and biochemistry, roles that have later become characteristic of systems biology: it has demonstrated the importance of (1) good definitions (Kholodenko et al. 1995), (2) quantitative approaches, (3) consensus or standardisation (Burns et al. 1985), (4) integration of experimental and theoretical work, and (5) principles emerging from molecular networking (Westerhoff et al. 2009a, b). MCA tells us that qualitative and unfounded statements such as the ones reviewed above in the context of defining metabolic control have no meaning in biology and thereby only add to confusion: flux control may well be distributed over all the steps in a pathway and has been shown to be distributed in almost all biological cases studied, with some steps having a higher degree of flux control than others.

Good definitions have more than one property, i.e. they should be unequivocal, operational, and understandable also to the nonexperts. In view of the latter, the magnitude of the flux control coefficients of a pathway step can be seen as a percentage of control exerted by that step over the flux of interest. It is also the percentage change in steady-state flux caused by a 1 % activation of only that step. In practice, pathway control is shared between all enzymes of the network (i.e. not only of the pathway), in proportions that differ between pathways. The attractive and thereby rather persistent, but flawed, concept of "the rate-limiting" step in a network process has been invalidated experimentally for metabolic (Groen et al. 1982) and gene expression (Jensen et al. 2000) pathways. For signal transduction pathways the subtleties in control are not less (Heinrich et al. 2002; Hornberg et al. 2005a).

Contrary to what is sometimes proposed as MCA is not a mere sensitivity analysis of fluxes and metabolite concentrations. Where a sensitivity analysis treats the sensitivities of a flux with respect to all parameters (e.g. temperature, enzyme activity, or Michaelis–Menten constant) equally, MCA starts from the complete set of sensitivities of a flux [or other state function (Westerhoff and Dam 1987)] for all process activities in the system. The corresponding sensitivity coefficients are called control coefficients. This focus enabled MCA to discover and prove laws that capture the dependencies of these control coefficients on each other. MCA also defines the subset of properties of the processes that matter most for the control properties of the system. This leads to the discovery of another set of laws relating systems properties to component properties. These laws are absent when the processes occur in isolation and are hence properties of the system only. They are among the first principles discovered by systems biology [review: (Westerhoff 2008)].

Together these laws enable the prediction of relative control exerted by each step (enzyme) on the system's variables (such as fluxes and metabolite concentrations) on the basis of certain kinetic properties of the component processes. As of the operational nature of the definitions of the control coefficients, this control can also be measured experimentally by applying a perturbation to the step being studied and measuring the effect on the variable of interest after the system has settled to a new steady state. By virtue of these properties, MCA is an example of systems biology.

In contrast to what its name suggests, metabolic control analysis is not limited to the topic of metabolism. Its principles and definitions also apply to gene expression and signal transduction pathways, where similar laws of "hierarchical control analysis" apply in addition (Kahn and Westerhoff 1991; Westerhoff 2008). As the methodology also applies to biology outside biochemistry, such as ecology (Getz et al. 2003), we shall refer to the generalised form of metabolic control analysis as control analysis.

3.3 The Control Coefficients

Control analysis relies on the proper definition of the system under investigation, i.e. on specifying what are the system's borders, which are the (dependent) variables, which are the parameters (=independent variables), and which are process activities (Westerhoff and Dam 1987). The dependencies of the variables on process activities are quantified in terms of control coefficients. A control coefficient is the relative measure of the change in a system variable (e.g. flux or concentration) upon perturbation of the activity of an elemental process (in - non-channelled pathways corresponding to the enzyme activities) (Westerhoff and Van Dam 1987). It is mathematically defined (Burns et al. 1985; Kholodenko et al. 1995) as

$$C_i^x \equiv \frac{\left(\frac{\mathrm{d}x}{\mathrm{d}p}\right)_{\text{global}}}{\left(\frac{\partial v_i}{\partial p}\right)_{\text{local}}} \cdot \frac{v_i}{x} = \frac{\frac{\mathrm{d}\ln x}{\mathrm{d}p}}{\frac{\partial \ln v_i}{\partial p}}, \tag{3.1}$$

where "x" is the dependent variable (e.g. flux J, concentration X or potential ψ) and v_i is a process activity such as the activity ($V_{\max}$ in both directions) of an enzyme

"i". p is a parameter that affects process i selectively. As indicated by the parentheses, the derivatives are both partial but in a different sense. The one in the numerator is that where only one of all process activities in the system is modulated, but all metabolite concentrations and hence process rates may vary as a consequence of that modulation; it is a systems property. With the proviso that all other process activities are *not* modulated, it can be seen as a total derivative. The one in the denominator modulates the process activity at constant values of all other metabolite concentrations that affect the process; like the elasticity coefficient (see below) it is a component property even though that component property may itself vary with the system's state.

The most common control coefficients quantify the control of fluxes or metabolite concentrations. However, any variable of the system will have a control coefficient defined by equations analogous to Eq. (3.1) (Westerhoff and Dam 1987). There is no need for the system to be at steady state (Westerhoff 2008); if it is not then x is a function of time, but until now the perturbation in process activity has been considered to be time independent. As many enzyme-catalysed reaction rates are proportional to the enzyme concentration (at least in a certain range of enzyme concentrations), control coefficients can often be written using total enzyme concentration E_i as parameter (Kacser and Burn 1973):

$$C_i^x = \frac{\mathrm{d}x}{\mathrm{d}E_i} \cdot \frac{E_i}{J} = \frac{\mathrm{d}\ln x}{\mathrm{d}\ln E_i}. \tag{3.2}$$

The definition of concentration control coefficient pertains equally to gene expression, signal transduction, as well as metabolic pathways, as well as to their integration (Kahn and Westerhoff 1991). In metabolic pathways fluxes are defined as the amount of a chemical element flowing down the pathway per unit time and unit biomass. In gene expression pathways, fluxes may be defined as the rate of synthesis of the mRNA, the rate of synthesis of the protein, and the chemical flux catalysed by that protein. In a signal transduction pathway it may make sense to define the ultimate flux as the steady rate of the process that is activated by the final signalling protein, which could well correspond to the concentration of the signalling state of that protein multiplied by a rate constant.

Figure 3.1 shows flux control coefficients of various enzymes in one, two, and three reactions' metabolic pathways. The rate-limiting enzymes in a metabolic pathway must have a flux control coefficient equal to 1. In this example, this only occurs in the one-enzyme pathway. Moreover, a rate-limiting enzyme does not remain rate-limiting upon overexpression. Indeed, most overexpression studies of enzymes have revealed that large increases of enzyme concentrations are not accompanied by equivalent increases in pathway flux. Usually, whilst one is increasing the amount of the hypothetically rate-limiting enzyme, its control over the pathway flux decreases until it approaches 0 eventually. Concentration control coefficients of various enzymes in three reactions' metabolic pathway are shown in Table 3.1.

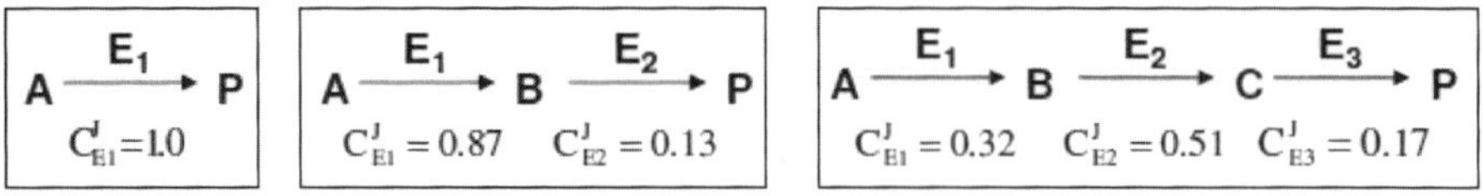

Fig. 3.1 One, two, and three reactions' metabolic pathways and exemplary flux control coefficients. A, P and enzymes E_1, E_2, and E_3 are present at fixed concentrations that may be modulated, whereas steady-state flux J and concentrations of B and C are dependent variables

Table 3.1 Concentration control coefficients in the three reactions' metabolic pathway shown in Fig. 3.1

	Enzymes		
Metabolites	E_1	E_2	E_3
B	2.65	−1.98	−0.67
C	0.18	0.29	−0.47

3.4 The Summation Laws

Control analysis has led to the identification of important properties of pathways at steady state pertaining to the summation of all the control coefficients of a pathway. For a steady state the sum of flux control coefficients of all the reactions in any pathway is equal to unity, using various procedures (Kacser and Burn 1973; Heinrich and Rapoport 1974; Westerhoff and Dam 1987; Giersch 1988; Westerhoff 2008) for the generalisation beyond metabolic pathways and steady state). This law can be derived by making a simultaneous small relative increase ($\delta_r x \equiv \delta x/x$) in all process rates of a metabolic system. Because the relative rates of production of each metabolite then increase by the same amount as the relative rates of its consumption, that metabolite's concentration remains unchanged and hence the steady-state condition is maintained. The metabolic flux through the pathway then increases exactly by the same relative amount "δ_r", whereas a transient time decreases by the fraction "δ_r". Mathematically this principle says that the flux is a homogeneous function of first degree, the metabolite concentrations are homogeneous functions of zero degree, and transient times are a function of degree minus 1, of all process activities: The summation laws are corollary of the Euler theorem for homogeneous functions which is also used in the derivation of the Gibbs–Duhem law (Westerhoff and Dam 1987; Giersch 1988; Westerhoff 2008). For flux control coefficients:

$$\sum_i C_i^J = 1 \tag{3.3}$$

for concentration control coefficients:

$$\sum_i C_i^{X_i} = 0 \tag{3.4}$$

for transient time, and area-under-the-curve control coefficients:

$$\sum_i C_i^{\tau_i} = -1 \tag{3.5}$$

where the summations are over all the reactions in the system. This may include not only the steps in the pathway of interest but also those of other pathways linking in. X_j may refer to any free variable in the system that does not have the dimension of time, such as the ratio of the phosphorylated to the non-phosphorylated form of a signal transduction protein, the ATP/ADP ratio, or the intracellular concentration of cAMP. Summation laws may be verified for the examples in Fig. 3.1 and Table 3.1.

The summation law proves that systems biology is essential for the understanding of life. Let us look at a linear pathway, where there is but a single steady-state flux and flux control coefficients are non-negative, and suppose that the flux control by the third enzyme equals 1. This implies that the first enzyme has no control on its own flux. The effect that enzyme has on life (which is through the flux that it enables) is not at all determined by the enzyme catalysing reaction 1, nor by the gene encoding that enzyme. Figure 3.1 gives other examples where no enzyme fully controls its own function, but the enzymes do this collectively.

An example from signal transduction can be found in the work of Kolodkin et al. (2010). They examined the control of the components of the signal transduction pathways of one nuclear hormone receptor (NR) on the flux through the signalling pathway of another nuclear hormone receptor. In contrast to cascade process signalling such as in the MAP kinase pathway (Hornberg et al. 2005a, b), in the case of nuclear hormone receptor signalling the signal flux equals a material flux, i.e. that of the nuclear hormone receptor itself. The nuclear hormone receptors serve as alternative cargos of a single transport system which uses a single transport channel (the nuclear pore complex, NPC). Hence the nuclear hormone receptors and their signal transduction pathways compete with one another (Fig. 3.2a). Kolodkin et al. calculated the extent to which a particular NR pathway flux would be controlled by both its own input, NPC, and output processes and those of the other NRs. If we first inspect the case with the lowest number of nuclear hormone receptors (i.e. $n = 2$ in Fig. 3.2b, c), then we see that the input process of NR1 controlled its own signal transduction flux for only some 25 %. The control by its own output and transport processes amounted to some 45 % and 25 %, respectively, the three controls adding up to 100 %, but all differing from zero. As the number of NRs using this same channel exceeded 6, the control by the input over its own signal flux control went up to some 45 % but never got even close to 100 %.

Much of the control, i.e. some −25 %, of the signal flux through the NR1 pathway was carried by the input reactions of the other NR when $n = 2$. This control was negative because the processes were competing. The transport and output processes of the second signal transduction route also controlled the flux through the first. The flux control coefficients all added up to 100 %, again suggesting that there was indeed full control but that this control is likely to be distributed, as is confirmed by the computations.

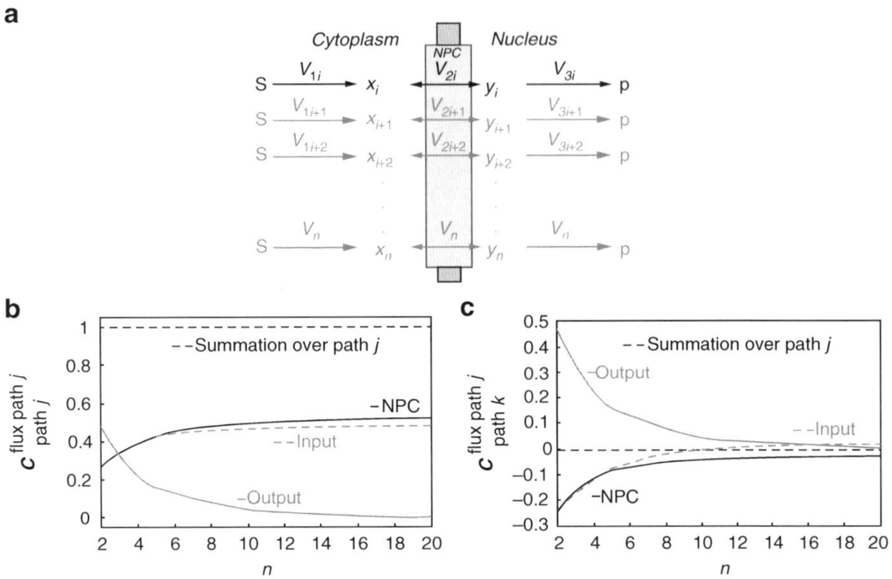

Fig. 3.2 (**a**) A number of parallel nuclear hormone receptor signal transduction pathways sharing the same nuclear pore complex (NPC). The pathways compete with each other for the NPC. (**b**) Signal flux control coefficients and their sums showing control of a particular nuclear receptor transportation flux (path *j*) by its own three pathway components, i.e. by the input (leading from the plasma membrane to close to the nuclear membrane), by the NPC itself, and by the output (the processes leading the nuclear receptor from the inside of the nuclear membrane to the establishment of the complex between DNA and the nuclear hormone). (**c**) The control of the flux through that same pathway, but now by the input, NPC, and output processes of a different nuclear hormone signal transduction pathway. Also here "Input" indicates the flux control coefficient by the input reaction of this particular nuclear receptor [the reaction between S and X_i, seen in (**a**)], "Output" indicates the flux control coefficient by the output reaction of this particular nuclear receptor (the reaction between Y_i and P), while "NPC" refers to the reactions through pore. "Summation over path *j*" gives the sum of all these three control coefficients (for many relevant details, see Kolodkin et al. 2010)

The cross control by any one other nuclear receptor pathway disappeared when the total number of nuclear hormone receptors using the transport protein exceeded 6: when the same transport channel is used by a number of cargoes, the signal transduction of any one NR systems is robust with respect to perturbations in the concentration of any single other signal transduction. When this number of active signal transduction pathways is lower, however, various signal transduction routes should be expected to be highly dependent on one another, to extents that depend on various factors, and often in nonintuitive ways. Control is 100 % but may be distributed.

The summation laws take the place of the intuition-based but in biology faltering concept of "the rate-limiting" step: For flux they confirm that there must be a complete rate limitation of 1, which is in line with intuition. However in biology this rate limitation need not reside in a single step, as confirmed experimentally by

(Groen et al. 1982). In signal transduction, the situation can become almost bizarre, with all kinases exerting a control of 1 on the steady metabolic flux and all phosphatases a control of -1 and the k_{cat} of the enzyme exerting the missing 1 (Kahn and Westerhoff 1991). For concentration they show that the concept that a single step limits a concentration is always wrong: if any step limits a concentration then there must be other steps that limit that concentration in the opposite direction. And for transient times, it reflects the opposite of limitation by a process activity: a step limits the magnitude of a transient time, e.g. in signal transduction, because it is too fast rather than too slow. Control of concentrations must be, and control of flux and transient time may well be, distributed over multiple steps hereby, but what determines this distribution? Enter the component properties. But which component properties?

3.5 The Elasticity Coefficients

The elasticity coefficients of an enzyme measure the most important component property, which is its immediate response to changes in its immediate environment. Said changes include changes in the concentrations of substrates, products, or allosteric effector, the concentration ratio of phosphorylated to non-phosphorylated form of signal transduction proteins, and the concentration of a growth factor such as EGF, pH, and temperature. Mathematically the elasticity coefficient is defined by (3.6):

$$\varepsilon_S^J = \left(\frac{\partial v_i}{\partial S}\right)_{\text{local}} \cdot \frac{S}{J} = \frac{\partial \ln v_i}{\partial \ln S}, \tag{3.6}$$

where v_i denotes the reaction rate and S denotes any metabolite concentration or other environmental factor.

3.6 Connections Between System-Level Control and Molecular Properties

The key systems biology property of control analysis is that it relates the kinetic properties of the component processes to the functional properties of the system. The connectivity (Kacser and Burn 1973; Westerhoff and Chen 1984) laws are such relationships. They relate the control coefficients to elasticity coefficients. The connectivity theorem for flux control coefficients (Kacser and Burn 1973) is valid for any free variable X (such as the concentration ratio of the phosphorylated to the non-phosphorylated form of a signal transduction protein) and any flux J in the network. It refers to the (multiplicative) products of the flux control coefficient of all steps by their elasticity coefficients towards X. It states that the sum of all of these is zero:

$$\sum C_i^J \cdot \varepsilon_X^i = 0. \tag{3.7}$$

This relationship also applies for the concentration control coefficients (Westerhoff and Chen 1984), as shown by (3.8) and (3.9):

$$\sum C_i^Y \cdot \varepsilon_X^i = 0, \quad X \neq Y \tag{3.8}$$

with the exception of

$$\sum C_i^X \cdot \varepsilon_X^i = -1 \tag{3.9}$$

The connectivity laws empower control analysis with the ability to describe how fluctuations in the concentrations of network components pathway propagate through the chain of reactions so as to be annihilated by the system's response (Westerhoff and Chen 1984). The kinetic properties of each process attenuate the perturbation to and from its immediate neighbours (Kholodenko et al. 2002).

The connectivity laws also reflect another characteristic of biological systems, i.e. that components are determined more by the collective of all processes than by themselves. For the signal transduction network of Fig. 3.3, one finds for the extent to which the kinase controls the phosphorylation state of E:

$$C_{\text{kinase}}^{\text{EP/E}} = -C_{\text{phosphatase}}^{\text{EP/E}} = \frac{1}{-\varepsilon_{\text{EP/E}}^{\text{kinase}} + \varepsilon_{\text{EP/E}}^{\text{phosphatase}}}. \tag{3.10}$$

This shows that it is impossible for kinase to be the only factor in control and that the control of the kinase depends on properties of both the kinase and the phosphatase. If the kinases and phosphatases are product insensitive, the dependence on kinase properties even drops from the equation, and the extent to which the kinase affects transcription does not depend on kinase properties. The importance of the phosphatase also reflects on the response of the transcription rate to changes in the concentration of the external signal:

$$R_{\text{signal}}^{\text{transcription}} = \frac{\varepsilon_{\text{EP/E}}^{\text{RNApolymerase}} \cdot \varepsilon_{\text{signal}}^{\text{kinase}}}{-\varepsilon_{\text{EP/E}}^{\text{kinase}} + \varepsilon_{\text{EP/E}}^{\text{phosphatase}}}. \tag{3.11}$$

The equations also generalise the case of zero-order ultrasensitivity which requires that both the kinase and phosphatase are insensitive to the fractional phosphorylation of the signal transduction protein E, leading to elasticity coefficients close to zero. Ortega et al. (2002) have shown that this situation is biochemically unlikely. Network topology and component elasticities together determine the magnitudes of the control coefficients in complex networks of signal transduction that may include gene expression and metabolism. In a sense systems

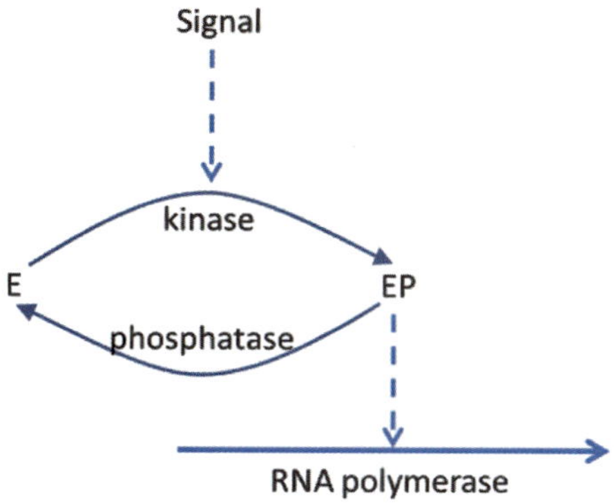

Fig. 3.3 A simple case of signal transduction with a single cascade of a kinase and a phosphatase effecting the phosphorylation of transcription factor *E*, which activates RNA polymerase into transcription. The external signal S activates the kinase

control is the inverse of local properties, but highly significantly it is a matrix inverse (inverse of the whole) rather than the collection of the individual inverses of the local properties (Westerhoff and Kell 1987; Kahn and Westerhoff 1991).

3.7 Determination of Control Coefficients

The obvious way of determining control coefficients experimentally is to follow their definition (see above), i.e. to perturb one step and let the system settle to a new steady state and then measure the change in the variable of interest. There are many ways of perturbing the rate of a reaction, each of them with its own advantages and disadvantages:

1. Alteration of enzyme concentration by genetic means [e.g. Flint et al. 1981; Dykhuizen et al. 1987; Fell 1992; Niederberger et al. 1992; Jensen et al. 1993) for further overview] has the disadvantage that other enzyme concentrations may also change. This then invokes a second type of control coefficient (Westerhoff and Workum 1990).
2. Titration with inhibitors (e.g. Groen et al. 1982) has the disadvantage that the in situ efficacy of the inhibitor needs to be assessed.
3. Titration with purified enzyme in vitro (e.g. Torres et al. 1986; Moreno-Sánchez et al. 2008) has the disadvantage that it only determines the control coefficient in the in vitro situation.

The perturbation should be specific. Should one desire a complete picture of the control of the variable in question, then the same procedure will have to be repeated for each step of the system. In addition, the perturbations should be small—because the steady state moves when the perturbations are finite, there is an error associated with large perturbations. However, the effects of small perturbations are usually also small and difficult to detect accurately. If one wants to use the enzyme concentration as the parameter to perturb, the rate of reaction must change proportionally with the enzyme concentration. If the relation between rate and enzyme

Table 3.2 Control coefficients of various levels kinases and phosphatases for the degree of phosphorylation at the next level down, for the signalling pathway shown in Fig. 3.4

Enzymes	Kinase 1	Ptase 1	Kinase 2	Ptase 2	Kinase 3	Ptase 3
$C^{X_{\text{levelA}}}_{i_{A+1,\text{SS}}}$	0.15	−0.15	0.21	−0.21	0.32	−0.32

Note: Ptase denotes Phosphatase

concentration is not linear such as when a process is catalysed by two proteins, or when proteins are substrates rather than catalysts such as in signal transduction, the protein concentration-based control coefficients will not conform to the above summation laws (Van Dam et al. 1993).

The alternative to experimental determination of the measurement of control coefficient is the computational determination, either from complete replica models of the network (e.g. Bakker et al. 1999; Conradie et al. 2010), or by inversion of the matrix of elasticity coefficients (Westerhoff and Kell 1987).

This indicates that the phosphorylation and dephosphorylations processes are equally important for steady-state signalling. This is clearest for a simple signal transduction cascade where there is only one kinase and one phosphatase per level:

$$C^{X_{i+1}}_{\text{phosphatase}} = -C^{X_{i+1}}_{\text{kinase}}. \tag{3.12}$$

3.8 Control and Signal Transduction

Kahn and Westerhoff (1991) developed HCA for signal transduction cascades at steady state, where extra control laws apply between the cascade levels: The steady-state degree of phosphorylation of a protein at any next level in the cascade is controlled by the protein, and the control by that kinase and that phosphatase adds up to zero:

$$\sum_{i_{A+1}=1}^{n_{A+1}} C^{X_{\text{level A}}}_{i_{A+1,\text{SS}}} = 0. \tag{3.13}$$

In Table 3.2 this summation law is demonstrated for the signalling pathway shown in Fig. 3.4.

This kind of signal transduction process is of interest in protein kinase cascades, such as the MAP kinase pathway. Here the first kinase activates the second by phosphorylation, which then activates the third, again by phosphorylation, and so on. The phosphorylated last kinase activates transcription. The above HCA implies that corresponding protein phosphatases deserve as much interest as the kinases for the ultimate steady-state levels, a conclusion that breaks with the research history, which focused on the kinases. Hornberg et al. (2005a, b) extended HCA mathematically to describe the maximum phosphorylation level obtained in the time-dependent functioning of the cascade. In this case the same summation law

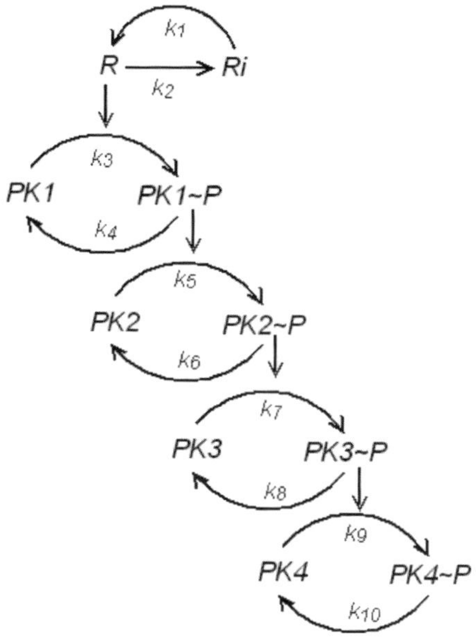

Fig. 3.4 Four steps signal transduction cascade of kinases and phosphatases

continued to apply to the maximum level, but now the kinases collectively were equally important as the phosphatases collectively, not individually. For all other points in the transient yet another law was derived mathematically, defining control by time itself (Westerhoff 2008).

This work had been preceded by an analysis of the cascades in terms of mass action kinetics by (Heinrich et al. 2002). In this description, the reaction rate for the phosphorylation of the next kinase in the pathway reads as follows:

$$v_{i+1} = k_{i+1}.x_i.e_i.(1 - x_{i+1}).e_{i+1} = \alpha_{i+1}.x_i.e_i.(1 - x_{i+1}), \tag{3.14}$$

where e_i represents the total concentration of the ith kinase in the pathway and x_i the fraction of the kinase that is in the phosphorylated, hence active, state. α_{i+1} is the pseudo-first-order rate constant. The corresponding phosphatase, at concentration f_{i+1}, would act at a rate as follows:

$$v_{-(i+1)} = k_{-(i+1)}.x_{i+1}.e_{i+1}.f_{i+1} = \beta_{i+1}.x_{i+1}.e_{i+1}. \tag{3.15}$$

The α's and the β's were proportional to the kinase and phosphatase activities (and concentrations) at the corresponding levels. For the steady state of permanent activation of the pathway, one then finds

$$x_{i+1} = \frac{k_{i+1}.x_i.e_i}{k_{i+1}.x_i.e_i + k_{-(i+1)}.f_{i+1}} = \frac{\alpha_{i+1}.x_i.e_i}{\alpha_{i+1}.x_i.e_i + \beta_{i+1}.e_{i+1}}. \tag{3.16}$$

Using the above equations for kinases and phosphatases in the consecutive steps in a signalling cascade, Heinrich et al. (2002) showed that signal amplification is possible particularly when the level of activation of the pathway (x_i in the above equation) is small. The signal amplification is demonstrated in Fig. 3.5a for the signalling cascade shown in Fig. 3.4. Signal dampening can also be achieved by such cascades (Fig. 3.5b). These equations again show, but now in a model-specific way, that kinases and phosphatases are equally important in determining amplification and indeed signal level.

3.9 Examples

3.9.1 *Galactose Signalling and Co-response on Galactose Metabolic Flux in* Saccharomyces cerevisiae

Figure 3.6 presents an example where control analysis of signal transduction and gene expression may lead. Much of this terrain is still unexplored.

3.9.2 *Regulatory Strength in Central Nitrogen Metabolism and Signal Transduction in* Escherichia coli

An in silico replica has been constructed for the complete central ammonium assimilation network of *E. coli* (Bruggeman et al. 2005). Ammonium assimilation is regulated by a variety of mechanisms, including interactions with metabolites (substrates, products, effectors), binding and release of regulatory proteins, and activities of modifier enzymes that covalently modify crucial proteins/enzymes. This latter part corresponds to signal transduction and a two-component regulatory system is involved. However, there are more regulatory links in the network, all of which, even the metabolic links, transduce signal. Using the replica model, the regulation of the two ammonium assimilating pathways, i.e. glutamine synthetase-glutamate synthase (GS-GOGAT) and glutamate dehydrogenase (GDH), was dissected quantitatively. Steady states and transient states of the entire network were examined, but special attention was given to GS, which is by far the most strongly regulated enzyme in the network. The wild-type and mutants lacking one of the important regulatory enzymes/proteins (GS, GOGAT, GDH, ATase, and UTase) were investigated in silico. Overall, the results of the kinetic model indicated that GS flux was tuned down at 1 mM ammonium, but that the extent of the decrease depended on the carbon and nitrogen status. Focusing on the transient short-term effects on GS activity upon a 0.05–1 mM ammonium upshift, the relative, average contribution of signal flux through the various direct regulators of GS (ADP, ATP, glutamine, glutamate, adenylylation state) could be calculated in terms of regulatory strengths. On a second timescale, signal flow through ADP was most regulating,

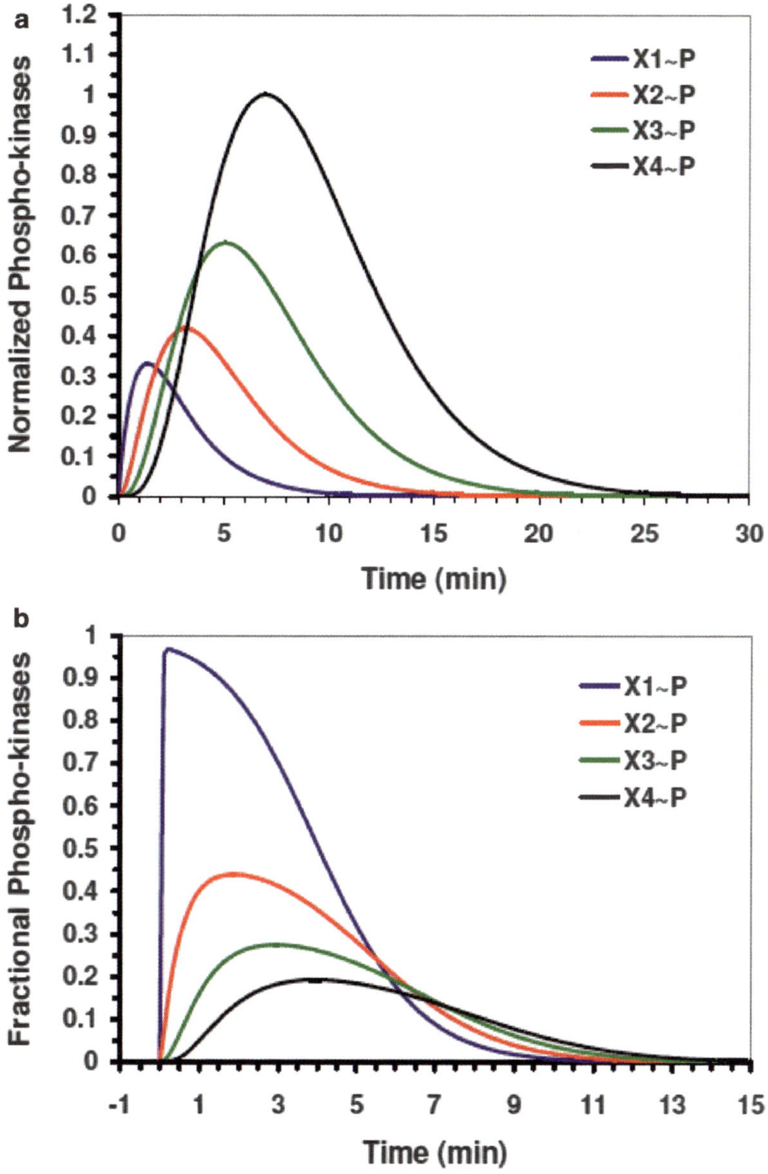

Fig. 3.5 (**a**) Signal amplification ahead a multicyclic signalling pathway composed of four cycles. The phosphorylation and dephosphorylation rate constants (k_s) used for simulations were: $k_{3,5,7,9} = 1$, $k_{4,6,8,10} = 0.5$, and $R_i = 0.1$. Activation of receptor concentration was modelled as reported by (Heinrich et al. 2002). Simulated fractional concentrations of activated kinases $X_i \sim P$ were plotted versus time. The activities of $X_i \sim P$ were normalised with respect to the amplitude of $X_4 \sim P$. (**b**) Signal repression along a signalling pathway in which rate constants were $k_{3,5,7,9} = 1$, $k_{4,6,8,10} = 0.5$, and $R_i = 50$

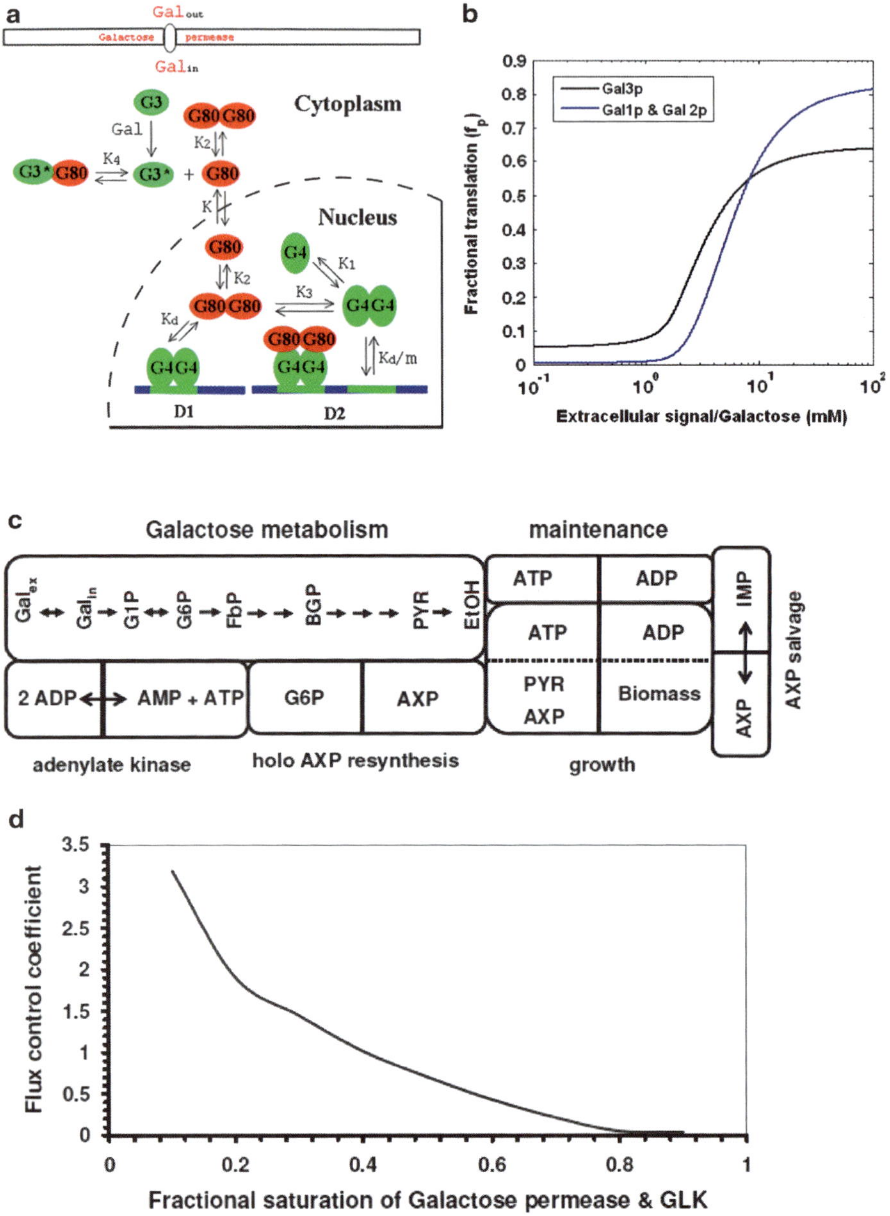

Fig. 3.6 (**a**) Galactose signalling for transcriptional regulation of *GAL* genes in *Saccharomyces cerevisiae* adopted from (Verma et al. 2003). *D*1 and *D*2 are genes having one and two binding sites for transcriptional factor in their promoter regions. *G*4, *G*80, *G*3, and *G*3* represent transcriptional activator Gal4p, transcriptional repressor Gal80p, signal transducer Gal3, and activated Gal3 protein. Gal4p and Gal80p form dimers and the Gal4p dimer binds to the binding sites of genes *D*1 and *D*2. Cooperative binding of Gal4p dimer to *D*2 is represented by parameter *m* when one site is already occupied by Gal4p dimer. The repressor Gal80p dimer binds with free Gal4p and also with the Gal4p bound with genes *D*1 and *D*2. *K* denotes distribution of Gal80p shuttling between the nucleus and cytoplasm. Activated Gal3p sequesters with Gal80p in

whereas on a second-to-minute timescale most signal flux into and out of GS turned to glutamate and the adenylylation state. On a minute timescale it was the adenylylation state signal flux into GS that dominated. On average for the whole transient period, some 60 % of the regulation was mediated by signal through the adenylylation state of GS and some 40 % by ADP, glutamate, glutamine, and ATP (Bruggeman et al. 2005). Again, i.e. also when it comes down to GS regulation, there is no single regulatory route, but all of the signal transducing routes contribute to different degrees on different timescales.

3.10 Medical Implications of Control Analysis

In medical applications the focus of control analysis applications has been to identify:

1. The enzymes having high flux control coefficients in the pathway influenced by disease; these are likely to be potent drug targets reducing metabolic flux through the pathway upon their inhibition.
2. The enzymes having high flux control coefficients in the pathogenic microorganism, for which the enzymes catalysing the homologous reaction in the complementary human metabolic pathway have a low flux control coefficient; these can be targeted to suppress the growth of microorganism with minimal effect on the humans.
3. Bottlenecks of processes important for nutrition; Rigoulet et al. (1988) calculated the redistribution of flux control coefficients in the TCA cycle in brain edema. In diabetes it is important to know which enzymes control the carbon flux into the gluconeogenic pathway. Groen et al. (1983) and Rigoulet et al. (1987) have shown that pyruvate carboxylase exerts much flux control and would, therefore, be a potent drug target.

MCA has been used to rank drug targets to suppress *Trypanosoma brucei*, the parasite causing sleeping disease. This was done by calculating which steps in glycolysis needed the least inhibition to achieve a certain inhibition of the glycolytic flux (Bakker et al. 1999). The glucose transporter in the *Trypanosoma brucei*

Fig. 3.6 (continued) cytoplasm to activate the transcription of *GAL* genes. K_i (i = 1–4) and K_d values represent dissociation constants for various protein–protein and DNA–protein interactions, respectively. All the parameters values, genes, and protein concentrations were taken from (Verma et al. 2003). (**b**) Simulated fractional protein expression from *GAL* genes having one and two binding sites for Gal4p dimer in response to galactose signalling. (**c**) Schematic of complete domino model for galactose metabolism in *Saccharomyces cerevisiae*, adopted from domino glucose metabolism (Verma et al. 2013). (**d**) Simulated profile of flux control coefficients for fractional saturation of galactose permease and galactokinase, enzymes catalysing reaction 1 and reaction 2, respectively, of domino galactose metabolism. Model parameters were replicated from domino glucose metabolism (Verma et al. 2013)

was identified to be the most promising target, followed by aldolase, glycerol-3-phosphate dehydrogenase, and glyceraldehyde-phosphate dehydrogenase. Eisenthal and Cornish-Bowden (1998) have suggested that metabolite would accumulate sufficiently to interfere with the action of competitive inhibitors. Following this reasoning, competitive inhibition of pyruvate export was found more adequate in their studies. The most promising target the glucose transporter (Bakker et al. 1999) should however be insensitive to this argument, as the extracellular glucose concentration will not be affected by such an inhibitor. Also glycolytic and energy metabolism in cancer has been studied from the control analysis perspective, leading to new drug targets (Moreno-Sanchez et al. 2010).

Letellier et al. (1998) detailed an important application of MCA in medicine, i.e. to enzyme deficiencies, in mitochondrial myopathies. MCA explained why flux is not considerably reduced above a certain threshold activity of the enzyme, and only after further inhibition, the flux decreased drastically. In case of enzyme deficiencies, control coefficients can only be used as a guideline. Therefore, Schuster and Holzhutter (1995) have used kinetic modelling to address the impact of enzyme deficiencies in erythrocytes.

One of the limitations with current chemotherapies is not that they fail to kill cancer cells but that they fail to destroy it at doses which are not harmful to normal cells. The therapeutic window between cytotoxicity in malignant cells versus normal tissue is too narrow. A key to finding targets where cancer may be differentially sensitive is to try and understand what are the differences in flux control between cancer and normal cells. Then drugs may be targeted to steps where the flux control is much higher in the tumour than in healthy tissues. These steps could be enzymes (Moreno-Sanchez et al. 2010) as well as signal transduction proteins (Hornberg et al. 2005a).

3.11 Software for Control Analysis

Several programmes are adopting the MCA approach and thus we avail of some useful tools for various aspects of MCA. Metabolic simulators such as COPASI (Hoops et al. 2006) are programmes for simulation of steady states and transient behaviour of biochemical pathways (including several compartments). They provide features to enable calculation of all the enzyme elasticities and flux and concentration control coefficients. JWS online (Olivier and Snoep 2004) is a public model repository enabling the use of realistic models through a web interface. Control coefficients become available at the clicking of a button and their dependence on network properties can be computed in close to realistic models, including those of signal transduction.

3.12 Perspectives

Molecular biological components and the systems in which they function (e.g. substrates, enzymes, metabolites, genes in a cell, tissue, or organism) and pathways mediating their functional outcome are many times more complex than networks and circuits such as the London Underground. Yet, the London Underground is already complex: making sure that one particular station functions efficiently such that each minute one train could depart does not guarantee that indeed one train will depart per minute. Most likely, many fewer trains will depart because further down the tube there are other stations that are less efficient, or because at some stations up the tube excessive numbers of people wish to board or leave the train, or because the train driver overslept. Network studies reside at the crossroads of disciplines, from mathematics (graph theory, combinatorics, probability theory) to physics (statistical thermodynamics, macromolecular crowding), and from computer science (network generating algorithms, combinatorial optimisation) to the life sciences (metabolic and regulatory networks between proteins and nucleic acids). The impact of network theory on understanding is strong in all natural sciences (Barabási and Albert 1999), especially in systems biology with gene networks (Alon 2007), metabolic networks (Schuster et al. 2002), plant systems biology, and even food webs (Getz et al. 2003). Yet, biological systems will not be understood by existing network theory alone. Their properties are much more complex than the properties of standard networks, for instance in that their networks adapt and change temporarily, are hierarchical in terms of space, time, and organisation, and have been optimised through evolution for multiple properties that we do not yet understand. New network theories are needed and will have to be more targeted towards understanding biological systems functionally. These will have to integrate strongly with genomics and molecular data, because different biological networks may need somewhat different theories, if only because their objective (evolutionary purpose) is different.

Acknowledgement HVW, MV, and SR thank the transnational program Systems Biology of MicroOrganisms (SysMO) and ERASysBioprogram and its funders EPSRC/BBSRC for supporting the following projects (R111828, MOSES project, ERASysBio, BB/F003528/1, BB/C008219/1, BB/F003544/1, BB/I004696/1, BB/I00470X/1, BB/I017186/1, EP/D508053/1, BB/I017186/1). HVW also thanks other contributing funding sources including the NWO, and EU-FP7 (EC-MOAN, UNICELLSYS, SYNPOL, ITFOM).

References

Alberghina L, Westerhoff HV (eds) (2005) Systems biology: definitions and perspectives. Springer, Berlin

Alon U (2007) Network motifs: theory and experimental approaches. Nat Rev Genet 8:450–461

Bakker BM, Michels PAM, Opperdoes FR, Westerhoff HV (1999) What controls glycolysis in bloodstream form Trypanosoma brucei? J Biol Chem 274:14551–9

Barabási A-L, Albert R (1999) Emergence of scaling in random networks. Science 286:509–12

Boogerd FC, Bruggeman FJ, Richardson RC, Stephan A, Westerhoff HV (2005) Emergence and its place in nature: a case study of biochemical networks. Synthese 145:131–64

Bruggeman FJ, Boogerd FC, Westerhoff HV (2005) The Multifarious short-term regualtion of ammonium assimilation of Escherichia coli: dissection using an in silico replica. FEBS J 272: 1965–85

Burns JA, Cornish-Bowden A, Groen AK, Heinrich H, Kacser H, Porteous JW et al (1985) Control of metabolic systems. Trends Biochem Sci 10:16

Caplan SR, Essig A (1969) Oxidative phosphorylation: thermodynamic criteria for the chemical and chemiosmotic hypotheses. Proc Natl Acad Sci USA 64:211–18

Conradie R, Bruggeman FJ, Ciliberto A, Csikász-Nagy A, Novák B, Westerhoff HV et al (2010) Restriction point control of the mammalian cell cycle via the cyclin E/Cdk2:p27 complex. FEBS J 277:357–67

Cortassa S, Aon MA, Westerhoff HV (1991) Linear nonequilibrium thermodynamics describes the dynamics of an autocatalytic system. Biophys J 60:794–803

Dykhuizen DE, Dean AM, Hartl DL (1987) Metabolic flux and fitness. Genetics 115:25–31

Eisenthal R, Cornish-Bowden A (1998) Prospects for antiparasitic drugs: the case of Trypanosoma brucei, the causative agent of African sleeping sickness. J Biol Chem 273:5500–05

Fell DA (1992) Metabolic control analysis: a survey of its theoretical and experimental development. Biochem J 286:313–30

Flint HJ, Tateson RW, Barthelmess IB, Porteous DJ, Donachie WD, Kacser H (1981) Control of the flux in the arginine pathway of Neurospora crassa. Modulations of enzyme activity and concentration. Biochem J 200:231–46

García-Contreras R, Vos P, Westerhoff HV, Boogerd FC (2012) Why in vivo may not equal in vitro – new effectors revealed by measurement of enzymatic activities under the same in vivo-like assay conditions. FEBS J 279:4145–59

Getz WM, Westerhoff HV, Hofmeyr J-HS, Snoep JL (2003) Control analysis of trophic chains. Ecol Model 168:153–71

Giersch C (1988) Control analysis of metabolic networks. Eur J Biochem 174:509–13

Groen AK, Wanders RJ, Westerhoff HV, van der Meer R, Tager JM (1982) Quantification of the contribution of various steps to the control of mitochondrial respiration. J Biol Chem 257: 2754–57

Groen AK, Vervoorn RC, Van der Meer R, Tager JM (1983) Control of gluconeogenesis in rat liver cells. I. Kinetics of the individual enzymes and the effect of glucagon. J Biol Chem 258:14346–53

Heinrich R, Rapoport TA (1974) A linear steady-state treatment of enzymatic chains. General properties, control, effector strength. Eur J Biochem 42:89–95

Heinrich R, Neel BG, Rapoport TA (2002) Mathematical models of protein kinase signal transduction. Mol Cell 9:957–70

Hendrickson DG, Hogan DJ, McCullough HL, Myers JW, Herschlag D, Ferrell JE et al (2009) Concordant regulation of translation and mRNA abundance for hundreds of targets of a human microRNA. PLoS Biol 7:e1000238

Hoops S, Sahle S, Gauges R, Lee C, Pahle J, Simus N et al (2006) COPASI-a COmplex PAthway SImulator. Bioinformatics 22:3067–74

Hornberg JJ, Binder B, Bruggeman FJ, Schoeberl B, Heinrich R, Westerhoff HV (2005a) Control of MAPK signalling: from complexity to what really matters. Oncogene 24:5533–42

Hornberg JJ, Bruggeman FJ, Binder B, Geest CR, de Vaate AJMB, Lankelma J et al (2005b) Principles behind the multifarious control of signal transduction. FEBS J 272:244–58

Jensen PR, Westerhoff HV, Michelsen O (1993) Excess capacity of H^+-ATPase and nverse respiratory control in Escherichia coli. EMBO J 12:1277–87

Jensen PR, van der Weijden CC, Jensen LB, Westerhoff HV, Snoep JL (2000) Extensive regulation compromises the extent to which DNA gyrase controls DNA supercoiling and growth rate of Escherichia col. Eur J Biochem 266:865–77

Kacser H, Burn JA (1973) The control of flux. Symp Soc Exp Biol 27:65–104

Kahn D, Westerhoff HV (1991) Control theory of regulatory cascades. J Theor Biol 153:255–85

Kholodenko BN, Molenaar D, Schuster S, Heinrich R, Westerhoff HV (1995) Defining control coefficients in "non-ideal" metabolic pathways. Biophys Chem 56:215–26

Kholodenko BN, Kiyatkin A, Bruggeman FJ, Sontag E, Westerhoff HV, Hoek JB (2002) Untangling the wires: a strategy to trace functional interactions in signaling and gene networks. Proc Natl Acad Sci USA 99:12841–46

Kolodkin AN, Bruggeman FJ, Plant N, Moné MJ, Bakker BM, Campbell MJ, van Leeuwen JP, Carlberg C, Snoep JL, Westerhoff HV (2010) Design principles of nuclear receptor signaling: how complex networking improves signal transduction. Mol Syst Biol 6(446):102

Kolodkin A, Simeonidis E, Westerhoff HV (2012) Computing life: add logos to biology and bios to physics. Prog Biophys Mol Biol 111(2–3):69–74

Letellier T, Malgat RR, Mazat JP (1998) Metabolic control analysis and mitochondrial pathologies. Mol Cell Biochem 184:409–17

Michaelis M, Menten ML (1913) Kinetics invertinwirkung. Biochem J 49:333–69

Mitchell P (1961) Coupling of phosphorylation to electron and hydrogen transfer by a chemi-osmotic type of mechanism. Nature 191:144–8

Moreno-Sánchez R, Encalada R, Marín-Hernández A, Saavedra E (2008) Experimental validation of metabolic pathway modeling. FEBS J 275:3454–69

Moreno-Sanchez R, Saavedra E, Rodriguez-Enriquez S, Gallardo-Perez JC, Quezada H, Westerhoff HV (2010) Metabolic control analysis indicates a change of strategy in the treatment of cancer. Mitochondrion 10:626–39

Niederberger P, Prasad R, Miozzari G, Kacser H (1992) A strategy for increasing an in vivo flux by genetic manipulations. The tryptophan system of yeast. Biochem J 287:473–79

Olivier BG, Snoep JL (2004) Web-based modelling using JWS-Online. Bioinformatics 20: 2143–44

Ortega F, Acerenza L, Westerhoff HV, Mas F, Cascante M (2002) Product dependence and bifunctionality compromise the ultrasensitivity of signal transduction cascades. Proc Natl Acad Sci USA 99:1170–75

Quinton-Tulloch MJ, Bruggeman FJ, Snoep JL, Westerhoff HV (2013) Trade-off of dynamic fragility but not of robustness in metabolic pathways in silico. FEBS J 280:160–73

Rigoulet M, Leverve XM, Plomp PAJ, Meijer AJ (1987) Stimulation by glucose of gluconeogenesis in hepatocytes isolated from starved rats. Biochem J 245:661–8

Rigoulet M, Averet N, Mazat JP, Guerin B, Cohadon F (1988) Redistribution of the flux-control coefficients in mitochondrial oxidative phosphorylations in the course of brain edema. Biochim Biophys Acta 932:116–23

Savageau MA (1976) Biochemical systems analysis: a study of function and design in molecular biology. Addison-Wesley, Reading, MA

Schuster R, Holzhutter HG (1995) Use of mathematical models for predicitng the metabolic effects of large-scale enzyme-activity alterations. Eur J Biochem 229:403–18

Schuster S, Hilgetag C, Woods JH, Fell DA (2002) Reaction routes in biochemical reaction systems: algebraic properties, validated calculation procedure and example from nucleotide metabolism. J Math Biol 45:153–81

Thiele I, Swainston N, Fleming RMT, Hoppe A, Sahoo S, Aurich MK et al (2013) A community-driven global reconstruction of human metabolism. Nat Biotechnol 31(5):419–25

Thomas S, Fell DA (1998) The role of multiple enzyme activation in metabolic flux control. Adv Enzyme Regul 38:65–85

Torres NV, Mateo F, Melendez-Hevia E, Kacser H (1986) Kinetics of metabolic pathways. A system in vitro to study the general control of flux. Biochem J 234:169–74

Van Dam K, Van der Vlag J, Kholodenko BN, Westerhoff HV (1993) The sum of the control coefficients of all enzymes on the flux through a group transfer pathway can be as high as two. Eur J Biochem 212:791–99

van Driel R, Fransz PF, Verschure PJ (2003) The eukaryotic genome: a system regulated at different hierarchical levels. J Cell Sci 116:4067–75

van Eunen K, Bouwman J, Daran-Lapujade P, Postmus J, Canelas AB, Mensonides FIC et al (2010) Measuring enzyme activities under standardized in vivo-like conditions for systems biology. FEBS J 277:749–60

Verma M, Bhat PJ, Venkatesh KV (2003) Quantitative analysis of GAL genetic switch of Saccharomyces cerevisiae reveals that nucleocytoplasmic shuttling of Gal80p results in a highly sensitive response to galactose. J Biol Chem 278:48764–69

Verma M, Zakhartsev M, Reuss M, Westerhoff HV (2013) 'Domino' systems biology and the 'A' of ATP. Biochim Biophys Acta 1827:19–29

Wagner A, Fell DA (2001) The small world inside large metabolic networks. Proc Biol Sci 7: 1803–10

Westerhoff HV (2008) Signalling control strength. J Theor Biol 252:555–67

Westerhoff HV, Chen Y-D (1984) How do enzyme activities control metabolite concentrations? Eur J Biochem 142:425–30

Westerhoff HV, Dam KV (1987) Thermodynamics and control of biological free-energy transduction. Elsevier, Amsterdam

Westerhoff HV, Kell DB (1987) Matrix method for determining steps most rate-limiting to metabolic fluxes in biotechnological processes. Biotechnol Bioeng 30:101–7

Westerhoff HV, Palsson BO (2004) The evolution of molecular biology into systems biology. Nat Biotechnol 22:1249–52

Westerhoff H, van Workum M (1990) Control of DNA structure and gene expression. Biomed Biochim Acta 49:839–53

Westerhoff HV, Aon MA, Kv D, Cortassa S, Kahn D, Mv W (1990) Dynamical and hierarchical coupling. Biochim Biophys Acta 1018:142–46

Westerhoff HV, Kolodkin A, Conradie R, Wilkinson S, Bruggeman F, Krab K et al (2009a) Systems biology towards life in silico: mathematics of the control of living cells. J Math Biol 58:7–34

Westerhoff HV, Winder C, Messiha H, Simeonidis E, Adamczyk M, Verma M et al (2009b) Systems biology: the elements and principles of life. FEBS Lett 583:3882–90

Chapter 4
MicroRNAs and Robustness in Biological Regulatory Networks. A Generic Approach with Applications at Different Levels: Physiologic, Metabolic, and Genetic

Jacques Demongeot, Olivier Cohen, and Alexandra Henrion-Caude

Abstract MicroRNAs have been discovered in the noncoding nuclear genome. They inhibit partly in a nonspecific manner the transcription of numerous genes, and the corresponding "inhibitory noise" prevents the weakest positive interactions of the genetic regulatory networks to be actually efficient, hence microRNAs control the number of attractors of these networks, e.g., by frequently forcing them to have only one or two possible behaviors for fulfilling a precise cell function (if we identify a network attractor with a precise differentiated cell state). More specifically, microRNAs have a great influence on the chromatin clock, which ensures the controlled mode of updating genetic regulatory networks. We analyze this influence as well as their impact on important functions like controlling the cell cycle, improving the defenses of a host against pathogens like viruses, and maintaining the homeostasis of energy metabolism. In the last case, we show the role of two types of microRNAs, both involved in the control of the mitochondrial genome: (1) nuclear microRNAs, called mitoMirs, inhibiting mitochondrial genes and (2) putative mitochondrial microRNAs located in the noncoding part of the mitochondrial genome that inhibit tRNAs function. We show the complex

J. Demongeot
AGIM FRE CNRS-UJF 3405, Faculty of Medicine, University J. Fourier of Grenoble, 38700 La Tronche, France

Escuela de Ingeniería Civil Informática y Departamento de Ingeniería Biomedica, Universidad de Valparaiso, Valparaiso, Chile
e-mail: Jacques.Demongeot@agim.eu

O. Cohen (✉)
AGIM FRE CNRS-UJF 3405, Faculty of Medicine, University J. Fourier of Grenoble, 38700 La Tronche, France
e-mail: Olivier.Cohen@ujf-grenoble.fr

A. Henrion-Caude
INSERM U 781, Université Paris Descartes, Hôpital Necker—Enfants Malades, 149 rue de Sèvres, 75015 Paris, France
e-mail: alexandra.caude@inserm.fr

M.A. Aon et al. (eds.), *Systems Biology of Metabolic and Signaling Networks*, Springer Series in Biophysics 16, DOI 10.1007/978-3-642-38505-6_4,

involvement of microRNAs in the ubiquitous p53 regulatory function of cell cycle control, then their global role in cell respiration homeostasis, in carcinogenesis, and finally we discuss the influence of microRNAs on the increase of robustness of genetic networks during evolution.

4.1 Introduction

The whole body physiology as well as cell metabolism are regulated by interaction networks that bring together as elementary nodes, cells or macromolecules like genes and their expression products, proteins, or other metabolites. This complex organization is made out of numerous weak interactions due to physicochemical forces like electrostatic or van der Waals forces, or to electrical and/or mechanical forces. The aim of this chapter is to show how mathematical theories like graph theory, discrete network theory, and dynamical systems theory are necessary to give a mechanistic description of how a cell works, a tissue or an organ grows, from the emergent properties of their constituents interacting at different levels of complexity. The corresponding regulatory networks made of elements (e.g., genes, proteins, cells,. . .) in interaction control important tissue functions like proliferation and differentiation and cellular functions like respiration or glycolysis. The dynamics of these networks depends highly on the relationships and delays between the kinetics of creation and/or transformation of their elements and then they need to be described in the framework of Systems Biology.

A system is a set of elements in interaction and the cell (resp. tissue) organization is a biological system, considered as a pyramid of components made of interacting macromolecules (resp. cells). Their observed spatio-temporal behavior (phenotype) can be explained through several loops of complexity from data acquisition to reconstruction of regulatory interaction networks (inverse problem) at different levels, allowing direct predictions by modeling and simulating them in silico. This complexity deals with kinetic rules (Henri–Michaelis–Menten, Hill, Monod–Wyman–Changeux, Thomas,. . .) prescribing how macromolecules, cells, and tissues are connected into integrated regulatory networks with architectural similarities both inside the cell and at tissue level. The dynamics allowed by these rules and the corresponding discrete or differential equations allow simulating trajectories to be compared with the temporal evolutions observed in experiments. These trajectories can be stable in different mathematical senses that we present in Sect. 4.2; a system being stable in all senses will be declared robust. In Sect. 4.3, we consider simple examples such as robust and non-robust systems. In Sect. 4.4, we examine the consequences of the role of microRNAs in the energetic system of the cell, at the physiologic vegetative level. In Sect. 4.5, we study the global role of microRNAs at the level of mitochondrial or chloroplast respiration, in the genesis of cancer and in the control of the defenses against infectious agents. In Sect. 4.6, we present a brief perspective about the role of microRNAs in maintaining robustness in genetic networks.

4.2 Definitions of the Notions of Stability and Robustness

There exist several definitions of the stability of a dynamical system and we will give hereafter the most useful. We define a trajectory $x(a,t)$ in a state space $E \subset \mathbb{R}^n$ as the set of all states observed as the time goes from initial value 0, corresponding to the state $x(a,0) = a$, to infinity. The set of states visited when t tends to infinity is called the limit set of the trajectory starting in a, and is denoted $L(a)$. If the initial states lie in a set A, then $L(A)$ is the union of all limit sets $L(a)$, for a belonging to A. Conversely, $B(A)$, called the attraction basin of A is the set of all initial conditions outside A, whose limit set $L(a)$ is included in A. We will call attractor in the following a set A such as (1) $A = L(B(A))$, (2) A is not contained in a wider set B, such as $\mathrm{d}(A,B\backslash A) = \inf_{a \in A,\ b \in B\backslash A} \mathrm{d}(a,b) = 0$ and verifying (1), and (3) A does not contain a strictly smaller subset C verifying (1) and (2). Such an attractor A associated to its basin $B(A)$ is the exact set of the states "attracting" the trajectories coming from $B(A)$ outside A.

4.2.1 *Definition of the Lyapunov (or Trajectorial) Stability*

A trajectory $x(a,t)$ is called Lyapunov stable, if no perturbation at any time t should be amplified: if $b = x(a,t) + \varepsilon$ denotes the perturbed state, then for any $s > t$, $\mathrm{d}(x(a,s), x(b,s-t)) \leq \varepsilon$.

4.2.2 *Definition of the Asymptotic Stability*

A trajectory $x(a,t)$ is called asymptotically stable, if any perturbation at any time t is asymptotically damped: if $b = x(a,t) + \varepsilon$ denotes the perturbed state, then $\lim_{s \to \infty} \mathrm{d}(x(a,s), x(b,s-t)) = 0$.

4.2.3 *Definition of the Structural Stability with Respect to a Parameter μ*

A dynamical system whose trajectories $x_p(a,t)$ depend on a parameter p is called structurally stable with respect to the parameter p, if no perturbation of p at any time can provoke a change in number or nature (fixed attractor, called steady state, or periodic attractor, called limit cycle) of its attractors. p may parametrize the state transition rule of the system or its architectural characteristics (number of elements, intensity of interactions).

4.2.4 Definition of the Structural Stability with Respect to Updating Modes

A dynamical system is called structurally stable with respect to updating modes, if no disturbance in the updating schedule, e.g., by passing from the sequential mode (in which nodes of the network are updated by the state transition rule one after the other in a given order) to the parallel one (in which all the nodes of the network are updated by the state transition rule at the same time), can change the number or nature of its attractors.

4.2.5 Definition of the Resistance to Boundary Perturbations (Resilience)

A dynamical system having a frontier separating them from its environment is resistant to boundary perturbations, if no perturbation in state or architecture of the environmental elements can provoke a change in number or nature of its attractors.

4.2.6 Definition of the Robustness (Resilience)

A dynamical system whose trajectories $x_p(a,t)$ depend on a parameter p is said to be robust (or resilient), if all of its trajectories are asymptotically stable and if it is boundary resistant and structurally stable with respect to any parameter or updating schedule perturbation. We will give in the next Section some examples of robust and non-robust biological networks at different levels, genetic, metabolic, and physiologic. These networks can be decomposed following their dynamical typology and we can distinguish between four categories of dynamics, whose definition will be given hereafter:

- Potential (or gradient, or purely dissipative)
- Hamiltonian (or conservative)
- Mixed potential-Hamiltonian (MPH)
- MPH with principal potential part.

4.2.7 Definition of a Potential Dynamics

A dynamical system has a potential dynamics if the velocity along its trajectories is equal to the gradient of a scalar potential P defined on the state space E. If the system is governed by a differential equation defining its state transition rule, we have: d$x(a,t)$/dt = $-\nabla P$ = $-$grad$P$ = $-\partial P/\partial x$, where $-\partial P/\partial x$ is the vector $(-\partial P/\partial x_1,\ldots,-\partial P/\partial x_n)$ and the state x is a vector of dimension n: $x = (x_1,\ldots, x_n)$. The system is called dissipative, because the potential P decreases along trajectories until attractors which are located on the minima of P.

4.2.8 Definition of a Hamiltonian dynamics

A dynamical system has a Hamiltonian dynamics if the velocity along its trajectories is tangent to the contour lines projected on E from the surface representative of an energy function H defined on E: d$x(a,t)$/dt = tang$H$. If the dimension of the system is 2, then the vector Tang$H$ is equal to $(\partial H/\partial x_2, -\partial H/\partial x_1)$. The system is said conservative, because the energy function H is constant along a trajectory.

4.2.9 Definition of a Mixed Potential-Hamiltonian Dynamics

A dynamical system has a mixed potential-Hamiltonian dynamics if the velocity along its trajectories can be decomposed into two parts, one potential and one Hamiltonian: d$x(a,t)$/dt = $-$grad$P$ + tang$H$. If the set of minima of P is a contour line of the surface H on E, then its connected components are attractors of the system.

4.2.10 Definition of a Principal Potential Part Dynamics

A mixed potential-Hamiltonian system has a principal potential part dynamics, if the ratio $\|\text{tangH}\|/\|\text{grad}P\|$ between the norms of the potential part and the Hamiltonian one tends to 0 when t tends to infinity.

4.3 Examples of Robust and Non-Robust Regulatory Networks

We will give as examples in the following some toy models coming from regulatory networks studied more in details in (Demongeot et al. 2000; Thellier et al. 2004; Cinquin and Demongeot 2005; Forest and Demongeot 2006; Forest et al. 2006; Jolliot and Prochiantz 2004; Demongeot et al. 2007a, b, 2008a, 2009a, 2010a, 2011a, b; Glade et al. 2007; Demongeot and Françoise 2006; Ben Amor et al. 2008;

Demongeot and Waku 2012a; Demongeot and Demetrius (submitted); van der Pol and van der Mark 1928; http://www.sciencesunivnantesfr_sites_genevieve_tulloue_index_fichiers_animflash.html).

4.3.1 Example of a Potential Metabolic Network in Morphogenesis

Metabolic networks used in plant or animal morphogenesis are fully connected with all interactions inhibitory, except positive auto-loops representing auto-catalytic processes. Such systems called n-switches (Demongeot et al. 2000; Thellier et al. 2004; Cinquin and Demongeot 2005) are driven by the differential equations ruling the concentrations x_i of growth hormones [as auxin for plants like Araucaria trees—see Fig. 4.1 and Forest and Demongeot (2006) and Forest et al. (2006)—or transduction peptides for animals (Jolliot and Prochiantz 2004)] in several locations i, the value $i = 0$ being reserved to the location of the first growth, 1 to the second growth location,... (apex leaves cells for plants and after first bud cells; medulla primitive cells for animal nervous system and after bulbar cells) following:

$$\forall i = 0, \ldots, n, \ \mathrm{d}X_i/\mathrm{d}t = \sigma + Va_iX_i^c/(1 + \Sigma_{j=0,\ldots,n}a_jX_j^c) - \mu_i x_i,$$

where σ, V, c, and a_i's are, respectively, a constant entry flux, the Hill's maximum velocity, cooperativity, and affinities, and μ_i's the degradation rates of hormones. The dynamics is ruled by the potential:

$$P(y) = -\sigma\sum\nolimits_i \mathrm{Log}y_i/2 - V\mathrm{Log}(1 + \Sigma_i a_i y_i^{2c})/4c + \Sigma_i \mu_i y_i^2/4, \ \text{where } y_i^2 = X_i$$

If we replace Hill's kinetics by allosteric Monod–Wyman–Changeux kinetics, the system remains potential (Demongeot et al. 2007a, b; Glade et al. 2007).

In the case of plant morphogenesis, the first attractor observed corresponds to the steady state values $x_0{}^*$ verifying $\sigma + Va_0x_0{}^{*c}/(1 + a_0x_0{}^{*c}) - \mu_0x_0{}^* = 0$ and $x_i{}^* = 0$, for $i > 0$. It corresponds to the growth of the apex, but after a certain time, due to the negative geotropic plant growth, auxin from apex leaves cells can no more sufficiently diffuse among the first bud of the plant for inhibiting its cells, then the second steady state can be expressed, verifying: $\sigma + Va_{0x1}{}^{**c}/(1 + a_0x_1{}^{**c}) - \mu_0x_1{}^{**} = 0$ and $x_i{}^{**} = 0$, for $i \neq 1$. An example of such successive bud growths is given by Araucaria tree (Fig. 4.1, middle).

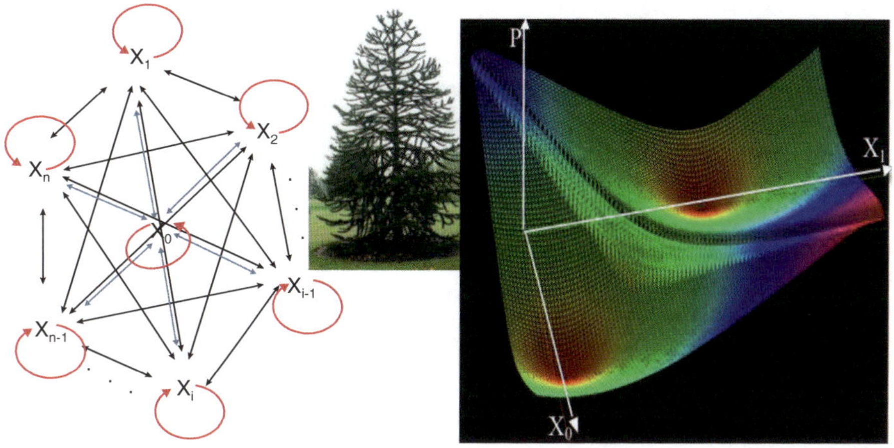

Fig. 4.1 *Left*: Representation of the interaction graph of an n-switch fully connected with negative interactions (*in black*) and positive ones (*in red*). *Middle*: *Araucaria* tree growth. *Right*: representation of the potential P in case $n = 1$: P is associated to the 2-switch defined by $\mu = 1$, $c = 2$, $V = 2$, $a_i = 0.1$, $\sigma = 1$. The surface of P is represented on the state space $E = (X_0 0 X_1)$

4.3.2 Example of a Hamiltonian Genetic Regulatory Network in Immunology

The network controlling the expression of the RAG (Recombination Activating Gene), responsible of the rearrangements of the $V(D)J$ region of the chromosome 14 in human, giving birth to the T-cell receptors alpha contains strong connected components (scc): (1) the subnetwork (given in Fig. 4.2) containing the gene GATA3 regulating the T helper cell maturation and made of two circuits, one positive of length 5 and another negative of length 3, tangent on the gene SOCS1 and (2) the subnetwork containing the gene PU.1 and made of two negative circuits of respective length 6 and 2 (Demongeot and Waku 2012a; Demongeot and Demetrius (submitted); Demongeot et al. 2011b). The circuits are Hamiltonian and the conservative energy is the discrete kinetic energy E defined on the circuit C by:

$$E(C) = \sum_{i \in C} (x_i(t) - x_i(t-1))^2/2,$$

where $x_i(t)$ is the Boolean state of the gene i at time t (equal to 1 if the gene is expressing its protein and 0 if it is in silence) and where the transitions on C are either the identity (symbol +) or the negation (symbol −). The number of attractors of the first scc is equal to 3 and of the second equal to 1 [cf. Demongeot and Waku (2012a), Demongeot and Demetrius (submitted), and Demongeot et al. (2011b) and numbers in red in Fig. 4.19 of the Mathematical Annex].

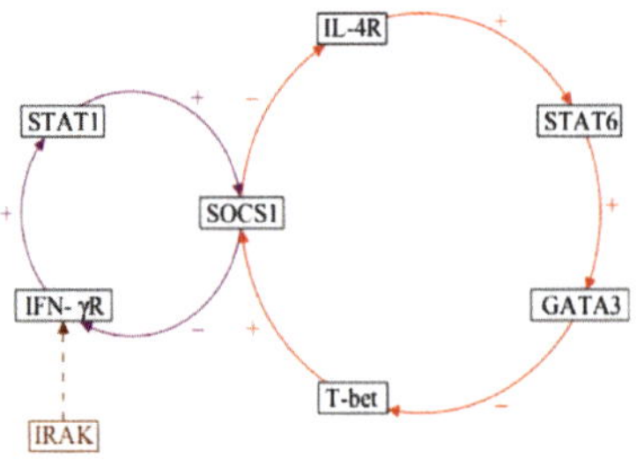

E (kinetic energy)	Attractor Nb	Attractor Length
0	2 fixed points	1
1	4	5
2	2	5

Fig. 4.2 *Left*: Immunetwork upstream the gene GATA3 regulating the T helper cell maturation (Demongeot et al. 2011b). *Right*: levels of the Hamiltonian discrete kinetic energy for the positive circuit of length 5 containing the gene GATA3 (*in red* on the *left*)

4.3.3 Example of a Mixed Potential-Hamiltonian Metabolic Network in Physiology

The van der Pol system has been used since about 80 years (van der Pol and van der Mark 1928; http://www.sciencesunivnantesfr_sites_genevieve_tulloue_index_fichiers_animflash.html) for representing the activity of cardiac cells (Fig. 4.3, bottom) and the differential equations representing its dynamics are defined by:

$$\mathrm{d}x/\mathrm{d}t = y,\ \mathrm{d}y/\mathrm{d}t = -x + \mu\left(1 - x^2\right)y$$

If $\mu = 0$, these equations are those of the simple pendulum and if $-x$ and $(1 - x^2)$ are replaced by polynomials in x of high order, they become Liénard systems (Demongeot et al. 2007a, b; Glade et al. 2007). It is possible to obtain a potential-Hamiltonian decomposition: $\mathrm{d}x/\mathrm{d}t = -\partial P/\partial x + \partial H/\partial y$, $\mathrm{d}y/\mathrm{d}t = -\partial P/\partial y - \partial H/\partial x$, with $H(x,y) = (x^2 + y^2)/2 \ -\mu xy(1 - x^2/4 + y^2/4)/2$. This decomposition is not unique and allows obtaining an approximation for the equation of its limit cycle in the form: $H(x,y) = c$ (Demongeot and Françoise 2006).

Figure 4.3 (top) shows a dynamical system in which H and P have a revolution symmetry and share contour lines of the corresponding surfaces especially a limit cycle in the state space $E = (x_1 0 x_2)$: the system can be assimilated, when the state has a norm sufficiently big, to the motion of the projection (in green) of a ball (in red) descending along the potential surface P until its minimal set, which is the contour line (in green) of a Hamiltonian surface (in red). Figure 4.3 (middle) shows the limit cycle for different values of the anharmonic parameter μ of the van der Pol equation. Figure 4.3 (bottom) shows that the rhythm of the potential of an isolated cardiac cell fits with the solution of a van der Pol equation, justifying the use of this equation to represent the whole heart kinetics (Van der Pol and van der Mark 1928; http://www.sciencesunivnantesfr_sites_genevieve_tulloue_index_fichiers_animflash.html).

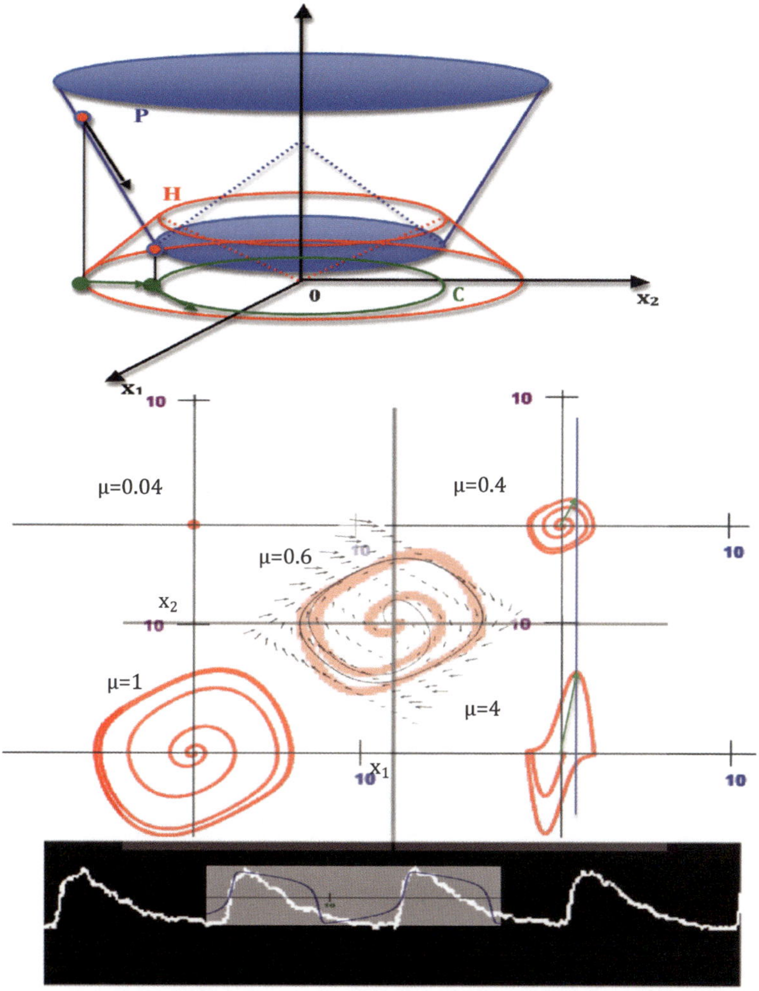

Fig. 4.3 Top: Dynamical system with the Hamiltonian and potential surfaces *H* and *P*, showing a revolution symmetry and sharing contour lines, especially the limit cycle C of the dynamical system (*in green*). *Middle*: simulations of van der Pol system $dx_1/dt = x_2$, $dx_2/dt = -x_1 + \mu(1-x_1^2)x_2$, for different values of the bifurcating anharmonic parameter μ (http://www.sciencesunivnantesfr_sites_genevieve_tulloue_index_fichiers_animflash.html). At the center, we have a representation of the velocity field for $\mu = 0.6$. *Bottom*: fit between the isolated cardiac cell rhythm (*in white*) and the solution of the van der Pol system (*in blue*)

4.3.4 Example of a Mixed Principal Potential Part Metabolic Network in High Glycolysis

The upper part of glycolysis contains two key enzymes, the PhosphoFructo-Kinase (PFK), which can be considered as following a Hill kinetics of order *n*, and the Aldolase, which follows a parabolic Hill (Demongeot and Doncescu

2009a). The substrates of the PFK and Aldolase are respectively fructose-6-phosphate and fructose-1,6-diphosphate, of which concentrations are x_1 and x_2, respectively. The corresponding 2-dimensional differential system is defined by the following equations:

$$\begin{aligned} \mathrm{d}x_1/\mathrm{d}t &= J - V_1 {x_1}^n/(K_1 + {x_1}^n), \mathrm{d}x_2/\mathrm{d}t = V_1 {x_1}^n/(K_1 + {x_1}^n) \\ &\quad - V_2 {x_2}^2/\left(K_2 + {x_2}^2\right), \text{or by changing the variables } y_i = (x_i)^2 \\ \mathrm{d}y_1/\mathrm{d}t &= \left[J - V_1 {y_1}^{2n}/\left(K_1 + {y_1}^{2n}\right)\right]/2y_1, \\ \mathrm{d}y_2/\mathrm{d}t &= \left[V_1 {y_1}^{2n}/\left(K_1 + {y_1}^{2n}\right) - V_2 {y_2}^4/\left(K_2 + {y_2}^4\right)\right]/2y_2 \end{aligned}$$

We can find a potential-Hamiltonian decomposition for this new differential system:

$$\mathrm{d}y_1/\mathrm{d}t = -\partial P/\partial y_1 + \partial H/\partial y_2 + R(y_1, y_2),\ \mathrm{d}y_2/\mathrm{d}t = -\partial P/\partial y_2 - \partial H/\partial y_1$$

where we have: $P = -J\ \mathrm{Log}(y_1)/2 + V_1 \mathrm{Log}(K_1 + {y_1}^{2n})/4n + V_2 \mathrm{Log}(K_2 + {y_2}^2)/4$, $H = -V_1 \int [{y_1}^{2n}/(K_1 + {y_1}^{2n})\mathrm{d}y_1]/2y_2$ and $R = -V_1 \int [{y_1}^{2n}/(K_1 + {y_1}^{2n})\mathrm{d}y_1]/2{y_2}^2$, H and R being negligible in the domain where ${y_2}^*$ is sufficiently large and ${y_1}^*$ sufficiently small, which corresponds for example to a large value of the V_{max} ratio V_1/V_2.

4.3.5 Example of a Robust Network, the Neural Hippocampus Network

In a toy model of the vegetative system, we can consider two simple networks, each of them having a regulon structure, i.e., two nodes in interaction with a negative circuit, one of them being auto-excitable (self-positive or auto-catalytic loop) [cf. Fig. 4.4, left and Elena et al. (2008) and Demongeot et al. (2002)]. The first regulon represents the vegetative control of the respiratory system with its inspiratory I and expiratory E neurons; the second one describes the vegetative control of the cardiac oscillator with the cardiomodulator bulbar node CM ruling the activity of the sinusal node S.

The corresponding dynamics is completely different if we consider the 2 regulons coupled or not, and noised or not. By modeling the transition of neuron states between time t and $(t + dt)$ by 2 coupled van der Pol differential equations: $\mathrm{d}x/\mathrm{d}t = y$, $\mathrm{d}y/\mathrm{d}t = -x + \mu(1 - x^2)y$, for the respiratory dynamics, where x represents the activity of neurons E, y the activity of neurons I, and μ is an anharmonic parameter, and $\mathrm{d}z/\mathrm{d}t = w$, $\mathrm{d}w/\mathrm{d}t = -z + \eta(1-z^2)w + k(y)y$, for the cardiac dynamics, where z represents the activity of pace-maker cells S, w the activity of neurons CM, η the anharmonic parameter, and $k(y)$ the coupling intensity parameter between inspiratory neurons I and cardiomodulator neurons

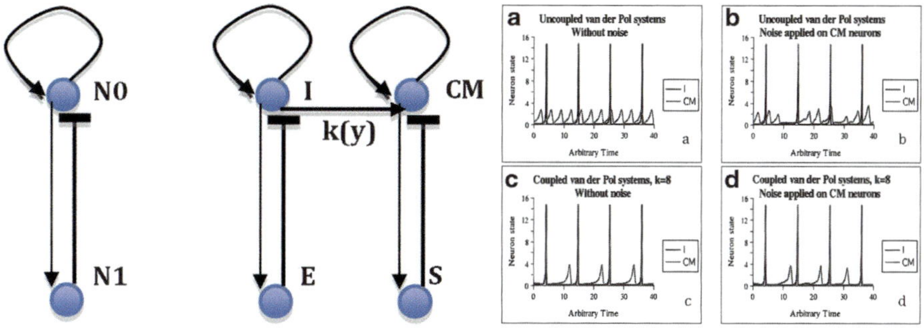

Fig. 4.4 *Left*: Structure of a single negative regulon with two nodes, N0 self-excitable and N1, and 2 regulons coupled between their respective self-excitable nodes I and CM by a directional edge with a coupling intensity $k(y)$. *Right*: Temporal series from simulations of the 2 negative regulons (neuron I in *black* and CM in *gray*), in four cases: (**a**) regulons are uncoupled without noise; (**b**) uncoupled with addition of noise to CM neurons; (**c**) coupled without noise; (**d**) coupled with noise. Parameters of the van der Pol equations are $\mu = 10$, $\eta = 1$, with $k(y) = 0$ when systems are uncoupled and $k(y) = 8$ when coupled

CM. Both respiratory and cardiac systems have indeed their own rhythm, but they are also coupled directionally: the cardiomodulator CM is coupled to the respiratory activity (inspiratory neurons I) via bulbar connections, causing a 1/1 harmonic entrainment in the case of coupling, perturbed in the case of uncoupling and robust in the case of coupling, by adding a Gaussian white noise to the second member of the van der Pol equations [cf. Fig. 4.4 Right and Elena et al. (2008) and Demongeot et al. (2002)].

4.3.6 *Example of an Asymptotically Stable and Structurally Instable Genetic Network Controlling Flower Morphogenesis*

The classical network controlling the flowering of *Arabidopsis thaliana* (Demongeot et al. 2010a; Mendoza and Alvarez-Buylla 1998) contains two strongly connected components and has five asymptotically stable attractors. The relative sizes of the five attraction basins corresponding to these attractors highly depend on the state of critical nodes as the gibberellin gene called RGA in Fig. 4.5 (plant hormone responsible of the flower growth).

4.3.7 *Example of a Non-Robust (Due to a Sensitivity to the Initial Conditions) Hamiltonian Population Dynamics Network*

V. Volterra introduced his famous differential system for interpreting the fluctuations observed in the struggle for life between a prey population of size x and a predator population of size y (Glade et al. 2007; Volterra 1926a, b, 1931) and

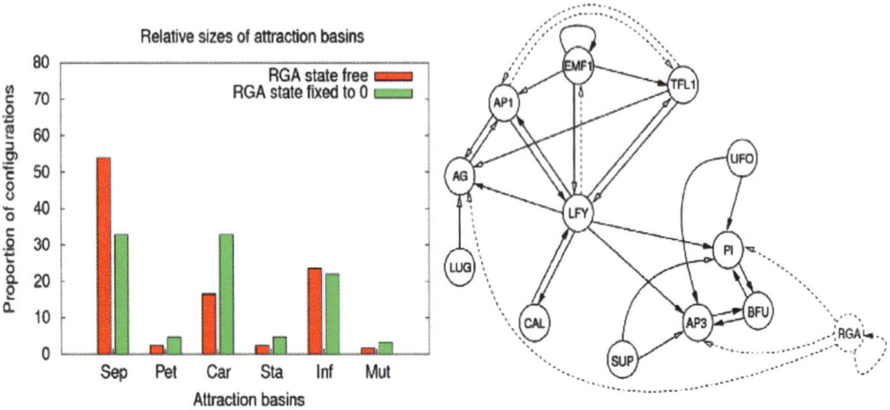

Fig. 4.5 *Left*: Relative sizes of the five attraction basins of the attractors of the genetic network controlling the flowering of *Arabidopsis thaliana*, in case of presence (*red*) or absence (*green*) of RGA regulation. *Right*: the network

A. Lotka (Lotka 1925) used the same system for representing the kinetics of bimolecular chemical reactions:

$$\mathrm{d}x/\mathrm{d}t = x(a - by),\ \mathrm{d}y/\mathrm{d}t = y(cx - d)$$

This system is purely Hamiltonian, by changing the variables $X = \mathrm{Log}x$ and $Y = \mathrm{Log}y$ (population «affinities») with a Hamiltonian function H equal to:

$$H(X, Y) = -ce^X + \mathrm{d}X - be^Y + aY,$$

the trajectories being exactly the contour lines of H. These trajectories are Lyapunov, but not asymptotically, stable. Hence, they are very sensitive to the initial conditions, which decide what will be the final shape of the temporal evolution of the system (cf. Fig. 4.6).

4.3.8 Example of a Non-Robust (Due to a Sensitivity to the Updating Mode) Genetic Network Controlling the Cell Cycle

The core of the genetic network controlling the cell cycle in mammals (Kohn 1999), the E2F box, has a strong connected component made of two intersecting positive circuits, one of length 4 and another of length 3 (Demongeot et al. 2008a, 2009a). The number and nature of its attractors depend both on the updating mode and on the state of a boundary node, the microRNA miRNA 159, which inhibits E2F. Then the system is not robust, essentially due to the occurrence of one periodic

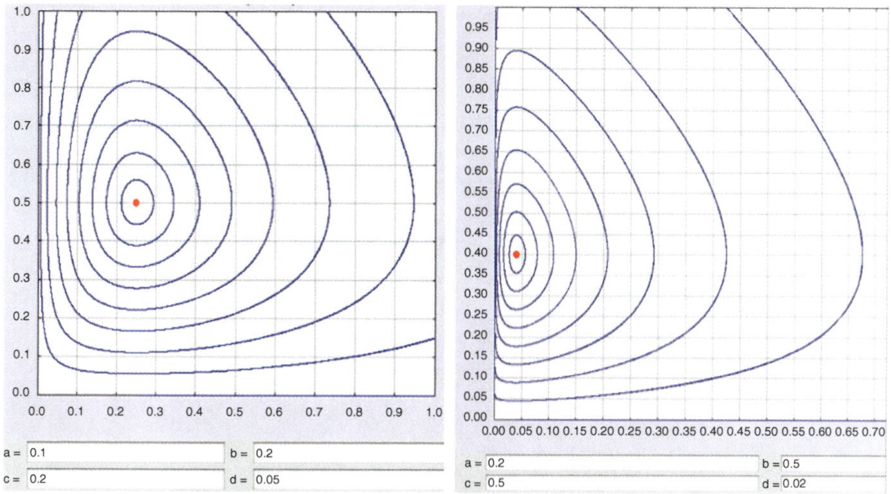

Fig. 4.6 Representation of the trajectories of a Lotka–Volterra system showing a collection of non-limit cycles (*in blue*) organized around a center (*in red*) for different values of the parameters *a*, *b*, *c*, *d* (http://www.medownloads.com/download-Lotka-Volterra-146411.htm)

attractor having an important Attraction Basin Relative Size (ABRS) in sequential updating mode when miRNA 159 is absent, and 4 periodic attractors when miRNA 159 is active and inhibits E2F (cf. Fig. 4.7).

4.4 MicroRNAs, MitomiRs, and ChloromiRs

4.4.1 Role of MicroRNAs in the General Architecture of Genetic Regulatory Networks

The microRNAs are parts of the nuclear genome which pertains to the noncoding genome. They have a partly unspecific role of inhibition, preventing the weakest part of the genetic regulatory networks to be expressed and hence the appearance of a too large number of attractors in these networks, i.e., forcing the network to have only few possible behaviors for fulfilling a precise function. MicroRNAs play an important role in both specific and nonspecific inhibition in many circumstances of the cell life, like chromatin clock control and have a big influence on many metabolic systems. We will first recall the origin of the microRNAs especially those acting on or pertaining to the mitochondrial (resp. chloroplast) genome called mitomiRs (resp. chloromiRs) (Bandiera et al. 2011, 2013; Demongeot et al. 2013a, b, c) and second present successively their role in controlling the energy system [and more generally homeostasis (Bernard 1865; Cannon 1932)], the chromatin clock, the morphogenesis and cell cycle, the immunologic system, and finally their possible influence in cancerogenesis and in robustness maintenance. For example,

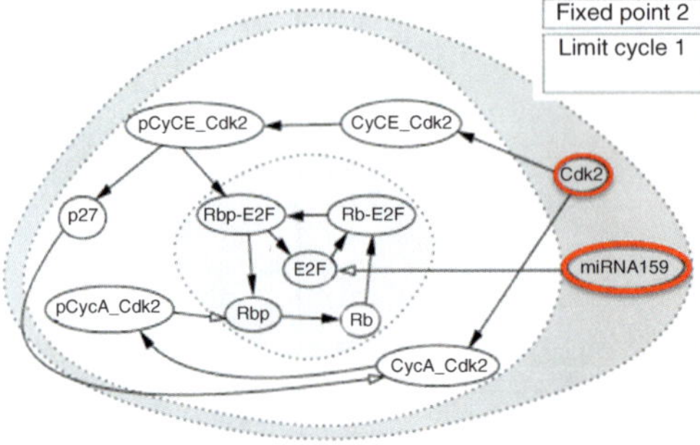

miRNA159=0	Sequential updating		Parallel updating		
Nature	Attractor	ABRS	Attractor	ABRS	AD
Fixed point 1	000000000000	6.25%	000000000000	0.5%	1.45
Fixed point 2	000000011111	56.25%	000000011111	99.5%	5.40
Limit cycle 1	000000001000 000000010111	37.5%	None		

miRNA 159 = 1	Parallel updating	
Nature	Attractor	ABRS
Fixed point 1	000010000000	0.12%
Fixed point 2	000010011011	0.54%
Limit cycle 1	000010001010 00001001000 1	10.6%
Limit cycle 2	000010010011 000010001011 000010011001 000010011010	26.6%
Limit cycle 3	000010010010 000010000011 000010001001 000010011000	32.4%
Limit cycle 4	000010000010 000010000001 000010001000 000010010000	23.7%

Fig. 4.7 *Left*: Interaction signed digraph modeling the genetic regulation network controlling the cell cycle in mammals (from Kohn 1999). *Black* (resp. *white*) *arrows* represent activations (resp. inhibitions), and the frontier genes are surrounded by a *red circle*. *Right*: network attractors, in absence (*top*) and presence (*bottom*) of the microRNA miRNA 159, with their Attraction Basin Relative Size (ABRS), equal to the proportion of initial conditions leading to the attractor, and in the case of fixed configurations, their Attraction Diameter (AD), equal to the maximal distance in the hypercube state space E between couples of states of their attraction basin, the gene state being represented by 0 (gene in silence) or 1 (gene in expression) in the following order: p27, Cdk2, pCyCE_Cdk2, CyCE_Cdk2, miRNA 159, pCycA_Cdk2, CycA_Cdk2, Rbp-E2F, Rb-E2F, E2F, Rbp, and Rb

the expression of the ubiquitous tumor suppressor p53, because involved in many cell functions (like cell cycle arrest, cellular senescence, apoptosis,. . .), plays a role in carcinogenesis, and many human tumors present defects in the p53 control pathway. Hence, we will examine the possible role of the microRNAs controlled by p53 and those controlling it. Eventually, we will discuss their role in maintaining the robustness of networks dedicated to a precise function during the evolution. MicroRNAs are present in almost all genetic regulatory networks acting as inhibitors targeting mRNAs, by hybridizing at most one of their triplets, hence acting as translation factors by preventing the protein elongation in the ribosome.

The interaction graph associated to a genetic regulatory network can be inferred from the experimental data and from the literature, e.g., from the genes co-expression data, whose correlation networks are built by using for example directional correlations or logical considerations about the observed fixed configurations. These correlation graphs are after pruned or completed by orienting, signing, and valuating their edges, hence creating new connected components. Interaction graphs architecture contains motifs having positive and negative circuits, a negative (positive) circuit being a closed path in the graph having an odd (even) number of inhibitions. These circuits are connected to tree structures:

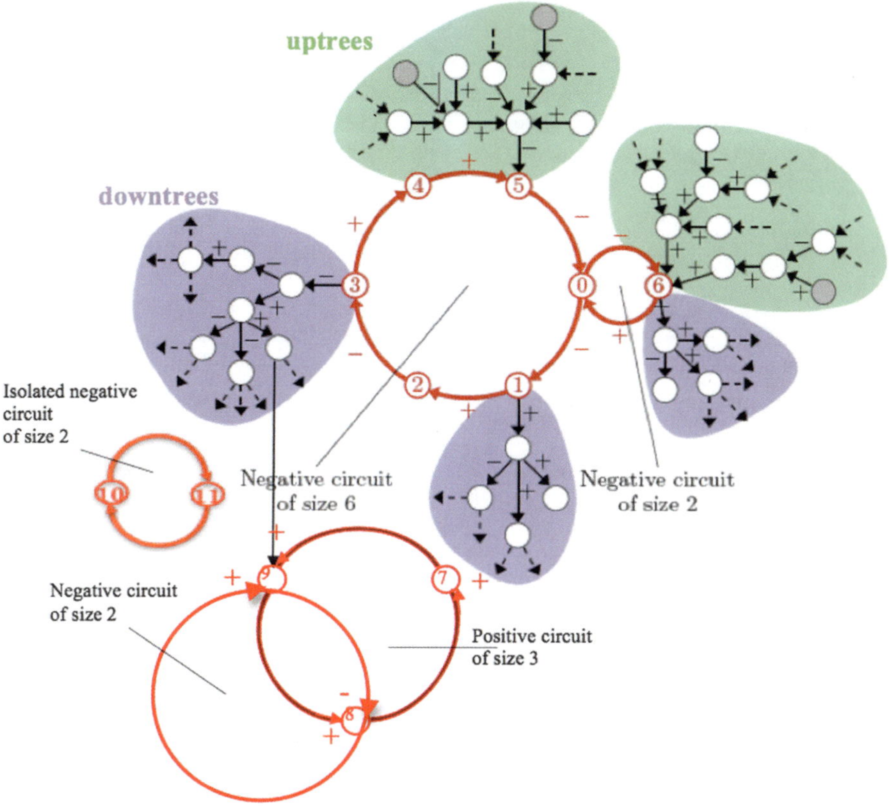

Fig. 4.8 The general architecture of a genetic regulatory network with up- and down-trees and circuits, positive or negative, isolated, tangential, or intersecting

(1) the upper trees control the circuits, their genes sources being often microRNAs and (2) the down-trees are controlled by the circuits until their final leaves, in general genes responsible of the final differentiation of a cell lineage (cf. Fig. 4.8).

The mitomiRs are localized in the nuclear noncoding genome and inhibit key genes involved in the energy system, like the ATPase or translocase genes, but they could also be present in the noncoding part of the mitochondrial genome, targeting the mitochondrial tRNAs, hence causing an unspecific inhibitory noise, which leaves expressed only the attractors of the dynamics of certain circuits on which this ubiquitous inhibition is compensated by a sufficient activation from the genes preceding in the trees or circuits the mitomiR target. In Fig. 4.9, we can see several situations in which the microRNA can or cannot prevent the attractor to behave as a periodic limit cycle. This last behavior (Fig. 4.9, right) is observed if the absolute value of the negative interaction weight related to the microRNA is less than the value of the preceding positive interaction. In this case, the microRNA has no influence on the network dynamics. This behavior is similar to the unspecific

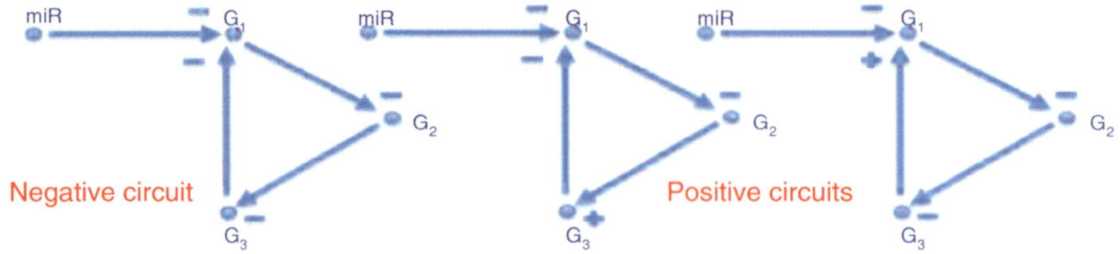

miR	G_1	G_2	G_3	miR	G_1	G_2	G_3	miR	G_1	G_2	G_3
0	1	1	1	0	1	1	1	0	1	1	1
0	0	0	0	0	0	0	1	0	1	0	0
0	1	1	1	0	0	1	0	0	0	0	1
0	0	0	0	0	1	1	1	0	1	1	1

miR	G_1	G_2	G_3	miR	G_1	G_2	G_3	miR	G_1	G_2	G_3
								miR- > G3 + (miR- < G3 +)			
1	1	1	1	1	1	1	1	1	1	1	1
1	0	0	0	1	0	0	1	1	0 (1)	0	0
1	0	1	1	1	0	1	1	1	0 (0)	1 (0)	0 (1)
1	0	1	0	1	0	1	1	1	0 (1)	1	0 (1)

Fig. 4.9 Dynamic behavior of circuits inhibited by the microRNA miR. *Left*: negative circuit whose limit cycle is canceled by miR and replaced by a fixed configuration. *Middle*: positive circuit whose limit cycle is canceled by miR and replaced by a fixed configuration. *Right*: positive circuit whose limit cycle is not canceled by miR

inhibitory activity the cortex exerts on the sub-cortical structures, which leaves effective only the sufficiently activated neural networks, as well as the anatomically identified subthalamic nuclei involved in motor control (Benabid et al. 1992), when they are co-activated by voluntary conscious and/or sensory unconscious inputs. Neural versus genetic metaphor seems pertinent because of the structural and functional analogy of their mathematical models.

We will use to compare microRNAs the normalized circular Hamming distance (equal to the number, divided by 22, of mismatches, i.e., the number of pairs different of A–U, U–A, C–G, G–C, G–U, and U–G, the classical pairs due to the Crick–Watson antisense hybridization, cf. Table 4.1) to the reference palindromic sequence AL. AL is close to the Archetypal Levin's tRNA loops sequence (Demongeot and Moreira 2007) and barycenter of a set of RNA rings of length 22 (like the microRNAs), whose the main characteristics is containing all the amino acids triplets, then constituting a "matrimonial agency" for amino acids favoring peptidic bonds as an ancestral ribosome (Hobish et al. 1995; Demongeot et al. 2009b, c). More, AL falls in the 5 % lowest tail part of the distribution of normalized circular Hamming distances to Rfam, a collection of multiple sequence alignments covering noncoding RNA families (Griffiths-Jones et al. 2005). Proximity to AL of microRNAs of Tables 4.1 and 4.2 is significant for a distance strictly

Table 4.1 RNA sequences with W, the observed number of subsequences of length 5 they shared with AL, its expectation $<W_R>$, the ratio $S = W/<W_R>$ with the significance of its difference with 1 and standard deviation σ_R (the number of stars corresponds to the number of standard deviations of the confidence interval not including the starred value)

RNA type	W	$<W_R>$	$S = W/<W_R>$	σ_R
tRNA conserved domains	3,498	1,262	2.77****	32
Introns	65,119	44,599	1.46***	202
snRNA (spliceosomal)	5,421	3,799	1.43***	59
snRNA	5,478	3,941	1.39***	59
rRNA	14,213	10,366	1.37***	97
miRNA	716	540	1.33***	22
tRNA	12,709	11,376	1.12**	105

Table 4.2 *Sequences alignment and distances to AL for* miR-20 [resp. miR-181b, miR-93 from (http://mirdborg/miRDB/; http://mirnamapmbcnctuedutw/)] and its inhibited targets, RNA-dependent Helicase P68 and Endonuclease CCR4 (resp. Helicase-DNA-binding protein KIAA1416, Exoribonuclease 2, and DNA Polymerase θ)

```
                 Sequences                         Identification                    Anti-matches

   UACUGCC  AUUCAAGAUGAAUGG 5'-3'AL                                                   14
3'_AUGGACG –UGAUAUUCGUGAAAU_5': hsa-miR-20
   . |.| .| : .|:| | | | ::.| | | | | | | |.
5'_CAGCAGTAATTATGGTGCACTTTt_3': Helicase P68 ENST00000225792 112-136

3'_AUGGACGU--------GAUAUUCG UGAAAU_5': hsa-miR-20
    | . | | . | | | . . .     | | | | | : | | | | | | | .
5'_TCCCAGCACCATTCTATAGGCAACTTTT_3': Endonuclease CCR4 ENST00000264903 3312-3340

   GUACUGCCAUUCAAGAUGAAUG 5'-3' AL                                                    15**
3'_UUGGGUGGCUGUCGUUACUUACAA_5': hsa-miR-181b
   | | | | : . .  | . | . .  | : | | | | | | | | | | .
5'_ AACCTTACAAGTGT AATGAATGTG_3': Helicase-DNA-binding protein KIAA1416 ENST00000333363 0-24

   CAUUCAAGAUGAAUGGUACUGC     5'-3' AL                                                15**
3'_GAUGGACGUGCUUGUCGUGAAA_5': hsa-miR-93
   : | | . : . . . | |       | | | | | | | | | |
5'_TTAATAAAAC –  TACAGTACTTT_3': Exoribonuclease 2 ENST00000262489 85-106

3'_GAUGGAC---- GUGCUUG--UCGUGAAA_5': hsa-miR-93
   . | : | : | | . . . . : | : | . | | . . | | | | | | | |
5'_ATGCTTGAAATATGCACTTAGCACTTT_3': DNA Polymerase θ ENST00000264233 293-320
```

Star number indicates the significance level l(s) for the anti-matches number to be greater than the expected value 12.6 + s (SupBinomial (22, 0.375) unilateral test: $s = 2{,}4$, $l(s) < 5\ 10^{-2} \equiv$**; s=3.4, $l(s)<10^{-2}\equiv$***; $s = 4.4$, $l(s)<10^{-3}\equiv$**** (Demongeot and Moreira 2007; Demongeot et al. 2009b))

more than 14, because the random variable equal to the circular Hamming distance between AL and a random sequence of length 22 is dominated by the Sup of 22 binomial variables B(22,0.375) having 12.6 as mean and 15 as 0.95-confidence upper-bound (Demongeot and Moreira 2007; Demongeot et al. 2009b).

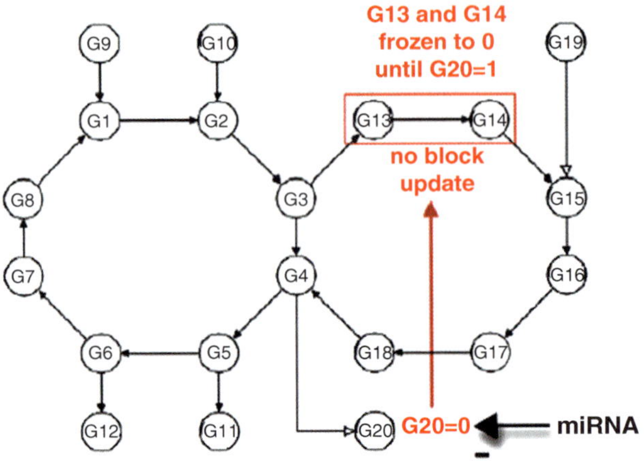

Fig. 4.10 Influence of the chromatin clock on the genetic network dynamics: the down-tree leaf G20 expresses or not a chromatin clock's enzyme, which authorizes or prevents the expression of the updating block G13-G14

In Table 4.1, the variable *W* equal to the observed number of subsequences of length 5 common to specified sets of RNA sequences and to the reference sequence AL is calculated for RNA sequences of the database *R*fam (Griffiths-Jones et al. 2005). $<W_R>$ is the expected mean of the random variable equal to the number of subsequences of length 5 common to the sets of RNA sequences and to a random ring of length 22. The quantity σ_R denotes the standard deviation of $<W_R>$. In each case, the value of the ratio $S = W/<W_R>$ exceeds 1, with a significance less than 10^{-3} for the tRNA conserved domains, which justifies the use of AL as reference sequence, especially for miRNAs and tRNAs. We will introduce in the following the notion of genetic threshold Boolean random regulatory network (getBren) with n genes, which is a set *N* of *n* random automata as defined in (Hopfield 1982; Hartwell et al. 1999; Weaver et al. 1999; Kauffman 1969; Thomas 1973; Demongeot et al. 2003) and in the present Mathematical Annex.

4.4.2 MicroRNAs and Chromatin Clock

The genes coding for enzymes involved in the chromatin clock as histone acetyltransferases, endonucleases, exonucleases, helicases, replicases, polymerases, etc., called clock genes, can be inhibited by many microRNAs preventing some blocks of genes to be co-expressed (cf. Fig. 4.10), e.g., RNA-dependent Helicase P68 and Endonuclease CCR4 are inhibited by the same miR-20, Helicase-DNA-binding protein KIAA1416 by miR181b, Exoribonuclease 2 and DNA Polymerase θ by the same miR-93. The influence of microRNAs on chromatin clock is partly ambiguous: Table 4.2 shows that (1) several steps like

Table 4.3 *Sequences alignment and distances to AL for* miR136, miR34a, and miR301 (from (http://mirdborg/miRDB/; http://mirnamapmbcnctuedutw/; http://ferrolab.dmi.unict.it/miro/)) and their inhibited targets, ATPase and Translocase

<table>
<tr><th>Sequences</th><th>Identification</th><th>Anti-matches</th></tr>
<tr><td colspan="3">ATPase SEQUENCE 1</td></tr>
<tr><td>GCCAU UCAAGAUG AAUGGUACU</td><td>5’-3’ AL</td><td>17****</td></tr>
<tr><td>3'_AGGUA-GUAGUUUUGU-UUACCUCA_5'</td><td>hsa-miR-136</td><td></td></tr>
<tr><td>| . : | | . : | | : . : . | | | . | : | | | | | | |</td><td></td><td></td></tr>
<tr><td>5'_TGTATGTATT--GTACATAGTGGAGT_3'</td><td>ENST00000276390_P 253-277</td><td></td></tr>
<tr><td colspan="3">ATPase SEQUENCE 2</td></tr>
<tr><td>CUGCCAUUCAAGAUGAAUGGUA</td><td>5’-3’ AL</td><td>15**</td></tr>
<tr><td>3'_UGUUGGUCGAUUCUGUGA-CGGU_5'</td><td>hsa-miR-34a</td><td></td></tr>
<tr><td>| . | : . | | | . | : | . | | | | | . | | | |</td><td></td><td></td></tr>
<tr><td>5'_AGAGGCAGATGACACACTGGCCA_3'</td><td>ENST00000248430 _P 2783-2806</td><td></td></tr>
<tr><td colspan="3">Translocase SEQUENCE 1</td></tr>
<tr><td>CUGCCAUUCAAGAUGAAUGGUA</td><td>5’-3’ AL</td><td>15**</td></tr>
<tr><td>3'_UGUUGGUCGA-UUCUGUGACGGU_5'</td><td>hsa-miR-34a</td><td></td></tr>
<tr><td>. . | | | | . . | | . | | | | | | | | | | : | |</td><td></td><td></td></tr>
<tr><td>5'_TGAACC----CTAAAGACACTGTCA_3'</td><td>ENST00000284320 53-74</td><td></td></tr>
<tr><td colspan="3">Translocase SEQUENCE 2</td></tr>
<tr><td>CCAUUCAAGAUGA AUGGUACUG</td><td>5’-3’ AL</td><td>16***</td></tr>
<tr><td>3'_CGAAACUGUUAUGA------UAACGUGAC_5'</td><td>hsa-miR-301</td><td></td></tr>
<tr><td>| : | . | | | . | | | : : | . . . . | . | | | | | | |</td><td></td><td></td></tr>
<tr><td>5'_GTTGTGAAAATGTTTGAAACTGCACTG_3'</td><td>ENST000002 84320 1999-2026</td><td></td></tr>
</table>

helicase function can be controlled by several microRNAs like mi-R20 and miR-181b and (2) several functions like helicase and endonuclease (resp. exoribonuclease and polymerase) are controlled by the same microRNA miR-20 (resp. miR-93).

4.4.3 *MicroRNas and Cellular Energetics: Oxidative Phosphorylation*

The cellular energy system of most of eukaryotic cells is essentially composed of glycolysis and aerobic oxidation. In eukaryotes, later stages of oxidative phosphorylation occur in mitochondria, with enzymatic steps like ATPase and translocase. For each of these genes, it is possible to find at least one microRNA inhibiting its activity (cf. Table 4.3). The microRNAs exert a translational repression preventing the enzyme synthesis in ribosomes. For example, in (Bandiera et al. 2011) are presented 2 microRNAs susceptible to hybridize with a perfect anti-match mitochondrial genes: hsa-miR-1974 and hsa-miR-1977 (Fig. 4.11) target indeed two mitochondrial tRNA genes, respectively, TRNE and TRNN, which code, respectively, for ATP8 and ND4L, and the hsa-miR-1978 targets a stretch of a mitochondrial rRNA sequence called RNR1. ATP8 is the ATP synthase protein 8, a subunit of the mitochondrial ATPase, and ND4L is a protein which provides instructions for making NADH. RNR1 is a subunit of the ribonucleotide reductase,

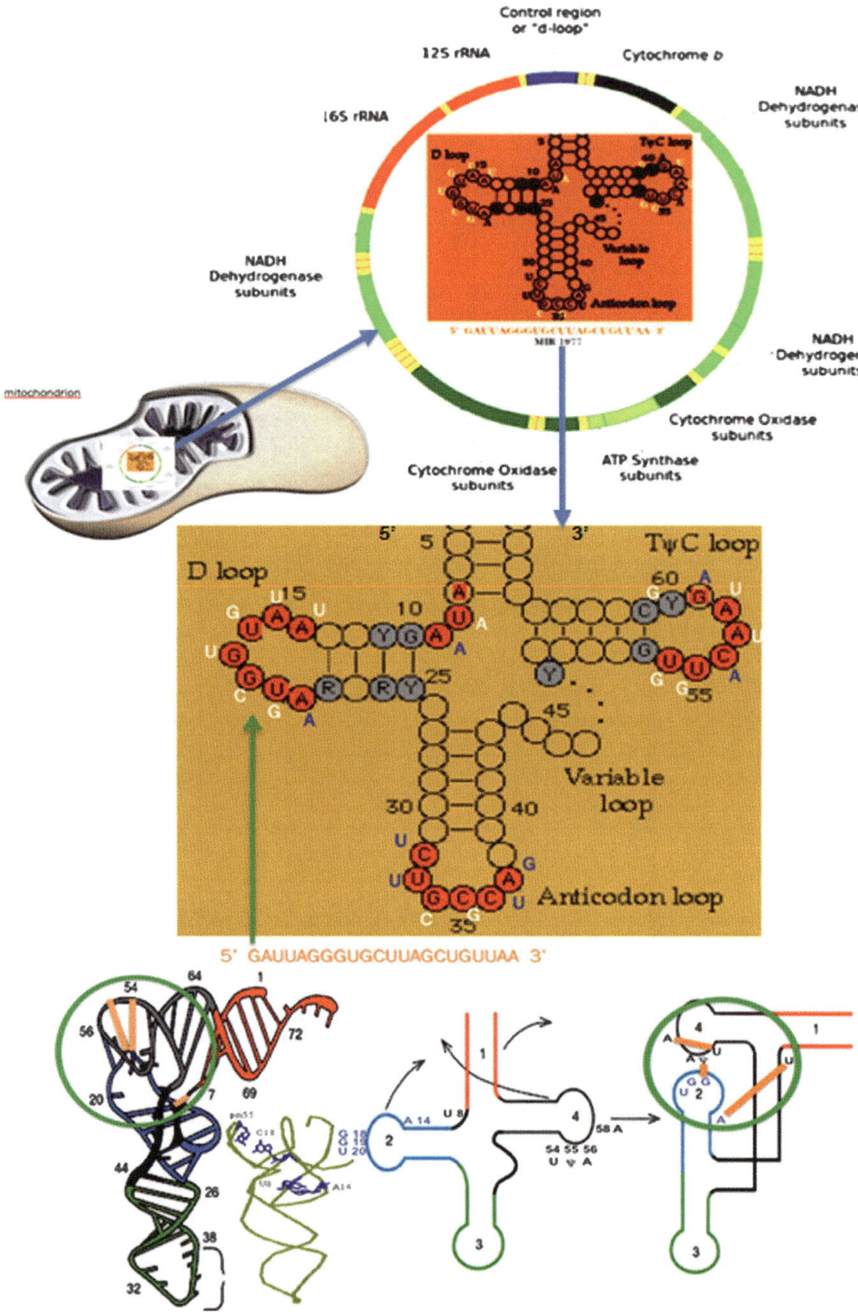

Fig. 4.11 Top: sequence of the nuclear mitomiR 1977: 5′_GAUUAGGGUGCUUAGCUG UUAA_3′ presenting 14 mismatches with the 3-loops sequence of the mitochondrial Gly-tRNA (Demongeot and Moreira 2007): 3′_YGAACUUACCGUCAUGGUAAAU_5′ made from the D-loop, Anticodon-loop, and TψC-loop. *Bottom*: 3D structure with U8-A14 reverse Hoogsteen pair, and tertiary pairs G18-ψ55, U54-A58 stabilizing, respectively, the sharp turn in D-loop and the tertiary T structure

Table 4.4 Alignment of AL with mitomiR 1974 from nuclear noncoding genome with indication of the D-loop and A14 turn bend in yellow (Bandiera et al. 2011; Demongeot et al. 2013a, b, c; http://mirdborg/miRDB/; http://mirnamapmbcnctuedutw/)

Sequences	Identification	Anti-matches
GUACUGCCAUUCAAGAUGAAUG 3'-5' AL 3'_AUAAGAGCGUGCCUGAUGUUGGU_5': hsa-mitomiR-1974		14

Table 4.5 Consensus sequences of mito²miRs from mitochondrial noncoding genome (Sbisa et al. 1997; Cui et al. 2007) and alignment with the Lewin ancestral tRNA D-loop and A14 turn bend in yellow (*R = puric==A or G; Y = pyrimidic=U, T, or C)*

Sequences	Identification	Anti-matches
3'_CGGGACUUCAUUCUUGGUCUA_5': hsa-mito²miR C 116		
5'_ RUUUCRARRURARYGGUAYUY_3': Lewin tRNA		16
CCAUUCAAGAUGAAUGGUACUG 3'-5' AL		18 matches
3'_CAGGACUUCAUUCUUGGUCUA_5':hsa-mito²miR CSBD353		
5'_RUUUCRARRURARYGGUAYUY_3': Lewin tRNA		16

an enzyme that catalyzes the formation of deoxyribonucleotides used in DNA synthesis from ribonucleotides, favoring the reduction of ATP and GTP, by increasing dNTP pools and, hence, decreasing the NTP pools necessary for a correct functioning of metabolisms like glycolysis.

A last but not least putative inhibitory mechanism comes from the regulation of the tRNA function by hybridizing by RNA sequences, the tRNA loops especially the tRNA D-loop: in Table 4.4 we see the possibility to hybridize tRNA loops (especially the D-loop and the A14 turn responsible of the tRNA tertiary structure; cf. Fig. 4.11) by the nuclear mitomiR 1974 and in Table 4.5, two possibilities of small mitochondrial RNAs called mito²miRs [C 116, CSBD 353 in Sbisa et al. (1997); Cui et al. (2007)] coming from the noncoding part of the mitochondrial genome, the Lewin's invariant part of tRNA secondary structure (Bandiera et al. 2011; Griffiths-Jones et al. 2005) serving as reference template for tRNA loops hybridization (Lewin et al. 2011; Turner et al. 2005; Sbisa et al. 1997; Cui et al. 2007; Bandiera et al. 2012). The unspecific inhibitory noise caused by this possible direct (inside the mitochondrion matrix) new regulation favors as indicated in the previous section the circuits with sufficiently strong interactions for resisting to the mitomiR inhibitory influence (Demongeot and Waku 2012b).

4.4.4 *MicroRNas and Cellular Energetics: Glycolysis/Oxidative Phosphorylation Coupling*

MicroRNAs inhibit all glycolytic steps (cf. Figs. 4.12 and 4.13, and Table 4.6) and depending on the intensity of these inhibitions favor at the level of pyruvate coupling, the entrance in the Krebs cycle, if translocase and ATPase are not

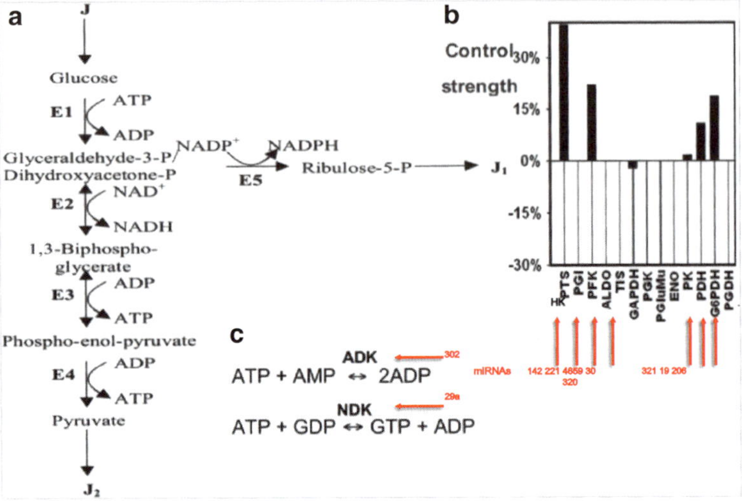

Fig. 4.12 (**a**) Glycolysis main steps with indication (**b**) of their enzyme control strengths and of their inhibitory microRNAs (*red arrows*). E1 denotes the four enzymes of the high glycolysis (hexokinase HK, phosphoglucose-isomerase PGI, phosphofructo-kinase PFK, and aldolase ALDO), E2 denotes the glyceraldehyde-3P-dehydrogenase, E3 denotes the three enzymes of the low glycolysis (phosphoglycerate-kinase, phosphoglycerate-mutase, enolase ENO), E4 denotes the pyruvate-kinase PK and E5 the three enzymes of the oxidative part of the pentose pathway (glucose-6P-dehydrogenase G6PDH, 6P-glucono-lactonase, and phosphogluconate-dehydrogenase PGDH, alternative to phospho-transferase system PTS). (**c**) Adenylate (guanylate) pool regulated by ADK (NDK)

inhibited and sufficiently present in the mitochondrial inner membrane [cf. Fig. 4.13 and Demongeot et al. (2007c)], or conversely privilege lactacte shuttle [like in brain astrocytes, in order to feed neurons in lactate (Aubert et al. 2005; 2007)].

Glycolysis (Fig. 4.12) has been modeled by several authors (Aubert et al. 2005, 2007; Boiteux et al. 1975; Demongeot and Seydoux 1979; Hervagault et al. 1983; Demongeot and Kellershohn 1983; Demongeot and Doncescu 2009b), especially its central allosteric step ruled by the phosphofructokinase (PFK), which is the key glycolytic enzyme, because of its highly nonlinear allosteric kinetics and of the presence as effectors of ATP and ADP (controlled by miR-298 and 29a in adenylates/guanylates pools regulated by ADK and DNPK, cf. Table 4.7) in a negative regulatory circuit causing, for critical values of the glucose entry flux J, oscillations for all glycolytic metabolites, with a period of several minutes (Aubert et al. 2005; Demongeot and Seydoux 1979). Let us define now by x_1, x_2, x_3, and x_4 the concentrations of the successive main metabolites of the glycolysis, respectively glucose, glyceraldehyde-3-P, 1,3-biphospho-glycerate, and phospho-enol-pyruvate. We assume that steps E2 and E3 of the glycolysis (Fig. 4.12a) are Michaelian and reversible. The complex E1 includes the allosteric irreversible kinetics of the phospho-fructo-kinase PFK with a cooperativity n [see Ovadi (1988), Reder (1988), Ritter et al. (2008), Thellier et al. (2006), and Demongeot

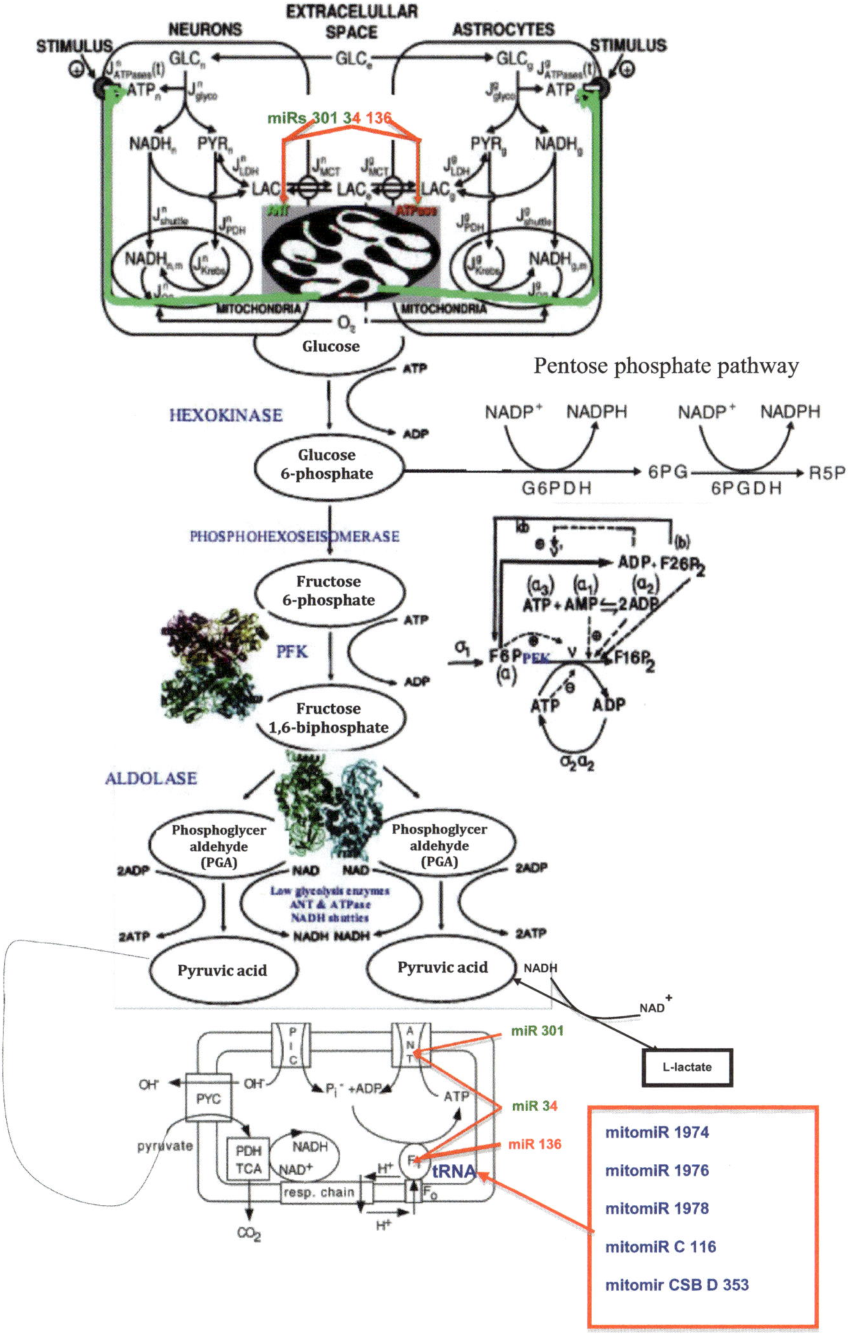

Fig. 4.13 *Top*: Connection between the glycolysis pathway of neurons and astrocytes through the lactate flux, controlled by ANT (green) and F_1 ATPase (*red*) inside the mitochondrial inner membrane. *Middle*: Glycolysis coupled to oxidative phosphorylation, with indication of the PFK kinetics with its effectors (activators and inhibitors). *Bottom*: indication of the main miRs and mitomiRs acting on ANT (*in green*), ATPase (*in red*), and mitochondrial tRNAs (*in blue*)

Table 4.6 Sequences alignment of nuclear miRNAs with glycolysis genes (Bandiera et al. 2011, 2013; Demongeot et al. 2013a, b, c)

Enzymes **Sequences** **Identification** **Anti -matches**

Hexokinase

```
 UUCAAGA      UGAAUGGUACUGCCA 5'-3' AL                                  16***
3'_AGGUAUU--UCAUCCUUUGUGAUGU_5': hsa-miR-142_3p
  | . : | | . | . | . . | | .  | : | | | | | : | .
5'_TGTATTATCACCAGCAGACACTGCC_3': Hexokinase Type I ENST00000317679 398-423
```

Triosephosphate isomerase

```
 UGAAGGUACUGUCCAUUCAAGA   5'-3' AL                                      15**
3'_AGCUACAUUGUCUGCUGGGUUUC_5': hsa-mir-221
     . | | . . . : : |      | . | | . : | : .
  5'_ AATAGCGTA-------GCCCAGAGT_3': Triosephosphate isomerase ENST00000229270_P 1972-1990
```

Phosphofructokinase

```
 AGAUGAAUGGUACUGCCAUUCA  5'-3'  AL                                      15**
3'_UUUCUUCUUAGACAUGGCAGCU_5': hsa-miR-4659b-3p
  : . | | | | | | | . . .   | | . | : . . : . .
5'_GTAGAAGAAAAGGTCCTTGTCC_3': Phosphofructokinase NCBI5208 2352-2374
```

Fructosebiphosphate aldolase

```
  GAAUGGUACUGCCAUUCAAGAU    5'-3' AL                                    15**
3'_UGGAGUGUGACAAUGGUGUUUG_5': hsa-miR-122
  | . . . . : | . . . . | . : | : | : : . | :
5'_AGG----GCCAAATAGCTATGCAG_3': Fructosebiphosphate aldolase C ENST00000226253 66-86
```

Deoxyribosephosphate aldolase

```
 GUACUGCCAUUCAAGAUGAAUG  5'-3' AL                                       13
3'_CGACGUUUGUAGGCUGACUUUC_5': hsa-miR-30a
  . . | | . | : : | | | | | . . | | | | | | | .
5'_TATG - AGGCATCC--ACTGAAAT_3': Deoxyribosephosphate aldolase   ENST00000025429_P 1925-1945
```

Lactate dehydrogenase

```
 AUGGUACUGCCAUUCAAGAUGA  5'-3' AL                                       13
 3'_GUGUCAACGGUCGACUCUAAU_5': hsa-miR-216
  | | . | | | . . . . .  | | | | : | | | | |
5'_CAAAGT-----------GCT GGGATTA_3': Lactate dehydrogenase  ENST00000300038 202-218
```

Pyruvate kinase

```
 GGUACUGCCAUUCAAGAUGAAU  5'-3' AL                                       13
3'_CUUGGGUGUUAGGGACCGAAU_5':hsa-miR-321
  | . . | | | | | | : | . . | | | | | | | .
5'_GCTCCCACAGT-ACTGGCTTT_3': Pyruvate kinase ENST00000271945_P  2622-2642
```

G6P deshydrogenase

```
 AGAUG  AAUGGUACUGCCAUUCA   5'-3' AL                                    13
3'_GGUGU- GUGAAGGAAUGUAAGGU_5': hsa-miR-206
  | | . | | . : : | . . | :  | . | | | | |  | | .
5'_CCTCAGTGCCACT  TGACATT CCT_3': G6P deshydrogenase ENST00000291567 417-440
```

Pyruvate dehydrogenase

```
 UCAA  GAUGAAUGGUACUGCCAU   5'-3' AL                                    13
3'_AGUCA-AAACGUAUCU-AAACGUGU_5': hsa-miR-19a
  . : | | | . . | | | | | | : . | . | | | | : | | |
5'_GTAGTAATTGCATGCAGTTTGTACA_3' Pyruvate dehydrogenase ENST00000254142 136-161
```

Table 4.7 Sequences alignment of nuclear miRNAs with adenylate/guanylate pool regulated by genes ADK and NDK (http://mirdborg/miRDB/; http://mirnamapmbcnctuedutw/; Sbisa et al. 1997; Cui et al. 2007; Bandiera et al. 2012)

<table>
<tr><th>Sequences</th><th>Identification</th><th>Anti-matches</th></tr>
<tr><td>GUACUGCCAUUCA AGAUGAAUG 5'-3' AL</td><td></td><td>13</td></tr>
<tr><td colspan="2">3'_AGUGGUUUUGUACC - UUCGUGAAU_5': hsa-miR-302</td><td></td></tr>
<tr><td colspan="2">. | | | | . . . | . : . | | | . | | | | | : | | :</td><td></td></tr>
<tr><td colspan="2">5'_TCACATCATTCTGGAAAGCATTTG_3': ADenylate Kinase ADK ENST00000284162_P 1322-1346</td><td></td></tr>
<tr><td>CAU UCAAGAUGAAUGGUACUGC 5'-3' AL</td><td></td><td>14</td></tr>
<tr><td colspan="2">3'_UUG-GCUAAAGUCUACCACGAUC_5': hsa-miR-29a</td><td>.</td></tr>
<tr><td colspan="2">| | | . : | | | | | | | . | . | | | | | . | .</td><td></td></tr>
<tr><td colspan="2">5'_AACATGATTTCATAAGGTGCAAA_3': Nucleoside Diphosphate Kinase NDK ENST00000218340 1541-1564</td><td></td></tr>
</table>

and Laurent (1983) for its complex kinetics], and we suppose that both pyruvate-kinase (E4) and dehydrogenases of the complex E5 are irreversible.

Numerous $V_{\max}$ V_i's can be diminished by the action of microRNAs inhibiting the synthesis of the corresponding enzymes (cf. Fig. 4.12). Let us consider now the differential system (S1) ruling the glycolysis and the pentose pathway until the ribulose-5-P:

$$\mathrm{d}x_1/\mathrm{d}t = J - V_1 x_1{}^n/(1 + x_1{}^n),\ \mathrm{d}x_2/\mathrm{d}t = V_1 x_1{}^n/(1 + x_1{}^n) - V_2 x_2/(1 + x_2) + L\alpha V_2 x_3/(1 + x_3),\ \mathrm{d}x_3/\mathrm{d}t = \alpha V_2 x_2/(1 + x_2) - V_3 x_3/(1 + x_3) - L\alpha V_2 x_3/(1 + x_3) + L' V_3 x_4/(1 + x_4), \mathrm{d}x_4/\mathrm{d}t = V_3 x_3/(1 + x_3) - V_4 x_4/(1 + x_4) - L' V_3 x_4/(1 + x_4)$$

Consider now the change of variables: $y_i = x_i{}^{1/2}$, $\mathrm{d}y_i/\mathrm{d}x_i = x_i{}^{-1/2}/2$, where the y_i's are ruled by the differential system (S2):

$$\begin{aligned}\mathrm{d}y_1/\mathrm{d}t &= \left[J - V_1 y_1{}^{2n}/\left(1 + y_1{}^{2n}\right)\right]/2y_1,\ \mathrm{d}y_2/\mathrm{d}t = \left[V_1 y_1{}^{2n}/\left(1 + y_1{}^{2n}\right) - \right. \\ &\quad \left. V_2 y_2{}^2/\left(1 + y_2{}^2\right) + L\alpha V_2 y_3{}^2/\left(1 + y_3{}^2\right)\right]/2y_2, \\ \mathrm{d}y_3/\mathrm{d}t &= \left[\alpha V_2 y_2{}^2/\left(1 + y_2{}^2\right) - (V_3 + L\alpha V_2) y_3{}^2/\left(1 + y_3{}^2\right)\right. \\ &\quad \left. + L' V_3 y_4{}^2/\left(1 + y_4{}^2\right)\right]/2y_3, \mathrm{d}y_4/\mathrm{d}t = \left[V_3 y_3{}^2/\left(1 + y_3{}^2\right)\right. \\ &\quad \left. - V_4 y_4{}^2/\left(1 + y_4{}^2\right) - L' V_3 y_4{}^2/\left(1 + y_4{}^2\right)\right]/2y_4\end{aligned}$$

When the attractor is a fixed point, the first enzymatic complex E1 has a stable stationary state defined by $V_1 y_1 *^{2n}/(1 + y_1 *^{2n}) = J$. If the steady state value y_1* is reached the first among the other y_j*'s, because V_1 is an order of magnitude higher than the other V_i's, the motion of y_1 is called rapid, and then (S2) becomes (S3):

$$\begin{aligned}\mathrm{d}y_2/\mathrm{d}t &= -\partial P^*/\partial y_2, \mathrm{d}y_3/\mathrm{d}t = -\partial P^*/\partial y_3 + \partial H^*/\partial y_3,\ \mathrm{d}y_4/\mathrm{d}t \\ &= -\partial P^*/\partial y_4 + \partial H^*/\partial y_4\end{aligned}$$

where:

$$P^* = -J\text{Log}y_2/2 + V_2\text{Log}\left(1 + {y_2}^2\right)/4 - L\alpha V_2{y_3}^2\text{Log}y_2/2\left(1 + {y_3}^2\right)$$
$$+ (V_3 + L\alpha V_2)\text{Log}\left(1 + {y_3}^2\right)/4 - V_3{y_3}^2\text{Log}y_4/2\left(1 + {y_3}^2\right)$$
$$+ (V_4 + L'V_3)\text{Log}\left(1 + {y_4}^2\right)/4$$

and

$H^* = -L\alpha V_2{y_3}^2\text{Log}y_2/4(1 + {y_3}^2) + L'V_3{y_4}^2\text{Log}y_3/2(1 + {y_4}^2) - V_3{y_3}^2\text{Log}y_4/2(1 + {y_3}^2)$.

In this case, parameters appearing only in P^*, like V_4, are modulating the localization of the fixed point; hence, the values of the stationary concentrations of the glycolytic metabolites [cf. Demongeot et al. (2007a, 2007b) and Glade et al. (2007) for a more general approach of the potential-Hamiltonian decomposition].

Let us suppose now that we measure the outflows J_1 and J_2 (cf. Fig. 4.12). Then from the system (S1) we can calculate the sharing parameter α (which regulates the pentose pathway and the low glycolysis dispatching) from the steady-state equations equalizing the in- and outflows at each step. By denoting the stationary state $x^* = \{x_i^*\}_{i\,=\,1,4}$, we have:

$$V_1{x_1^*}^n/(1 + {x_1^*}^n) = J,\ \ V_2x_2^*/(1 + x_2^*) - L\alpha V_2x_3^*/(1 + x_3^*) = J$$
$$(1 - \alpha)V_2x_2^*/(1 + x_2^*) = J_1, V_3x_3^*/(1 + x_3^*)$$
$$= J_2(V_4 + L'V_3)/V_4$$

Hence, we can calculate α by using the following formula:

$$\alpha^2J_2LV_2(V_4 + L'V_3) + \alpha(JV_4 - J_2LV_2(V_4 + L'V_3)) + J_1V_4 - J$$
$$= 0 \text{ or } \alpha^2 - \alpha(1 - K) + K' = 0,$$

where we have denoted:

$$K = JV_4/J_2LV_2(V_4 + L'V_3) \text{ and } K' = (J_1V_4 - J)/J_2LV_2(V_4 + L'V_3).$$

When the flux of the kth step in a metabolic network has reached its stable stationary value Φ_k, then the notion of control strength C_{ki} exerted by the metabolite x_i on this flux Φ_k is defined by (Kaczer and Burns 1973; Wolf and Heinrich 2000):

$$C_{ki} = \partial\text{Log}\Delta\Phi_k/\partial\text{Log}\Delta x_i$$

and we have:

$$\forall k = 1, n,\ \ \Sigma_{i=1,n}C_{ki} = 1.$$

If each metabolite x_i is controlled by a set of m microRNAs and let M_{ij} denote the control strength of the microRNA j with concentration m_j on x_i, i.e., we have: $M_{ij} = \partial\text{Log}\Delta x_i/\partial\text{Log}\Delta m_j$, then we get that the product of matrices $P = \text{CM}$ is a matrix whose the sum of coefficients on a line is 1, summarizing the effects of the microRNAs on the metabolic network. If we calculate the stationary flux increment $\Delta\Phi_k$ resulting from a perturbation Δm from the value $m = 0$, as:$\Delta\Phi_k(\Delta m) = \Phi_k(0)$ $[\exp(\Sigma_j v_{kj}\Delta m_j)-1] \approx \Phi_k(0)\Sigma_j v_{kj}\Delta m_j$, then the general term of P, i.e., the control strength P_{kj} exerted by the microRNA j on the flux Φ_k, can be calculated, if all Δm_j are small as:

$$P_{kj} = \partial\text{Log}\Delta\Phi_k/\partial\text{Log}\Delta m_j \approx v_{kj}\Delta m_j/\Sigma_j v_{kj}\Delta m_j, \quad \text{where } v_{kj} \approx [\partial\Delta\Phi_k/\partial\Delta m_j]/\Phi_k(0)$$

The directed signed graph associated to the incidence (or adjacency) matrix P is the same as the up-tree part of the interaction graph of a genetic network corresponding to inhibitions by microRNAs, where the interaction weight of the microRNA j on the gene expressing the enzyme k is equal to v_{kj} (cf. Mathematical Annex), which justifies the use of similar tools like the entropy of the network and the subdominant eigenvalue of the Markovian matrix P, for characterizing the stability and the robustness of the metabolic network. The molecules controlled by microRNAs through the expression of their genes are proteins, enzymes, carriers, or membrane receptors, and the control strength equation can be used to prove that the most regulated molecules in glycolysis are enzymes like hexokinase, PFK, GPDH, and ADK, which rule the pool of the energetic molecules, ATP, ADP, and AMP, which conversely are mainly produced by the glycolysis (Wolf and Heinrich 2000; Ruoff et al. 2003; Bier et al. 1996; Mourier et al. 2010). In Fig. 4.12, we see that the most sensitive steps of glycolysis are hexokinase, PFK, and GPDH, inhibited in human by the microRNAs hsa-miR-19, 4659/320, and 142, respectively.

When oscillations occur, we can use the variables T_{ki} (resp A_{ki}) to quantify the control by Δx_i of the period τ_k (resp. intensity amplitude I_k) of the kth step flux (Baconnier et al. 1993), Δx_i being the perturbation of the concentration of the ith "pacemaker" effector (i.e., parameter causing oscillations), from its bifurcation values x_i:

$$T_{ki} = \partial\text{Log}\tau_k/\partial\text{Log}\,\Delta x_i \text{ and } A_{ki} = \partial\text{Log}I_k/\partial\text{Log}\Delta x_i.$$

If ξ is the eigenvalue of the Jacobian matrix of the differential system for which the stationary state has bifurcated in a limit cycle (Hopf bifurcation), then $\tau_k = 2\pi/\text{Im}\xi$, and if we consider a 2D potential-Hamiltonian example like:

$$\mathrm{d}x_1/\mathrm{d}t = -\partial P/\partial x_1 + \partial H/\partial x_2, \quad \mathrm{d}x_2/\mathrm{d}t = -\partial P/\partial x_2 - \partial H/\partial x_1,$$

then we have:

$$\mathrm{Im}\xi = \left|(\Delta P)^2 - 4(C(P) + C(H) + \partial^2 P/\partial x_1 \partial x_2 (\partial^2 H/\partial {x_2}^2 - \partial^2 H/\partial {x_1}^2) + \partial^2 H/\partial x_1 \partial x_2 (\partial^2 P/\partial {x_1}^2 . \partial^2 P/\partial {x_2}^2))\right|^{1/2}/2,$$

where $\Delta P = \partial^2 P/\partial {x_1}^2 + \partial^2 P/\partial {x_2}^2$ is the Laplacian of P and $C(P) = \partial^2 P/\partial {x_1}^2 \partial^2 P/\partial {x_2}^2 - (\partial^2 P/\partial x_1 \partial x_2)^2$ is the mean Gaussian curvature of the surface P, both taken at the stationary state of the differential system.

Concerning the energy balance in neurons, if the apparent V_{max} of the PFK is diminished in astrocytes due to a lack of ATP (used to inactivate via phosphorylation the neurotransmitter receptors), then the oscillatory behavior is less frequent and the production rate of lactate from pyruvate is more important than in neuron, creating a flux of lactate to neurons. The neurons consume the lactate coming from the extracellular space, partially replenished by the astrocytes production. That gives to neurons an ATP level higher than in astrocytes, with an extra-pyruvate production from lactate, and extra-oxygen and glucose consumption theoretically predicted and experimentally observed.

This ATP level depends on ATPase and Translocase concentrations, which are in human under the negative control of 3 miRNAs, the hsa-miR-136, 34 (common), and 301 (cf. Table 4.3) acting as boundary control nodes (http://microrna.sanger.ac.uk) by regulating the neuronal oxidative system efficacy. These enzymes are located on the mitochondrial inner membrane surface whose protein content can generate the ATP/GTP turnover (Bier et al. 1996), as well as the proton leak (Mourier et al. 2010).

4.4.5 MicroRNAs, Morphogenesis, and Cell Cycle

The Engrailed gene is required for the proper segmentation and maintenance of the posterior compartment of the Drosophila embryo (Almeida and Demongeot 2012), but also the Engrailed gene efficiently activates the gene Elk, necessary for the control of K^+ ion channels in excitable cells, and the human Engrailed homologues encode homeo-domains containing proteins and are implicated in the control of the pattern formation during the development of the central nervous system. On the top of Fig. 4.14, we have represented the action of the gene Elk, which controls for example positively the ability of CA3 cells Yi's to express their negative feedback upon the CA1 cells Xi's inside the hippocampus (Tonnelier et al. 1999), showing that the same set of genes can be involved from the development phase until the control of a high level function as the hippocampus memory ability.

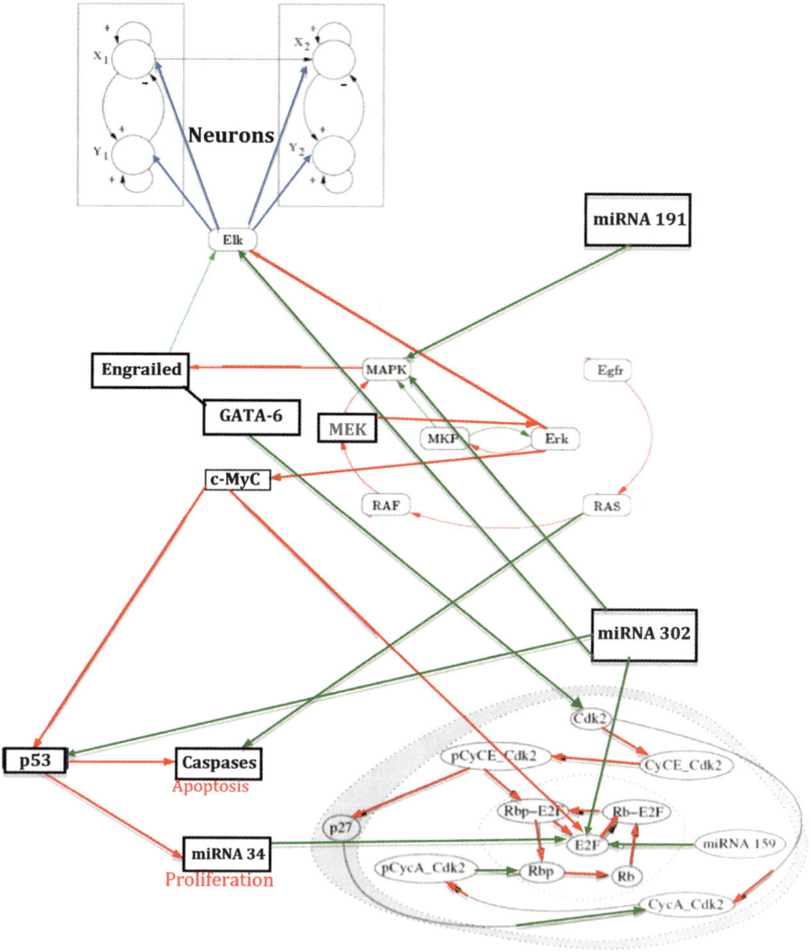

	Sequential updating		Parallel updating		
Nature	Attractor	ABRS	Attractor	ABRS	AD
Fixed point 1	000000000000	6.25%	000000000000	0.5%	1.45
Fixed point 2	000000011111	56.25%	000000011111	99.5%	5.40
Limit cycle 1	000000001000 000000010111	37.5%	None		

Fig. 4.14 *Top*: Connection between hippocampus neural network and p53 through Engrailed regulation network, with balance (activations *in red*, inhibitions *in green*) between apoptosis and proliferation. *Bottom*: Sequential and parallel attractors of the cell cycle network (E2F box)

We see in Fig. 4.14 that the gene Engrailed is not isolated, but is located at the intersection of four important networks: the Engrailed control network centered on MAPK and controlled by miRNA191, the p53 regulation network (Fig. 4.15), the

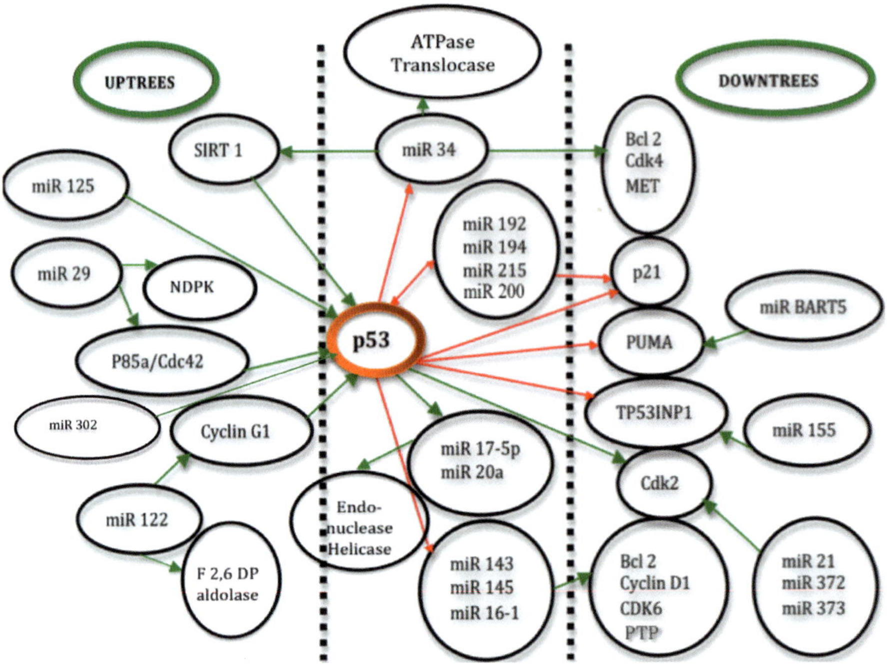

Fig. 4.15 Regulation of protein p53 showing relationships with numerous microRNAs with inhibiting (*green*) or activating (*red*) interactions

cytoskeleton network connected to the Myosin network (in green in Fig. 4.16) responsible of the control of morphogenetic processes like the gastrulation (Almeida and Demongeot 2012), the apoptosis and proliferation networks controlled by the ubiquitous protein p53, inhibited by miRNA302; p53 is a transcription factor of miRNA 34 (Raver-Shapira et al. 2007), which is a translation inhibitor of E2F in the mammal proliferation box (Kohn et al. 2006) and of ATPase and Translocase: p53 is indeed the center of a complex regulatory subnetwork, in activatory or inhibitory relationship with numerous other microRNAs (Fig. 4.15).

The E2F box, the core of the proliferation box (Kohn et al. 2006), presents several attractors both in sequential and parallel updating modes (cf. Fig. 4.14 bottom), especially the periodic behavior (000000001000, 000000010111), when its nodes are ordered as follows: p27, Cdk2, pCyCE_Cdk2, CyCE_Cdk2, miRNA 159, pCycA_Cdk2, CycA_Cdk2, Rbp-E2F, Rb-E2F, E2F, Rbp, and Rb. The triple action (accelerate, stop, and slow down the cell cycle) on the proliferation process is exerted negatively by the gene GATA-6, which is inhibited 1 time out of 2 by MAPK, and successively positively and negatively (through p53) by the gene c-MyC which is activated 1 time out of 2 by Erk. Then, the limit cycle of order 4 brought by the negative circuit of size 2 (MKP/Erk) leads the genes MKP, Erk,

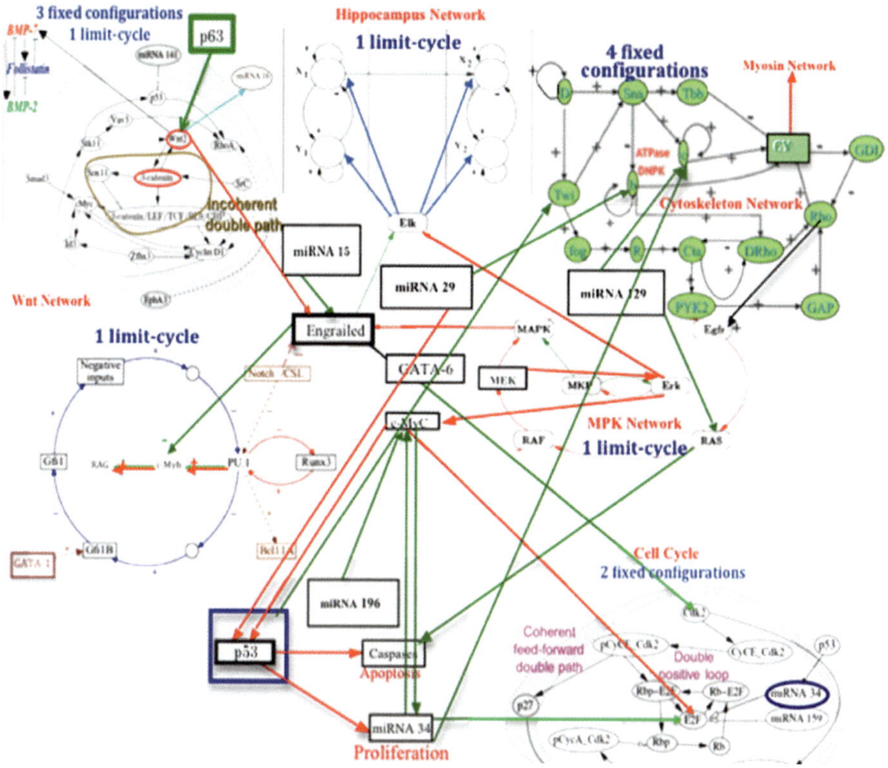

Fig. 4.16 Engrailed control network with its regulatory environment of up- or down-subnetworks. The number and nature of attractors of subnetworks are indicated in *blue*. *Green* or *white*-headed *arrows* correspond to inhibitions, *red* or *black*-headed to activations

MAPK, Engrailed, GATA-6, c-MyC, p53, miRNA34, Cdk2, E2F, RAS, and caspases, through the microRNAs miR34 and miR302, to enter in the limit cycle:

$$(000000110110,\ 011000010010,\ 111101000011,\ 100111100011).$$

Then, the E2F/proliferation box is activated 1 time out of 4 and the apoptosis box 1 time out of 2, such as we observe the following cyclic behavior: 4 cells, 4 cells, 3 cells, 2 cells, 4 cells, etc. This dynamic behavior allows the exponential growth of proliferation compensating exactly in a tissue the linear growth of apoptosis—two daughter cells replacing two dead cells during one period of the limit cycle, hence ensuring the conservation of the tissue volume and function. If the double dual control (acceleration and brake) by c-Myc is no more ensured, a pathologic proliferation (like in cancerogenesis) or excessive cell death (like in aging) can occur (Table 4.8).

Table 4.8 Sequences alignment of nuclear miRNAs with c-Myc and Engrailed genes (http://mirdborg/miRDB/; http://mirnamapmbcnctuedutw/; Raver-Shapira et al. 2007)

<table>
<tr><th>Sequences</th><th>Identification</th><th>Anti-matches</th></tr>
<tr><td>UCAAGAUGAAUGGUACUGCCAU</td><td>5’-3’ AL</td><td>14</td></tr>
<tr><td>3'_GGUUGUUGUACUUUGAUGGAU_5'</td><td>hsa-miR-196</td><td></td></tr>
<tr><td>. | . | : | | : : | . | | | | | : | | | .</td><td></td><td></td></tr>
<tr><td>5'_GCCATAATGT-AAACTGCCTC_3'</td><td>c-Myc ENST00000259523 125-145</td><td></td></tr>
<tr><td>UUCAAGAUGAAUGGUACUGCCA</td><td>5’-3’ AL</td><td>14</td></tr>
<tr><td>3'_GUGUUU-GGUAAUACACGACGAU_5'</td><td>hsa-miR-15a</td><td></td></tr>
<tr><td>| | . | : | . : : . | | | | . | | | | | | . .</td><td></td><td></td></tr>
<tr><td>5'_CAAAGATTTT TTAT- TGCTGCAT_3'</td><td>Engrailed ENST00000297375 369-391</td><td></td></tr>
</table>

Table 4.9 Comparison between cellular oxygen consumption of mammals (Hainaut and Hollstein 2000)

	V_{O2max}/M (ml.s^{-1}.kg^{-1})	V/M (ml/kg)
Dog	2.29	40.6
Goat	0.95	13.8
Horse	2.23	30.0
Steer	0.85	11.6

Table 4.10 Number of common subsequences of lengths 6 and 7 between the mitochondrial tRNA loops and the noncoding mitochondrial genomes of bos (total length 8,493 bases) and equus (total length 4,535 bases) (http://megasunbchumontrealca/ogmp/projects/other/cp_list.html)

Subsequences	Observed bos	Expected bos	Observed equus	Expected equus
TACCAC	2	2	0	1
TACCAT	1	2	0	1
ACCGTT	2	2	0	1
ACCATT	1	2	0	1
CTTGAA	1	2	0	1
ATTTGAA	1	0.5	0	0.25
Total	8	10.5 ± 6.5 ($m \pm 2\sigma$)	0	5.25 ± 4.6 ($m \pm 2\sigma$)

4.5 Global Influence of microRNAs

4.5.1 Physiological Role in Cell Energetics

The unspecific inhibitory noise from the microRNAs has a global resultant on the intermediary metabolism in, e.g., higher mammals. Their role is essential in the brake/acceleration balance of the network dynamics preventing artificial attractors to be expressed, which could lead either to pathologic rhythms, or to steady states with too high concentrations of certain metabolites.

Following the comparative physiology study about cell consumption of oxygen by mammals done in (Weibel et al. 1991), let us consider V, the total mitochondrial volume of muscle and V_{O2max}, the O_2 consumption rate per unit mass of whole body mass M of the mitochondria of these mammals. Table 4.9 shows that nonruminant

Table 4.11 Number of common subsequences of length 6 and 7 between the mitochondrial tRNA loops and the noncoding mitochondrial genomes of capra (total length 1,531 bases) and canis (total length 3,239 bases) (http://megasunbchumontrealca/ogmp/projects/other/cp_list html)

Subsequences	Observed capra	Expected capra	Observed canis	Expected canis
TACCAC	1	0.4	0	0.8
TACCAT	0	0.4	0	0.8
ACCGTT	1	0.4	0	0.8
ACCATT	1	0.4	0	0.8
CTTGAA	0	0.4	0	0.8
ATTTGAA	0	0.1	0	0.2
Total	3	2.1 ± 2.9	0	4.2 ± 4.1

Table 4.12 Number of common subsequences of length 6 and 7 between the mitochondrial tRNA loops and the noncoding mitochondrial genomes of ruminants (total length 10,024 bases) and nonruminants (total length 7,774 bases) (http://megasunbchumontrealca/ogmp/projects/other/cp_list html). The stars correspond to the statistical significance of the difference between the ruminants and non ruminants totals of matches

Subsequences	Observed ruminants	Expected ruminants	Observed nonruminants	Expected nonruminants
TACCAC	3	2.4	0	1.9
TACCAT	1	2.4	0	1.9
ACCGTT	3	2.4	0	1.9
ACCATT	2	2.4	0	1.9
CTTGAA	1	2.4	0	1.9
ATTTGAA	1	0.6	0	0.5
Total	11**	12.7 ± 7.1	0**	10 ± 6.3

**p<0.01.

animals like horse (equus) and dog (canis) have an efficacy in O_2 consumption about 2.5 times more than ruminants like steer (bos) and goat (capra). We have then studied in the noncoding (called d-loop) mitochondrial genome of the nonruminants if there are less mitomiRs candidates than in those of ruminants.

Table 4.10 shows that such mitomiRs exist and contain long subsequences (whose length is between 6 and 7 bases) common with the mitochondrial tRNA loops. On the contrary, these subsequences are totally absent in the noncoding mitochondrial genome of the nonruminants. These genomes have been extracted for these different animals from the classical genetic databases dedicated to the whole mitochondrial (http://megasunbchumontrealca/ogmp/projects/other/cp_list.html) and chloroplast (http://www.ncbinlmnihgov/nuccore/EF115542; http://www.ncbinlmnihgov/nuccore/AB042240) genomes: Megasun, from the University of Montreal (http://megasunbchumontrealca/ogmp/projects/other/cp_list.html), and Nuccore, from the NCBI (http://www.ncbinlmnihgov/nuccore/EF115542; http://www.ncbinlmnihgov/nuccore/AB042240).

From Tables 4.10, 4.11, and 4.12, we see that the absence of mitochondrial small subsequences, we have called mito^{2}miRs, in the noncoding (called d-loop) mitochondrial genomes of nonruminants (for a total length of 10,024 bases) susceptible to hybridize the mitochondrial tRNAs inside the mitochondrion matrix is highly

Table 4.13 Comparison between cellular CO_2 diurnal consumption between 2 cereals, sorghum and wheat (Tang et al. 2012)

	V_{CO2max}/S (mmol s^{-1} m^{-2})
Sorghum bicolor	41.1 ± 2.8
Triticum æstivum	27.9 ± 1

Table 4.14 Number of common subsequences of length 6 and 7 between the chloroplast tRNA loops and the noncoding chloroplast genomes of sorghum (total length 15,852 bases) and triticum (total length 16,014 bases)

Subsequences	Observed sorghum	Expected sorghum	Observed triticum	Expected triticum
TACCACT	0	1	0	1
TACCATT	4	1	4	1
TACCGCT	0	1	0	1
TACCGTT	1	1	1	1
ATTTGAA	3	1	4	1
GTTTGAA	1	1	5	1
ATTCGAA	3	1	5	1
GTTCGAA	3	1	3	1
ACTTGAA	0	1	2	1
GCTTGAA	0	1	1	1
ACTCGAA	3	1	3	1
GCTCGAA	1	1	1	1
Total	19**	11.7 ± 6.8 ($m \pm 2\sigma$)	29**	11.6 ± 6.8 ($m \pm 2\sigma$)

significant (observed number equal to 0 and mean expected number 10 with a standard deviation equal to 3.2). A classical normal test of equality between empirical means shows that the number of ruminant mito^{2}miRs is significantly (with a threshold of 10^{-3}) superior to that of nonruminant. The absence of possibility of inhibition of the tRNA transcription inside the nonruminant mitochondria can participate to the better efficacy of the cell respiration in these nonruminants with respect to the ruminants.

In the same vein, we have analyzed data coming from noncoding chloroplast genome (http://www.ncbinlmnihgov/nuccore/EF115542; http://www.ncbinlmnihgov/nuccore/AB042240) of cereals, one belonging to the C_4 plants, the sorghum (Sorghum bicolor), and the other to the C_3 plants, the wheat (Triticum æstivum) which represent the most important cereals in temperate countries. The C_4 plants are known to have a better efficacy of their chloroplast diurnal respiration than the C_3 plants, as we can see on Table 4.13 (Byrd et al. 1992; Farquhar et al. 1980), where the sorghum is shown to be 1.5 times more efficient that the wheat, by comparing the CO_2 consumption rate per unit surface of whole surface S of the leaves of these cereals.

In these cereals, the CO_2 assimilation is usually limited by the capacity of photosynthetic electron transport to supply ATP and NADPH to regenerate RuBP (Byrd et al. 1992; Farquhar et al. 1980). Then we can compare as for the animal mitochondria the presence of subsequences common to the noncoding chloroplast genome and to the chloroplast tRNA loops.

From the Table 4.14, we see that the number of chloroplast small subsequences, we are calling chloromiRs, in the noncoding chloroplast genome of a C_4 plant, the sorghum (Sorghum bicolor, for a total length of 15,852 bases) susceptible to hybridize the chloroplast tRNAs inside the chloroplast matrix is highly significantly less than the same number observed in a C_3 plant, the wheat (Triticum æstivum, for a total length of 16014 bases).

A classical normal test of equality between empirical means shows that the number of the C_4 plant chloromiRs is significantly (with a threshold of 10^{-3}) less than that of the C_3 plant. The diminution of possibility of inhibition of the tRNA transcription inside the chloroplast of the sorghum can participate to the better efficacy of the cell diurnal respiration in this C_4 plant with respect to the less efficient C_3 plant, the wheat.

4.5.2 Pathological Role in Cancer

For example, we can notice in patients with melanoma the presence of microRNAs like miR-221 reducing the isomerase activity in melanoma cells (Segura et al. 2012), hence increasing the high glycolysis rate [and favoring the pentose phosphate pathway and the Warburg effect (Tennant et al. 2010; Diaz-Ruiz et al. 2011; Demetrius and Simon 2012; Davies et al. 2012)] as well as the proliferative growth rate of the melanoma cells by targeting c-kit, p27 and p57, and in the patients with prostate cancer the presence of microRNAs like miR-34a, repressing the oxidative phosphorylation through the enzymes Translocase and ATPase (favoring the Warburg effect) and the cell cycle through E2F and CD44, inhibits the prostate cancer stem cells and prevents metastasis (Liu et al. 2011).

In the numerous papers devoted to microRNAs and cancer, we can notice the possibility of an under-expression of the microRNA hsa-miR-320 which reduces the PFK activity (Tang et al. 2012), favoring the high glycolysis rate. Associated like in the melanoma data with a blockage of the oxidative phosphorylation or with an over-expression of hsa-miR-19a which inhibits the Pyruvate Dehydrogenase, we have another example reinforcing the Warburg hypothesis concerning the energetic origin of cancer. The last example of miR-dependent cancer is the Hepatocellular Carcinoma (HCC), the major primary liver cancer. The glypican-3 (GPC3) is one of the most abnormally expressed genes in this cancer, which could play a role in liver carcinogenesis. In (Maurel et al. 2012; Huang et al. 2012) using a functional screening, the authors found that miR-96, miR-129-1-3p, miR-219-5p miR-1271, miR-1291, and miR-1303 differentially control the GPC3 expression in HCC cells. More precisely, miR-219-5p exerts its tumor-suppressive effect in hepatic carcinogenesis through its negative regulation of GPC3 expression.

Eventually, we can notice the role of the tumor suppressors like p53, p63, and p73, which are highly controlled by and are controlling numerous micro-RNAs [cf. Figs. 4.15 and 4.17 and Boominathan (2010a) and Boominathan (2010b)]. The

miR-143	Dengue	2	3-	10	1	10	147	138	tgagctacag	tgagctacag
miR-143	Dengue	2	3-	11	12	22	280	270	gcttcatctca	gcttcatctca
miR-143	Dengue	3	3-	7	8	14	273	267	cagtgct	cagtgct
miR-143	Dengue	4	3-	9	2	10	82	74	gagctacag	gagctacag
miR-143	Japanese encephalitis	-	5+	7	14	20	6	12	ttcatct	ttcatct
miR-143	Japanese encephalitis	-	3-	10	12	21	252	243	gcttcatctc	gcttcatctc
miR-143	Japanese encephalitis	rAT	3-	7	3	9	558	552	agctaca	agctaca
miR-143	Japanese encephalitis	-	5+	7	7	13	63	69	acagtgc	acagtgc

Fig. 4.17 Sequences viewing perfect matches between nuclear microRNA hsa-miR-143 with viral genomes

Table 4.15 Sequences showing anti-matches between nuclear microRNA hsa-miR-143 with viral genomes

Sequences	Identification	Anti-matches
5'-AGAUGAAUGGUACUGCCAUUCA-3'	AL	14
3'-ACUCGAUGUCACGAAGUAGAGU_5':	hsa-miR-143	

Table 4.16 Sequences alignment of nuclear miRNAs with pb1-5 H1N1 influenza A gene (Raver-Shapira et al. 2007; Kohn et al. 2006; Weibel et al. 1991)

Sequences	Identification	Anti-matches
3' UACCGUCAUGGUAAGUAGAACU 5'	AL	13
5' AGUGGGGAACCCUUCCAUGAGG 3':	hsa-miR-491	
3' UACCACCUUGUCUAGAAGUACU 5':	pb1-5	13
3' UAGAACUUACCGUCAUGGUAAG 5'	AL	13
5' AUGGUGGGCCGCAGAACAUGUG 3':	hsa-miR-654	
3' UACCACCUUGUCUAGAAGUACU 5':	pb1-5	14

role we have already commented of p53 in the dual regulation of the apoptosis versus proliferation processes is probably the main origin of its major influence in carcinogenesis and tumor-suppression, depending on its up- or downregulation by microRNAs.

4.5.3 Pathological Role in Infectious Diseases

MicroRNAs are closely related to viral genomes (He et al. 2009; Song et al. 2010; Li et al. 2011; Koparde and Singh 2010; Cullen 2010; Wang et al. 2009) due to a long co-evolution between hosts, vectors, and pathogens. In Table 4.15 and

Table 4.17 The conserved binding site (in blue) of AL with pb1gene compared across a selection of influenza viral strains (collected from NCBI new release 715 [22/01/2010])

Virus NCBI Type, Nb	5'-3' Sequences
	14 anti-matches AL 3'GUAGAACUUACCGUCAUGGUAA5'
H1N1, FLAP1MB	GAGGAGTTCACTGAGATCATGAAGATCTGTTCCACCATTG
H2N1, FJ686720	GAGGAGTTTGCTGAGATCATGAAGATCTGTTCCACCATTG
H3N1, CY039618	GGAGGAGTTTGCTGAGATCATGAAGATCTGTTCCACCATTG
H4N1, CY034651	GGAGGAATTTGCTGAGATCATGAAGATCTGTTCCACCATTG
H4N4, CY004720	AGAGGAGTTCGCTGAGATCATGAAGATCTGTTCCACCATTG
H5N1, FJ573466	AGGGTTTGCTGAGATCATGAAGATCTGTTCCACCATTGA
H6N1, CY042630	GGAAGAGTTTGCTGAGATCATGAAGATCTGTTCCACCATTG
H7N1, AB268552	AGAAGAGTTTGCTGAGATCATGAAGATCTGTTCCACCATTG
H9N1, CY005104	AGAGGAGTTTGCTGAGATCATAAAGATCTGTTCCACCAT
H10N1, CY032866	AGGGGAGTTTGCTGAGATCATGAAGATCTGTTCCACCATTG
H11N1, FJ517280	GGAGGAGTTTGCTGAGATCATGAAGATCTGTTCCACCATTG
H12N1, CY005349	AGAGGAGTTTGCTGAGATCATAAAGATCTGTTCCACCATTGA
H1N2, GQ405858	GAAGACTTCGCTGAGATCATGAAGATCTGTTCCACCATTG
H1N3, CY042123	GAGGAGTTTGCTGAGATCATGAAGATCTGTTCCACCATTGA
H1N4, CY042541	GAGGAGTTTGCTGAGATCATGAAGATCTGTTCCACCATTGA
H1N5, CY020867	GAGGAGTTTGCTGAGATCATGAAGATCTGTTCCACCATTGA
H1N7, CY042483	GAAGAGTTTGCTGAGATCATGAAGATCTGTTCCACCATTG
H2N8, CY004560	GAGGAGTTCGCTGAGATCATGAAGATCTGTTCCACCATTG
H1N9, CY042159	GAGGAGTTTGCTGAGATCATGAAGATCTGTTCCACCATTGA

Fig. 4.17, an example shows that microRNA hsa-miR-143 presents numerous perfect matches with Hamming distance 0 between long subsequences (whose length is between 7 and 11 bases) it shares with genomes of Dengue and Japanese encephalitis viruses (Demongeot et al. 2009b, c).

In Table 4.16, several microRNAs (miR-491 and 654) are inhibiting viral H1N1 influenza A proteins like pb1, which is a critical RNA polymerase subunit of the virus H1N1 influenza A, necessary for the virus replication both in vitro and in vivo (He et al. 2009; Song et al. 2010; Li et al. 2011).

The microRNAs like miR-491 can also hybridize many strains of several influenza viruses in conserved viral strains (Table 4.17), proving the possibility of an unspecific immunologic defense, acting before the immune cell response and preventing a too rapid early viral replication in human host. It could be interesting in the future to see the influence of microRNAs on the global dynamics of the metabolic networks involving proteins like pb1, when their interaction graphs will be better known (Koparde and Singh 2010; Cullen 2010; Wang et al. 2009).

We will give indications in Sect. 4.6 about the way to study the attractors of the networks reduced to their strong connected components, which are circuits of genes isolated, tangential or intersected, on which calculations about the number and nature of their attractors can be possible.

4.6 Perspective: The General Architecture of a Genetic Regulatory Network and the Problem of Its Robustness

The general architecture of a genetic regulatory network is given like in Fig. 4.8 or in Fig. 4.16 by a digraph framework in which the nodes sources are often micro-RNAs exerting their partly unspecific basic inhibitory influence, and the nodes sink are important genes controlling vital functions like the RAG (Recombinase Activating Gene) responsible for the TCR building in the immune system [cf. Fig. 4.16 and Demongeot and Waku (2012a), Pasqual et al. (2002), Baum et al. (2004), and Thuderoz et al. (2010)].

The network robustness (or resilience) (Demongeot and Waku 2012a; Demongeot and Demetrius (submitted); Blanchini and Franco 2011; Lesne 2008; Gunawardena 2010; Waddington 1940; Thom 1972; Cinquin and Demongeot 2002) can be defined as the capacity to return at its ordinary asymptotic dynamical behavior (called attractor) after endogenous or exogenous perturbations affecting:

- The state of certain of its genes, e.g., in case of specific silencing by micro-RNAs
- The state of its boundary, notably the appearance of new regulations from a mutation in the noncoding genome giving birth to new micro-RNAs
- Its architecture, by creating new links between proteins needing to take into account nonlinear interactions (Demongeot and Sené 2011).

A study about the influence of the microRNAs on the robustness of a network needs the exact counting of the dynamical attractors of this network, then to know the reduction or on the contrary the amplification factor caused, for example, by the circuit opening due to the effective inhibition of a gene (until its possible knockout) by a microRNA, or by the appearance of new genes, which corresponds to important architectural perturbations. A parameter devoted to represent the robustness of a biological network is its evolutionary entropy [cf. Mathematical Annex below and Demongeot et al. (2008b), Demongeot and Sené (2008), Demetrius (1983, 1997), Kühn (2010), Fogelman Soulié et al. (1989), Cosnard and Goles (1977), and Demongeot et al. (2012)], defined as the Kolmogorov–Sinaï entropy of the Markov process underlying the gene states updating. The microRNAs can have a double opposite influence on this parameter, causing its increase (and hence that of the network robustness) when they create an unspecific inhibitory "noise", dispatching into the gene state space the probability to have more possible phenotypes, and they can also provoke a decrease of the robustness, by opening circuits of the strongly connected components of the network, hence by diminishing the number of positive circuits, responsible of the attractor number of the whole network.

The example in Fig. 4.16 shows that we can count the attractor number of the main subnetworks of a functional network like that organized around the gene Engrailed. These attractor numbers are given in blue in Fig. 4.16, as well as their nature (steady state or limit cycle). The global attractor of the whole network is

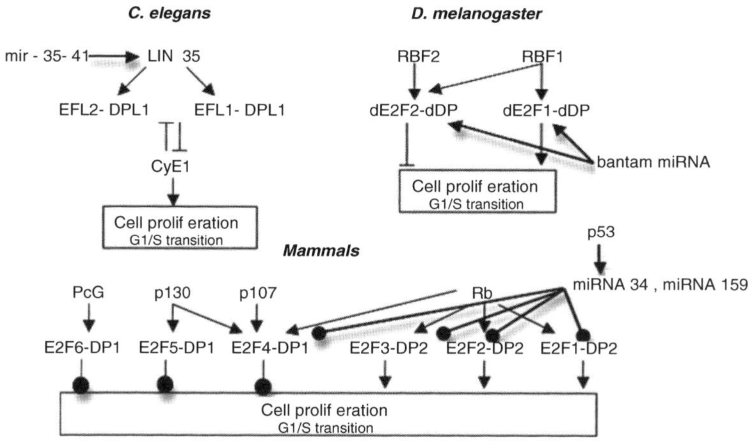

Fig. 4.18 Different control mechanisms of the G1/S transition in different species, *Caenorhabditis elegans*, *Drosophila melanogaster*, and mammals (after Caraguel et al. 2010)

more difficult to obtain, in particular when two subnetworks create a new strong connected component, like in the case of the MPK and Cytoskeleton subnetworks in Fig. 4.16, which creates a new level of complexity that needs further theoretical research.

In Fig. 4.18, a last example shows that the same function (cell proliferation) is regulated with progressively more frontier nodes of the regulatory networks, when passing from *C. elegans* and *D. melanogaster* to mammals (Caraguel et al. 2010; Kohn et al. 2006; Massirer et al. 2012; Herranz et al. 2010). The progressive appearance during the evolution of many upstream controllers until the mammals microRNAs and p53 provides to the cell cycle network a robust Rb-E2F control at the G1/S transition.

4.7 Conclusion

The sensitivity of real biological networks to endogenous or exogenous perturbations appears dominant in the case of the inhibitory actions exerted by the microRNAs and we have noticed this influence for many cases. More systematic studies have to be performed in order to confirm the dominant influence of boundary negative interactions, thanks to which the Hopfield-like regulatory interaction networks seem to become more robust, and also to make more precise their influence on the number of attractors, which is conjectured to diminish, when microRNAs are multiple on the boundary of the interaction graph of the network.

For example, MAPK in Fig. 4.14 is inhibited not only by miR-191 but also by miR-350 (http://mirdborg/miRDB/; http://mirnamapmbcnctuedutw/), which is also involved in the memory storage and evocation (Smalheiser et al. 2011;

Ben Amor et al. 2010a, b), and the role of this inhibition could lead to a restricted number of dynamical behaviors susceptible to be stored and evoked inside the subnetworks of hippocampus. Many microRNAs are also controlling morphogenesis of teeth (where miR-140, miR-31, miR-875-5p, and miR-141 were expressed mainly during tooth morphogenesis), feathers or hairs, with miRs like miR-30 activating Wnt through WWP1 [cf. Fig. 4.16 and Michon et al. (2012) and Michon et al. (2008)] or miR-16 inhibited by Wnt/ß-catenin signaling and inhibiting Cdk2 in the proliferation box [cf. Fig. 4.16 and Takeshita et al. (2010) and Martello et al. (2007)].

More generally, the passage from a random network structure to a small world (Duchon et al. 2006; Demongeot et al. 2010b, 2011c) during evolution, by increasing the microRNAs number in case of networks controlling the same function (like the cell cycle), could have contributed to increase network robustness. Eventually, considering the role of a large inhibition by microRNAs in genetic networks modeling the chromatin clock, could allow understanding the role of the state-dependent expression schedule. These two last perspectives would lead to further investigations and constitute challenges for future work.

Acknowledgments We are indebted to A. Doncescu, M. Noual, and S. Sené for many helpful discussions. This work has been supported by the EC NoE VPH.

Mathematical Annex

In mathematical modeling, a real genetic regulatory network is called a genetic threshold Boolean regulatory network (denoted in the following getBren). A getBren N can be considered as a set of random automata, defined by:

1. Any random automaton i of the getBren N owns at time t a state $x_i(t)$ valued in $\{0,1\}$, 0 (resp. 1) meaning that gene i is inactivated or in silence (resp. activated or in expression). The global state of the getBren at time t, called configuration in the sequel, is then defined by: $x(t) = (x_i(t))_{i \in \{1,n\}} \in \Omega = \{0,1\}^n$
2. a getBren N of size n is a triplet (W,Θ,P), where:
   - W is a matrix of order n, where the coefficient $w_{ij} \in \mathbb{R}$ represents the interaction weight gene j has on gene i. Sign(W) $= (\alpha_{ij} = \text{sign}(w_{ij}))$ is the adjacency (or incidence) matrix of the interaction graph G.
   - Θ is an activation threshold vector of dimension n, its component θ_i being the activation threshold attributed to automaton i
   - M: $\mathbf{P}(\Omega) \rightarrow [0,1]^{m \times m}$ (where $\mathbf{P}(\Omega)$ is the set of all subsets of Ω and $m = 2^n$) is a Markov transition matrix, built from local probability transitions P_i giving the new state of the gene i at time t+1 according to W, Θ, and configuration $x(t)$ of N at time t such that:

$\forall g \in \{0,1\}, \beta \in \Omega, P_{i,g}{}^{\beta}(\{x_i(t+1) = g|x(t) = \beta\}) = \exp[g(\Sigma_{j\in Ni} w_{ij}\beta_j - \theta_i)/T]/Z_i$, where $Z_i = [1 + \exp[(\Sigma_{j\in Ni} w_{ij}\beta_j - \theta_i)/T]$, N_i is the neighborhood of the gene i in the getBren N, i.e., the set of genes j (including possibly i) such that $w_{ij} \neq 0$, and $P_{i,g}{}^{\beta}$ is the probability for the gene i of passing to the state g at time t+1, from the state β at time t on N_i. M denotes the transition matrix built from the $P_{i,g}{}^{\beta}$, s. M depends on the update mode chosen for changing the states of the getBren automata. For the extreme values of the randomness parameter T, we have:

1. if $T = 0$, the getBren becomes a deterministic threshold automata network and the transition can be written as:

$$x_i(t+1) = h(\Sigma_{j\in Ni} w_{ij} x_j(t) - \theta_i),$$

where h is the Heaviside function: $h(y) = 1$, if $y > 0$;

$h(y) = 0$, if $y < 0$,

except for the case $\Sigma_{j\in Ni} w_{ij} x_j(t) - \theta_I = 0$, for which, if necessary, 1 and 0 are both chosen with probability ½

2. When T tends to infinity, then $P_{i,g}{}^{\beta} = 1/2$ and each line of M is the uniform distribution on Ω.

If the mode of updating the states of the n genes is sequential, the invariant measure μ expressing the asymptotic state of N is the Gibbs measure on Ω defined by (Demongeot and Waku 2012a; Demongeot and Demetrius (submitted); Demongeot et al. 2011b, 2008b, 2012; Demongeot and Sené 2008, 2011; Fogelman Soulié et al. 1989; Cosnard and Goles 1977):

$$\forall x \in \Omega, \mu(\{x\}) = \exp((\Sigma_{i\in x,\, j\in Ni} w_{ij} x_i x_j - \theta_i)/T)/Z,$$

where $Z = \Sigma_{y\in\Omega} \exp((\Sigma_{j\in y, k\in Nj} w_{jk} y_j y_k - \theta_j)/T)$.

We define the random energy U (Demongeot and Sené 2008, 2011; Demongeot et al. 2008b) and random frustration F (Demongeot and Waku 2012a; Demongeot and Demetrius (submitted)) of a getBren N by:

$$\forall x \in \Omega, U(x) = \Sigma_{i,j\in\{1,n\}} \alpha_{ij} x_i x_j = Q_+(N) - F(x),$$

where $Q_+(N)$ is the number of positive edges in the interaction graph G of the network N and $F(x)$ the global frustration of x, i.e., the number of pairs (i,j) where the values of x_i and x_j are contradictory with the sign α_{ij} of the interaction between genes i and j: $F(x) = \Sigma_{i,j\in\{1,n\}} F_{ij}(x)$, where F_{ij} is the local frustration of the pair (i,j) defined by:

$$\begin{aligned} F_{ij}(x) = 1,\ &\text{if } \alpha_{ij} = 1, x_j = 1 \text{ and } x_i = 0, \quad \text{or } x_j = 0 \text{ and } x_i = 1,\\ &\text{and if } \alpha_{ij} = -1,\ x_j = 1 \quad \text{and } x_i = 1, \quad \text{or } x_j = 0 \text{ and } x_i = 0, \end{aligned}$$

$F_{ij}(x) = 0$, elsewhere.

Eventually, we define the random global dynamic frustration D by:

$$D(x(t)) = \Sigma_{i,j \in \{1,n\}} D_{ij}(x(t)),$$

where D_{ij} is the local dynamic frustration of the pair (i,j) defined by:

$$D_{ij}(x(t)) = 1, \text{ if } \alpha_{ij} = 1, x_i(t) \neq h(\Sigma_{j \in Ni} w_{ij} x_j(t) - \theta_i) \text{ or } \alpha_{ij} = -1,$$
$$x_i(t) = h(\Sigma_{j \in Ni} w_{ij} x_j(t) - \theta_i),$$

$D_{ij}(x(t)) = 0$, elsewhere.

Then we prove the following Proposition 1:

Proposition 1. *Let us consider the random energy U and the random frustration F of a getBren N having a constant absolute value w for its interaction weights, null threshold Θ, temperature T equal to 1 and being sequentially updated, then:*

1. $U(x) = \Sigma_{i,j \in \{1,n\}} \alpha_{ij} x_i x_j = Q_+(\mathrm{N}) - \mathrm{F(x)}$, *where $Q_+(N)$ is the number of positive edges in the interaction graph G of the network*
2. $E_\mu(U) = \partial \log Z / \partial w$, *where the free energy* $\log Z$ *is equal to the quantity* $\log(\Sigma_{y \in \Omega} \exp(\Sigma_{j \in y, k \in y} w_{ij} y_j y_k))$ *and μ is the invariant Gibbs measure defined by:* $\forall\, x \in \Omega$, $\mu(\{x\}) = \exp(\Sigma_{i \in x, j \in x} w_{ij} x_i x_j)/Z$
3. $Var_\mu U = Var_\mu F = -\partial E_\mu / \partial \log w$, *where* $E_\mu = -\Sigma_{x \in \Omega} \mu(\{x\}) \log(\mu(\{x\})) = \log Z - wE(U)$ *is the entropy of μ, maximal among entropies corresponding to all probability distributions ν for the U's having the same given expectation* $E_\nu(U) = E_\mu(U)$.

Proof. 1. It is easy to check that: $U(x) = Q_+(N) - F(x)$,

2. The expectation of U, denoted $E_\mu(U)$, is given by:

$$E_\mu(U) = \Sigma_{x \in \Omega} \Sigma_{i \in x, j \in x} \alpha_{ij} x_i x_j \exp\left(\Sigma_{i \in x, j \in x} w x_i x_j\right)/Z = \partial \log Z / \partial w$$

3. Following Demongeot and Waku (2012a, b and submitted), we have $\mathrm{Var}_\mu U = \mathrm{Var}_\mu F = -\partial E_\mu / \partial \log w$, and according to Demongeot et al. (2008b), Demongeot and Sené (2008), Demetrius (1983, 1997), Kühn (2010), and Fogelman Soulié et al. (1989), E_μ is maximal among the proposed set of entropies.∎

Proposition 2. *Let us consider the getBren N with T = 0, sequentially or synchronously updated, defined from a potential P defined by:*

$$\forall x \in \Omega, P(x) = \Sigma_k ({}^t x A_k x) x_k + {}^t x W x + \Theta x,$$

where A, W, Θ are, respectively, integer tensor, matrix, and line vector. Let suppose also that:

$$\forall i = 1, \ldots, n, \Delta x_i \in \{-1, 0, 1\}.$$

If h denotes the Heaviside function, consider now the potential automaton i defined by:

$$x_i(t+1) = h(-\Delta P/\Delta x_i + x_i(t)),$$

*and by the condition $x_i(t + 1) \geq 0$, if $x_i(t) = 0$, such that the flow remains in Ω. Then, if the tensor A is symmetrical with vanishing diagonal (*i.e., *if we have the equalities: $\forall$ i,j,k = 1,...,n, $a_{ijk} = a_{ikj} = a_{kij} = a_{jki} = a_{jik} = a_{kji}$, and $a_{iik} = 0$), and if each sub-matrix (on any subset J of indices in {1,...,n}) of A_k and W are non-positive with vanishing diagonal, P decreases on the trajectories of the potential automaton, for any mode of implementation of the dynamics (sequential, block sequential, and parallel). Hence, the stable fixed configurations of the automaton correspond to the minima of its potential P.*

Proof. We have, for a discrete function *P* on *Ω*:

$$\Delta P(x)/\Delta x_i = [P(x_1, \ldots, x_i + \Delta x_i, \ldots, x_n) - P(x_1, \ldots, x_i, \ldots, x_n)]/\Delta x_i$$

and the proof, based on the existence of a Lyapunov function proved in Ben Amor et al. (2008), Demongeot et al. (2006, 2008a), Kühn (2010), Fogelman Soulié et al. (1989), and Cosnard and Goles (1977), results from the Proposition 1 of (Cosnard and Goles 1977). ∎

Proposition 3. *Let us consider the Hamiltonian getBren which is a circuit with constant absolute value w for its interaction weights, null threshold Θ, and temperature T equal to 0, sequentially or synchronously updated, whose Hamiltonian H is defined by:*

$$\begin{aligned} H(x(t)) &= \sum\nolimits_{i=1,\ldots,n} (x_i(t) - x_i(t-1))^2/2 \\ &= \sum\nolimits_{i=1,\ldots,n} (h(w_{i(i-1)} x_{i-1}(t-1) - x_i(t-1))^2/2, \end{aligned}$$

then H equals the total discrete kinetic energy and the half of the global dynamic frustration D(x(t)). The result remains available if the automata network is a circuit in which transition functions are Boolean identity or negation.

Proof. It is easy to check that: $H(x(t)) = D(x(t))/2$. ∎

Table 4.18 Values of the global frustration D, attractor numbers and periods for positive circuits of order 8 with Boolean transitions identity or negation

D (frustration)	Attractor number	Attractor period
0	2	1
2	7	8
4	3	4
4	16	8
6	7	8
8	1	2

Proposition 1 is used to estimate the evolution of the robustness of a network, because from Demongeot and Waku (2012a, b) and Demongeot and Demetrius (submitted) it results that the quantity $E = E_{\mu}—E_{\text{attractor}}$, called the evolutionary entropy serves as robustness parameter (Ben Amor et al. 2008; Demongeot et al. 2009a, b, 2010a; Demongeot and Waku 2012a, b; Elena et al. 2008; Lesne 2008; Gunawardena 2010), being related to the capacity a getBren has to return to μ, the equilibrium measure, after endogenous or exogenous perturbation. $E_{\text{attractor}}$ can be evaluated by the quantity:

$$E_{\text{attractor}} = -\Sigma_{k=1,m\leq 2}n\mu(C_k)\log\mu(C_k),$$

where m is the number of attractors and $C_k = B(A_k)\cup A_k$ is the union of the attractor A_k and of its attraction basin $B(A_k)$. A systematic calculation of $E_{\text{attractor}}$ allows quantifying the increase in complexity of a network ensuring a dedicated regulatory function in different species: for example, the increase of the inhibitory sources with multiple targets in up-trees converging on a conserved subgraph of a genetic network (e.g., the core regulating the cell cycle in *C. elegans*, *D. melanogaster,* and mammals, cf. Fig. 4.18) causes a decrease of its attractor number (Demongeot and Waku 2012a; Demongeot and Demetrius (submitted); Caraguel et al. 2010), hence an increase of its evolutionary entropy, showing that the robustness of a network is in this case positively correlated with its connectivity (i.e., the ratio between the numbers of interactions and genes in the network).

Propositions 2 and 3 give examples of extreme cases, where the networks are either discrete (Cinquin and Demongeot 2005) or continuous (Cinquin and Demongeot 2002) potential (or gradient) systems, generalizing previous works on continuous or discrete networks in which authors attempt to explicit Waddington and Thom chreode's potential-like (Waddington 1940; Thom 1972), or Hamiltonian (Demongeot and Demetrius (submitted); Demongeot et al. 2011b): in (Demongeot et al. 2012) for example, it is proposed a method for calculating the number of attractors in case of circuits with Boolean transitions identity or negation. These results about attractors counting constitute a partial response to the discrete version of the XVIth Hilbert's problem and can be approached by using the Hamiltonian energy levels. For example, for a positive circuit of order 8, it is easy to prove that, in case of parallel updating, we have only even values for the global frustration D (they are odd for a negative circuit), corresponding to different values

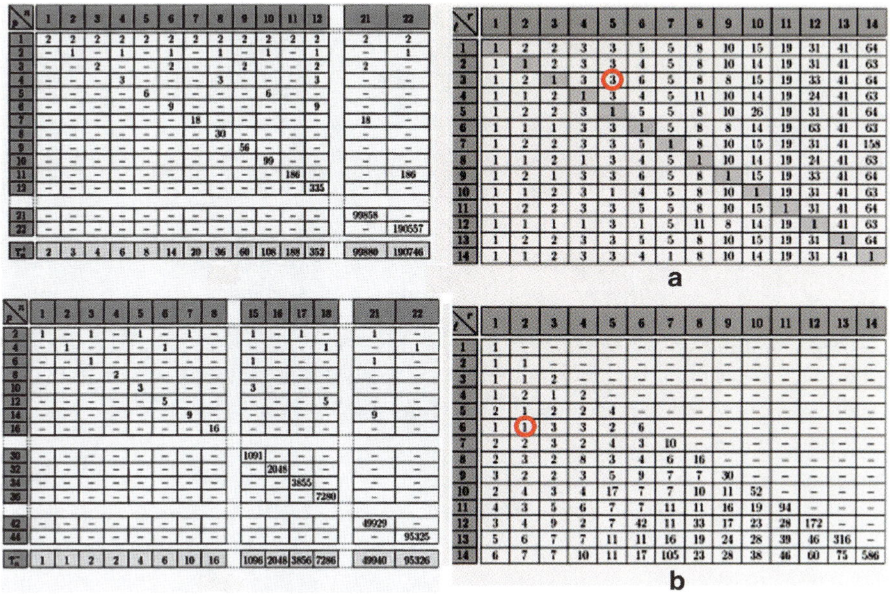

p \ n	1	2	3	4	5	6	7	8	9	10	11	12	21	22
1	2	2	2	2	2	2	2	2	2	2	2	2	2	2
2	–	1	–	1	–	1	–	1	–	1	–	1	–	1
3	–	–	2	–	–	2	–	–	2	–	–	2	2	–
4	–	–	–	3	–	–	–	3	–	–	–	3	–	–
5	–	–	–	–	6	–	–	–	–	6	–	–	–	–
6	–	–	–	–	–	9	–	–	–	–	–	9	–	–
7	–	–	–	–	–	–	18	–	–	–	–	–	18	–
8	–	–	–	–	–	–	–	30	–	–	–	–	–	–
9	–	–	–	–	–	–	–	–	56	–	–	–	–	–
10	–	–	–	–	–	–	–	–	–	99	–	–	–	–
11	–	–	–	–	–	–	–	–	–	–	186	–	–	186
12	–	–	–	–	–	–	–	–	–	–	–	335	–	–
21	–	–	–	–	–	–	–	–	–	–	–	–	99858	–
22	–	–	–	–	–	–	–	–	–	–	–	–	–	190557
T_n^+	2	3	4	6	8	14	20	36	60	108	188	352	99880	190746

p \ n	1	2	3	4	5	6	7	8	15	16	17	18	21	22
2	1	–	1	–	1	–	1	–	1	–	1	–	1	–
4	–	1	–	–	–	1	–	–	–	–	–	1	–	1
6	–	–	1	–	–	–	–	–	1	–	–	–	1	–
8	–	–	–	2	–	–	–	–	–	–	–	–	–	–
10	–	–	–	–	3	–	–	–	3	–	–	–	–	–
12	–	–	–	–	–	5	–	–	–	–	–	5	–	–
14	–	–	–	–	–	–	9	–	–	–	–	–	9	–
16	–	–	–	–	–	–	–	16	–	–	–	–	–	–
30	–	–	–	–	–	–	–	–	1091	–	–	–	–	–
32	–	–	–	–	–	–	–	–	–	2048	–	–	–	–
34	–	–	–	–	–	–	–	–	–	–	3855	–	–	–
36	–	–	–	–	–	–	–	–	–	–	–	7280	–	–
42	–	–	–	–	–	–	–	–	–	–	–	–	49929	–
44	–	–	–	–	–	–	–	–	–	–	–	–	–	95325
T_n^-	1	1	2	2	4	6	10	16	1096	2048	3856	7286	49940	95326

a

l \ r	1	2	3	4	5	6	7	8	9	10	11	12	13	14
1	1	2	2	3	3	5	5	8	10	15	19	31	41	64
2	1	1	2	3	3	4	5	8	10	14	19	31	41	63
3	1	2	1	3	3	6	5	8	8	15	19	33	41	64
4	1	1	2	1	3	4	5	11	10	14	19	24	41	63
5	1	2	2	3	1	5	5	8	10	26	19	31	41	64
6	1	1	1	3	3	1	5	8	8	14	19	63	41	63
7	1	2	2	3	3	5	1	8	10	15	19	31	41	158
8	1	1	2	1	3	4	5	1	10	14	19	24	41	63
9	1	2	1	3	3	6	5	8	1	15	19	33	41	64
10	1	1	2	3	1	4	5	8	10	1	19	31	41	63
11	1	2	2	3	3	5	5	8	10	15	1	31	41	64
12	1	1	1	1	3	1	5	11	8	14	19	1	41	63
13	1	2	2	3	3	5	5	8	10	15	19	31	1	64
14	1	1	2	3	3	4	1	8	10	14	19	31	41	1

b

l \ r	1	2	3	4	5	6	7	8	9	10	11	12	13	14
1	1	–	–	–	–	–	–	–	–	–	–	–	–	–
2	1	1	–	–	–	–	–	–	–	–	–	–	–	–
3	1	1	2	–	–	–	–	–	–	–	–	–	–	–
4	1	2	1	2	–	–	–	–	–	–	–	–	–	–
5	2	1	2	2	4	–	–	–	–	–	–	–	–	–
6	1	1	3	3	2	6	–	–	–	–	–	–	–	–
7	2	2	3	2	4	3	10	–	–	–	–	–	–	–
8	2	3	2	8	3	4	6	16	–	–	–	–	–	–
9	3	2	2	3	5	9	7	7	30	–	–	–	–	–
10	2	4	3	4	17	7	7	10	11	52	–	–	–	–
11	4	3	5	6	7	7	11	11	16	19	94	–	–	–
12	3	4	9	2	7	42	11	33	17	23	28	172	–	–
13	5	6	7	7	11	11	16	19	24	28	39	46	316	–
14	6	7	7	10	11	17	105	23	28	38	46	60	75	586

Fig. 4.19 *Left*: Number of attractors of period *p* for positive (*top*) and negative (*bottom*) circuits of order *n*, in case of parallel updating mode (after Demongeot et al. 2012). *Right*: Total number of attractors in case of tangent circuits, where (**a**) the left circuit of order *l* is negative and the right circuit of order *r* is positive and (**b**) both side circuits are negative

of the period of the attractors (fixed configurations, if the period is 1, or limit cycles of configurations, if the period is strictly more than 1, cf. Table 4.18).

For example, if the common gene PU.1 of the example of the Fig. 4.16 is split into 2 genes in order to incorporate the negative circuit of order 2 inside the negative circuit of order 6, it would give birth to a positive circuit of order 8 having 8 attractors, and only one attractor in the case of tangency of the two negative circuits (Table 4.18 and Fig. 4.19). More generally, the number of observed attractors (fixed points and cell cycles) in circuits with Boolean transitions identity or negation can be calculated in case of parallel updating by using Tables given in (Demongeot et al. 2012) for the cases of isolated, tangential, or intersecting circuits, generalizing the results of Fig. 4.16 (from Demongeot et al. 2012). We observe a rapid increase of the number of cell cycles, when the order of the circuits increases in the network, corresponding in general to an increase of their evolutionary entropy *E*. A way to diminish this number is to introduce up-stream inhibitors like microRNAs.

A last important feature of the getBren dynamics, as we have seen in Sect. 4.4.2, is the existence of genes influencing directly the opening of the DNA inside the chromatin, hence allowing or not the gene expression. If these genes are controlled by microRNAs (Demongeot et al. 2013a, b, c), it is necessary to generalize the getBren structure by considering that the possibility to update a block of genes at iteration *t* is depending on the state of *r* "clock" genes (i.e., involved in the chromatin updating clock) $k_1,\ldots, k_r$ (like histone acetyltransferase, endonucleases, exonucleases, helicase, replicase, polymerases) depending on s microRNAs, $l_1,\ldots, l_s$. Then the transition for a gene i, such as i does not belong to $\{k_1,\ldots, k_r\}$, could be written as:

$\forall g\in\{0,1\}, \beta\in\{0,1\}^n$, if $\forall j = 1,\ldots,r$, $x_{kj}(t) = 1$, then:

1. $P_{i,g}{}^{\beta}(\{x_i(t+1) = g|x(t) = \beta\}) = \exp[g(\Sigma_{j\in Ni} w_{ij}\beta_j - \theta_i)/T]/[1+ \exp[(\Sigma_{j\in Ni} w_{ij} \beta_j - \theta_i)/T]$, if microRNAs $l_1,\ldots,l_s$ are silent
   and
2. $P_{i,0}{}^{\beta}(\{x_i(t+1) = 0|x(t) = \beta\}) = 1$, if not.

We can remark that the case (1) implies that: $\forall j = 1,\ldots,s$, $x_{lj}(t-1) = 0$. To make the transition rule more precise, we can for the sake of simplicity, decide that the indices $k_1,\ldots,k_r$ of the r "clock" genes are $1,\ldots,r$ and then we have the three possible following behaviors:

1. If $y(t) = \Pi_{i=1,\ldots,r}\, x_i(t) = 1$, then the rule (2) is available
2. If $y(t) = 0$ and $\Sigma_{s=t,\ldots,t-c}\, y(s) > 0$, then $x(t+1) = x(t-s^*)$, where s^* is the last time before t, where $y(s^*) = 1$
3. If $y(t) = 0$ and $\Sigma_{s=t,\ldots,t-c}\, y(s) = 0$, then $x(t+1) = 0$ (by exhaustion of the pool of genes still in expression)

The dynamical system remains autonomous (with respect to the time t, i.e., depends on t only through the set of state variables $\{x(t-c),\ldots, x(t-1)\}$), but a theoretical study of its attractors (like in Demongeot et al. 2012), with this state-dependent updating schedule, is very difficult to perform and will be investigated further by the community of biomathematicians.

References

Almeida L, Demongeot J (2012) Predictive power of "a minima" models in biology. Acta Biotheor 60:3–19

Aubert A, Costalat R, Magistretti PJ (2005) Brain lactate kinetics: modeling evidence for neuronal lactate uptake upon activation. Proc Natl Acad Sci USA 102:16448–53

Aubert A, Pellerin L, Magistretti PJ, Costalat R (2007) A coherent neurobiological framework for functional neuroimaging provided by a model integrating compartmentalized energy metabolism. Proc Natl Acad Sci USA 104:4188–93

Baconnier P, Pachot P, Demongeot J (1993) An attempt to generalize the control coefficient concept. J Biol Syst 1:335–47

Bandiera S, Rüberg S, Girard M, Cagnard N, Hanein S, Chrétien D, Munnich A, Lyonnet S, Henrion-Caude A (2011) A nuclear outsourcing of RNA interference components to human mitochondria. PLoS ONE 6:e20746

Bandiera S, Ruberg S, Hanein S, Munnich A, Demongeot J, Lyonnet S, Henrion-Caude A (2012) Achieving a novel dynamic in mitochondrial diseases: identification of mitochondrial microRnas. In: Actes 6èmes Assises de Génétique Humaine et Médicale. FAGHEM, Paris, p 6

Bandiera SR, Mategot RJ, Demongeot J, Henrion-Caude A (2013) MitomiRs: delineating the intracellular localization of microRNAs at mitochondria. Free Radic Biol Med. doi:10.1016/j.freeradbiomed.2013.06.013

Baum TP, Pasqual N, Thuderoz F, Hierle V, Chaume D, Lefranc MP, Jouvin-Marche E, Marche P, Demongeot J (2004) IMGT/GeneInfo: enhancing V(D)J recombination database accessibility. Nucleic Acids Res 32:51–4

Ben Amor H, Demongeot J, Elena A, Sené S (2008) Structural sensitivity of neural and genetic networks. Lect Notes Comp Sci 5317:973–86

Ben Amor H, Glade N, Lobos C, Demongeot J (2010a) The isochronal fibration: characterization and implication in biology. Acta Biotheor 58:121–42

Ben Amor H, Glade, Demongeot J (2010) Mnesic evocation: an isochron-based analysis. In: IEEE AINA'10 & BLSMC'10. IEEE Proceedings, Piscataway, pp 745–50

Benabid AL, Lavallée S, Hoffmann D, Cinquin P, Le Bas JF, Demongeot J (1992) Computer suport for the Talairach system. In: Kelly PJ (ed) Computers in stereotactic surgery. Blackwell, Cambridge, pp 230–45

Bernard C (1865, 1923 reed.) Introduction à la Médecine Expérimentale. J Gibert, Paris

Bier M, Teusink B, Kholodenko BN, Westerhoff HV (1996) Control analysis of glycolytic oscillations. Biophys Chem 62:15–24

Blanchini F, Franco E (2011) Structurally robust biological networks. BMC Syst Biol 5:74

Boiteux A, Goldbeter A, Hess B (1975) Control of oscillating glycolysis of yeast by stochastic periodic and steady source of substrate: a model and experimental study. Proc Natl Acad Sci USA 72:3829–33

Boominathan L (2010a) The tumor supressors p53 p63 and p73 are regulators of MicroRNA processing complex. PLoS One 5:e10615

Boominathan L (2010b) The guardians of the genome (p53 TA-p73 and TA-p63) are regulators of tumor supressor miRNAs network. Cancer Metastasis Rev 29:613–39

Byrd GT, Sage RF, Brown RH (1992) A comparison of dark respiration between C3 and C4 plants. Plant Physiol 100:191–8

Cannon WB (1932) The wisdom of the body. WW Norton, New-York

Caraguel F, Tayyab M, Giroud F, Demongeot J (2010) Evolution of the genetic regulatory networks: the example of the cell cycle control network from gastrulation modelling to apocatagenesis. In: Chang E, Barolli L (eds) IEEE AINA' 10. IEEE Proceedings, Piscataway, pp 767–74

Cinquin O, Demongeot J (2002) Positive and negative feedback: striking a balance between necessary antagonists. J Theor Biol 216:229–41

Cinquin O, Demongeot J (2005) High-dimensional switches and the modeling of cellular differentiation. J Theor Biol 233:391–411

Cosnard M, Goles E (1977) Discrete states neural networks and energies. Neural Netw 10:327–34

Cui P, Ji R, Ding F, Qi D, Gao H, Meng H, Yu J, Hu S, Zhang H (2007) A complete mitochondrial genome sequence of the wild two-humped camel (Camelus bactrianus ferus): an evolutionary history of camelidae. BMC Genomics 8:241–4

Cullen BR (2010) Five questions about viruses and MicroRNAs. PLoS Pathog 6:e1000787

Davies PC, Demetrius L, Tuszynski JA (2012) Cancer as a dynamical phase transition. Theor Biol Med Model 8:30

Demetrius L (1983) Statistical mechanics and population biology. J Stat Phys 30:709–53

Demetrius L (1997) Directionality principles in thermodynamics and evolution. Proc Natl Acad Sci USA 9:3491–98

Demetrius LA, Simon DK (2012) An inverse-Warburg effect and the origin of Alzheimer's disease. Biogerontology 13(6):583–94

Demongeot J, Demetrius L (submitted) Complexity and stability in biological systems. Acta Biotheoretica

Demongeot J, Waku J (submitted) Robustness in genetic regulatory networks: mathematical approach and biological applications. Neural Networks

Demongeot J, Doncescu A (2009) Modelling the glycolysis: an inverse problem approach. In: IEEE AINA '09 & BLSMC '09. IEEE Proceedings, Piscataway, pp 930–5
Demongeot J, Doncescu (2009) A Modelling the glycolysis: an inverse problem approach. In: IEEE AINA' 09 & BLSMC' 09. IEEE Proceedings, Piscataway, pp 930–5
Demongeot J, Françoise JP (2006) Approximation for limit cycles and their isochrons. Comptes Rendus Biologies 329:967–70
Demongeot J, Hazgui H (In Press) MicroRNAs: unspecific inhibitory regulation in immunologic control and in mitochondrial respiration. In: Barolli L et al. (eds) IEEE AINA '13. IEEE Proceedings, Piscataway
Demongeot J, Kellershohn N (1983) Glycolytic oscillations: an attempt to an "in vitro" reconstitution of the higher part of glycolysis. Lect Notes Biomaths 49:17–31
Demongeot J, Laurent M (1983) Sigmoidicity in allosteric models. Math Biosci 67:1–17
Demongeot J, Moreira A (2007) A circular RNA at the origin of life. J Theor Biol 249:314–24
Demongeot J, Sené S (2008) Asymptotic behavior and phase transition in regulatory networks II Simulations. Neural Netw 21:971–9
Demongeot J, Sené S (2011) The singular power of the environment on nonlinear Hopfield networks. In: CMSB'11 ACM Proceedings, New York, pp 55–64
Demongeot J, Seydoux F (1979) Oscillations glycolytiques: modélisation d'un système minimum à partir des données physiologiques et moléculaires. In: Delattre P, Thellier M (eds) Elaboration et justification de modèles. Maloine, Paris, pp 519–36
Demongeot J, Waku J (2012a) J Robustness in biological regulatory networks. II Application to genetic threshold Boolean random regulatory networks (getBren). Comptes Rendus Mathématique 350:225–8
Demongeot J, Waku J (2012b) Robustness in biological regulatory networks. III Application to genetic networks controlling the morphogenesis. Comptes Rendus Mathématique 350:289–92
Demongeot J, Thomas R, Thellier M (2000) A mathematical model for storage and recall functions in plants. CR Acad Sci Sciences de la Vie 323:93–7
Demongeot J, Virone G, Duchêne F, Benchetrit G, Hervé T, Noury N, Rialle V (2002) Multi-sensors acquisition data fusion knowledge mining and alarm triggering in health smart homes for elderly people. Comptes Rendus Biologies 325:673–82
Demongeot J, Aracena J, Thuderoz F, Baum TP, Cohen O (2003) Genetic regulation networks: circuits, regulons and attractors. Comptes Rendus Biologies 326:171–88
Demongeot J, Elena A, Weil G (2006) Potential-Hamiltonian decomposition of cellular automata. Application to degeneracy of genetic code and cyclic codes III. Comptes Rendus Biologies 329:953–62
Demongeot J, Glade N, Forest L (2007a) Liénard systems and potential-Hamiltonian decomposition I Methodology. Comptes Rendus Mathématique 344:121–6
Demongeot J, Glade N, Forest L (2007b) Liénard systems and potential-Hamiltonian decomposition II Algorithm. Comptes Rendus Mathématique 344:191–4
Demongeot J, Glade N, Hansen O, Moreira A (2007c) An open issue: the inner mitochondrial membrane (IMM) as a free boundary problem. Biochimie 89:1049–57
Demongeot J, Elena A, Sené S (2008a) Robustness in neural and genetic networks. Acta Biotheoretica 56:27–49
Demongeot J, Jezequel C, Sené S (2008b) Asymptotic behavior and phase transition in regulatory networks I Theoretical results. Neural Netw 21:962–70
Demongeot J, Ben Amor H, Gillois P, Noual M, Sené S (2009a) Robustness of regulatory networks a generic approach with Applications at different levels: physiologic metabolic and genetic. Int J Mol Sci 10:4437–73
Demongeot J, Drouet E, Moreira A, Rechoum Y, Sené S (2009b) MicroRNAs: viral genome and robustness of the genes expression in host. Phil Trans Roy Soc A 367:4941–65
Demongeot J, Glade N, Moreira A, Vial L (2009c) RNA relics and origin of life. Int J Mol Sci 10:3420–41

Demongeot J, Goles E, Morvan M, Noual M, Sené S (2010a) Attraction basins as gauges of environmental robustness in biological complex systems. PloS ONE 5:e11793
Demongeot J, Noual M, Sené S (2010) On the number of attractors of positive and negative Boolean automata circuits. In: Chang E, Barolli L (Eds) IEEE AINA' 10. IEEE Proceedings, Piscataway, pp 782–9
Demongeot J, Henrion-Caude A, Lontos A, Promayon E (2011a) General architecture of a genetic regulation network Applications to embryologic and immunologic control. In: Lenaerts T, Giacobini M, Bersini H, Bourgine P, Dorigo M, Doursat R (eds) ECAL'11 advances in artificial life. MIT Press, Cambridge MA, pp 1–8
Demongeot J, Elena A, Noual M, Sené S, Thuderoz F (2011b) "Immunetworks" attractors & intersecting circuits. J Theor Biol 280:19–33
Demongeot J, Elena A, Noual M, Sené S (2011) Random Boolean networks and attractors of their intersecting circuits. In: Pllana S, Barolli L, Xhafa F (Eds) IEEE AINA '11. IEEE Proceedings, Piscataway, pp 483–7
Demongeot J, Noual M, Sené S (2012) Combinatorics of Boolean automata circuits dynamics. Discrete Appl Math 160:398–415
Demongeot J, Cohen O, Doncescu A, Henrion-Caude A (2013a) MitomiRrs and energetic regulation. In: Barolli L et al. (eds) IEEE AINA' 13. IEEE Proceedings, Piscataway, pp 1501–1508
Demongeot J, Hazgui H, Vuillerme N (2013b) MicroRNAs: unspecific inhibitory regulation in immunologic control and in mitochondrial respiration. In: IEEE AINA' 13 and BLSMC' 13, IEEE proceedings, Piscataway, pp 1509–16
Demongeot J, Hazgui H , Bandiera S, Cohen O, Henrion-Caude A (2013c) MitomiRs, ChloromiRs and general modelling of the microRNA inhibition. Acta Biotheor 61. doi:10.1007/s10441-013-9190-8
Diaz-Ruiz R, Rigoulet M, Devin A (2011) The Warburg and Crabtree effects: On the origin of cancer cell energy metabolism and of yeast glucose repression. Biochim Biophys Acta 1807:568–76
Duchon A, Hanusse N, Lebhar E, Schabanel N (2006) Could any graph be turned into a small-world? Theor Comput Sci 355:96–103
Elena A, Ben-Amor H, Glade N, Demongeot J (2008) Motifs in regulatory networks and their structural robustness. In: BIBE 2008. IEEE Proceedings, Piscataway, pp 1–6
Farquhar GD, von Caemmerer S, Berry JA (1980) A biochemical model of photosynthetic CO2 assimilation in leaves of C3 species. Planta 149:78–90
Fogelman Soulié F, Goles E, Martinez S, Mejia C (1989) Energy function in neural networks with continuous local functions. Complex Syst 3:269–93
Forest L, Demongeot J (2006) Cellular modelling of secondary radial growth in conifer trees: application to *Pinus radiata*. Bull Math Biol 68:753–84
Forest L, Martinez S, Padilla F, Demongeot J, San Martin J (2006) Modelling of auxin transport affected by gravity and differential radial growth. J Theor Biol 241:241–51
Glade N, Forest L, Demongeot J (2007) Liénard systems and potential-Hamiltonian decomposition III Applications in biology. Comptes Rendus Mathématique 344:253–8
Griffiths-Jones S, Marshall M, Khanna A, Eddy SR, Bateman A (2005) Rfam: annotating non-coding RNAs in complete genomes. Nucleic Acids Res 33:121–4
Gunawardena S (2010) The robustness of a biochemical network can be inferred mathematically from its architecture. Biol Syst Theor 328:581–2
Hainaut P, Hollstein M (2000) p53 and human cancer: the first ten thousand mutations. Adv Cancer Res 77:81–137
Hartwell LH, Hopfield JJ, Leibler S, Murray AW (1999) From molecular to modular cell biology. Nature 402:47–52
He T, Feng G, Chen H, Wang L, Wang Y (2009) Identification of host encoded microRNAs interacting with novel swine-origin influenza A (H1N1) virus and swine influenza virus. Bioinformation 4:112–8

Herranz H, Hong X, Pérez L, Ferreira A, Olivieri D, Cohen SM, Milán M (2010) The miRNA machinery targets Mei-P26 and regulates Myc protein levels in the Drosophila wing. EMBO J 29:1688–98
Hervagault JF, Duban MC, Kernevez JP, Thomas D (1983) Multiple steady states and oscillatory behavior of a compartmentalized phosphofructokinase system. Proc Natl Acad Sci USA 80:5455–59
Hobish MK, Wickramasinghe NSMD, Ponnamperuma C (1995) Direct interaction between amino-acids and nucleotides as a possible physico-chemical basis for the origin of the genetic code. Adv Space Res 15:365–75
Hopfield JJ (1982) Neural networks and physical systems with emergent collective computational abilities. Proc Natl Acad Sci USA 79:2554–58
http://ferrolabdmiunictit/miro/
Huang N, Lin J, Ruan J, Su N, Qing R, Liu F, He B, Ly C, Zheng D, Luo R (2012) MiR-219-5p inhibits hepatocellular carcinoma cell proliferation by targeting glypican-3. FEBS Lett 586:884–91
Jolliot A, Prochiantz A (2004) Transduction peptides: from technology to physiology. Nat Cell Biol 6:189–96
Kaczer H, Burns JA (1973) The control of flux. Symp Soc Exp Bot 28:65–104
Kauffman SA (1969) Metabolic stability and epigenesis in randomly constructed genetic nets. J Theor Biol 22:437–67
Kohn KW (1999) Molecular interaction map of the mammalian cell cycle control and DNA repair systems. Mol Biol Cell 10:2703–34
Kohn KW, Aladjem MI, Weinstein JN, Pommier Y (2006) Molecular interaction map of bioregulatory networks: a general rubric for systems biology. Mol Biol Cell 17:1–13
Koparde P, Singh S (2010) Avian influenza and micro RNA: role of bioinformatics. J Bioinform Sequence Anal 3:11–22
Kühn R (2010) Equilibrium analysis of complex systems. Lecture Notes 7CCMCS03, King's College, London
Lesne A (2008) Robustness: confronting lessons from physics and biology. Biol Rev Cambridge Philos Soc 83:509–32
Lewin B, Krebs JE, Kilpatrick ST, Goldstein ES (2011) Genes X. Jones & Bartlett, Sudbury, MA
Li W, Yang X, Jiang Y, Wang B, Yang Y, Jiang Z, Li M (2011) Inhibition of influenza A virus replication by RNA interference targeted against the PB1 subunit of the RNA polymerase gene. Arch Virol 156:1979–87
Liu C, Kelnar K, Liu B, Chen X, Calhoun-Davis T, Li H, Patrawala L, Yan H, Jeter C, Honorio S, Wiggins JF, Bader AG, Fagin R, Brown D, Tang DG (2011) The microRNA miR-34a inhibits prostate cancer stem cells and metastasis by directly repressing CD44. Nat Med 17:211–5
Lotka A (1925) Elements of physical biology. Williams & Wilkins, Baltimore
Martello G, Zacchigna L, Inui M, Montagner M, Adorno M, Mamidi A, Morsut L, Soligo S, Tran U, Dupont S, Cordenonsi M, Wessely O, Piccolo S (2007) MicroRNA control of nodal signalling. Nature 449:183–8
Massirer KB, Perez SG, Mondol V, Pasquinelli AE (2012) The miR-35-41 family of MicroRNAs regulates RNAi sensitivity in *Caenorhabditis elegans*. PLoS Genet 8:e1002536
Maurel M, Jalvy S, Ladeiro Y, Combe C, Vachet L, Sagliocco F, Bioulac-Sage P, Pitard V, Jacquemin-Sablon H, Zucman-Rossi J, Laloo B, Grosset CF (2012) A functional screening identifies five miRNAs controlling glypican-3: role of miR-1271 down-regulation in hepatocellular carcinoma. Hepatology 57:195–204
Mendoza L, Alvarez-Buylla E (1998) Dynamics of the genetic regulatory network for *Arabidopsis thaliana* flower morphogenesis. J Theor Biol 193:307–19
Michon F, Forest L, Collomb E, Demongeot J, Dhouailly D (2008) BMP2 and BMP7 play antagonistic roles in feather induction. Development 135:2797–805
Michon F, Tummers M, Kyyrönen M, Frilander MJ, Thesleff I (2012) Tooth morphogenesis and ameloblast differentiation are regulated by micro-RNAs. Dev Biol 340:355–68

Mourier A, Devin A, Rigoulet M (2010) Active proton leak in mitochondria: a new way to regulate substrate oxidation. Biochim Biophys Acta 1797:255–61
Ovadi J (1988) Old pathway-new concept: control of glycolysis by metabolite-modulated dynamic enzyme associations. Trends Biochem Sci 13:486–90
Pasqual N, Gallagher M, Aude-Garcia C, Loiodice M, Thuderoz F, Demongeot J, Ceredig R, Marche PN, Jouvin-Marche E (2002) Quantitative and qualitative changes in ADV-AJ rearrangements during mouse thymocytes differentiation: implication for a limited TCR ALPHA chain repertoire. J Exp Med 196:1163–74
Raver-Shapira N, Marciano E, Meiri E, Spector Y, Rosenfeld N, Moskovits N, Bentwich Z, Oren M (2007) Transcriptional activation of miR-34a contributes to p53-mediated apoptosis. Mol Cell 26:731–43
Reder C (1988) Metabolic control theory: a structural approach. J Theor Biol 135:175–201
Ritter JB, Genzela Y, Reichl U (2008) Simultaneous extraction of several metabolites of energy metabolism and related substances in mammalian cells: Optimization using experimental design. Anal Biochem 373:349–69
Ruoff P, Christensen MK, Wolf J, Heinrich R (2003) Temperature dependency and temperature compensation in a model of yeast glycolytic oscillations. Biophys Chem 106:179–92
Sbisa E, Tanzariello F, Reyes A, Pesole G, Saccone C (1997) Mammalian mitochondrial D-loop region structural analysis: identification of new conserved sequences and their functional and evolutionary implications. Gene 205:125–40
Segura MF, Greenwald HS, Hanniford D, Osman I, Hernando E (2012) MicroRNA and cutaneous melanoma: from discovery to prognosis and therapy. Carcinogenesis 33:1823–32
Smalheiser NR, Lugli G, Thimmapuram J, Cook EH, Larson J (2011) Endogenous siRNAs and non-coding RNA-derived small RNAs are expressed in adult mouse hippocampus and are up-regulated in olfactory discrimination training. RNA 17:166–81
Song L, Liu H, Gao S, Jiang W, Huang W (2010) Cellular MicroRNAs inhibit replication of the H1N1 influenza A virus in infected cells. J Virol 84:8849–60
Takeshita F, Patrawala L, Osaki M, Takahashi R, Yamamoto Y, Kosaka N, Kawamata M, Kelnar K, Bader AG, Brown D, Ochiya T (2010) Systemic delivery of synthetic MicroRNA-16 inhibits the growth of metastatic prostate tumors via downregulation of multiple cell-cycle genes. Mol Ther 18:181–7
Tang H, Lee M, Sharpe O, Salamone L, Noonan EJ, Hoang CD, Levine S, Robinson WH, Shrager JB (2012) Oxidative stress-responsive microRNA-320 regulates glycolysis in diverse biological systems. FASEB J 26:4710–21
Tennant DA, Dur RV, Gottlieb E (2010) Targeting metabolic transformation for cancer therapy. Nat Rev Cancer 10:267–77
Thellier M, Demongeot J, Guespin J, Ripoll C, Norris V, Thomas R (2004) A logical (discrete) formulation model for the storage and recall of environmental signals in plants. Plant Biol 10:1055–75
Thellier M, Legent G, Amar P, Norris V, Ripoll C (2006) Steady-state kinetic behaviour of functioning-dependent structures. FEBS J 273:4287–99
Thom R (1972) Structural stability and fluctuations and morphogenesis. Benjamin, Reading, MA
Thomas R (1973) Boolean formalisation of genetic control circuits. J Theor Biol 42:563–85
Thuderoz F, Simonet MA, Hansen O, Dariz A, Baum TP, Hierle V, Demongeot J, Marche PN, Jouvin-Marche E (2010) From the TCRAD rearrangement quantification to the computational simulation of the locus behavior. PloS Comp Biol 6:e1000682
Tonnelier A, Meignen S, Bosch H, Demongeot J (1999) Synchronization and desynchronization of neural oscillators: comparison of two models. Neural Netw 12:1213–28
Turner P, McLennan A, Bates A, White M (2005) Molecular biology. Taylor & Francis, New York
http://www.medownloads.com/download-Lotka-Volterra-146411.htm
van der Pol B, van der Mark M (1928) Le battement du cœur considéré comme oscillation de relaxation et un modèle électrique du cœur (The beating of the heart considered as relaxation oscillation and an electric model of the heart). L'onde électrique 7:365–92
Volterra V (1926a) Variazioni e uttuazioni del numero d'individui in specie animali conviventi. Mem Acad Lincei 2:31–113

Volterra V (1926b) Fluctuations in the abundance of a species considered mathematically. Nature 118:558–60

Volterra V (1931) Leçons sur la Théorie Mathématique de la Lutte pour la Vie. Gauthier-Villars, Paris

Waddington CH (1940) Organizers and genes. Cambridge University Press, Cambridge UK

Wang Y, Brahmakshatriya V, Zhu H, Lupiani B, Reddy SM, Yoon BJ, Gunaratne PH, Kim JH, Chen R, Wang J, Zhou H (2009) Identification of differentially expressed miRNAs in chicken lung and trachea with avian influenza virus infection by a deep sequencing approach. BMC Genomics 10:512

Weaver DC, Workman CT, Stormo GD (1999) Modeling regulatory networks with weight matrices. Pac Symp Biocomput 4:112–23

Weibel ER, Taylor CR, Hopeler H (1991) The concept of symmorphosis: a testable hypothesis of structure–function relationship. Proc Natl Acad Sci USA 88:10357–61

Wolf J, Heinrich R (2000) Effect of cellular interaction on glycolytic oscillations in yeast: a theoretical investigation. Biochem J 345:321–34

Chapter 5
Dynamics of Mitochondrial Redox and Energy Networks: Insights from an Experimental–Computational Synergy

Sonia Cortassa and Miguel A. Aon

Abstract Functionally, a cell comprises spatially distributed and compartmentalized subsystems, the dynamics of which occurs on several temporal scales. Interactivity in complex spatiotemporally organized cellular systems is fundamental to their counterintuitive behavior and one of the main reasons why their study needs mathematical modeling. But models alone are not enough; what we ultimately require is a combined experimental–theoretical approach in order to validate our models as rigorously as possible.

We explore in a detailed example the success of experimental–modeling synergy leading to the elucidation of the mechanisms involved in synchronized mitochondrial oscillations in the heart, and the discovery there of new related mechanisms. This work involves successive and iterative reciprocal potentiation of the loop via experiments and computational modeling: simulation–validation and prediction–experimentation thereby alternate so as to provide a deeper understanding of complex biological phenomena.

The concept of network has become central in systems biology. Conceptually, networks can be approached from different angles. One is *morphological*, in which mitochondrial spatial organization corresponds to a network because they exhibit a spatial arrangement with a defined pattern that *topologically* connects them in a certain way, for example, in a lattice as in cardiac cells, or reticular random as in neurons or cancer cells. However, underlying these networks there is another vast network of metabolic reactions with nodes represented by substrates, products, ion gradients, and links by enzymes catalyzing reactions between substrates and products, or transporters and channels modulating the passage of ions and metabolites across membranes. The metabolic network distributes among

S. Cortassa • M.A. Aon (✉)
Division of Cardiology, Johns Hopkins University, Baltimore, MD 21205, USA
e-mail: maon1@jhmi.edu

M.A. Aon et al. (eds.), *Systems Biology of Metabolic and Signaling Networks*, Springer Series in Biophysics 16, DOI 10.1007/978-3-642-38505-6_5,

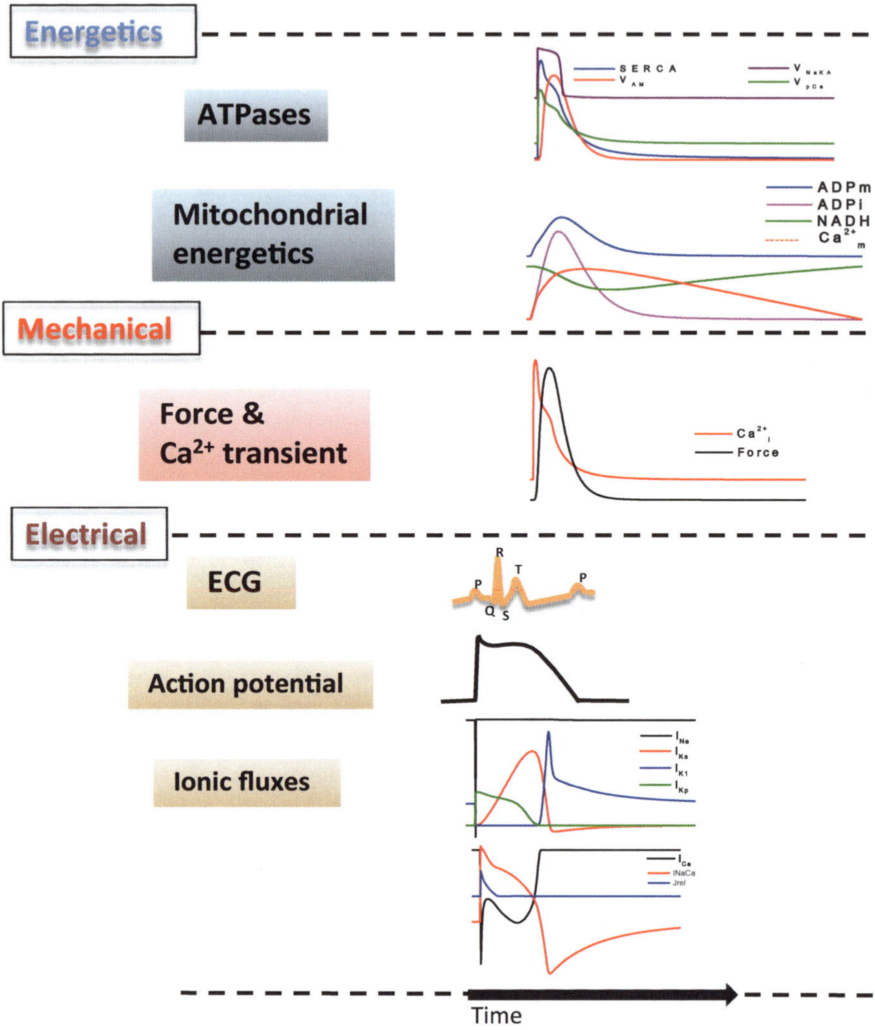

Fig. 5.1 Temporal unfolding of electrical, mechanical, and energetic processes during a single cardiomyocyte beat as computed by the Excitation–Contraction coupling Mitochondrial Energetic (ECME) model. The depolarization and repolarization phases of the action potential (AP) take place during the first 200 ms when the ECME model is stimulated every 2 s. This calculation reveals the multiple temporal scales involved. The *gray shaded* area is meant to highlight the earlier occurrence of electrical as compared with mechanical (contraction–relaxation) and metabolic/energetic processes. Mitochondrial energy fuels the electrical and contractile machinery of the heart cell on a slower time scale (few seconds) compared with the electrical processes (milliseconds), which are followed by mechanical events associated with the force of contraction and the Ca^{2+} transient

different compartments whose interaction is mediated by transport processes. If we zoom in within a set of processes we will find even more detailed networks, and the dynamics is reflected by fluxes occurring in different compartments, e.g., by channels and pumps in the plasma membrane, sarcoplasmic reticulum, myofibrils, and mitochondria.

The molecular view analyzes interactions in biomolecular networks involving different components which result in varied functional outputs: protein–protein, genetic expression (multi arrays), regulatory (protein–DNA interactions, combinatorial transcription factors), and signaling (signal transduction pathways through protein–protein and protein–small molecule interactions) (Cortassa et al. 2012).

From a temporal perspective a highly tuned response exists in energy supply by mitochondria to the demand by electromechanical processes in the heart operating in the millisecond range (e.g., action potentials, calcium transients) (Fig. 5.1). This tight match between energy supply and demand can be more readily fulfilled by the highly synchronized and robust action of mitochondrial networks. In the heart, mitochondria constitute an extensive subcellular network, which occupies ~30 % of the heart cell volume, and appears to be wrapped by the sarcoplasmic reticulum and in close vicinity with the myofilaments and t-tubules. During maximal workload, the whole ATP pool in the heart cell is turned over in a few seconds, while ~2 % of that pool is consumed in each heartbeat. Both constancy and flexibility are required from the mitochondrial network in response to the changing metabolic demand for supplying a steady output of ATP to fuel contraction, and to adapt the rate of energy provision. Whereas under normal physiological conditions the availability of energy is fine-tuned to match changes in energy demand, under stress this is not the case.

The idea that mitochondria may function as a coordinated network of oscillators emerged from studies on living cardiomyocytes subjected to metabolic stress. The network behavior of mitochondria depends on local as well as global coordination in the cell, and ROS-induced ROS release is a mechanism that was shown to exert both local and cell-wide influence on the network. Mitochondrial network organization may be also essential for the temporal organization of the heart rhythm.

Mitochondrial network energetics, or the functioning of mitochondria as networks, represents an advantageous behavior for its coordinated action, under normal physiology, provides overall and usual robustness despite occasional failure in a few nodes, and improves energy supply during a swiftly changing demand (Aon and Cortassa 2012). Mitochondrial network energetics along with its remarkable nonlinear properties together with those of the whole heart itself set the stage for the appearance of critical phenomena and bifurcations leading to self-organized, emergent behavior. An amazing example of the latter is given by the existence, at critical points (mitochondrial criticality), of emergent macroscopic self-organized behavior escalating from the subcellular to the whole heart, eventually leading to the death of the animal. The demonstration of the involvement of mitochondrial oscillations in reperfusion-related arrhythmias after ischemic injury, and of their

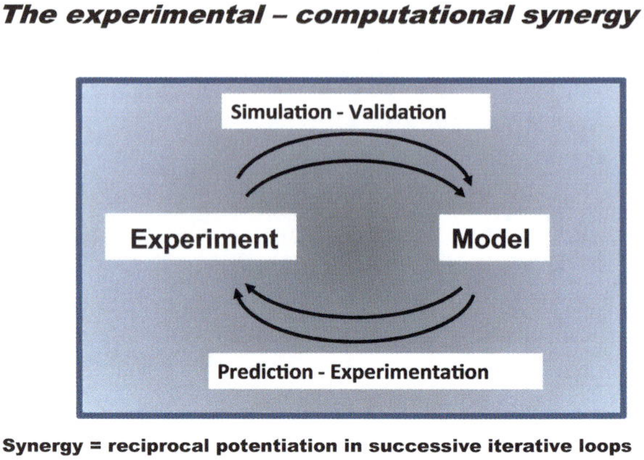

Fig. 5.2 The experimental—computational synergy

pharmacological reversion, blunting oscillations, and stabilizing the action potential in different animal models, is strong proof of the involvement of the network behavior of mitochondrial energetics.

5.1 Experimental–Computational Synergy

In the next sections we will delineate the powerful synergy arising from a combined experimental and theoretical approach to unravel mechanisms underlying complex dynamic behavior. Specifically, we will describe how the interaction of computational modeling and experimental work led to the vision of mitochondria behaving as networks in architectural, topological, and dynamic senses. We also explore the physiological and pathophysiological consequences of the network behavior of mitochondrial function.

Figure 5.2 shows an overall flow diagram of the experimental–computational synergy. The main driving force underlying the synergy is the continual interaction between experiment and computational model which gives rise to iterative loops. The intrinsic dynamic of these loops works as follows: a model is validated as can be judged from its ability to simulate experimental results; this triggers model prediction of new, unexpected behavior which elicits experimental verification, followed by the discovery of novel properties revealed by the experiments which in turn feedback on the model that can be tested again in its ability to simulate them. If the model is unable to simulate the new behavior, then modifications (changes either in the structure of the model, or in the rate expressions or just in parameter values) are introduced. These changes may take the form of model upgrades to account for new processes, or different functional relationships between the model components already in place.

The iterative dynamic of the *experiment* ↔ *model* loop represents the synergy through theoretical prediction that works as a hypothesis-driven experimental test,

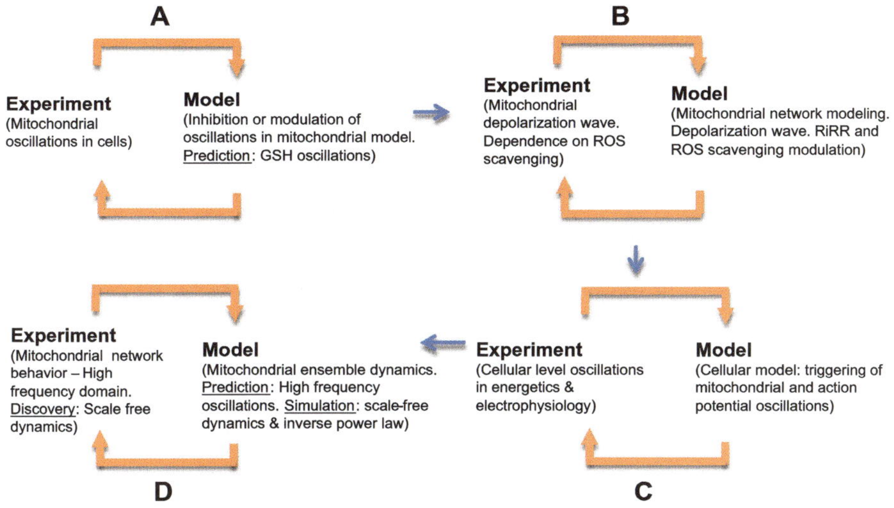

Fig. 5.3 Overview of the experimental—computational synergy as applied to the elucidation of the mechanisms involved in mitochondrial oscillations in the heart, and the discovery of scale-free dynamics in the mitochondrial network

and the results of which then become a new functional/mechanistic insight into the behavior of the experimental system. The loop can then reiterate from novel behavior as experimentally verified, and this can subsequently be further investigated in order to see if the model in its present form can account for the new phenomena. If not, then the model is further refined and upgraded.

5.2 Oscillatory Phenomena in Cardiomyocytes: A Case Study of Experimental–Modeling Synergy

In order to show how the experimental–computational synergy can be used to address a specific biological problem, we analyze the experimental demonstration of cell-wide mitochondrial oscillations in living cardiomyocytes (Figs. 5.3 and 5.4). To understand the mechanism underlying the oscillations, we developed a computational model of the mitochondrial oscillator (Cortassa et al. 2004). One of the aims of this modeling was to investigate the role of ROS in the mitochondrial oscillations described in living cardiomyocytes subjected to oxidative stress (Aon et al. 2003).

In the heart, under normal physiological conditions, the availability of energy is fine-tuned to match changes in energy demand. However, under stress this is not the case. Metabolically stressful conditions such as substrate deprivation, or oxidative stress, represent a pathophysiological situation under which mitochondrial

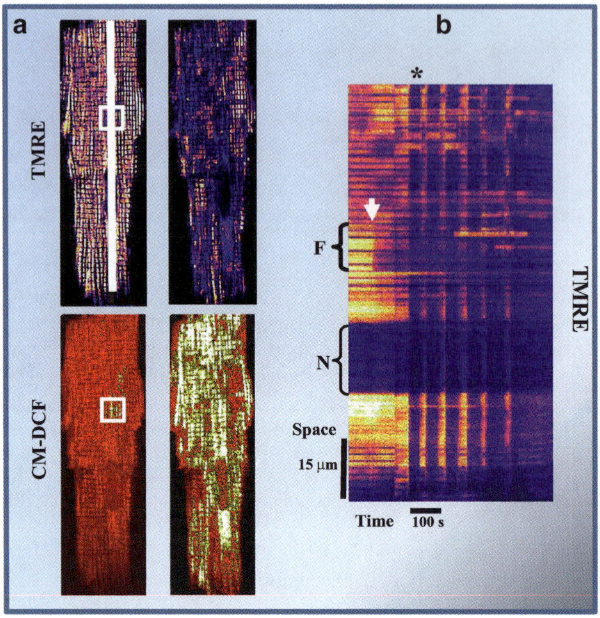

Fig. 5.4 Cell-wide synchronized mitochondrial oscillations after local generation of ROS. (**a**) Cardiomyocyte loaded at 37 °C with tetramethylrhodamine ethyl ester (TMRE, $\Delta\Psi_m$ indicator, *upper images*) and 5-(-6)-chloromethyl-2, 7-dichlorohydrofluorescein diacetate (CM-H_2DCFH, ROS-sensitive, lower images). By using two-photon laser excitation, and after 10–20 control images were collected, a small region of a cardiac myocyte (20 × 20 pixels) was excited in a single flash resulting in rapid loss of mitochondrial membrane potential, $\Delta\Psi_m$ (**a**, *white square* in *upper left*; **b**, *white arrow*) and local generation of ROS (**a**, *white square* in *lower left*). Thereafter, $\Delta\Psi_m$ remained depolarized in the *flashed area* throughout the experiment (see **b**). The *right images* in **a** show the first whole-cell $\Delta\Psi_m$ depolarization (**b**, *asterisk*) after a delay time. (**b**) Time-line image of TMRE created by analyzing a line drawn along the longitudinal axis of the cell (shown in **a**, *upper left*). The *arrow* points out the timing of the flash and the brackets point out the flash region (*Upper*) and the nucleus (*Lower*). The synchronous $\Delta\Psi_m$ mitochondrial oscillations are evident as vertical *blue bands*. The mitochondria that do not belong to the spanning cluster remained visibly polarized. Reproduced from Aon MA, Cortassa S, and O'Rourke B. (2004) PNAS 101, 4447–4452

energetics oscillates. Under these conditions the performance mode of mitochondrial function becomes a key arbiter of life and death at cellular and organ levels (Aon et al. 2006a; Gustafsson and Gottlieb 2008; O'Rourke et al. 2005). The idea that mitochondria could function as a coordinated network of oscillators emerged from studies in living cardiomyocytes subjected to metabolic stress (Aon et al. 2003, 2004a, 2006b).

5.2.1 Experimental Studies: A Brief Phenomenological Description of Mitochondrial Oscillations

The mitochondrial oscillator was first described experimentally under pathophysiological conditions of metabolic stress, e.g., substrate deprivation (Romashko et al. 1998), or oxidative stress (Aon et al. 2003). The mechanisms underlying the synchronization and propagation of mitochondrial oscillations in intact cardiomyocytes were explored in detail employing two-photon laser scanning fluorescence microscopy (Aon et al. 2003).

Experimentally, oscillations were triggered in a reproducible manner, a key for studying the underlying mechanisms under controlled conditions. Two-photon microscopy gives a detailed spatial picture of the mitochondrial network as can be seen in Fig. 5.4, in which a freshly isolated cardiomyocyte loaded with membrane potential and ROS sensors is shown. The fluorescence spatiotemporal dynamics along a line drawn throughout the longitudinal axis of an individual cell can be obtained. A time-line image of fluorescence intensity of TMRM or CM-DCF results in 2D plots that contain the whole spatial and temporal information of the stack of images. In these pseudo-color plots the blue bars correspond to mitochondrial membrane potential ($\Delta\Psi_{m}$,) depolarization, and the yellow zones in between to $\Delta\Psi_m$ repolarization. These 2D plots clearly show that while the oscillations affect the whole cell, the flashed zone remains depolarized with high ROS (Fig. 5.4), and oxidized NADH (not shown) (Aon et al. 2003). After about 1 min, whole-cell mitochondrial oscillations are triggered whereby both $\Delta\Psi_m$ and the reduced state of NADH are synchronized into phase; with each $\Delta\Psi_m$ depolarization an associated burst in the rate of ROS production occurs.

5.2.2 Modeling Studies: A Brief Description of the Mitochondrial Oscillator

A model describing mitochondrial energetics and Ca^{2+} handling (Cortassa et al. 2003) was extended to describe the key features of the proposed mechanism of mitochondrial oscillations based on our experimental findings (Cortassa et al. 2004). The addition to this model of a leak of electrons from the respiratory chain to produce the free radical superoxide, $O_2^{\cdot-}$, as previously proposed for an outwardly rectifying inner membrane anion channel (IMAC) modeled after the centum pS channel in which conductance is $O_2^{\cdot-}$ activated (Borecky et al. 1997), and a cytoplasmic ROS scavenging system in the cytoplasm, was sufficient to support limit-cycle oscillations within certain parametric domains of our model (Cortassa et al. 2004). The normal anion permeability of IMAC would permit the passage of $O_2^{\cdot-}$ from the matrix to the cytoplasmic side of the inner membrane. In addition, the IMAC opening probability was assumed to be increased by $O_2^{\cdot-}$ at an external site.

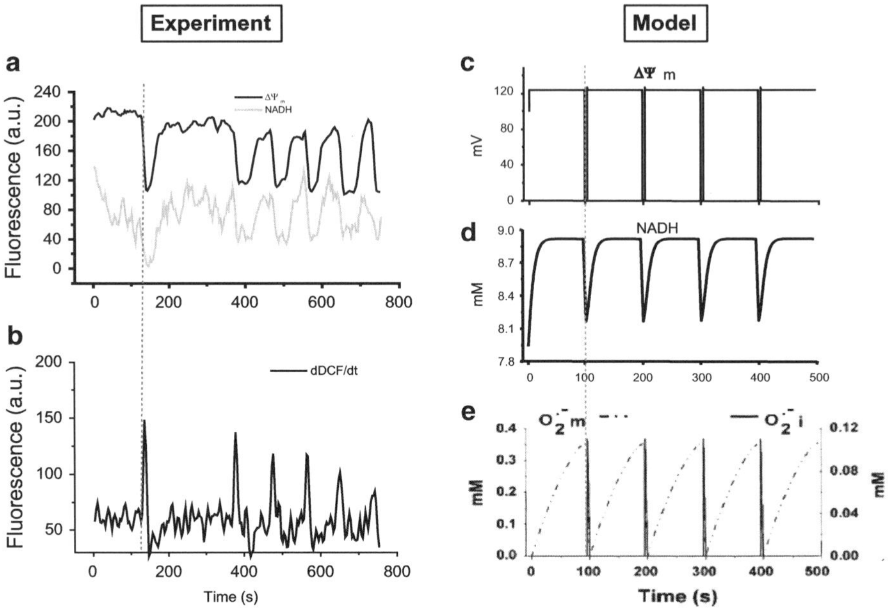

Fig. 5.5 Dynamic behavior of energetic and redox variables during mitochondrial oscillations (**a**, **b**) Shown is the simultaneous recording of the average whole-cell fluorescence of TMRE (a probe of $\Delta\Psi_m$) and NADH (autofluorescence) (**a**), and the rate of ROS accumulation (first derivative, dDCF/dt, of the CM-DCF signal) (**b**). (**c**–**e**) These *panels* display the simulation of the experimentally observed oscillations in $\Delta\Psi_m$, (**c**) NADH (**d**), and ROS (**e**, superoxide, $O_2^{\cdot-}$, in the model) with a mitochondrial oscillator model. The dashed lines are meant to emphasize the phase relationship between the different signals in the experiment (**a**, **b**) and variables in the simulation (**c**–**e**). Notice that, in the experiment as well as in the simulation, the peak in ROS accumulation occurs concomitantly with $\Delta\Psi_m$ depolarization and NADH oxidation during the initial phase of the oscillation. The model simulation further shows that the spike of $O_2^{\cdot-}$ corresponds to the release of the free radical accumulated in the mitochondrial matrix. Panels **a**, **b** were modified from Aon, Cortassa, Marban, O'Rourke (2003) J Biol Chem 278, 44735–44. Panels **c**–**e** were reproduced from Cortassa, Aon, Winslow, O'Rourke (2004) Biophys J 87, 2060–73

According to the postulated mechanism, under oxidative stress, the enhanced ROS production from the electron transport chain leads to accumulation in the mitochondrial matrix to critical levels (Aon et al. 2003, 2004a), thereby triggering the opening of IMAC in a positive feedback loop (Cortassa et al. 2004). According to this model, a burst in cytoplasmic ROS accompanies $\Delta\Psi_m$ depolarization (Fig. 5.5) as a result of a mixed process of accelerated $O_2^{\cdot-}$ production occurring concomitantly with a sudden increase in mitochondrial respiration. A pulse of cytoplasmic $O_2^{\cdot-}$ is released (Aon et al. 2007a; Cortassa et al. 2004).

The mitochondrial oscillator behaves as a relaxation oscillator, composed of both slow (ROS accumulation in the mitochondrial matrix) and fast (the IMAC opening and rapid ROS release) processes (Fig. 5.5e).

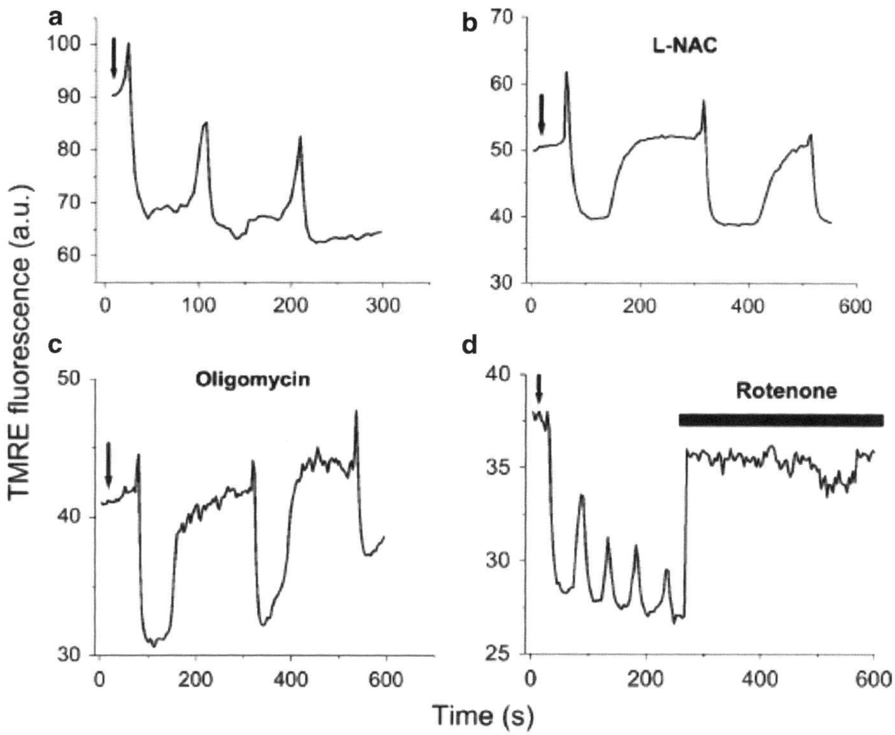

Fig. 5.6 Effects of scavengers or inhibitors of ROS production on mitochondrial oscillations. Recordings of the TMRE signal of myocytes showing cell-wide mitochondrial oscillations after a laser flash (*arrows*) in the absence (**a**) or in the presence of 4 mM *N*-acetyl-L-cysteine (L-NAC) for 30 min (**b**), or 10 μg/ml of oligomycin for 60 min (**c**), or after the acute addition of rotenone (15 μM) (**d**). Oscillations in mitochondrial metabolism were triggered as described in Fig. 5.4 (see also text). Reproduced from Cortassa, Aon, Winslow, O'Rourke (2004) Biophys J 87, 2060–73

5.3 First Iterative Loop: Cell-wide Mitochondrial Oscillations

Cell-wide synchronized oscillations in mitochondrial NADH, $\Delta\Psi_m$, and ROS could be reproducibly triggered by a focalized laser flash affecting only a few mitochondria (Fig. 5.4). We used two-photon scanning laser fluorescence microscopy and a laser flash to induce localized oxidative stress in a few mitochondria (~50). After a few seconds (~40 s on average), a synchronized cell-wide $\Delta\Psi_m$ depolarization occurred that extended and prolonged into cell-wide oscillations of the mitochondrial network in all tested energetic/redox variables ($\Delta\Psi_m$, NADH, ROS, GSH) (Figs. 5.4, 5.5, and 5.7).

We then concentrated on the cellular redox balance affecting the ROS production and ROS scavenging capacity of cells (Fig. 5.6). The oscillations could be slowed by increasing the ROS scavenging capacity of the cell (Fig. 5.6b), or by

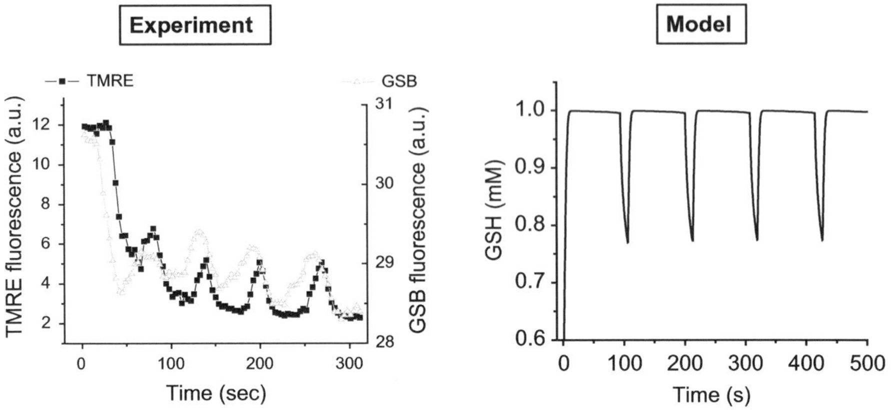

Fig. 5.7 Glutathione oscillations. (**a**) Experimental demonstration of GSH oscillations (70 s period) recorded simultaneously with $\Delta\Psi_m$. Oscillations were triggered after a localized laser flash as described in Fig. 5.4 in freshly isolated cardiomyocytes loaded with 100 nM tetramethylrhodamine methyl ester (TMRM) and 50 μM monochlorobimane, MCB (Cortassa et al. 2004). Reproduced from Cortassa, Aon, Winslow, O'Rourke (2004) Biophys J 87, 2060–73

blocking the mitochondrial ATP synthase (Fig. 5.6c). Mitochondrial oscillations could be interrupted by acute inhibition of mitochondrial ROS production (Fig. 5.6d), or by blockage of mitochondrial IMAC (Aon et al. 2003); these were later found to be activated under moderate oxidative stress (Aon et al. 2007b).

In the respiratory chain, oscillations were suppressed when we inhibited electron transfer complex I (rotenone), III (antimycin A/myxothiazol), or IV (cyanide), or the phosphorylation machinery F_1F_0 ATPase (oligomycin) or adenine nucleotide translocator (ANT, bongkrekic acid). On the other hand, we reinforced the scavenging system of cardiac cells by adding a superoxide dismutase mimetic or N-acetyl cysteine (Aon et al. 2008a). Although we could suppress the mitochondrial oscillations by preincubation in the presence of ROS scavengers, it took much longer than respiratory inhibitors to see the effect (1–3 min vs. 1 h) (Aon et al. 2003). The scavengers also slowed the initial $\Delta\Psi_m$ depolarization wave (Aon et al. 2004a; Cortassa et al. 2004).

Model simulations showed the ability to reproduce the ~100 s oscillatory period and the phase relationship between $\Delta\Psi_m$ and NADH observed experimentally (Fig. 5.5). The model was also able to simulate other major experimental findings, including (1) the requirements of ROS to cross a threshold so as to trigger fast $\Delta\Psi_m$ depolarization, (2) the suppressive effect of inhibitors of the electron transport chain, adenine nucleotide translocator (ANT) and the F_1F_0 ATPase on ROS production and $\Delta\Psi_m$ oscillation (Cortassa et al. 2004), (3) the effects of anion channel inhibitors, and (4) the sensitivity of the oscillator to the level of ROS scavengers.

After validation, the model led to predictions that were tested in the experimental system. The model anticipated oscillatory behavior of reduced glutathione (GSH), and that was experimentally demonstrated as well as its phase relationship with $\Delta\Psi_m$ (Fig. 5.7) (Cortassa et al. 2004).

5.4 Second Iterative Loop: Mitochondrial Criticality and Network Redox Energetics During Oscillations

5.4.1 *Mitochondrial Criticality*

Under oxidative stress mitochondrial behavior reaches a critical point that we called *mitochondrial criticality* (Aon et al. 2004a), an emergent macroscopic response manifested as a generalized $\Delta\Psi_m$ collapse followed by synchronized oscillation in the mitochondrial network under stress (Fig. 5.4). As the mitochondria approach criticality, two main questions arise (Aon et al. 2006a): (1) how does the signal propagate throughout the network? and (2) how does $\Delta\Psi_m$ depolarization occur almost simultaneously in distant regions of the cell?

Applying percolation theory to the problem (see Box 5.1) we found that, prior to the first global $\Delta\Psi_m$ depolarization, approximately 60 % of the mitochondria had accumulated ROS to a level roughly 20 % above baseline (Aon et al. 2004a), which was the threshold for activation of the oscillator at the whole cell level. This critical density of mitochondria (60 %) was consistent with that predicted by percolation theory (Box 5.1). Moreover, the spatial distribution of mitochondria at the threshold exhibits a fractal dimension in agreement with theory (Aon et al. 2003, 2004b).

Beyond criticality, self-sustained oscillations in $\Delta\Psi_m$ continue as a consequence of a bifurcation in the dynamics of the system (Cortassa et al. 2004). However, the spatial pattern of subsequent depolarization of the network will typically follow that of the original percolation cluster, with some mitochondria always remaining outside the cluster. Another important feature of the percolation model is that the global transition can be prevented if the $O_2^{\cdot -}$ concentration reaching the neighboring mitochondrion is decreased below threshold, either by decreasing $O_2^{\cdot -}$ production (e.g., by inhibiting respiration), decreasing $O_2^{\cdot -}$ release (e.g., by inhibiting IMAC), or increasing the local ROS scavenging capacity (e.g., by increasing the GSH pool) (Aon et al. 2004a, 2007b; Cortassa et al. 2004).

Box 5.1: Standard 2D Percolation Theory as Applied to Explain Mitochondrial Criticality

Percolation describes how local neighbor–neighbor interactions among elements in a lattice can scale to produce a macroscopic response spanning from one end of the array to the other (Stauffer and Aharony 1994). Such a "spanning cluster" forms when there is a critical density of elements close to the threshold for a transition (the percolation threshold). Experimentally, the "spanning cluster" involved ~60 % of the mitochondrial lattice with increased levels of ROS (Aon et al. 2004a). This value was consistent with a critical density of mitochondria at the percolation threshold (*pc*), which, for a square lattice in percolation theory, is equal to 0.593 or ~59 % (Feder 1988; Stauffer and Aharony 1994).

(continued)

Box 5.1 (continued)

Another signature feature of percolation processes at *pc* is that they are organized as fractals. This property implies that local processes can scale to produce macroscopic behavior. At *pc*, the mass of the spanning cluster increases with the size of the lattice, *L*, as a power law, L^{Df}, with D_f as the fractal dimension (Feder 1988; Mandelbrot 1977; Stauffer and Aharony 1994). Fractal box counting analysis of our data yielded a fractal dimension of $D_f \sim 1.82$, close to that exhibited by percolation clusters and cytoskeletal lattices at p_c ($D_f \sim 1.90$) (Aon and Cortassa 1994, 1997; Aon et al. 2003; Feder 1988; Stauffer and Aharony 1994).

Several interesting properties of the mitochondrial response can be explained when the network is considered as a percolation cluster. First, the question of the limited diffusivity and lifetime of $O_2^{\cdot -}$ as the triggering molecule is answered, since the only relevant diffusion distance is the inter-mitochondrial spacing (~1 μm). As long as there are enough neighboring mitochondria belonging to the spanning cluster (i.e., they have accumulated enough $O_2^{\cdot -}$ to approach the percolation threshold) an universal phase transition will occur (Feder 1988; Schroeder 1991; Stauffer and Aharony 1994) and mitochondria will depolarize for the first time throughout the cell (Aon et al. 2003, 2004a).

5.4.2 Modeling Mitochondrial Network Redox Energetics

The mitochondrial oscillator model utilized in the first iterative loop corresponds to an isolated mitochondrion representing the average behavior of the mitochondrial population, actually (in the case of the cardiomyocyte or cardiac cell) organized as a network. Therefore, accounting for the spatial relationships between individual mitochondria within the network became crucial to simulating the initial depolarization wave that signals the energetic collapse of the mitochondrial network. The importance of this seminal event in the escalation of failures, from mitochondria propagating to cells and groups of them, and finally attaining the whole heart, made it worthwhile (and in fact crucial) to unravel the fundamental mechanisms involved. More specifically, we wanted to explore whether a reaction–diffusion mechanism could be responsible of the spreading of failure of individual organelles to the whole cell. According to our model, wave propagation comprises the nonlinear dependence of an IMAC opening on $O_2^{\cdot -}$ accumulation in the matrix, and the free radical autocatalytic release and spreading in the network; we hypothesized that mechanistically this could be enough to reproduce the wave phenomenon and the underlying ROS-induced ROS release (RIRR).

In order to achieve this goal we developed a mathematical model of RIRR based on reaction–diffusion (RD-RIRR) in one- and two-dimensional mitochondrial

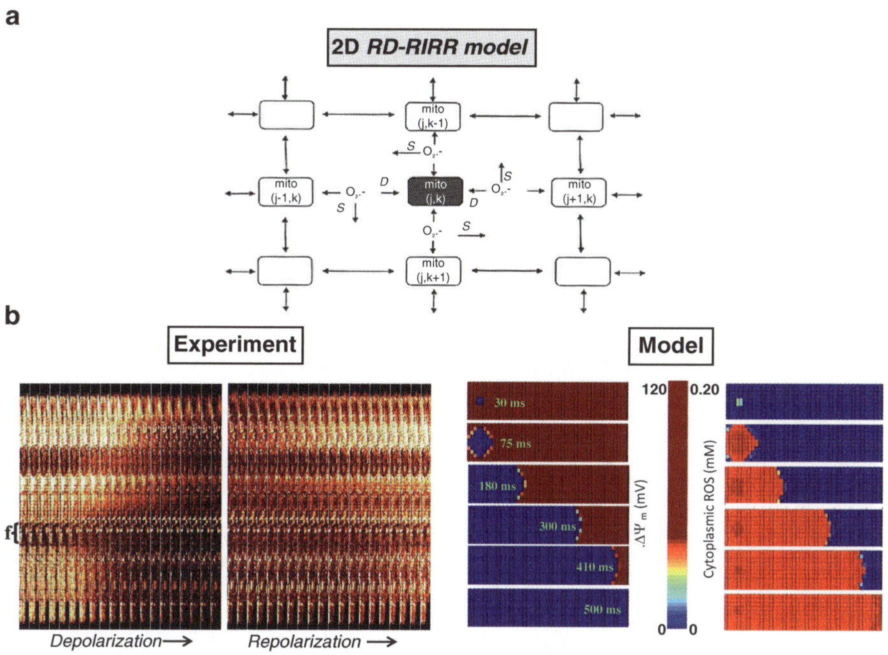

Fig. 5.8 Scheme of the RD-RIRR mitochondrial network model. In the two-dimensional RD-RIRR model neighboring mitochondria are chemically coupled with each other through superoxide anion, $O_2^{\cdot -}$, diffusion. *Light* and *dark gray* indicate polarized and depolarized mitochondria, respectively. *Arrows* indicate release of $O_2^{\cdot -}$ and its effect on mitochondrial neighbors. D stands for $O_2^{\cdot -}$ diffusion, and S for $O_2^{\cdot -}$ scavenging by Cu,Zn SOD and catalase. Reproduced from Zhou, Aon, Almas, Cortassa, Winslow, O'Rourke (2010) PLoS Computational Biology 6(1): e1000657. doi:10.1371/journal.pcbi.1000657

networks (see Box 5.2). The nodes of the RD-RIRR network are comprised of models of individual mitochondria that include a mechanism of ROS-dependent oscillation based on the interplay between ROS production, transport, and scavenging, and incorporating the tricarboxylic acid (TCA) cycle, oxidative phosphorylation, and Ca^{2+} handling. Local mitochondrial interaction is mediated by $O_2^{\cdot -}$ diffusion and the $O_2^{\cdot -}$-dependent activation of IMAC (Fig. 5.8a).

In a 2D network composed of 500 mitochondria, model simulations reveal $\Delta\Psi_m$ depolarization waves similar to those observed when isolated guinea pig cardiomyocytes are subjected to local laser flash or antioxidant depletion (Fig. 5.8b).

Box 5.2: Accounting for ROS-Induced ROS Release Based on Reaction–Diffusion: Computational Model of a Network Comprised by 500 Mitochondria Based on the ROS-Dependent Mitochondrial Oscillator

The thin optical sectioning ability of two-photon laser scanning fluorescence microscopy can be used to examine the behavior of the mitochondrial network in a single plane of a cardiomyocyte (Fig. 5.4) (Aon et al. 2003, 2004a). To compare the experimental results obtained with optical imaging experiments in cardiomyocytes subjected to oxidative stress, a mitochondrial reaction–diffusion ROS-induced ROS release (RD-RIRR) computational model was developed. Each non-boundary node (mitochondrion) in the network was considered to have four nearest neighbors for $O_2^{\cdot-}$ interaction (Fig. 5.8). At each node (*j*, *k*) of the 2D network, $O_2^{\cdot-}$ dynamics is described by the mass balance equation based on $O_2^{\cdot-}$ reaction and diffusion (Zhou et al. 2010):

$$\begin{aligned} &\frac{\partial C_{O_2-i}(x,y,t)}{\partial t} = D_{O_2-i}\left(\frac{\partial^2 C_{O_2-i}(x,t)}{\partial x^2} + \frac{\partial^2 C_{O_2-i}(y,t)}{\partial y^2}\right) + f(C_{O_2-i},t) \\ &\text{Boundary conditions}: \frac{\partial C_{O_2-i}(0,t)}{\partial x} = 0; \frac{\partial C_{O_2-i}(X,t)}{\partial x} = 0 \\ &\frac{\partial C_{O_2-i}(0,t)}{\partial y} = 0; \frac{\partial C_{O_2-i}(Y,t)}{\partial y} = 0 \\ &\text{Initial conditions}: C_{O_2-i}(x,y,0) = g(x,y) \end{aligned} \tag{5.1}$$

where $D_{O_{2.}}$- is the cytoplasmic $O_2^{\cdot-}$ diffusion coefficient, *X* and *Y* indicate the total lengths in the dimensions *x* and *y*, respectively, and $f\ (C_{O2.-i}\ ,\ t) = Vt_{O2.-i}\ (t) - VSOD_{O_{2.}}\text{-}i\ (t)$. $Vt_{O2.-i}$ is the rate of $O_2^{\cdot-}$ transport (release) from the mitochondrion (via IMAC), and $VSOD_{O2.-i}$, the $O_2^{\cdot-}$ scavenging rate by Cu, Zn superoxide dismutase (SOD). The function *g*(*x*,*y*) describes the distribution of $O_2^{\cdot-}$ at time 0 (the initial condition). The spatial coordinates, *x* and *y*, are subjected to discretization to numerically solve the system by the finite difference method. Non-flux boundary conditions were used.

To solve this large nonlinear network consisting of 500 (50 × 10) mitochondria (each node described by 15 state variables), a high-performance parallel computer was used. To be suitable for parallel computation, Eq. (5.1) was rewritten in the matrix form using forward Euler method to approximate the time derivative of $C_{O2.-i}$ at each node (*j*, *k*):

$$C_{O_2-i}(j,k,t+\Delta t) = C_{O_2-i}(j,k,t) + [\text{Diff}_{O_2-i}(j,k,t) + f(C_{O_2-i}(j,k,t))]\,\Delta t$$

(continued)

Box 5.2 (continued)

Reaction–diffusion theory (pioneered by Turing (1952)), as a basis for pattern formation in biological or chemical systems, emphasizes the importance of two components; an autocatalytic reaction producing a local product (mediator or 'morphogen' as originally defined by Turing), and the transport of this product by diffusion away from the source. This process can give rise to spontaneous symmetry-breaking and the appearance of self-organized spatial patterns including waves and oscillations (Aon et al. 1989; Cortassa et al. 1990; Meinhardt 1982; Nicolis and Prigogine 1977). With respect to the present model, the reaction consists of the reduction of O_2 to produce ROS (specifically $O_2^{\cdot-}$) driven by mitochondrial electron donors (e.g., NADH). The local concentration of $O_2^{\cdot-}$ around the mitochondrion is shaped by several others factors, including buffering by the antioxidant reactions and transport of $O_2^{\cdot-}$ across the mitochondrial membrane. Diffusion of the $O_2^{\cdot-}$ to neighboring mitochondria is shaped by the $O_2^{\cdot-}$ diffusion coefficient, $D_{O2.-}$, and the amount of the $O_2^{\cdot-}$ scavenger enzyme superoxide dismutase present, which consequently determines the rate of propagation of $\Delta\Psi_m$ depolarization through the network. As expected, increasing SOD concentration slowed down the depolarization wave.

The rate of propagation of the depolarization wave in the model corresponded to 26 μm s^{-1} with low $D_{O2.-}$ (of the order of 10^{-14} cm^2 s^{-1}), which compares well with the experimentally determined 22 μm s^{-1} at 37 °C (Aon et al. 2004a). A restricted diffusion range of $O_2^{\cdot-}$ in cells is consistent with experimental data; however, the actual diffusion coefficient of $O_2^{\cdot-}$ in cells (with antioxidant systems disabled) has not been determined and is likely to be influenced by local reactions with other molecules and molecular crowding around mitochondria, which would decrease the effective volume, increase the viscosity of the medium, and increase collision probability. This assumption of restricted diffusion is represented by the low $D_{O2.-}$ in the model.

5.4.3 Experimental–Theoretical Test of a Main Mechanistic Assumption of Mitochondrial Oscillations: Superoxide as a Trigger of $\Delta\Psi_m$ Depolarization

The original ROS-dependent mitochondrial oscillator model (Cortassa et al. 2004) considered cytoplasmic $O_2^{\cdot-}$ as the primary ROS that would increase IMAC open probability in an autocatalytic process. H_2O_2 was ruled out because it had been shown experimentally that superoxide dismutase mimetics, which should enhance H_2O_2 accumulation, in fact suppressed the oscillations in $\Delta\Psi_m$ (Aon et al. 2003). To test the assumption that $O_2^{\cdot-}$ could directly trigger IMAC opening, we applied increasing concentrations of potassium superoxide (KO_2, an $O_2^{\cdot-}$ donor) to partially permeabilized myocytes. Increasing the exogenous cytoplasmic KO_2

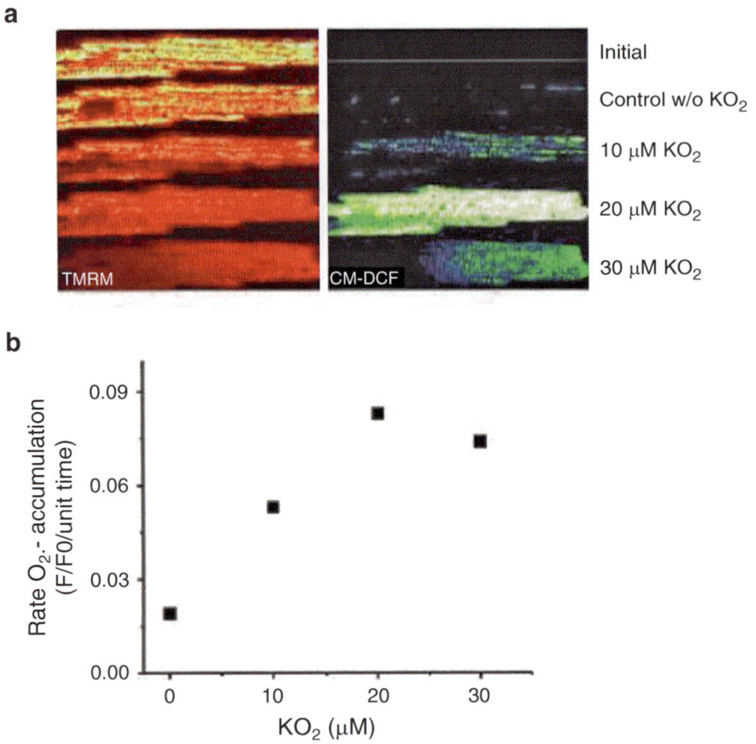

Fig. 5.9 Mitochondrial $O_2^{\cdot-}$ and $\Delta\Psi_m$ in response to increased exogenous $O_2^{\cdot-}$ Myocytes were loaded with TMRM (100 nM) and CM-H_2DCFDA (2 μM) for at least 20 min and imaged using two-photon laser scanning fluorescence microscopy. After loading, the excess dye was washed out and the cells were briefly superfused with a permeabilizing solution (saponin) (Aon et al. 2007). After permeabilization, the myocytes were continuously perfused with an intracellular solution containing GSH:GSSG at a ratio of 300:1. The TMRM was included in the medium to avoid depletion of the probe during depolarization–repolarization cycles. (**a**) The TMRM and CM-DCF images of a permeabilized cardiomyocyte at time zero after loading and before (*top row image*) or after permeabilization and 5 min imaging under control conditions (Control, *second row*) or the presence of KO_2, an $O_2^{\cdot-}$ donor (10 μM, third row; 20 μM, fourth row; 30 μM, fifth row), after 3 min equilibration in each case. RIRR-mediated $\Delta\Psi_m$ depolarization without a permeability transition occurs at the two lower concentrations, while loss of the CM-DCF probe (~500 MW) from the mitochondrial matrix due to PTP opening occurs at 30 μM KO_2. (**b**) The rates of $O_2^{\cdot-}$ accumulation as a function of KO_2 concentration. Slopes were calculated when the linear rate of change of the CM-DCF signal stabilized under each condition. Reproduced from Zhou, Aon, Almas, Cortassa, Winslow, O'Rourke (2010) PLoS Computational Biology 6(1): e1000657. doi:10.1371/journal.pcbi.1000657

concentration, from 10 to 20 μM, elicited progressive $\Delta\Psi_m$ depolarization and increased the rate of mitochondrial $O_2^{\cdot-}$ accumulation (Fig. 5.9b). Exposure of the cell to 30 μM KO_2 induced an irreversible collapse of $\Delta\Psi_m$, accompanied by the complete release of the $O_2^{\cdot-}$ sensor, indicative of permeability transition pore (PTP) opening (Fig. 5.9a).

5.4.4 *Propagation of Depolarization Through ROS-Induced ROS Release (RIRR)*

RIRR as the basic mechanism of propagation of $\Delta\Psi_m$ depolarization and $O_2^{\cdot-}$ release was also demonstrated in the 2D RD-RIRR mitochondrial network model (Zhou et al. 2010). Consisting of five hundred (10 × 50) mitochondria, the 2D RD-RIRR model was parametrically initialized to represent a condition of high oxidative stress to simulate a mitochondrial network at criticality. Approximately 1 % (6 out of 500) of the mitochondrial network was induced to undergo depolarization (Fig. 5.8b). A local increase in $O_2^{\cdot-}$ concentration and depolarization of $\Delta\Psi_m$ in this area was evident, similar to those processes observed in experiments in which we applied a localized laser flash to a fraction of the mitochondrial network (Fig. 5.4). $\Delta\Psi_m$ depolarization propagated outward in all directions from the six perturbed mitochondria and then appeared as a longitudinal wave as the edges of the array were encountered (Fig. 5.8b, model left panel). Importantly, a wave of increased $O_2^{\cdot-}$ accompanied the $\Delta\Psi_m$ depolarization wave (Fig. 5.8b, model right panel).

Moreover, the model further contributed to our understanding by showing that (1) local gradients of cytoplasmic $O_2^{\cdot-}$, determined by diffusion and scavenger capacity, play a significant role in determining the rate of propagation of the $\Delta\Psi_m$ depolarization and repolarization waves; and (2) by uncovering a novel aspect of the synchronization mechanism, i.e., that clusters of mitochondria that are in a state characteristic of the oscillatory domain of the parametric space can entrain mitochondria that would otherwise display stable dynamics (Zhou et al. 2010).

While focusing on a specific mechanism of RIRR (i.e., IMAC-mediated), the model results provided general theoretical support for mitochondrial communication occurring by way of $O_2^{\cdot-}$ diffusion. The RD-RIRR model simulations confirm that $O_2^{\cdot-}$ diffusion occurring locally between neighboring mitochondria over a distance of a few microns is sufficient for propagation and synchronization of $\Delta\Psi_m$ depolarization over a larger distance (Aon et al. 2004a; Zhou et al. 2010).

5.5 Third Iterative Loop: Mitochondrial Network Redox Energetics Escalates from Subcellular to Cellular Level

The experimental studies had shown that laser flash-induced cell-wide mitochondrial oscillations in isolated cardiomyocytes produced shortening of the cellular action potential (AP) (Akar et al. 2005; Aon et al. 2003). The collapse of $\Delta\Psi_m$

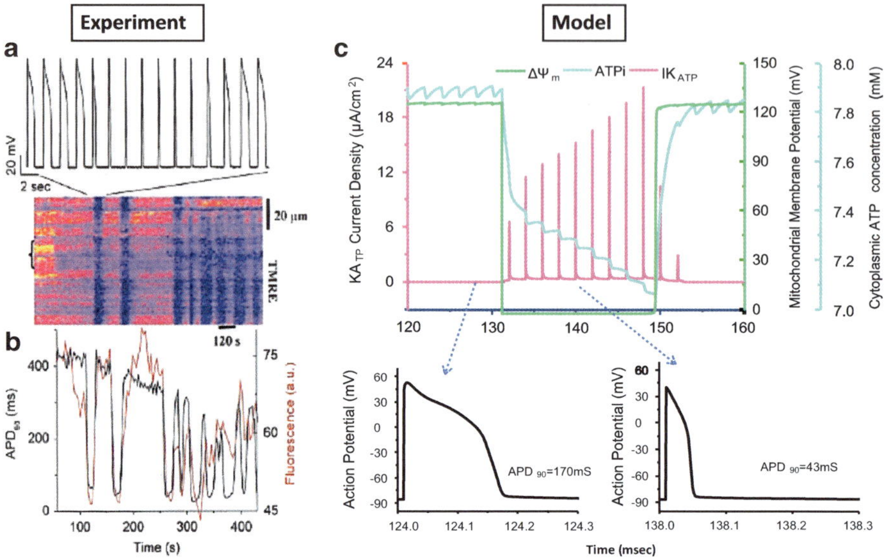

Fig. 5.10 Effects of mitochondrial oscillation on the electrical excitability of the cardiomyocyte (**a**) Action potentials (AP, *upper panel*) evoked by brief current injections were recorded in current clamp mode during whole-cell patch clamp while simultaneously imaging $\Delta\Psi_m$ with TMRE (*lower panel*). During a synchronized cell-wide depolarization–repolarization cycle, the AP shortened in synchrony with fast mitochondrial depolarization, and the cell became unexcitable in the fully depolarized state (remaining upward spikes are from the stimulus only). Recovery of $\Delta\Psi_m$ coincided with restoration of AP. (**b**) Temporal correlation between the AP duration at 90 % repolarization and $\Delta\Psi_m$. The current–voltage relationship of the oscillatory membrane current fits the profile of the sarcolemmal K_{ATP} current (Aon et al. 2003). (**c**) Simulations of $\Delta\Psi_m$, cytoplasmic ATP concentration ([ATP]i) and sarcolemmal K_{ATP} current during oscillations triggered by oxidative stress, and comparison of APs during polarized and depolarized states. Modified from Aon, Cortassa, Marban, O'Rourke (2003) J Biol Chem 278, 44735–44, and Zhou, Cortassa, Wei, Aon, Winslow, O'Rourke (2009) Biophys J 97, 1843–1852

triggered by ROS was shown experimentally to be coupled to the opening of sarcolemmal ATP-sensitive potassium (K_{ATP}) channels, contributing to electrical dysfunction during ischemia–reperfusion. K_{ATP} channels have a low open probability under physiological conditions, but are rapidly activated during ischemia or metabolic inhibition (Lederer et al. 1989; Noma 1983).

Figure 5.10 shows that during the depolarized phase of the oscillation, the AP interval shortens as a consequence of the activation of sarcolemmal K_{ATP} currents rendering the myocyte electrically inexcitable during the nadir of $\Delta\Psi_m$ (Akar et al. 2005; Aon et al. 2003). The fact that these effects were mediated by $\Delta\Psi_m$ depolarization was indicated by inhibition of the IMAC-mediated mitochondrial oscillations with 4′-chlorodiazepam (4′-Cl-DZP) an intervention that concomitantly reestablished and stabilized the sarcolemmal AP.

Highly nonlinear interactions are involved in the mitochondrial oscillation-driven inexcitability of the cardiac cell, including communication through transport between different subcellular compartments and processes of different nature, i.e., metabolic and electromechanical (Fig. 5.1). A comprehensive mathematical model represents an invaluable tool for addressing this problem. Coupling between the mitochondrial energy state and electrical excitability mediated by the sarcolemmal K_{ATP} current (IK,ATP) was incorporated into a computational model of excitation–contraction coupling linked to mitochondrial bioenergetics, and accounting for mitochondrial RIRR.

The model at this stage produced results that were very similar to the experimental observations. During the phase of mitochondrial depolarization, the AP shortened by almost 75% (Fig. 5.10) and the intracellular Ca^{2+} transient amplitude and force production decreased (Zhou et al. 2009). Mitochondrial depolarization accelerates K_{ATP} current activation because the decreased $\Delta\Psi_m$ causes the ATP synthase to run in reverse, thus consuming cytoplasmic ATP and decreasing the phosphorylation potential. Tight coupling between the mitochondrial energy state and the sarcolemmal K_{ATP} current is facilitated by the high-energy phosphoryl transfer reactions of the cytoplasm (Cortassa et al. 2006; Dzeja and Terzic 2003; Sasaki et al. 2001; Zhou et al. 2009). Most importantly, whole-cell model simulations demonstrated that increasing the fraction of oxygen diverted from the respiratory chain to ROS production triggers limit-cycle oscillations of $\Delta\Psi_m$, redox potential, and mitochondrial respiration through the activation of the ROS-sensitive IMAC.

5.6 Fourth Iterative Loop: Scale-Free Dynamics of Mitochondrial Network Redox Energetics

Another model prediction concerned the oscillatory frequency as a function of the superoxide dismutase (SOD) activity. This prediction resulted from stability analysis of the model that showed amplitude and frequency modulation of oscillations when the balance of ROS was changed by increasing SOD (Fig. 5.11). In the absence of metabolic stress, cardiomyocytes loaded with tetramethylrhodamine methyl ester (TMRM, $\Delta\Psi_m$ fluorescent reporter) display stable $\Delta\Psi_m$ for more than an hour. Using two-photon scanning laser fluorescence microscopy, cells were imaged every 110 ms and the average fluorescence from the whole or part of the network, calculated.

Applying power spectral analysis (PSA) and relative dispersional analysis (RDA) to $\Delta\Psi_m$ time series we found that, collectively, cardiac mitochondria behave as a highly correlated network of oscillators (see Box 5.3). PSA revealed that the ensemble of oscillators represented by the mitochondrial network exhibits many frequencies across temporal scales, spanning from milliseconds to minutes. This behavior corresponds to scale-free dynamics, mathematically characterized by an inverse power law when represented in a log–log plot of power versus frequency (Fig. 5.12; see also Boxes 5.3 and 5.4). The power spectrum followed a homogeneous inverse power law of the form $1/f^{\beta}$ with $\beta \sim 1.7$ (Aon et al. 2006b).

The origins of the inverse power law behavior exhibited by mitochondrial network dynamics

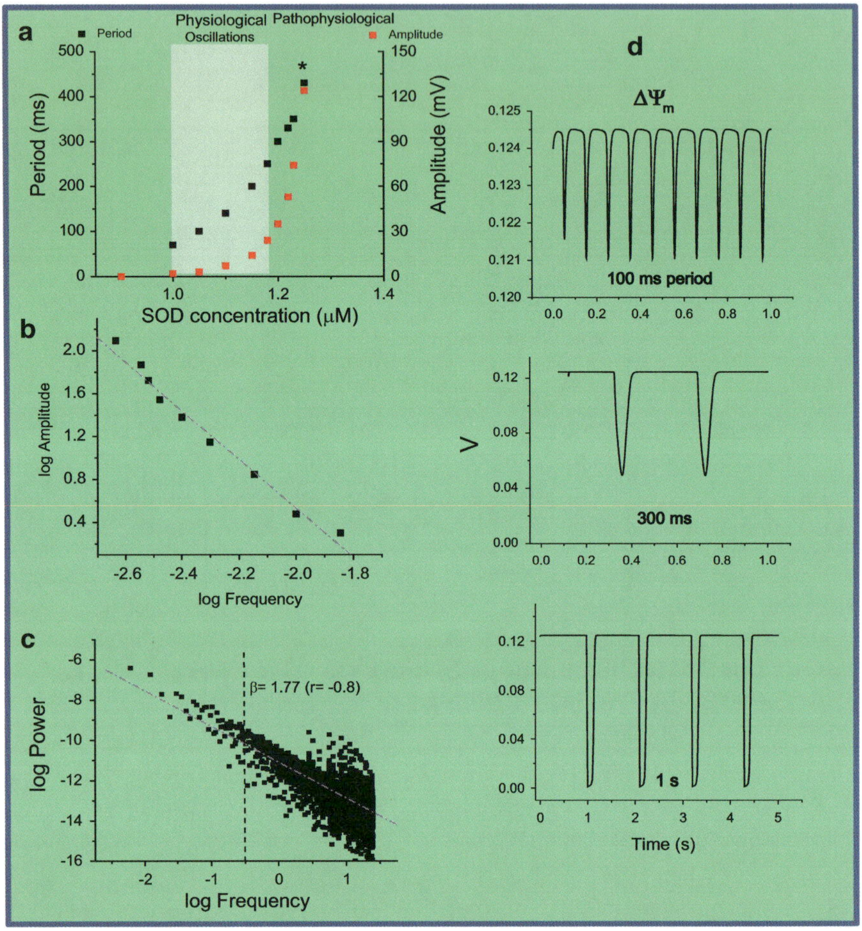

Fig. 5.11 Modulation of the oscillation period by the rate of ROS scavenging through SOD, and inverse power law behavior of the amplitude versus frequency relationship exhibited by the mitochondrial oscillator. (**a**) Oscillations with different periods and amplitude in $\Delta\Psi_m$ were simulated with our computational model of the mitochondrial oscillator by changing SOD concentration (Cortassa et al. 2004). (**b**) The double log graph of the amplitude versus frequency (1/period) was plotted from $\Delta\Psi_m$ oscillations with amplitudes in the range of 2–124 mV and periods ranging from 70 to 430 ms, respectively (see Aon et al. 2006, and their Supplemental Material for more details). (**c**, **d**) From the simulations, we selected five oscillatory periods in the high-frequency domain (between 70 and 300 ms) and one from the low-frequency (1-min period) domain and attributed each one of them proportionally to a network composed by 500 mitochondria as described in Aon et al. (2006). A matrix containing a total of 500 columns (mitochondria) and 6,000 rows was constructed. The time steps represented by the rows correspond to a fixed integration step of 20 ms for the numerical integration of the system of ordinary differential equations. We applied RDA and PSA to the average value of each row of the matrix at, e.g., time 1, T1, that represents the experimental average value of fluorescent intensity of the $\Delta\Psi_m$ probe (corresponding to mV) obtained every 110 ms from 500 mitochondria (on average) from each image of our stack

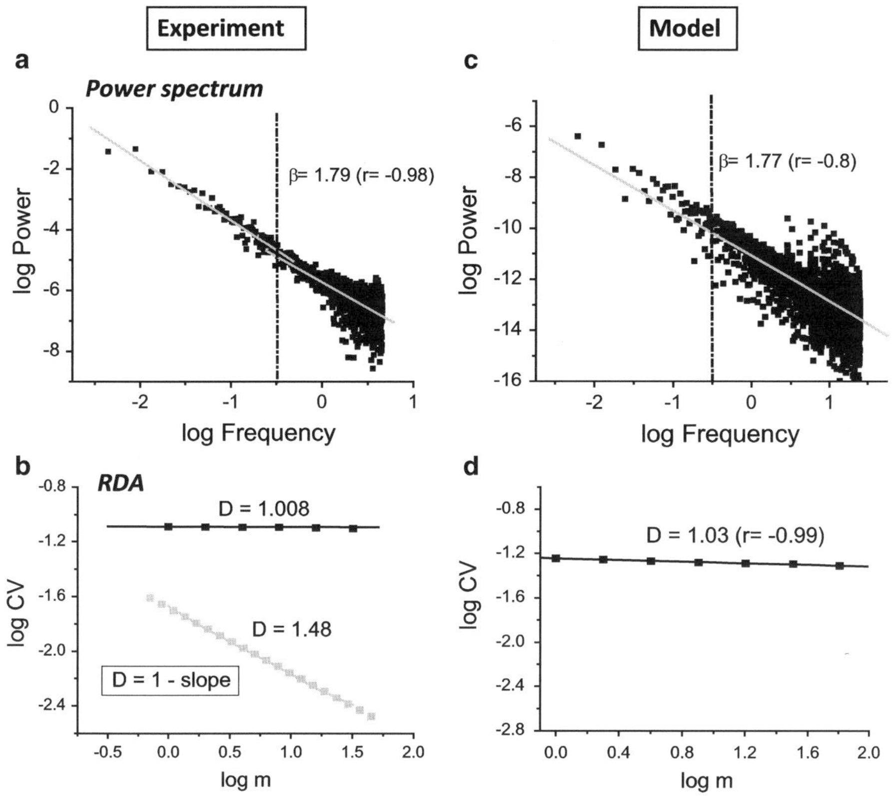

Fig. 5.12 Experimental evidence and model simulation of the inverse power law behavior observed after RDA or PSA of the fluorescence time series of TMRM. Mathematical procedures and simulations were performed as described in Boxes 5.3 and 5.4, and Fig. 5.11

RDA unveiled the existence of long-term temporal correlation ("memory") among oscillators in the network (Fig. 5.12; see also Box 5.3). This led us to conclude that the behavior in the physiological domain is also oscillatory but with low-amplitude high-frequency oscillations. These results also indicated that the oscillators are weakly coupled by low levels of mitochondrial ROS in the high-frequency domain. We considered the behavior under these conditions to belong to the "physiological" state, because fluctuations at high-frequency (but restricted amplitude) range imply depolarization only of between microvolts to a few millivolts (Fig. 5.11) (Aon et al. 2006b). Decreasing mitochondrial ROS production at the level of the respiratory chain, or blocking the ROS-induced ROS release (R-IRR) mechanism by inhibiting the mitochondrial benzodiazepine receptor in the physiological domain, consistently diminished the extent of correlated behavior of the mitochondrial network in the high-frequency domain (Aon et al. 2006b).

Under metabolic stress, however, an imbalance between production and scavenging can take ROS over a threshold resulting in strong coupling between mitochondria through RIRR (Figs. 5.4 and 5.8) (Aon et al. 2003; Zorov et al. 2000). When subjected to these challenging stressful conditions, the mitochondrial network spontaneously organizes into a synchronized cluster with a dominant low-frequency high-amplitude oscillation that spans the whole cell (Aon et al. 2004a, 2006b). The RIRR mechanism is effective locally, but long-range synchronization is due to the attainment of criticality by ~60 % of mitochondria that form a cluster, the "spanning cluster", across the cardiomyocyte. Mitochondria belong to the "spanning cluster" when ROS (more specifically, $O_2^{\cdot -}$) attain a threshold in the matrix, after which $\Delta\Psi_m$ depolarization ensues, triggering a similar response in neighboring mitochondria.

Underlying the inverse power law behavior observed experimentally in the power spectrum is the inverse relationship found in the model simulations of the double log plot of amplitude versus frequency (Fig. 5.11). Two key factors contribute to this dependence—the superoxide dismutase (SOD) activity and the balance between the rate of ROS production and scavenging. In the oscillatory domain, an increase in the SOD rate results in longer periods and higher amplitude oscillations (Fig. 5.11).

We hypothesized that if, according to the experimental results, the mitochondrial network were exhibiting a mixture of frequencies, then we should be able to simulate the inverse power law behavior obtained by either PSA or RDA (see Box 5.4). To test this hypothesis we simulated five different oscillatory periods ranging from 70 ms to 300 ms and one long period (1 min) oscillation (Fig. 5.11). A combination of 80 % short-period and 20 % long-period oscillations allowed us to simulate the inverse power law behavior observed experimentally by either PSA or RDA (Fig. 5.12). This result demonstrated that mixing a relatively few (six) periods of limit-cycle type of oscillation is enough to explain our experimental data. Using a similar approach, we were also able to simulate the transition from physiological to pathophysiological behavior (Aon et al. 2006b). This transition, according to simulations, and in agreement with experimental data, is effected when at least 60 % of the mitochondrial network dynamics is dominated by the long period, high-amplitude $\Delta\Psi_m$ oscillations (Fig. 5.12) ((Aon et al. 2006b), and their Supplementary Material).

Box 5.3: Fast Fourier Transform (FFT) Is a Key Mathematical Procedure Utilized in Power Spectral Analysis (PSA)

FFT of a time series enables the statistical determination of the power (equivalent to the amplitude squared) of each frequency component of a signal. When this analytical procedure was applied to the time series of $\Delta\Psi_m$ from the mitochondrial network, a large number of frequencies in multiple timescales became evident (Aon et al. 2006b, 2008b). Thus,

(continued)

Box 5.3 (continued)
statistically, the mitochondrial network rather than exhibiting a single "characteristic" frequency shows multiple frequencies, typical of *dynamic fractals*.

Fractals possess a two-sided nature both as geometric (spatial) and as dynamic objects. This trait enables techniques commonly applied for characterization and quantification of fractals in the spatial domain to be applied to describe dynamic behavior. The different spatial and temporal scales displayed by an object (be it a shape or a time series) can be quantified by lacunarity, a mass-related statistical parameter quantified by the coefficient of variation or relative dispersion, RD (= standard deviation (SD)/mean). RD is a strong function of scale (Aon and Cortassa 2009) that in the case of self-similar time series or "dynamic fractals" remains constant (i.e., the object looks the same at all scales) (West 1999). The determination of RD at successively larger intervals from a time series constitutes the basic mathematical procedure for applying relative dispersional analysis (RDA). RDA of $\Delta\Psi_m$ time series from the mitochondrial network revealed long-term temporal correlations ("memory").

Box 5.4: Simulation of the Inverse Power Law Behavior Exhibited by the Mitochondrial Network of Cardiomyocytes
From the simulations, we selected five oscillatory periods in the high-frequency domain (between 70 and 300 ms) and one from the low-frequency (1 min period) domain and attributed each one of them proportionally to a network composed of 500 mitochondria (i.e., every 100 mitochondria will oscillate with the same period). This number of mitochondria is similar to that present in a single optical slice of a cardiomyocyte (~1 μm focal depth) that we analyze by two-photon laser scanning microscopy with 110 ms/frame time resolution. Our experimental results could be precisely simulated with a mixture of 80 % short-period and 20 % long-period oscillations.

According to this protocol, we then constructed a matrix: with mitochondria in columns, and time, Ti, on rows. The final matrix contained a total of 500 columns and 6,000 rows. The time steps represented by the rows correspond to a fixed integration step of 20 ms for the numerical integration of the system of ODEs. The fixed integration step of 20 ms was chosen for the simulation of all periods within the range of 70–300 ms and 1 min in order to avoid aliasing effects.

	Mito1	Mito2	Mito3	...$Mito_n$
T_0				
T_1				
T_2				
...T_n				

5.7 Mitochondrial Oscillations in the Intact Heart: Testing the Consequences of Mitochondrial Criticality

We determined if the nonlinear oscillatory phenomena described in single cells can be observed at the level of the whole heart. Studies performed in permeabilized cardiomyocytes revealed that the critical state can be induced by partial depletion of the GSH pool and that the reversible (IMAC-mediated) and irreversible (PTP-mediated) depolarization of $\Delta\Psi_m$ can be distinguished from the cytoplasmic glutathione redox status. IMAC-mediated $\Delta\Psi_m$ oscillation was triggered at a GSH/GSSG ratio of 150:1–100:1, whereas PTP opening is triggered at a GSH/GSSG of 50:1 (Aon et al. 2007b). These results pointed out that GSH and probably also the glutathione redox potential are the main cellular variables that determine the approach of the mitochondrial network to criticality through an increase in oxidative stress by the overwhelming of the antioxidant defenses.

Extending the mechanistic findings in permeabilized cardiomyocytes (Aon et al. 2007b) to the mitochondrial ROS-dependent oscillator described in living cardiac myocytes (Aon et al. 2003, 2004a), and computational models (Cortassa et al. 2004) to the level of the myocardial syncytium, we showed that mitochondrial $\Delta\Psi_m$ oscillations could be triggered by ischemia/reperfusion (I/R) or GSH depletion in intact perfused hearts using two-photon scanning laser microscopy (Slodzinski et al. 2008). These results confirmed that the appearance of oscillatory behavior is not restricted to isolated cardiomyocytes but also happens in the epicardium of intact hearts (either flash-triggered or GSH depletion elicited), in both $\Delta\Psi_m$ and NADH (Fig. 5.13).

An important prediction of the percolation model utilized to explain the mechanism of mitochondrial synchronization at criticality is that the global transition can be prevented if the $O_2^{\cdot -}$ concentration reaching the neighboring mitochondrion is decreased below threshold. This can be accomplished either by decreasing $O_2^{\cdot -}$ production (inhibiting respiration), decreasing $O_2^{\cdot -}$ release (inhibiting IMAC), or increasing the local ROS scavenging capacity (increasing the GSH pool) (Aon et al. 2004a, 2007b; Cortassa et al. 2004). Within this rationale, $\Delta\Psi_m$ depolarization induced by depleting the GSH pool could induce cardiac arrhythmias even under normoxic conditions (Aon et al. 2007b, 2009). Indeed, Brown et al. (2010) demonstrated that systematic oxidation of the GSH pool with diamide in Langendorff-perfused guinea pig hearts elicited ventricular fibrillation under normoxia (Fig. 5.14b, d). Experimental evidence further indicating the involvement of IMAC was noted when the arrhythmias induced by GSH depletion (Brown et al. 2010) or H_2O_2-elicited oxidative stress (Biary et al. 2011) were prevented with the IMAC blocker 4'-Cl-DZP (Fig. 5.14c, e).

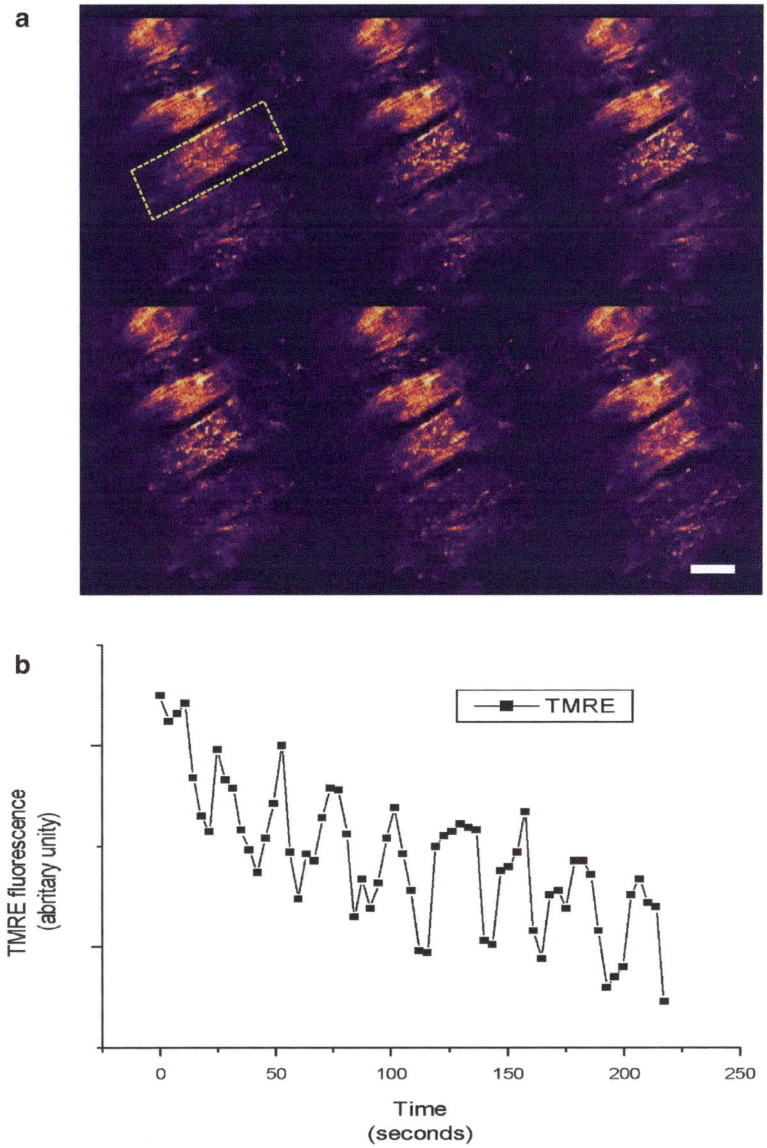

Fig. 5.13 Mitochondrial membrane potential oscillations in myocytes of the intact heart. (**a**, **b**) After 2–3 min of normoxic perfusion with Ca^{2+}-free Tyrode's buffer, the $\Delta\Psi_m$ signal (TMRE fluorescence) became unstable and spontaneous oscillations were observed. Sustained oscillations in $\Delta\Psi_m$ were observed for several minutes in the cell indicated by the *yellow dashed line* in panel **a**. Both intracellular and intercellular heterogeneity of $\Delta\Psi_m$ is evident in the epicardial optical sections. *Scale bar* in **a** equals 20 μm. Reproduced from Slodzinski, Aon, O'Rourke (2008) J Mol Cell Cardiol 45, 650–660

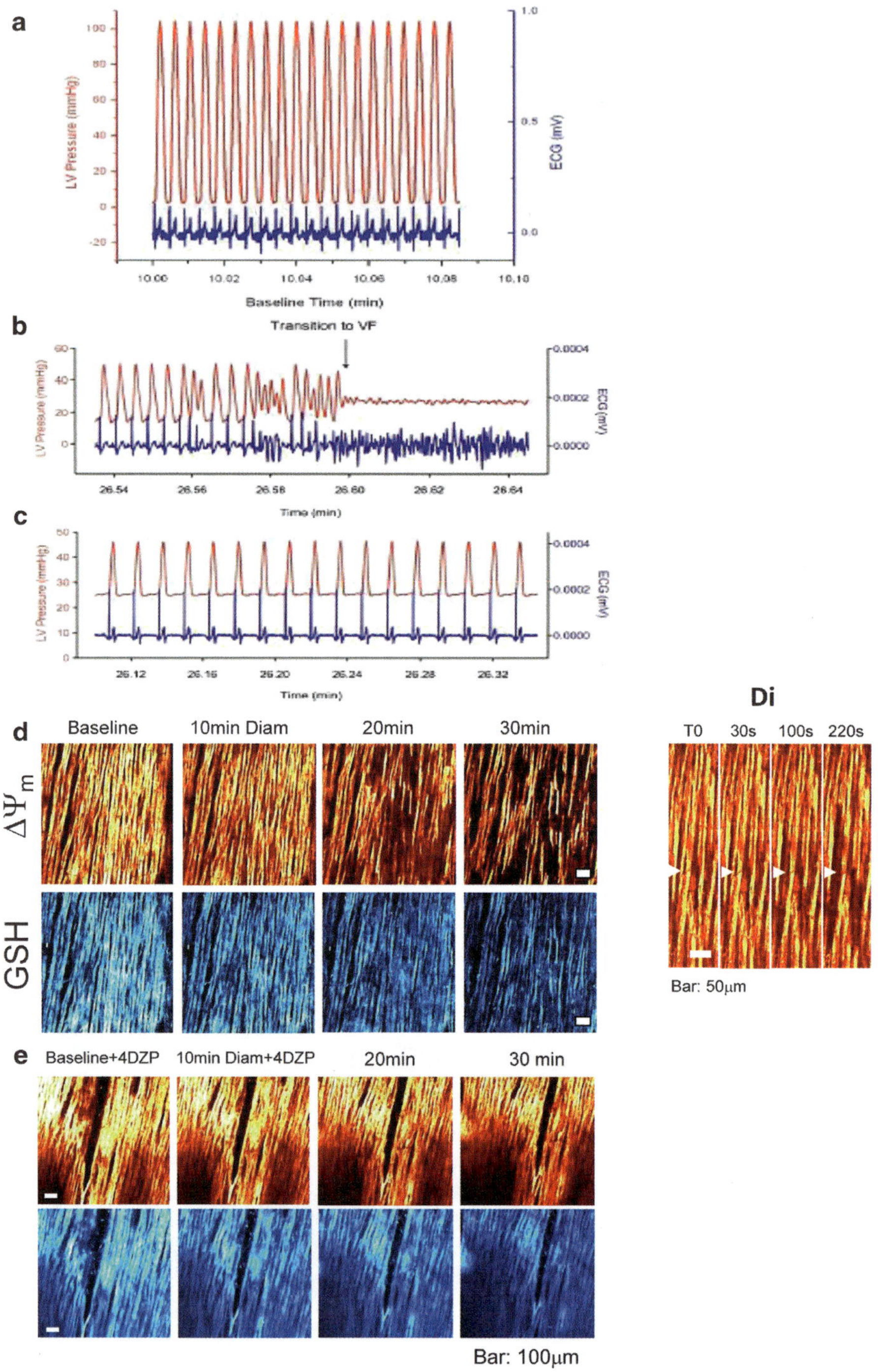

Fig. 5.14 Representative LV pressure and ECG from guinea pig heart. (**a**) Control heart after 10 min of baseline perfusion with simultaneous left ventricular (LV), pressure (*red*), and

5.8 New Modeling Developments Along the Experimental–Computational Synergy

The latest version of the isolated mitochondrion model—the mitochondrial energetic-redox (ME-R) model—includes all four main redox couples NADH/NAD^+, NADPH/$NADP^+$, GSH/GSSG, and $Trx(SH)_2$/TrxSS together with a complete array of antioxidant defenses. All four variables are considered as present in two compartments: matrix and extra-mitochondrial; the latter compartment comprising intermembrane space and cytoplasm (Kembro et al. 2013). Also taken into account are the $NADP^+$-dependent isocitrate dehydrogenase (IDH2) in the TCA cycle, and transhydrogenase (THD), two of the three main NADPH sources in mitochondria. The Trx system involves thioredoxin reductase and peroxiredoxin whilst the glutaredoxin system accounts for the recovery of glutathionylated proteins (using GSH as cofactor), superoxide dismutases (SOD) (matrix-located MnSOD and extra-mitochondrial Cu,ZnSOD), and catalase activity also in the extra-mitochondrial compartment.

The model by Kembro et al. (2013) has been formulated on the basis of our mitochondrial energetics version that included pH regulation, ion dynamics (H^+, Ca^{2+}, Na^+, Pi), respiratory fluxes from complex I and II, tricarboxylic acid cycle (TCA cycle) dynamics, adenine nucleotide exchange (ANT), and ATP synthesis (Wei et al. 2011).

The qualitative dynamic behavior exhibited by the new ME-R model reveals that, as in former versions (Cortassa et al. 2004; Zhou et al. 2009), the underlying oscillatory mechanism involves ROS imbalance determined by the interplay between ROS production and scavenging as the main trigger of oscillations. This happens irrespective of the bi-compartmental nature of the ME-R model, accounting for ROS scavenging in both the matrix and extra-mitochondrial space.

5.9 Conclusions

Experimental–computational synergy involves the reciprocal potentiation of the loop involving experimental work and mathematical modeling that operates iteratively via the multiple simulation–validation and prediction–experimentation

Fig. 5.14 (continued) electrocardiogram (ECG, blue). (**b**) LV pressure and ECG in a heart during diamide treatment showing the transition to ventricular fibrillation (VF, *blue*) with concomitant loss of pump function (*red*). (**c**) LV pressure and ECG in a heart during diamide treatment plus 64 μM 4'-ClDzp. (**d**, **e**) Simultaneous imaging of $\Delta\Psi_m$ (*top*) and GSH (*bottom*) in intact guinea pig hearts using two-photon microscopy after exposure to diamide (D) and diamide + 4′ chlorodiazepam, 4′-ClDzp (**e**). The inset Di shows in detail the propagation of $\Delta\Psi_m$ depolarization in the syncytium at the cardiomyocyte level. Reproduced from Brown, Aon, Frasier, Sloan, Maloney, Anderson, O'Rourke (2010) J Mol Cell Cardiol 48, 673–679

loops. When applied systematically and rigorously the synergetic loop of experiment-model enables a deeper understanding of complex biological phenomena as shown in this chapter. Thorough application of this strategy both enabled us to formulate the following new concepts and to unravel complex emergent phenomena exhibited by networks of metabolic and electromechanical processes in the cardiac cell:

- Mitochondria are organized as a dynamic network of oscillators in the cardiac myocyte
- The mitochondrial network is synchronized by ROS to different degrees of coupling strength, either weakly or strongly during physiological or pathophysiological behavior, respectively.
- The mitochondrial network is embedded within other metabolic networks of the cardiac cell, thereby interacting with metabolic, electrical, and mechanical processes.
- Emergent phenomena take place in these networks in the form of physiologically normal long-term correlations involving signaling processes, or as failures that can propagate from the subcellular to the whole heart producing potentially catastrophic arrhythmias (see Chap. 10).
- Striking similarities between the cardiac redox control systems and those in the evolutionarily distant organism, baker's yeast (*Saccharomyces cerevisiae*), indicate an ancient commonality of central core metabolic mechanisms (see Chap. 12), and how a mutually enhanced understanding of these can be gleaned from comparative studies of different biological systems (Aon et al. 2007c, 2008b; Lemar et al. 2007; Lloyd et al. 2012).

Acknowledgments This work was performed with the financial support of R21HL106054 and R01-HL091923 from NIH.

References

Akar FG, Aon MA, Tomaselli GF, O'Rourke B (2005) The mitochondrial origin of postischemic arrhythmias. J Clin Invest 115:3527–35

Aon MA, Cortassa S (1994) On the fractal nature of cytoplasm. FEBS Lett 344:1–4

Aon MA, Cortassa S (1997) Dynamic biological organization. Fundamentals as applied to cellular systems. Chapman & Hall, London

Aon MA, Cortassa S (2009) Chaotic dynamics, noise and fractal space in biochemistry. In: Meyers R (ed) Encyclopedia of complexity and systems science. Springer, New York

Aon MA, Cortassa S (2012) Mitochondrial network energetics in the heart. Wiley Interdiscip Rev Syst Biol Med 4:599–613

Aon M, Thomas D, Hervagault JF (1989) Spatial patterns in a photobiochemical system. Proc Natl Acad Sci USA 86:516–9

Aon MA, Cortassa S, Marban E, O'Rourke B (2003) Synchronized whole cell oscillations in mitochondrial metabolism triggered by a local release of reactive oxygen species in cardiac myocytes. J Biol Chem 278:44735–44

Aon MA, Cortassa S, O'Rourke B (2004a) Percolation and criticality in a mitochondrial network. Proc Natl Acad Sci USA 101:4447–52
Aon MA, O'Rourke B, Cortassa S (2004b) The fractal architecture of cytoplasmic organization: scaling, kinetics and emergence in metabolic networks. Mol Cell Biochem 256–257:169–84
Aon MA, Cortassa S, Akar FG, O'Rourke B (2006a) Mitochondrial criticality: a new concept at the turning point of life or death. Biochim Biophys Acta 1762:232–40
Aon MA, Cortassa S, O'Rourke B (2006b) The fundamental organization of cardiac mitochondria as a network of coupled oscillators. Biophys J 91:4317–27
Aon MA, Cortassa S, Lemar KM, Hayes AJ, Lloyd D (2007a) Single and cell population respiratory oscillations in yeast: a 2-photon scanning laser microscopy study. FEBS Lett 581:8–14
Aon MA, Cortassa S, Maack C, O'Rourke B (2007b) Sequential opening of mitochondrial ion channels as a function of glutathione redox thiol status. J Biol Chem 282:21889–900
Aon MA, Cortassa S, O'Rourke B (2007a) On the network properties of mitochondria. Wiley-VCH, p 111–35
Aon MA, Cortassa S, O'Rourke B (2008a) Mitochondrial oscillations in physiology and pathophysiology. Adv Exp Med Biol 641:98–117
Aon MA, Roussel MR, Cortassa S, O'Rourke B, Murray DB, Beckmann M, Lloyd D (2008b) The scale-free dynamics of eukaryotic cells. PLoS One 3:e3624
Aon MA, Cortassa S, Akar FG, Brown DA, Zhou L, O'Rourke B (2009) From mitochondrial dynamics to arrhythmias. Int J Biochem Cell Biol 41:1940–8
Biary N, Xie C, Kauffman J, Akar FG (2011) Biophysical properties and functional consequences of reactive oxygen species (ROS)-induced ROS release in intact myocardium. J Physiol 589:5167–79
Borecky J, Jezek P, Siemen D (1997) 108-pS channel in brown fat mitochondria might Be identical to the inner membrane anion channel. J Biol Chem 272:19282–9
Brown DA, Aon MA, Frasier CR, Sloan RC, Maloney AH, Anderson EJ, O'Rourke B (2010) Cardiac arrhythmias induced by glutathione oxidation can be inhibited by preventing mitochondrial depolarization. J Mol Cell Cardiol 48:673–9
Cortassa S, Sun H, Kernevez JP, Thomas D (1990) Pattern formation in an immobilized bienzyme system. A morphogenetic model. Biochem J 269:115–22
Cortassa S, Aon MA, Marban E, Winslow RL, O'Rourke B (2003) An integrated model of cardiac mitochondrial energy metabolism and calcium dynamics. Biophys J 84:2734–55
Cortassa S, Aon MA, Winslow RL, O'Rourke B (2004) A mitochondrial oscillator dependent on reactive oxygen species. Biophys J 87:2060–73
Cortassa S, Aon MA, O'Rourke B, Jacques R, Tseng HJ, Marban E, Winslow RL (2006) A computational model integrating electrophysiology, contraction, and mitochondrial bioenergetics in the ventricular myocyte. Biophys J 91:1564–89
Cortassa S, Aon MA, Iglesias AA, Aon JC, Lloyd D (2012) An introduction to metabolic and cellular engineering, 2nd edn. World Scientific, Singapore
Dzeja PP, Terzic A (2003) Phosphotransfer networks and cellular energetics. J Exp Biol 206: 2039–47
Feder J (1988) Fractals. Plenum, New York
Gustafsson AB, Gottlieb RA (2008) Heart mitochondria: gates of life and death. Cardiovasc Res 77:334–43
Kembro JM, Aon MA, Winslow RL, O'Rourke B, Cortassa S (2013) Integrating mitochondrial energetics, redox and ROS metabolic networks: a two-compartment model. Biophys J 104: 332–343
Lederer WJ, Nichols CG, Smith GL (1989) The mechanism of early contractile failure of isolated rat ventricular myocytes subjected to complete metabolic inhibition. J Physiol 413:329–49
Lemar KM, Aon MA, Cortassa S, O'Rourke B, Muller CT, Lloyd D (2007) Diallyl disulphide depletes glutathione in Candida albicans: oxidative stress-mediated cell death studied by two-photon microscopy. Yeast 24:695–706

Lloyd D, Cortassa S, O'Rourke B, Aon MA (2012) What yeast and cardiomyocytes share: ultradian oscillatory redox mechanisms of cellular coherence and survival. Integr Biol (Camb) 4:65–74

Mandelbrot BB (1977) The fractal geometry of nature. W.H. Freeman, New York

Meinhardt H (1982) Models of biological pattern formation. Academic, London

Nicolis G, Prigogine I (1977) Self-organization in nonequilibrium systems : from dissipative structures to order through fluctuations. Wiley, New York, p 491

Noma A (1983) ATP-regulated K+ channels in cardiac muscle. Nature 305:147–8

O'Rourke B, Cortassa S, Aon MA (2005) Mitochondrial ion channels: gatekeepers of life and death. Physiology (Bethesda) 20:303–15

Romashko DN, Marban E, O'Rourke B (1998) Subcellular metabolic transients and mitochondrial redox waves in heart cells. Proc Natl Acad Sci USA 95:1618–23

Sasaki N, Sato T, Marban E, O'Rourke B (2001) ATP consumption by uncoupled mitochondria activates sarcolemmal K(ATP) channels in cardiac myocytes. Am J Physiol Heart Circ Physiol 280:H1882–8

Schroeder M (1991) Fractals, chaos, power laws. Minutes from an infinite paradise. W.H. Freeman and Company, New York

Slodzinski MK, Aon MA, O'Rourke B (2008) Glutathione oxidation as a trigger of mitochondrial depolarization and oscillation in intact hearts. J Mol Cell Cardiol 45:650–60

Stauffer D, Aharony A (1994) Introduction to percolation theory. Taylor & Francis, London

Turing AM (1952) The chemical basis of morphogenesis. Phil Trans R Soc Lond B237:37–72

Wei AC, Aon MA, O'Rourke B, Winslow RL, Cortassa S (2011) Mitochondrial energetics, pH regulation, and Ion dynamics: a computational-experimental approach. Biophys J 100: 2894–903

West BJ (1999) Physiology, promiscuity and prophecy at The Millennium: A tale of tails. World Scientific, Singapore

Zhou L, Cortassa S, Wei AC, Aon MA, Winslow RL, O'Rourke B (2009) Modeling cardiac action potential shortening driven by oxidative stress-induced mitochondrial oscillations in guinea pig cardiomyocytes. Biophys J 97:1843–52

Zhou L, Aon MA, Almas T, Cortassa S, Winslow RL, O'Rourke B (2010) A reaction–diffusion model of ROS-induced ROS release in a mitochondrial network. PLoS Comput Biol 6: e1000657

Zorov DB, Filburn CR, Klotz LO, Zweier JL, Sollott SJ (2000) Reactive oxygen species (ROS)-induced ROS release: a new phenomenon accompanying induction of the mitochondrial permeability transition in cardiac myocytes. J Exp Med 192:1001–14

Chapter 6
Adenylate Kinase Isoform Network: A Major Hub in Cell Energetics and Metabolic Signaling

Song Zhang, Emirhan Nemutlu, Andre Terzic, and Petras Dzeja

Abstract The adenylate kinase isoform network is integral to the cellular energetic system and a major player in AMP metabolic signaling circuits. Critical in energy state monitoring and stress response, the dynamic behavior of the adenylate kinase network in governing intracellular, nuclear, and extracellular nucleotide signaling processes has been increasingly revealed. New adenylate kinase mutations have been identified that cause severe human disease phenotypes such as reticular dysgenesis associated with immunodeficiency and sensorineural hearing loss and primary ciliary dyskinesia characteristic of chronic obstructive pulmonary disease. The adenylate kinase family comprises nine major isoforms (AK1–AK9), and several subforms with distinct intracellular localization and kinetic properties designed to support specific cellular processes ranging from muscle contraction, electrical activity, cell motility, unfolded protein response, and mitochondrial/nuclear energetics. Adenylate kinase and AMP signaling is necessary for energetic communication between mitochondria, myofibrils, and the cell nucleus and for metabolic programming facilitating stem cell cardiac differentiation and mitochondrial network formation. Moreover, it was discovered that during cell cycle, the AK1 isoform translocates to the nucleus and associates with the mitotic spindle to provide energy for cell division. Furthermore, deletion of *Ak2* gene is embryonically lethal, indicating critical significance of catalyzed phosphotransfer in the crowded mitochondrial intracristae and subcellular spaces for ATP export and intracellular distribution. Taken together, new evidence highlights the importance of the system-wide adenylate kinase isoform network and adenylate kinase-mediated phosphotransfer and AMP signaling in cellular energetics, metabolic sensing, and regulation of nuclear and cell cycle processes which are critical in tissue homeostasis, renewal, and regeneration.

S. Zhang • A. Terzic • P. Dzeja (✉)
Center for Regenerative Medicine, Division of Cardiovascular Diseases, Department of Medicine, Mayo Clinic, Rochester, MN 55905, USA
e-mail: dzeja.petras@mayo.edu

E. Nemutlu
Analytical Chemistry, Hacettepe University, Ankara, Turkey

M.A. Aon et al. (eds.), *Systems Biology of Metabolic and Signaling Networks*, Springer Series in Biophysics 16, DOI 10.1007/978-3-642-38505-6_6,

6.1 Adenylate Kinase Isoform Network

Adenylate kinase, a ubiquitous enzyme with a unique property to catalyze the reaction 2ADP↔ATP + AMP, is indispensable for nucleotide biosynthesis and a sensitive reporter of the cellular energy state, translating small changes in the balance between ATP and ADP into relative large changes in AMP concentration, so that enzymes and metabolic sensors that are affected by AMP can respond with high sensitivity and fidelity to stress signals (Dzeja and Terzic 2009; Dzeja et al. 1998; Noda 1973; Noma 2005). Moreover, adenylate kinase, via a series of spatially linked enzymatic reactions, can facilitate propagation of nucleotide signals in the intracellular, extracellular, and mitochondrial intracristal spaces, thus coordinating energy transfer events and the response of metabolic sensors and nucleotide/nucleoside receptor signaling (Carrasco et al. 2001; Dzeja et al. 1985, 1998, 2002, 2007a, b). A significant progress has been made in defining dynamics of conformational transitions that are functionally important in adenylate kinase catalysis (Daily et al. 2010, 2012; Henzler-Wildman et al. 2007).

Distributed throughout the cell and cellular compartments, the adenylate kinase isoform network delivers β- and γ-high-energy phosphoryls of ATP to ATPases and monitors ATP/ADP metabolic imbalances (Dzeja and Terzic 2009). In response to stress adenylate kinase generates AMP signals which are delivered to metabolic sensors to adjust energy metabolism and cell functions according to changes in physiological state and energetic environment (Dzeja and Terzic 2003, 2009; Janssen et al. 2004; Noma 2005; Pucar et al. 2002). Adenylate kinase-mediated metabolic monitoring and downstream AMP signaling AK → AMP → AMP sensors (including AMPK, K-ATP, and AMP-sensitive metabolic enzymes) network is increasingly recognized as a major homeostatic hub, which is critical in regulation of diverse cellular processes (Dzeja and Terzic 2009; Noma 2005). So far, up to nine distinct adenylate kinase isoforms and a number of subforms with different intracellular localization and energetic-metabolic signaling roles have been identified (Fig. 6.1) (Amiri et al. 2013; Collavin et al. 1999; Dzeja and Terzic 2009; Dzeja et al. 2011a; Janssen et al. 2000, 2004; Noma 2005; Panayiotou et al. 2011; Ren et al. 2005; Ruan et al. 2002). The energetic signaling role of adenylate kinase has gained particular significance after discovery that this enzyme, through a chain of sequential reactions, facilitates the transfer and utilization of both β- and γ- phosphoryls of the ATP molecule, thereby doubling the energetic potential of ATP and cutting in half the cytosolic diffusional resistance for energy transmission (Dzeja et al. 1985, 1998). Turnover of ATP α-, β-, and γ-phosphoryls can be followed by ^{18}O-assisted ^{31}P NMR, a versatile technique for measurement of intracellular dynamics of energy metabolism (Fig. 6.1) (Nemutlu et al. 2012b). Isotope labeling studies indicate that in intact tissues the highest adenylate kinase-catalyzed ATP β-phosphoryl turnover is in the kidney, which approximates 98 % of γ-ATP turnover, followed by the liver (80 %), the heart (15–40 %), and contracting (10–17 %) or resting (3–5 %) skeletal muscles suggesting a centralized role of adenylate kinase in tissue energy homeostasis (Dzeja and Terzic 2003, 2007, 2009).

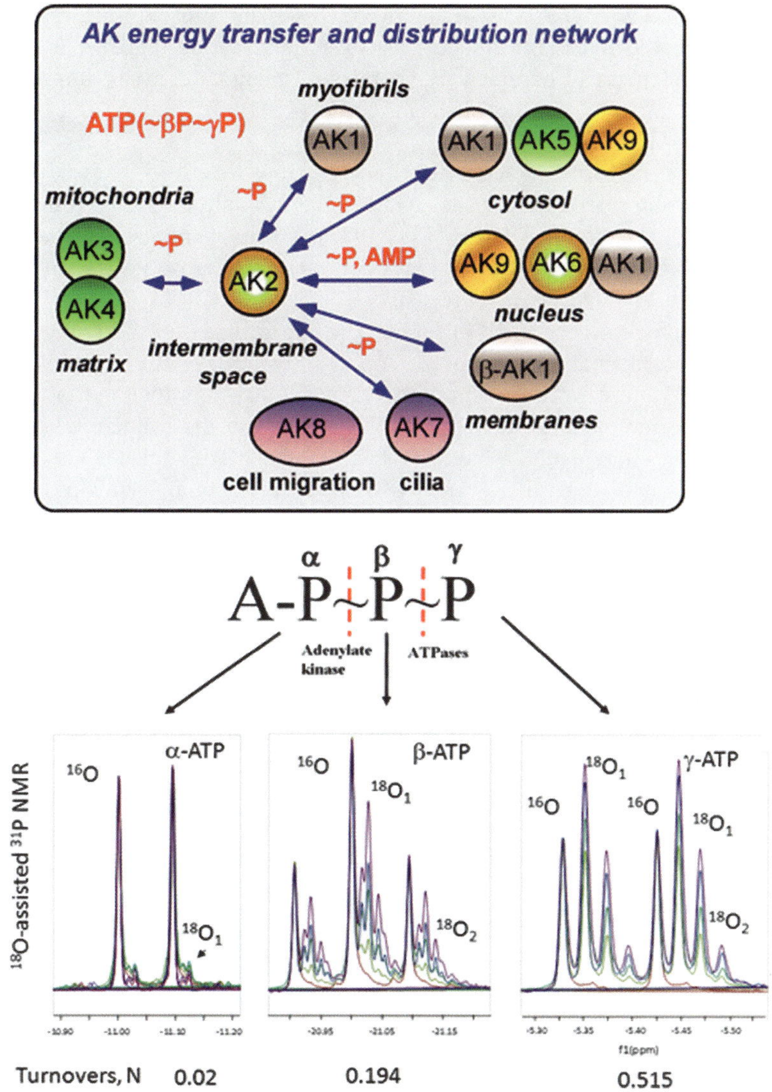

Fig. 6.1 Adenylate kinase isoform metabolic signaling, energy transfer, and distribution network (*upper panel*). Distributed throughout cellular compartments, adenylate kinases facilitate nucleotide exchange, ATP delivery to ATPases, and AMP signal to metabolic sensors. AK2 positions in the "bottleneck" of the network and its deficiency is embryonically lethal, slowing down the flow of nucleotides back and forth from cytosol to mitochondria. Adenylate kinase catalyzed turnover of ATP β-phosphoryls can be followed by ^{18}O-labeling detected by ^{18}O-induced shift in ^{31}P NMR spectra of ATP (*lower panel*) (Nemutlu et al. 2012a, b). Different colors indicate different times (2, 5, 10, and 15 min) of ^{18}O-labeling of ATP α-, β-, and γ-phosphoryls. Turnovers of each ATP phosphoryl were calculated as described (Nemutlu et al. 2012b)

The highest total adenylate kinase activity is in skeletal muscle followed by heart, brain, kidney, and liver (Borglund et al. 1978), suggesting that flux and enzyme activity are independent parameters. Therefore, functional activity imposed metabolic flux cannot be predicted from activity measurements or from model calculations (Nemutlu et al. 2012b).

Three major isoforms, AK1, AK2, and AK3 are localized in the cytosol, the mitochondrial intermembrane space, and the matrix, facilitating transcellular nucleotide exchange (Dzeja et al. 1985; Noda 1973). The existence of the whole family of adenylate kinases with different expression profiles, substrate specificities, and kinetic properties provides evidence of specialized functions in specific cellular processes (Panayiotou et al. 2011). The major cytosolic isoform AK1 is expressed at high levels in skeletal muscle, brain, and heart, whereas AK2 is expressed in the mitochondrial intermembrane space of tissues rich in mitochondria like liver, kidney, heart, and skeletal muscle. AK3 is located in the mitochondrial matrix, with high expression levels in the liver, heart, and skeletal muscle. Tissue specific adenylate kinase isoforms AK4 and AK5, with preferential localization in mitochondrial matrix and cytosol, respectively, have been cloned (Miyoshi et al. 2009; Noma 2005; Panayiotou et al. 2010; Van Rompay et al. 1999; Yoneda et al. 1998). AK4 contains an N-terminal mitochondrial import sequence and is expressed at low levels in brain, kidney, liver, and heart tissues (Panayiotou et al. 2010). AK5 is mostly cytosolic or both cytosolic and nuclear depending on the transcript variants. With two separate active functional domains, AK5 is expressed almost exclusively in brain, although there is evidence that it exists in other tissues too (Solaroli et al. 2009; Van Rompay et al. 1999). AK4 protein levels are increased in cultured cells exposed to hypoxia and in animal models of neurodegenerative diseases (Liu et al. 2009). Although AK4 might be enzymatically less active, it retains nucleotide binding capability, interacts with the mitochondrial ADP/ATP translocator, and serves a stress responsive function, promoting cell survival and proliferation (Liu et al. 2009). Both *AK4* and *AK3* are among hypoxia-inducible factor 1 (HIF-1) regulated genes promoting cell survival (Hu et al. 2006; Semenza 2000). AK5 is detected in human pancreatic beta-cells and implicated in the regulation of the K-ATP channel (Stanojevic et al. 2008), while appearance of autoantibodies to AK5 in refractory limbic encephalitis patients carrying poor prognosis (Tuzun et al. 2007).

More recently, the existence of an additional AK1 gene product, the p53-inducible membrane-bound myristoylated AK1β, has been reported and implicated in p53-dependent cell cycle arrest and nucleotide exchange in the submembrane space (Collavin et al. 1999; Janssen et al. 2004; Ruan et al. 2002). In this context, the gene encoding AK1 is downregulated during tumor development, which could be associated with lower AK1β levels and cell cycle disturbances (Vasseur et al. 2005). AK1β also has been demonstrated to be associated with the nuclear envelope (Janssen et al. 2004) and proteome studies identify AK1β in epithelium microvilli (Bonilha et al. 2004), suggesting a role in energy support of nuclear and epithelia transport processes.

The minor AK6 isoform, also known as transcription factor TAF9 and hCINAP, is localized in the cell nucleus along with AK5 which associates with centrosomes; both isoforms are required for cell growth (Noma 2005; Ren et al. 2005; Zhai et al. 2006). Human AK5 was identified to have two enzymatically active adenylate kinase domains which could catalyze sequential phosphoryl transfer (Solaroli et al. 2009). Another isoform, AK7, has a tissue-specific expression pattern and its activity has been associated with cilia function (Fernandez-Gonzalez et al. 2009). The next AK8 isoform is the second known human adenylate kinase with two complete and active catalytic domains within its polypeptide chain, a feature that previously was shown also for AK5 (Panayiotou et al. 2011). AK8 has nuclear, nucleoli, and mitochondrial localization (see http://www.proteinatlas.org/ENSG00000165695). Both AK7 and the full length AK8 have high affinity for AMP and are more efficient in AMP phosphorylation as compared to the major cytosolic isoform AK1 (Panayiotou et al. 2011). This property may be advantageous for energetic circuits supporting ciliary motility and cell migration (Dzeja and Terzic 1998). Recently, a new member of the adenylate kinase family has been identified and named AK9 (Amiri et al. 2013). This isoform has both cytosolic and nuclear localization and its significance remains to be determined. Also, the exact substrate specificity of the AK9 needs to be resolved with pure enzyme preparations (Amiri et al. 2013).

The significance of the adenylate kinase isoform network and AMP signaling is highlighted by recent studies indicating that mutations in AK1, AK2, and AK7 genes are associated with severe human disease and that AK1 and AK2 isoforms are critical in stem cell differentiation as well as for unfolded protein stress response (Burkart et al. 2010; Dzeja et al. 2011a; Fernandez-Gonzalez et al. 2009; Inouye et al. 1998; Lagresle-Peyrou et al. 2009; Pannicke et al. 2009). Protein knockdown using siRNAs indicates that AK1, AK2, and AK5 isoforms are involved in cardiomyocyte differentiation, mitochondrial biogenesis, and network formation and for development of cardiac beating area and contractile performance (Dzeja et al. 2011a). Deficiency of AK2, which is localized to the mitochondrial intermembrane space at crossroads of high-energy phosphoryl flux, arrests hematopoietic stem cell differentiation in humans (Lagresle-Peyrou et al. 2009; Pannicke et al. 2009) and is embryonically lethal in *Drosophila* and mice (Fujisawa et al. 2009; Noma 2005; Zhang et al. 2010b). Absence or reduction of AK2 protein would interfere with mitochondrial bioenergetics and mitochondria–nucleus energetic communication that could compromise implementation of the leukocyte developmental program (Lagresle-Peyrou et al. 2009; Pannicke et al. 2009). Indeed, preliminary data indicate that *AK2−/−* mouse embryonic fibroblasts (MEFs) have severely disrupted mitochondrial cristae structure, and display low growth and proliferation potential (Zhang et al. 2010b). *Ak2* deficiency could compromise evolvement of the mitochondrial network and establishment of metabolic circuits that are part of developmental programming and execution of cell differentiation sequences (Chung et al. 2007, 2008, 2010; Dzeja et al. 2011a). Indeed, AK2 regulates cell growth, viability, and proliferation in insect growth and development (Chen et al. 2012). Moreover, it was discovered that adenylate kinase and associated AMPK constitute a major metabolic signaling axis guiding energetics of cell cycle, stem cell differentiation, and lineage specification; the defects in AMP metabolic signaling could lead to cardiac malformations (Dzeja et al. 2011a). Thus, AK2 deficiency disrupts the mitochondrial–cytosolic–nuclear flow of energy and the developmental metabolic information governing cell differentiation.

The mitochondrial AK2 isoform has the highest affinity (lowest Km) for AMP ($\leq$10 μM) among AMP metabolizing enzymes and is highly concentrated in the narrow intermembrane space (Dzeja and Terzic 2009; Walker and Dow 1982). Virtually, all the AMP reaching mitochondria is converted to ADP and channeled into oxidative phosphorylation maintaining a low cytosolic AMP concentration. In such a way, adenylate kinase tunes cytosolic AMP signals and guards the cellular adenine nucleotide pool (Dzeja and Terzic 2009). During intense physical activity or metabolic stress, such as ischemia, AMP concentration rises, turning on other AMP-metabolizing enzymes, such as AMP deaminase and 5′-nucleotidase, producing IMP and adenosine. In this regard, a marked elevation of mitochondrial AK2 activity has been demonstrated in hypertrophy in response to increased energy demand and the necessity to maintain the cellular adenine nucleotide pool (Seccia et al. 1998).

Muscles of *Ak1* knockout mice, with one less phosphotransfer chain, display lower energetic efficiency, slower relaxation kinetics, and a faster drop in contractility upon ischemia associated with compromised myocardial–vascular crosstalk (Dzeja et al. 2007b; Pucar et al. 2000, 2002). A mechanistic basis for thermodynamically efficient coupling of cell energetics with cellular functions lies in the unique property of adenylate kinase catalysis which transfers both β- and γ-phosphoryls of ATP, doubling the energetic potential of ATP as an energy-carrying molecule (Dzeja et al. 1985, 1998, 2002; Dzeja and Terzic 2003, 2009).

More recently, it was demonstrated that cytoskeleton-based cell motility can be modulated by spatial repositioning of AK1 enzymatic activity providing local ATP supplies and "on-site" fueling of the actomyosin-machinery (van Horssen et al. 2009). Another study suggests that intracellular and extracellular adenylate kinase play an important role in nucleotide energetic signaling, regulating actin assembly–disassembly involved in cell movement and chemotaxis (Kuehnel et al. 2009). Such integrated energetic and metabolic signaling roles place adenylate kinase in a unique position within the cellular metabolic regulatory network.

New studies published within the past years have uncovered much wider involvement of adenylate kinase in energy support, metabolic monitoring, and managing cellular functions with mutations in this enzyme associated with human diseases (Dzeja and Terzic 2009; Lagresle-Peyrou et al. 2009; Panayiotou et al. 2011; van Horssen et al. 2009). In addition to mutations of adenylate kinase isoform 7, AK7, gene which has been found to be associated with primary ciliary dyskinesia which leads to chronic obstructive pulmonary disease (COPD) in humans (Fernandez-Gonzalez et al. 2009), new mutations in both adenylate kinases 7 and 8, AK7 and AK8, genes were found associated with ciliary defects and congenital hydrocephalus (Vogel et al. 2012). Also, recent studies have uncovered that dysregulation of adenylate kinase AK1-, AK2-, and AK5-mediated energetic and AMP metabolic signaling is associated with Alzheimer's and Huntington neurodegenerative disorders (Park et al. 2012; Seong et al. 2005) and the pathophysiological process of temporal lobe epilepsy (Peng et al. 2012). The level of AK4 is decreased in Parkinson's disease models in response to Parkin expression (Narendra et al. 2012). Moreover, AK4 was discovered to be a progression-associated gene in human lung cancer that promotes metastasis (Jan et al. 2012). In this regard, new findings indicate that development of feto-placental unit and reproductive efficiency of women as well as blood glucose and glycated hemoglobin levels depends on the genetic variability of Ak(1)1 and Ak(1)2 phenotypes (Fulvia et al. 2011; Gloria-Bottini et al. 2011, 2012, 2013). The therapeutic potential of modulating adenylate kinase activity is demonstrated by the finding that decreased expression of AK1 in myocardial infarction is restored by resveratrol treatment, contributing to increase myocardial energetic efficiency (Lin et al. 2011). Taken together, these new data suggest very fundamental roles of adenylate kinase isoforms in various human organs to support vital physiological functions, while dysregulation of adenylate kinase-mediated energetic and metabolic signaling precipitates disease phenotypes.

6.2 Adenylate Kinase Isoforms in Nuclear and Cell Cycle Energetics

Many nuclear processes, including DNA replication and cell cycle events such as mitotic spindle movement, chromosome disjunction, and karyokinesis, also ATP-dependent chromatin remodeling and gene transcription, as well as initiation of developmental and regenerative programming, require robust energy supply (Dzeja and Terzic 2003, 2007, 2009; Folmes et al. 2011a, b, 2012a; Morettini et al. 2008; Rosenfeld et al. 2009). However, the nucleus is separated from the cytosolic energetic system, and energy supply routes to nuclear processes, including cell cycle and cytokinesis machinery, are largely unknown (Dzeja et al. 2002; Dzeja and Terzic 2007). In this regard, cells and nuclei of dividing and regenerating cells in tissues are enriched in energetic and phosphotransfer enzymes to support high energy needs of genetic reprogramming and cell division cycle (Dzeja and Terzic 2009; Hand et al. 2009; Noda 1973; Ottaway and Mowbray 1977; Rosenfeld et al. 2009). Cell life and renewal rate of a cell population depend on balance between cell division, cell cycle arrest, differentiation, and apoptosis (Mandal et al. 2010; Walsh et al. 2010). During evolution, cells have developed a conserved metabolic signaling mechanism controlling cell division, proliferation, and regeneration depending on nutrient and energy resources (Collavin et al. 1999; Folmes et al. 2012a, b; Nakada et al. 2010; Romito et al. 2010). When energy resources are plenty, cells can grow, proliferate, and regenerate. When energy is low, metabolic signaling turns on a cascade which activates cell cycle metabolic checkpoints preventing cell division (Dzeja and Terzic 2009; Gan et al. 2010; Mandal et al. 2010). The regulation of these processes is central for maintaining tissue homeostasis, whereas dysregulation may lead to diseases such as cancer, diabetes, and neurodegenerative disorders (Dzeja and Terzic 2009; Motoshima et al. 2006). It has been suggested that dynamics of nuclear–cytosolic protein localization, phosphorelays of energetic and signaling cascades, and spatially and temporally controlled proteolysis with protein modification events are overlayed on the network that controls cell cycle progression and differentiation (Domian et al. 1997; Ptacin and Shapiro 2013).

Phosphotransfer enzyme adenylate kinase facilitates mitochondrial–nuclear energetic communication (Dzeja et al. 2002) and within the nucleus, embedded into organized enzymatic complexes of nucleotide-metabolizing and phosphotransfer enzymes, is involved in maintaining proper nuclear dNTP ratios and facilitates channeling of nucleotides into DNA replication machinery (Kim et al. 2005). Imbalances in dNTP ratios affect the fidelity of DNA replication and repair leading to acquired mutations (Dzeja and Terzic 2007; Kim et al. 2005). In yeast AK(Adk1p) is bound to chromatin throughout the cell cycle, physically interacts with pre-replicative complexes, including Orc3p and Cdc14p proteins, and is required for DNA replication initiation and cell cycle control (Cheng et al. 2011). Thus, adenylate kinase in the nucleus surveys nucleotide ratios necessary for error-free DNA replication, while another nuclear protein Rad50, harboring or having associated adenylate kinase activity, participates in DNA double-strand break repair (Bhaskara et al. 2007). As AK1 has no known nuclear localization sequence, the molecular mechanisms of AK translocation to the nucleus are still obscure (Strobel et al. 2002).

In recent years, nuclear adenylate kinase isoforms have been characterized in several organisms, such as *Drosophila melanogaster* (Meng et al. 2008), *Saccharomyces cereviciae* (FAP7) (Juhnke et al. 2000), *Caenorhabditis elegans* (Zhai et al. 2006), and *Homo sapiens* (hCINAP) (Malekkou et al. 2010; Ren et al. 2005). Moreover, the human isoform (hCINAP) is involved in Cajal body organization, gene transcription, and cell cycle progression (Santama et al. 2005; Zhang et al. 2010a). Recent study of nuclear shuttling of adenylate kinase in *Trypanosoma cruzi* (TcADKn) has identified its noncanonical nuclear localization signal, being one of the few atypical NLS that involves the catalytic site of the protein (Walker domain or P-loop) (Camara Mde et al. 2013). It is postulated that TcADKn enters the nucleus in an unfolded conformation, being the nuclear localization signal within the P-loop, and once it enters the nucleus, it folds correctly regarding the active site inside the protein. TcADKn could be forming a complex with other proteins, which are recognized by the importin and then enters the nucleus, or it could be recognized directly by the importing complex (Camara Mde et al. 2013). TcADKn nuclear export depends on the nuclear exportation adapter CRM1, and its nuclear shuttling is regulated by nutrient availability, ribosome biogenesis, DNA integrity, and oxidative stress. Posttranslational modification of AK isoforms (acetylation, myristoylation, glutathionylation, and S-nitrosylation) can play a role in directing AK to nuclear and other compartments, according to local energetic needs

(Fratelli et al. 2002; Ginger et al. 2005; Janssen et al. 2004; Klier et al. 1996; Rahlfs et al. 2009; Shi et al. 2008).

The AK6 isoform was identified to be localized to the cell nucleus, where energy provision and nucleotide channeling into DNA synthesis play critical roles in processing genetic information (Dzeja et al. 2002; Ren et al. 2005; Zhai et al. 2006). However, there is still controversy regarding the AK6 isoform, which is also known as TAF9 RNA polymerase II possessing ATPase activity (Santama et al. 2005), suggesting that other adenylate kinase isoforms (AK1 and AK5) can also subserve nuclear energetic needs (Dzeja et al. 2002; Janssen et al. 2004; Noda 1973). AK5 may associate with centrosomes (unpublished). Knockdown of AK6 slows growth and development of *C. elegans* (Zhai et al. 2006), while in yeast, a point mutation in the *Fap7* gene, an analog of AK6, reduces growth on glucose (Juhnke et al. 2000). Through biochemical purification and mass spectrometry analysis, a putative homolog of the *S. cerevisiae* Rps14 protein was identified as a partner of AK6 (Feng et al. 2012). Most importantly, *aak6* T-DNA insertion mutants had decreased stem growth compared with wild-type plants. These data indicate that AAK6 exhibits adenylate kinase activity and is an essential growth factor in *Arabidopsis* (Feng et al. 2012). In this regard, TcADKn is involved in ribosome biogenesis, being involved in the processing of the 18S precursor at site D by directly interacting with TcRps14 (Camara Mde et al. 2013). Another nuclear protein Rad50, a member of DNA repair RAD50/MRE11/NBS1 protein complex (MRN), which is essential for sensing and signaling from DNA double-strand breaks, in addition to ATP binding and hydrolysis, may have intrinsic or associated adenylate kinase activity required for efficient tethering between different DNA molecules (Bhaskara et al. 2007; Randak et al. 2012). A mutation affecting ATPase/adenylate kinase activity of Rad50, necessary for DNA tethering, also abolishes the formation of viable spores (Bhaskara et al. 2007). Thus, adenylate kinase and adenylate kinase activity-possessing proteins play a significant role in the energetics of the nucleus which is separated from major ATP generating processes in the cytosol.

Recently, it was demonstrated that AK1 translocates to the nucleus during cell division and associates with the mitotic spindle to provide energy for chromosome disjunction (Fig. 6.2) (Dzeja et al. 2011a). The discovery of nuclear translocation of AK1 in metaphase is in line with the adenylate kinase role in energy support of motility of cilia and flagella which have 9 + 2 microtubular structures similar to those of mitotic spindle (Cao et al. 2006; Wirschell et al. 2004). In mitotic spindles, AK1 is expected to associate with motor or anchoring proteins as it does with the Oda5 protein of the dynein complex in flagella to provide "on-site" ATP fueling capacity (van Horssen et al. 2009; Wirschell et al. 2004). In yeasts, AK1 interacts with TEM1, a component of the whole network of cell cycle regulation, including AMN1, BUB2, ARP2, TORC2, BFA1, and many others (Szappanos et al. 2011; Tarassov et al. 2008). Beside TEM1, AK1 interacts with CLB2, Rad53, SPL2, and PHO85, which are involved in cell cycle regulation and others (Ho et al. 2002). Our studies revealed several new candidate proteins for the AK1 and AK2 interactomes that include cell motility, cell cycle, and cytokinesis-associated proteins, like

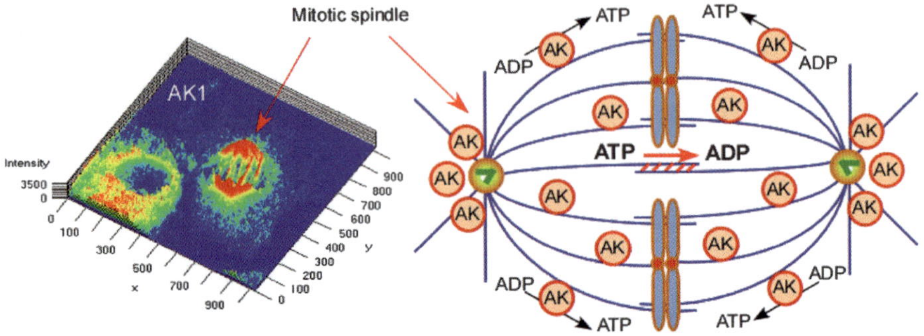

Fig. 6.2 Adenylate kinase provides energy and metabolic signaling for cell division cycle. Adenylate kinase translocates to the cell nucleus and associates with the mitotic spindle during cell division cycle (Dzeja et al. 2011a). Such adenylate kinase redistribution is advantageous for delivery and on-site regeneration of ATP, thus furnishing high energy needs of mitotic spindle movements required for chromosome disjunction and completion of cell division cycle through cytokinesis. *AK* adenylate kinase

myosin-9, tubulin, TEM1, DUO1, CBL2, PHO85, IPL1, MAD2, and others. In this regard, the AK2 isoform, which when mutated causes leukocyte developmental arrest, in yeasts interacts with DUO1, a component of Dam1 complex, and with a network regulating cell cycle including IPL1 (Aurora kinase subunit), MAD2, NAP1, SLJ15, STU2, and others (Tarassov et al. 2008).

The second *AK1* gene product, the p53-inducible membrane-bound myristoylated AK1β isoform, is implicated in p53-dependent cell cycle arrest and nucleotide exchange in the submembrane space (Collavin et al. 1999; Janssen et al. 2004; Ruan et al. 2002). Since gene encoding AK1 is downregulated during tumor development, it could result in lower AK1β levels and cell cycle disturbances (Vasseur et al. 2005). AK1β also associates with the nuclear envelope (Janssen et al. 2004) and epithelium microvilli (Bonilha et al. 2004), suggesting a role in energy support of nuclear and epithelia transport processes.

Developmental deployment and upregulation of the adenylate kinase/AMPK signaling axis provides a nucleo-cytosolic energetic and metabolic signaling vector integral to execution of stem cell cardiac differentiation (Dzeja et al. 2011a). Other studies indicate that adenylate kinase isoforms in the brain may contribute to neuronal maturation and regeneration (Inouye et al. 1998; Yoneda et al. 1998). Thus, targeted redistribution of the adenylate kinase-AMPK circuit associated with cell cycle and asymmetric cell division uncovers a new regulator for cardiogenesis and heart tissue regeneration.

Recently, using metabolomic technologies, we have discovered that in adult hearts increased expression of adenylate kinase isoforms (AK1, AK2, and AK1β) and high energy and AMP signal dynamics, measured by ^{18}O-labeling, is misread by AMPK as a "low energy" state, inducing blockade of cell cycle metabolic checkpoint and cardiomyocyte proliferation and renewal (Mandal et al. 2005; Zhang et al. 2012). Similarly, an excess AMP signaling in the periphery, induced by both leptin and fructose, convey to the brain false "low energy" signals forcing it to increase food consumption (Dzeja and Terzic 2009). The uncovered AMP metabolic flux-sensing mechanism opens new avenues for targeted regulation of the heart regenerative potential, which is critical for repair of injured hearts. We are developing approaches to regulate AMP signaling and increase AK1 entry into the nucleus of adult cardiomyocytes to facilitate nuclear energetics, reentry into cell cycle, proliferative potential, and regeneration (Dzeja et al. 2011a; Rubart and Field 2006; Walsh et al. 2010).

Accordingly, nuclear translocation of AK1, and in some cases AK2, upon association with mitotic spindle and cytokinesis apparatus provides local ATP regenerating capacity (~P) necessary for energy-dependent mitosis and cytokinesis. AMPK plays a critical role in phosphorylating (−P) and orchestrating cell cycle components (Dzeja et al. 2011a). At the same time, AK–AMP–AMPK signaling axis controls the p53/p21/cyclin metabolic cell cycle checkpoint (Mandal et al. 2010). In adult heart, high energy metabolism, AMP signaling dynamics, and relatively high AMPK activity might be interpreted as conditions unfavorable for undergoing mitosis. p53 also regulates membrane-associated AK1β expression which can provide local signals to membrane-associated AMPK and exert additional control on the G1/S transition (Collavin et al. 1999). LKB1 is an important regulator of AMPK and downstream metabolic checkpoint events associated with cell proliferation (Gurumurthy et al. 2010; Nakada et al. 2010). Moreover, the AK–AMP–AMPK axis controls GSK3β which inhibits cell cycle progression; conversely inhibition of GSK3β restarts cardiomyocyte cell cycle (Campa et al. 2008; Woulfe et al. 2010). Defects in AK1 nuclear translocation and AMP/LKB1/AMPK signaling could result in polyploidy and insufficient energy, and signaling to cytokinesis machinery could lead to multinucleation (Jansen et al. 2009; Liu et al. 2010; Nakada et al. 2010; Walsh et al. 2010). Another phosphotransfer enzyme thymidine kinase, the activity of which is critical for dTTP production and DNA replication, is downregulated in adult cardiomyocytes limiting regenerative potential (Gillette and Claycomb 1974). Thus, energy supply to nuclear processes is an emerging field in cellular physiology with the potential to advance our understanding of metabolic requirements and energetic costs of executing cell division cycle and developmental and regenerative programming.

In summary, accumulating evidence suggests a fundamental role for adenylate kinase and phosphotransfer enzymes in general in cellular energetics and metabolic signaling processes. Their importance is highlighted by the wide range of metabolic rearrangements and compensations in energy and metabolic networks induced by genetic deficiencies (Dzeja et al. 2011b; Janssen et al. 2000), while uncompensated deficiency (AK2) is embryonically lethal (Fujisawa et al. 2009; Zhang et al. 2010b).

Acknowledgments This work was supported by National Institutes of Health, Marriott Heart Disease Research Program, Marriott Foundation, and Mayo Clinic. We thank Mayo Graduate School student Jennifer Shing for reading and editing the manuscript.

References

Amiri M, Conserva F, Panayiotou C, Karlsson A, Solaroli N (2013) The human adenylate kinase 9 is a nucleoside mono- and diphosphate kinase. Int J Biochem Cell Biol 45(5):925–31

Bhaskara V, Dupre A, Lengsfeld B, Hopkins BB, Chan A, Lee JH, Zhang X, Gautier J, Zakian V, Paull TT (2007) Rad50 adenylate kinase activity regulates DNA tethering by Mre11/Rad50 complexes. Mol Cell 25:647–61

Bonilha VL, Bhattacharya SK, West KA, Sun J, Crabb JW, Rayborn ME, Hollyfield JG (2004) Proteomic characterization of isolated retinal pigment epithelium microvilli. Mol Cell Proteomics 3:1119–27

Borglund E, Brolin SE, Agren A (1978) Adenylate kinase activity in various organs and tissues of mice with the obese-hyperglycemic syndrome (gene symbol Ob/Ob). J Histochem Cytochem 26:127–30

Burkart A, Shi X, Chouinard M, Corvera S (2010) Adenylate kinase 2 links mitochondrial energy metabolism to the induction of the unfolded protein response. J Biol Chem 286:4081–89

Camara Mde L, Bouvier LA, Canepa GE, Miranda MR, Pereira CA (2013) Molecular and functional characterization of a Trypanosoma cruzi nuclear adenylate kinase isoform. PLoS Negl Trop Dis 7:e2044

Campa VM, Gutierrez-Lanza R, Cerignoli F, Diaz-Trelles R, Nelson B, Tsuji T, Barcova M, Jiang W, Mercola M (2008) Notch activates cell cycle reentry and progression in quiescent cardiomyocytes. J Cell Biol 183:129–41

Cao W, Haig-Ladewig L, Gerton GL, Moss SB (2006) Adenylate kinases 1 and 2 are part of the accessory structures in the mouse sperm flagellum. Biol Reprod 75:492–500

Carrasco AJ, Dzeja PP, Alekseev AE, Pucar D, Zingman LV, Abraham MR, Hodgson D, Bienengraeber M, Puceat M, Janssen E, Wieringa B, Terzic A (2001) Adenylate kinase phosphotransfer communicates cellular energetic signals to ATP-sensitive potassium channels. Proc Natl Acad Sci USA 98:7623–28

Chen RP, Liu CY, Shao HL, Zheng WW, Wang JX, Zhao XF (2012) Adenylate kinase 2 (AK2) promotes cell proliferation in insect development. BMC Mol Biol 13:31

Cheng X, Xu Z, Wang J, Zhai Y, Lu Y, Liang C (2011) ATP-dependent pre-replicative complex assembly is facilitated by Adk1p in budding yeast. J Biol Chem 285:29974–80

Chung S, Dzeja PP, Faustino RS, Perez-Terzic C, Behfar A, Terzic A (2007) Mitochondrial oxidative metabolism is required for the cardiac differentiation of stem cells. Nat Clin Pract Cardiovasc Med 4(Suppl 1):S60–S67

Chung S, Dzeja PP, Faustino RS, Terzic A (2008) Developmental restructuring of the creatine kinase system integrates mitochondrial energetics with stem cell cardiogenesis. Ann N Y Acad Sci 1147:254–63

Chung S, Arrell DK, Faustino RS, Terzic A, Dzeja PP (2010) Glycolytic network restructuring integral to the energetics of embryonic stem cell cardiac differentiation. J Mol Cell Cardiol 48:725–34

Collavin L, Lazarevic D, Utrera R, Marzinotto S, Monte M, Schneider C (1999) wt p53 dependent expression of a membrane-associated isoform of adenylate kinase. Oncogene 18:5879–88

Daily MD, Phillips GN Jr, Cui Q (2010) Many local motions cooperate to produce the adenylate kinase conformational transition. J Mol Biol 400:618–31

Daily MD, Makowski L, Phillips GN Jr, Cui Q (2012) Large-scale motions in the adenylate kinase solution ensemble: coarse-grained simulations and comparison with solution X-ray scattering. Chem Phys 396:84–91

Domian IJ, Quon KC, Shapiro L (1997) Cell type-specific phosphorylation and proteolysis of a transcriptional regulator controls the G1-to-S transition in a bacterial cell cycle. Cell 90:415–24

Dzeja PP, Terzic A (1998) Phosphotransfer reactions in the regulation of ATP-sensitive K^+ channels. FASEB J 12:523–29

Dzeja PP, Terzic A (2003) Phosphotransfer networks and cellular energetics. J Exp Biol 206:2039–47
Dzeja PP, Terzic A (2007) Mitochondria-nucleus energetic communication: role for phosphotransfer networks in processing cellular information. In: Gibson G, Dienel G (eds) Brain energetics: integration of molecular and cellular processes. Springer, New York, pp 641–66
Dzeja P, Terzic A (2009) Adenylate kinase and AMP signaling networks: metabolic monitoring, signal communication and body energy sensing. Int J Mol Sci 10:1729–72
Dzeja P, Kalvenas A, Toleikis A, Praskevicius A (1985) The effect of adenylate kinase activity on the rate and efficiency of energy transport from mitochondria to hexokinase. Biochem Int 10:259–65
Dzeja PP, Zeleznikar RJ, Goldberg ND (1998) Adenylate kinase: kinetic behavior in intact cells indicates it is integral to multiple cellular processes. Mol Cell Biochem 184:169–82
Dzeja PP, Bortolon R, Perez-Terzic C, Holmuhamedov EL, Terzic A (2002) Energetic communication between mitochondria and nucleus directed by catalyzed phosphotransfer. Proc Natl Acad Sci USA 99:10156–61
Dzeja P, Chung S, Terzic A (2007a) Integration of adenylate kinase, glycolytic and glycogenolytic circuits in cellular energetics. In: Saks V (ed) Molecular system bioenergetics: energy for life. Wiley-VCH, Weinheim, Germany, pp 265–301
Dzeja PP, Bast P, Pucar D, Wieringa B, Terzic A (2007b) Defective metabolic signaling in adenylate kinase AK1 gene knock-out hearts compromises post-ischemic coronary reflow. J Biol Chem 282:31366–72
Dzeja PP, Chung S, Faustino RS, Behfar A, Terzic A (2011a) Developmental enhancement of adenylate kinase-AMPK metabolic signaling axis supports stem cell cardiac differentiation. PLoS One 6:e19300
Dzeja PP, Hoyer K, Tian R, Zhang S, Nemutlu E, Spindler M, Ingwall JS (2011b) Rearrangement of energetic and substrate utilization networks compensate for chronic myocardial creatine kinase deficiency. J Physiol 589:5193–211
Feng X, Yang R, Zheng X, Zhang F (2012) Identification of a novel nuclear-localized adenylate kinase 6 from Arabidopsis thaliana as an essential stem growth factor. Plant Physiol Biochem 61:180–6
Fernandez-Gonzalez A, Kourembanas S, Wyatt TA, Mitsialis SA (2009) Mutation of murine adenylate kinase 7 underlies a primary ciliary dyskinesia phenotype. Am J Respir Cell Mol Biol 40:305–13
Folmes CD, Nelson TJ, Martinez-Fernandez A, Arrell DK, Lindor JZ, Dzeja PP, Ikeda Y, Perez-Terzic C, Terzic A (2011a) Somatic oxidative bioenergetics transitions into pluripotency-dependent glycolysis to facilitate nuclear reprogramming. Cell Metab 14:264–71
Folmes CD, Nelson TJ, Terzic A (2011b) Energy metabolism in nuclear reprogramming. Biomark Med 5:715–29
Folmes CD, Dzeja PP, Nelson TJ, Terzic A (2012a) Metabolic plasticity in stem cell homeostasis and differentiation. Cell Stem Cell 11:596–606
Folmes CD, Nelson TJ, Dzeja PP, Terzic A (2012b) Energy metabolism plasticity enables stemness programs. Ann N Y Acad Sci 1254:82–9
Fratelli M, Demol H, Puype M, Casagrande S, Eberini I, Salmona M, Bonetto V, Mengozzi M, Duffieux F, Miclet E, Bachi A, Vandekerckhove J, Gianazza E, Ghezzi P (2002) Identification by redox proteomics of glutathionylated proteins in oxidatively stressed human T lymphocytes. Proc Natl Acad Sci USA 99:3505–10
Fujisawa K, Murakami R, Horiguchi T, Noma T (2009) Adenylate kinase isozyme 2 is essential for growth and development of Drosophila melanogaster. Comp Biochem Physiol B Biochem Mol Biol 153:29–38
Fulvia GB, Antonio P, Anna N, Patrizia S, Ada A, Egidio B, Andrea M (2011) Adenylate kinase locus 1 polymorphism and feto-placental development. Eur J Obstet Gynecol Reprod Biol 159:273–75

Gan B, Hu J, Jiang S, Liu Y, Sahin E, Zhuang L, Fletcher-Sananikone E, Colla S, Wang YA, Chin L, Depinho RA (2010) Lkb1 regulates quiescence and metabolic homeostasis of haematopoietic stem cells. Nature 468:701–4

Gillette PC, Claycomb WC (1974) Thymidine kinase activity in cardiac muscle during embryomic and postnatal development. Biochem J 142:685–90

Ginger ML, Ngazoa ES, Pereira CA, Pullen TJ, Kabiri M, Becker K, Gull K, Steverding D (2005) Intracellular positioning of isoforms explains an unusually large adenylate kinase gene family in the parasite Trypanosoma brucei. J Biol Chem 280:11781–89

Gloria-Bottini F, Antonacci E, Cozzoli E, De Acetis C, Bottini E (2011) The effect of genetic variability on the correlation between blood glucose and glycated hemoglobin levels. Metabolism 60:250–5

Gloria-Bottini F, Nicotra M, Amante A, Pietropolli A, Neri A, Bottini E, Magrini A (2012) Adenylate kinase genetic polymorphism and spontaneous abortion. Am J Hum Biol 24:186–8

Gloria-Bottini F, Neri A, Pietropolli A, Bottini E, Magrini A (2013) Ak(1) genetic polymorphism and season of conception. Eur J Obstet Gynecol Reprod Biol 166:161–3

Gurumurthy S, Xie SZ, Alagesan B, Kim J, Yusuf RZ, Saez B, Tzatsos A, Ozsolak F, Milos P, Ferrari F, Park PJ, Shirihai OS, Scadden DT, Bardeesy N (2010) The Lkb1 metabolic sensor maintains haematopoietic stem cell survival. Nature 468:659–63

Hand NJ, Master ZR, Eauclaire SF, Weinblatt DE, Matthews RP, Friedman JR (2009) The microRNA-30 family is required for vertebrate hepatobiliary development. Gastroenterology 136:1081–90

Henzler-Wildman KA, Lei M, Thai V, Kerns SJ, Karplus M, Kern D (2007) A hierarchy of timescales in protein dynamics is linked to enzyme catalysis. Nature 450:913–6

Ho Y, Gruhler A, Heilbut A, Bader GD, Moore L, Adams SL, Millar A, Taylor P, Bennett K, Boutilier K, Yang L, Wolting C, Donaldson I, Schandorff S, Shewnarane J, Vo M, Taggart J, Goudreault M, Muskat B, Alfarano C, Dewar D, Lin Z, Michalickova K, Willems AR, Sassi H, Nielsen PA, Rasmussen KJ, Andersen JR, Johansen LE, Hansen LH, Jespersen H, Podtelejnikov A, Nielsen E, Crawford J, Poulsen V, Sorensen BD, Matthiesen J, Hendrickson RC, Gleeson F, Pawson T, Moran MF, Durocher D, Mann M, Hogue CW, Figeys D, Tyers M (2002) Systematic identification of protein complexes in Saccharomyces cerevisiae by mass spectrometry. Nature 415:180–3

Hu CJ, Iyer S, Sataur A, Covello KL, Chodosh LA, Simon MC (2006) Differential regulation of the transcriptional activities of hypoxia-inducible factor 1 alpha (HIF-1alpha) and HIF-2alpha in stem cells. Mol Cell Biol 26:3514–26

Inouye S, Seo M, Yamada Y, Nakazawa A (1998) Increase of adenylate kinase isozyme 1 protein during neuronal differentiation in mouse embryonal carcinoma P19 cells and in rat brain primary cultured cells. J Neurochem 71:125–33

Jan YH, Tsai HY, Yang CJ, Huang MS, Yang YF, Lai TC, Lee CH, Jeng YM, Huang CY, Su JL, Chuang YJ, Hsiao M (2012) Adenylate kinase-4 is a marker of poor clinical outcomes that promotes metastasis of lung cancer by downregulating the transcription factor ATF3. Cancer Res 72:5119–29

Jansen M, Ten Klooster JP, Offerhaus GJ, Clevers H (2009) LKB1 and AMPK family signaling: the intimate link between cell polarity and energy metabolism. Physiol Rev 89:777–98

Janssen E, Dzeja PP, Oerlemans F, Simonetti AW, Heerschap A, de Haan A, Rush PS, Terjung RR, Wieringa B, Terzic A (2000) Adenylate kinase 1 gene deletion disrupts muscle energetic economy despite metabolic rearrangement. EMBO J 19:6371–81

Janssen E, Kuiper J, Hodgson D, Zingman LV, Alekseev AE, Terzic A, Wieringa B (2004) Two structurally distinct and spatially compartmentalized adenylate kinases are expressed from the AK1 gene in mouse brain. Mol Cell Biochem 256–7:59–72

Juhnke H, Charizanis C, Latifi F, Krems B, Entian KD (2000) The essential protein fap7 is involved in the oxidative stress response of Saccharomyces cerevisiae. Mol Microbiol 35:936–48

Kim J, Shen R, Olcott MC, Rajagopal I, Mathews CK (2005) Adenylate kinase of Escherichia coli, a component of the phage T4 dNTP synthetase complex. J Biol Chem 280:28221–29

Klier H, Magdolen V, Schricker R, Strobel G, Lottspeich F, Bandlow W (1996) Cytoplasmic and mitochondrial forms of yeast adenylate kinase 2 are N-acetylated. Biochim Biophys Acta 1280:251–6

Kuehnel MP, Reiss M, Anand PK, Treede I, Holzer D, Hoffmann E, Klapperstueck M, Steinberg TH, Markwardt F, Griffiths G (2009) Sphingosine-1-phosphate receptors stimulate macrophage plasma-membrane actin assembly via ADP release, ATP synthesis and P2X7R activation. J Cell Sci 122:505–12

Lagresle-Peyrou C, Six EM, Picard C, Rieux-Laucat F, Michel V, Ditadi A, Demerens-de Chappedelaine C, Morillon E, Valensi F, Simon-Stoos KL, Mullikin JC, Noroski LM, Besse C, Wulffraat NM, Ferster A, Abecasis MM, Calvo F, Petit C, Candotti F, Abel L, Fischer A, Cavazzana-Calvo M (2009) Human adenylate kinase 2 deficiency causes a profound hematopoietic defect associated with sensorineural deafness. Nat Genet 41:106–11

Lin JF, Wu S, Huang SS, Lu BY, Lin SM, Tsai SK (2011) Resveratrol protects left ventricle by increasing adenylate kinase and isocitrate dehydrogenase activities in rats with myocardial infarction. Chin J Physiol 54:406–12

Liu R, Strom AL, Zhai J, Gal J, Bao S, Gong W, Zhu H (2009) Enzymatically inactive adenylate kinase 4 interacts with mitochondrial ADP/ATP translocase. Int J Biochem Cell Biol 41:1371–80

Liu Z, Yue S, Chen X, Kubin T, Braun T (2010) Regulation of cardiomyocyte polyploidy and multinucleation by CyclinG1. Circ Res 106:1498–1506

Malekkou A, Lederer CW, Lamond AI, Santama N (2010) The nuclear ATPase/adenylate kinase hCINAP is recruited to perinucleolar caps generated upon RNA pol.II inhibition. FEBS Lett 584:4559–64

Mandal S, Guptan P, Owusu-Ansah E, Banerjee U (2005) Mitochondrial regulation of cell cycle progression during development as revealed by the tenured mutation in Drosophila. Dev Cell 9:843–54

Mandal S, Freije WA, Guptan P, Banerjee U (2010) Metabolic control of G1-S transition: cyclin E degradation by p53-induced activation of the ubiquitin-proteasome system. J Cell Biol 188:473–9

Meng G, Zhai R, Liu B, Zheng X (2008) Identification of a novel nuclear-localized adenylate kinase from Drosophila melanogaster. Biochemistry (Mosc) 73:38–43

Miyoshi K, Akazawa Y, Horiguchi T, Noma T (2009) Localization of adenylate kinase 4 in mouse tissues. Acta Histochem Cytochem 42:55–64

Morettini S, Podhraski V, Lusser A (2008) ATP-dependent chromatin remodeling enzymes and their various roles in cell cycle control. Front Biosci 13:5522–32

Motoshima H, Goldstein BJ, Igata M, Araki E (2006) AMPK and cell proliferation–AMPK as a therapeutic target for atherosclerosis and cancer. J Physiol 574:63–71

Nakada D, Saunders TL, Morrison SJ (2010) Lkb1 regulates cell cycle and energy metabolism in haematopoietic stem cells. Nature 468:653–8

Narendra D, Walker JE, Youle R (2012) Mitochondrial quality control mediated by PINK1 and Parkin: links to parkinsonism. Cold Spring Harb Perspect Biol 4

Nemutlu E, Juranic N, Zhang S, Ward LE, Dutta T, Nair KS, Terzic A, Macura S, Dzeja PP (2012a) Electron spray ionization mass spectrometry and 2D (31)P NMR for monitoring (18) O/(16)O isotope exchange and turnover rates of metabolic oligophosphates. Anal Bioanal Chem 403(3):697–706

Nemutlu E, Zhang S, Gupta A, Juranic NO, Macura SI, Terzic A, Jahangir A, Dzeja P (2012b) Dynamic phosphometabolomic profiling of human tissues and transgenic models by 18O-assisted 31P NMR and mass spectrometry. Physiol Genomics 44(7):386–402

Noda L (1973) Adenylate kinase. In: Boyer P (ed) The enzymes, 3rd edn. Academic, New York, pp 279–305

Noma T (2005) Dynamics of nucleotide metabolism as a supporter of life phenomena. J Med Invest 52:127–36

Ottaway JH, Mowbray J (1977) The role of compartmentation in the control of glycolysis. Curr Top Cell Regul 12:107–208

Panayiotou C, Solaroli N, Johansson M, Karlsson A (2010) Evidence of an intact N-terminal translocation sequence of human mitochondrial adenylate kinase 4. Int J Biochem Cell Biol 42:62–9

Panayiotou C, Solaroli N, Xu Y, Johansson M, Karlsson A (2011) The characterization of human adenylate kinases 7 and 8 demonstrates differences in kinetic parameters and structural organization among the family of adenylate kinase isoenzymes. Biochem J 433:527–34

Pannicke U, Honig M, Hess I, Friesen C, Holzmann K, Rump EM, Barth TF, Rojewski MT, Schulz A, Boehm T, Friedrich W, Schwarz K (2009) Reticular dysgenesis (aleukocytosis) is caused by mutations in the gene encoding mitochondrial adenylate kinase 2. Nat Genet 41:101–5

Park H, Kam TI, Kim Y, Choi H, Gwon Y, Kim C, Koh JY, Jung YK (2012) Neuropathogenic role of adenylate kinase-1 in Abeta-mediated tau phosphorylation via AMPK and GSK3beta. Hum Mol Genet 21:2725–37

Peng X, Wang L, Chen G, Wang X (2012) Dynamic expression of adenylate kinase 2 in the hippocampus of pilocarpine model rats. J Mol Neurosci 47:150–7

Ptacin JL, Shapiro L (2013) Chromosome architecture is a key element of bacterial cellular organization. Cell Microbiol 15:45–52

Pucar D, Janssen E, Dzeja PP, Juranic N, Macura S, Wieringa B, Terzic A (2000) Compromised energetics in the adenylate kinase AK1 gene knockout heart under metabolic stress. J Biol Chem 275:41424–29

Pucar D, Bast P, Gumina RJ, Lim L, Drahl C, Juranic N, Macura S, Janssen E, Wieringa B, Terzic A, Dzeja PP (2002) Adenylate kinase AK1 knockout heart: energetics and functional performance under ischemia-reperfusion. Am J Physiol Heart Circ Physiol 283:H776–H782

Rahlfs S, Koncarevic S, Iozef R, Mailu BM, Savvides SN, Schirmer RH, Becker K (2009) Myristoylated adenylate kinase-2 of Plasmodium falciparum forms a heterodimer with myristoyltransferase. Mol Biochem Parasitol 163:77–84

Randak CO, Ver Heul AR, Welsh MJ (2012) Demonstration of phosphoryl group transfer indicates that the ATP-binding cassette (ABC) transporter cystic fibrosis transmembrane conductance regulator (CFTR) exhibits adenylate kinase activity. J Biol Chem 287:36105–10

Ren H, Wang L, Bennett M, Liang Y, Zheng X, Lu F, Li L, Nan J, Luo M, Eriksson S, Zhang C, Su XD (2005) The crystal structure of human adenylate kinase 6: an adenylate kinase localized to the cell nucleus. Proc Natl Acad Sci USA 102:303–8

Romito A, Lonardo E, Roma G, Minchiotti G, Ballabio A, Cobellis G (2010) Lack of sik1 in mouse embryonic stem cells impairs cardiomyogenesis by down-regulating the cyclin-dependent kinase inhibitor p57kip2. PLoS One 5:e9029

Rosenfeld SS, van Duffelen M, Behnke-Parks WM, Beadle C, Corrreia J, Xing J (2009) The ATPase cycle of the mitotic motor CENP-E. J Biol Chem 284:32858–868

Ruan Q, Chen Y, Gratton E, Glaser M, Mantulin WW (2002) Cellular characterization of adenylate kinase and its isoform: two-photon excitation fluorescence imaging and fluorescence correlation spectroscopy. Biophys J 83:3177–87

Rubart M, Field LJ (2006) Cardiac regeneration: repopulating the heart. Annu Rev Physiol 68:29–49

Santama N, Ogg SC, Malekkou A, Zographos SE, Weis K, Lamond AI (2005) Characterization of hCINAP, a novel coilin-interacting protein encoded by a transcript from the transcription factor TAFIID32 locus. J Biol Chem 280:36429–441

Seccia TM, Atlante A, Vulpis V, Marra E, Passarella S, Pirrelli A (1998) Mitochondrial energy metabolism in the left ventricular tissue of spontaneously hypertensive rats: abnormalities in both adeninenucleotide and phosphate translocators and enzyme adenylate-kinase and creatine-phosphokinase activities. Clin Exp Hypertens 20:345–58

Semenza GL (2000) HIF-1 and human disease: one highly involved factor. Genes Dev 14:1983–91
Seong IS, Ivanova E, Lee JM, Choo YS, Fossale E, Anderson M, Gusella JF, Laramie JM, Myers RH, Lesort M, MacDonald ME (2005) HD CAG repeat implicates a dominant property of huntingtin in mitochondrial energy metabolism. Hum Mol Genet 14:2871–80
Shi Q, Feng J, Qu H, Cheng YY (2008) A proteomic study of S-nitrosylation in the rat cardiac proteins in vitro. Biol Pharm Bull 31:1536–40
Solaroli N, Panayiotou C, Johansson M, Karlsson A (2009) Identification of two active functional domains of human adenylate kinase 5. FEBS Lett 583:2872–76
Stanojevic V, Habener JF, Holz GG, Leech CA (2008) Cytosolic adenylate kinases regulate K-ATP channel activity in human beta-cells. Biochem Biophys Res Commun 368:614–9
Strobel G, Zollner A, Angermayr M, Bandlow W (2002) Competition of spontaneous protein folding and mitochondrial import causes dual subcellular location of major adenylate kinase. Mol Biol Cell 13:1439–48
Szappanos B, Kovacs K, Szamecz B, Honti F, Costanzo M, Baryshnikova A, Gelius-Dietrich G, Lercher MJ, Jelasity M, Myers CL, Andrews BJ, Boone C, Oliver SG, Pal C, Papp B (2011) An integrated approach to characterize genetic interaction networks in yeast metabolism. Nat Genet 43:656–62
Tarassov K, Messier V, Landry CR, Radinovic S, Serna Molina MM, Shames I, Malitskaya Y, Vogel J, Bussey H, Michnick SW (2008) An in vivo map of the yeast protein interactome. Science 320:1465–70
Tuzun E, Rossi JE, Karner SF, Centurion AF, Dalmau J (2007) Adenylate kinase 5 autoimmunity in treatment refractory limbic encephalitis. J Neuroimmunol 186:177–80
van Horssen R, Janssen E, Peters W, van de Pasch L, Lindert MM, van Dommelen MM, Linssen PC, Hagen TL, Fransen JA, Wieringa B (2009) Modulation of cell motility by spatial repositioning of enzymatic ATP/ADP exchange capacity. J Biol Chem 284:1620–27
Van Rompay AR, Johansson M, Karlsson A (1999) Identification of a novel human adenylate kinase. cDNA cloning, expression analysis, chromosome localization and characterization of the recombinant protein. Eur J Biochem 261:509–17
Vasseur S, Malicet C, Calvo EL, Dagorn JC, Iovanna JL (2005) Gene expression profiling of tumours derived from rasV12/E1A-transformed mouse embryonic fibroblasts to identify genes required for tumour development. Mol Cancer 4:4
Vogel P, Read RW, Hansen GM, Payne BJ, Small D, Sands AT, Zambrowicz BP (2012) Congenital hydrocephalus in genetically engineered mice. Vet Pathol 49:166–81
Walker EJ, Dow JW (1982) Location and properties of two isoenzymes of cardiac adenylate kinase. Biochem J 203:361–9
Walsh S, Ponten A, Fleischmann BK, Jovinge S (2010) Cardiomyocyte cell cycle control and growth estimation in vivo–an analysis based on cardiomyocyte nuclei. Cardiovasc Res 86:365–73
Wirschell M, Pazour G, Yoda A, Hirono M, Kamiya R, Witman GB (2004) Oda5p, a novel axonemal protein required for assembly of the outer dynein arm and an associated adenylate kinase. Mol Biol Cell 15:2729–41
Woulfe KC, Gao E, Lal H, Harris D, Fan Q, Vagnozzi R, DeCaul M, Shang X, Patel S, Woodgett JR, Force T, Zhou J (2010) Glycogen synthase kinase-3beta regulates post-myocardial infarction remodeling and stress-induced cardiomyocyte proliferation in vivo. Circ Res 106:1635–45
Yoneda T, Sato M, Maeda M, Takagi H (1998) Identification of a novel adenylate kinase system in the brain: cloning of the fourth adenylate kinase. Brain Res Mol Brain Res 62:187–95
Zhai R, Meng G, Zhao Y, Liu B, Zhang G, Zheng X (2006) A novel nuclear-localized protein with special adenylate kinase properties from *Caenorhabditis elegans*. FEBS Lett 580:3811–17
Zhang J, Zhang F, Zheng X (2010a) Depletion of hCINAP by RNA interference causes defects in Cajal body formation, histone transcription, and cell viability. Cell Mol Life Sci 67:1907–18

Zhang S, Nemutlu E, Dzeja P (2010b) Metabolomic profiling of adenylate kinase AK1−/− and AK2+/− transgenic mice: effect of physical stress. Circulation 122, A20435

Zhang S, Nemutlu E, Terzic A, Dzeja P (2012) Adenylate kinase phosphotransfer and AMP signaling regulate cardiomyocyte cell cycle and heart regenerative capacity. Circulation 126, A18852

Part III
Systems Biology of Cellular Structures and Fluxes

Chapter 7
Moonlighting Function of the Tubulin Cytoskeleton: Macromolecular Architectures in the Cytoplasm

Judit Ovádi and Vic Norris

Abstract Cells face the enormous challenge of generating a single phenotype that must be coherent with myriad internal and external conditions. For such phenotypes to have multifarious but meaningful outputs entails the sensing, and integration of a wide variety of chemical and physical information, hence the coordination of metabolic and signaling processes. This sensing, integration, and coordination are carried out by the complex ultrastructural arrays and moonlighting functions of the cytoskeletal network. In the cellular context, the direction and potency of sensing are determined by the structure-related responses of the cytoskeletal network to the activity of individual macromolecules in conjunction with associated metabolites and nucleotides. These responses comprise the binding (hetero-association) of these macromolecules to the cytoskeleton and the consequences of this binding on the behavior of both partners, among them the stability and dynamics of the cytoskeleton, and the catalytic and regulatory properties of the individual proteins (and/or their specific complexes). The latter is of specific importance in regulation at a high level of organization via the formation of microcompartments in linear pathways or at metabolic crossroads. In addition, key players in many metabolic and signaling pathways are nucleotides such as ATP and GTP that have a crucial role in cytoskeleton-mediated events. These issues are illustrated with examples, and the sensing power of dynamic macromolecular associations is discussed.

J. Ovádi (✉)
Institute of Enzymology, Research Centre for Natural Sciences, Hungarian Academy of Sciences, Karolina út 29, Budapest 1113, Hungary
e-mail: ovadi@enzim.hu

V. Norris
Faculty of Science, University of Rouen, Mont Saint Aignan Cedex, France

M.A. Aon et al. (eds.), *Systems Biology of Metabolic and Signaling Networks*, Springer Series in Biophysics 16, DOI 10.1007/978-3-642-38505-6_7,

7.1 Macromolecular Associations

Cell compartments are crowded with solutes, soluble macromolecules such as enzymes, nucleic acids, and structural proteins, and membranes to create complex structures that continue to be discovered (Fridman et al. 2012). The high protein density within the large compartments of the cells predominantly determines major characteristics of cellular environment such as viscosity, diffusion, and heterogeneity. The fact that the solvent viscosity of cytoplasm is not substantially different from that of water is explained by intracellular structural heterogeneity: macromolecular density is relatively low within the interstitial voids in the cell because many soluble enzymes are apparently integral parts of the insoluble cytomatrix and/or cytoskeletal ultrastructures. In other words, macromolecules are not distributed homogeneously and the cell should not be considered as a "bag." Indeed, most of the cytoplasmic proteins have nondiffusing forms or diffuse on the timescale of hours as evidenced by high voltage electron microscopy [(Ovadi and Saks 2004) and refs therein].

The eukaryotic cytoskeleton, which consists of three filamentous systems, has many physiological and pathological functions. The dynamic reorganizing ability of the filamentous structures is highly variable in different cells and tissues; the precise regulation of microtubular dynamics is critical for cell cycle progression, cell signaling, intracellular transport, cell polarization, and organismal development. The dynamic instability of microtubules in living cells has been analyzed by microscopy of microinjected or expressed fluorescent tubulin, time-lapse microscopy, and analysis of time-dependent, microtubule length changes (Kamath et al. 2010). The multiple functions of these superstructures are achieved by the static and dynamic associations of macromolecules and ligands, and by posttranslational modifications. In fact, macromolecular associations resulting from crowding (Swaminathan et al. 1997) create intracellular superstructures such as the microtrabecular lattice which is formed by the organization of the microtubule systems while the soluble enzymes are integral parts of the cytomatrix, a cytoplasmic association of enzymes (Minton and Wilf 1981; Porter et al. 1983). The full range of ultrastructures and functions characteristic of the cytoskeleton and, in particular, the microtubule system has only been found so far in eukaryotes. However, different types of filamentous structures have been identified in prokaryotes (Ingerson-Mahar and Gitai 2012).

One of the best understood of prokaryotic filaments is the tubulin-like FtsZ protein. This protein assembles into a ring-like structure, the Z-ring, at the site of cell division (Errington et al. 2003; Romberg and Levin 2003). FtsZ can also assemble into helices, of unknown function, that are highly dynamic (Thanedar and Margolin 2004). The FtsZ protein has a structural homology to tubulin (Lowe and Amos 1998). In vitro, FtsZ can form a wide variety of polymeric structures, some of which may resemble eukaryotic microtubules, depending on the presence and concentrations of lipids, divalent ions, GTP, and other proteins (Alexandre et al. 2002; Gonzalez et al. 2005; Popp et al. 2010; Gundogdu et al. 2011; Hsin et al. 2012; Martin-Garcia et al. 2012; Hou et al. 2012). As in the case of eukaryotic tubulin, the assembly of FtsZ into filaments can be mediated by accessory proteins such as ZipA. This prokaryotic protein in particular is considered to resemble typical MAPs (Amos et al. 2004). In addition, direct interaction between FtsZ and enzymes involved in glucose metabolism the glucosyltransferase UgtP occurs when the intracellular concentration of UDP-glucose is high and leads to a partial inhibition of division that is thought to constitute metabolic sensing (Weart et al. 2007). Direct interaction between FtsZ and thioredoxin, which helps maintain the intracellular redox potential (along with its moonlighting activities), may also permit metabolic sensing (Kumar et al. 2004); such interaction occurs in chloroplasts too (Balmer et al. 2003). The bacterial actin, MreB, which also interacts with thioredoxin, forms a helix that changes its location depending on FtsZ (Figge et al. 2004). Interactions also occur during cell division between FtsZ and a dozen other proteins involved, for example, in the synthesis of lipids and peptidoglycan (Norris et al. 2007). Intriguingly, production in a bacterium of S100B, a human protein that undergoes a calcium-dependent conformational change to bind to tubulin, results in its colocalization with FtsZ and inhibition of cell division (Ferguson and Shaw 2004).

Specificity is an important criterion in evaluating the physiological relevance of enzyme interactions; specific recognition of one protein by another is based on their surface complementarity. This recognition may depend on the conformational state of both partners as influenced by ligands and by additional macromolecules. The ensemble of these effects on interactions determines the functions of both the microtubule system and the proteins/enzymes involved in the metabolic and signaling pathways.

The associations of the glycolytic kinases with the microtubular system are oligomeric and isoform specific (Partikian et al. 1998), e.g., in the case of the dissociable muscle PFK isoform, the inactive dimeric form of the enzyme binds to MTs whilst in the case of brain the active tetrameric form does not bind (Verkman 2002) (Fig. 7.1). Some of these interactions are modulated by key metabolites and nucleotides, like fructose-phosphates and ATP (Wágner et al. 2001) indicating that the intra- and intermolecular forces between enzyme subunits and enzyme/MT, respectively, might control the dynamism and superstructure of the microtubular network in neuronal or mitotic cells. Such dynamic macromolecular associations could constitute an innovative control mechanism for energy production via glycolysis. In addition, the involvement of these associations in many physiological functions of the cell makes it likely that when they are perturbed pathological, e.g., neurodegenerative, processes result.

7.2 Moonlighting Proteins Display Multiple Functions

Decoding the human genome has made it clear that the straightforward *"one gene—one enzyme"* law of classical molecular biology is inadequate since many more proteins have been identified than protein-coding sequences which account for only

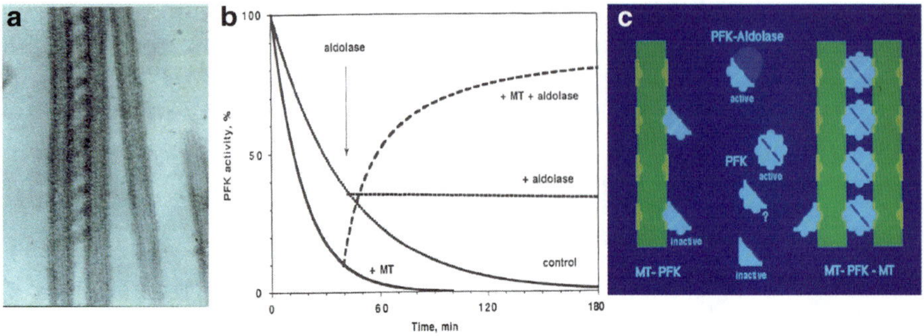

Fig. 7.1 PFK and MT: ultrastructure-dependent functions. (**a**): Periodical cross-linking of MTs by tetrameric PFK formed by dimer–dimer association of PFK bound primarily to MTs. (**b**): the addition of aldolase to the partial dissociated PFK prevents further dilution-induced dissociation, thus the inactivation of PFK. (**c**): Schematic presentation of multiple interactions in MT–PFK–aldolase system

a very small fraction of the genome (http://www.ornl.gov/sci/techresources/Human_Genome/faq/genenumber.shtml).

The molecular functions of many of the products of these genes have yet to be assigned; in addition, the network of all macromolecular interactions at gene and protein levels is far from being worked out. It requires the integration of many different kinds of experimental and computational methods. A functional genomic strategy using metabolome data revealed the phenotypes conferred by silent mutations (M/2) and the complex mechanisms leading from a single gene mutation to a highly variable phenotype (Raamsdonk et al. 2001).

The existence of moonlighting proteins is generally acknowledged. Their multiple functions do not result from gene fusions, splice variants, or posttranslational modifications (Jeffery 2003) but rather from the same protein having different functions depending on its location in a particular compartment, on its oligomerization state, or on its interaction with other molecular partners (Fig. 7.2). These features are not coded in the genome sequences, thus making the prediction of the moonlighting characteristics of gene products almost impossible. This has not precluded data on moonlighting proteins accumulating and being categorized (Sriram et al. 2005).

An excellent example of a classic metabolic enzyme displaying moonlighting characteristics is glyceraldehyde-3-phosphate dehydrogenase (GAPDH). This glycolytic enzyme has distinct functions that depend on its oligomerization state and location within the cell: the tetramer catalyzes the conversion of 3-phosphoglyceraldehyde in the cytosol, whereas the monomer catalyzes the release of uracil from DNA in the nucleus (Meyer-Siegler et al. 1991). The oligomerization state of this enzyme is modulated by the cellular concentrations of adenosine triphosphate (ATP) and nicotinamide-adenine-dinucleotide (NAD^+), and by its interactions with potential partners such as function-related glycolytic enzymes (Olah et al. 2008). The truncated toxic polypeptide derived from the β-amyloid

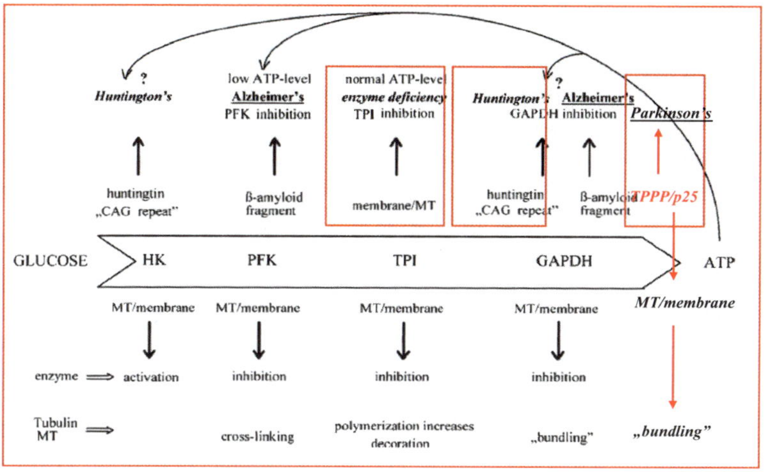

Fig. 7.2 Connection between glycolysis and neurodegenerative disorders: interaction of glycolytic enzymes attached to microtubule/membrane with pathological proteins resulting in mutual functional effects (Modified Fig. 5 from Ovadi et al. 2004)

precursor protein affects the functions of the dehydrogenase (Mazzola and Sirover 2003) (cf. Fig. 7.2). Cytoskeletal filaments are also targets of GAPDH and involve microfilaments or microtubules depending on the cell type (Ovadi and Srere 2000). GAPDH also associates with nucleic acids in the nucleus as well as in the cytoplasm. In summary, the different functions of GAPDH can be switched on by multiple factors so as to activate apparently unrelated pathways (Jeffery 1999).

Unlike classical cases in which moonlighting proteins display distinct physiological functions, there are situations that involve their functions being perverted, a pathological switch that may put these proteins into a specific subset of moonlighting proteins termed *neomoonlighting* proteins (Jeffery 2011). This pathological switch is based on different mechanisms and is characteristic of structurally unfolded (disordered) proteins with extensive capacity to adopt different conformations upon binding distinct partners.

Many of the neomoonlighting proteins do not have stable, well-defined 3D structures; consequently, they are liable to form pathological ultrastructures, such as fibrils, oligomers, or aggregates with distinct toxicity. Although the molecular events that initiate formation of these ultrastructures, such as aberrant protein–protein interactions, are similar in many "disordered diseases," the outcome as the clinical symptoms can be very different (Ovadi and Orosz 2009). The conformational instability of the neomoonlighting proteins can contribute extensively to the complex phenotype of a given disorder. Thus, a full characterization of the proteome is essential for understanding the pathological mechanisms of human diseases.

An established example is the Tubulin Polymerization Promoting Protein, TPPP/p25, a disordered protein that is abundant in the pathological human brain

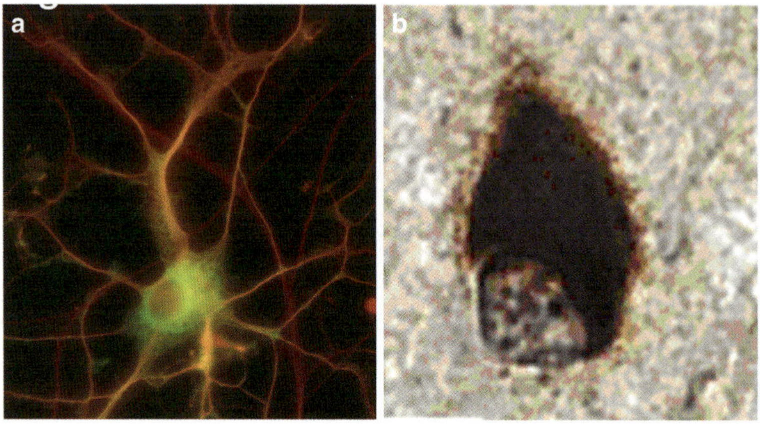

Fig. 7.3 *Moonlighting* function of TPPP/p25. (**a**): colocalization of EGFP-TPPP/p25 with the microtubule network (*red*) detected by immunofluorescence; (**b**): co-enrichment of TPPP/p25 and α-synuclein in human pathological inclusion visualized by immunohistochemistry of human pathological brain tissue for α-synuclein

tissue characteristic of synucleinopathies (Kovacs et al. 2004). This protein is primarily expressed in oligodendrocytes where the tubulin polymerization promoting and microtubule bundling activity of TPPP/p25 are crucial for the development of projections in the course of differentiation of the progenitor cells leading to myelinization, requiring the ensheathment of axons for their normal function in the central nervous system (CNS) (Fig. 7.3). However, its function is entirely different when it is co-localized and concentrated in neurons or glia cells with α-synuclein forming pathological inclusions such as Lewy bodies (Ovadi 2011).

Moonlighting proteins display autonomous, markedly different, functions such as structural and catalytic functions that are sometimes unrelated. Their interactions are particularly characteristic of disordered proteins. One of their characteristic features is that they use regions outside the active site for other functions, such as regulatory and structural functions (Khersonsky and Tawfik 2010). Moonlighting proteins are present in diverse organisms (Huberts and van der Klei 2010), and the implications of their different functions should be considered when trying to explain the multiple symptoms of single-gene disorders or to predict the consequences of metabolic engineering (Flores and Gancedo 2011)

Diseases resulting from disordered proteins require unconventional approaches that do not rely on gene knockouts and that lead to medicines without toxic side effects. Therefore, certain moonlighting proteins, such as those involved in macromolecular ultrastructures and in metabolic and signaling pathways, are potentially attractive targets for drug therapies. This is because it may be possible to find drugs that inhibit their pathological but not their physiological functions hence with very limited, toxic effects.

7.3 Sensor Potency of Enzyme-Decorated Microtubules

Cells are confronted with the challenge of generating a single phenotype that must be coherent with a huge number of combinations of internal and external conditions. Generating the meaningful outputs required for these phenotypes entails sensing and integrating a wide diversity of chemical and physical information, and hence the coordination of metabolic and signaling processes. We have proposed that this sensing, integration, and coordination is achieved by the complex structures and moonlighting functions of the cytoskeletal network (Norris et al. 2010; Norris et al. 2013).

In this proposal, the direction and potency of the sensing would be determined by the structure-related changes of the cytoskeletal network in response to the activity of individual macromolecules—along with associated metabolites and nucleotides—inside the living cell (Fig. 7.4). These responses comprise the binding of these macromolecules to the cytoskeleton and the consequences of this binding on the behavior of both partners, i.e., cytoskeleton dynamics and the catalytic and regulatory properties of the individual proteins (and/or their specific complexes). The binding of macromolecules to the cytoskeleton is of specific importance in the regulation at a higher level via the formation of microcompartments in linear pathways or at metabolic crossroads. In addition, nucleotides such as ATP and GTP, which can both influence and respond to cytoskeleton-mediated events, play key roles in many metabolic and signaling pathways.

If the binding by the cytoskeleton of an enzyme increases its activity, it may well be that an enzyme that is catalytically active has a higher probability of binding to the cytoskeleton. Microtubule binding to glycolytic enzymes is known to alter the catalytic and regulatory properties of several enzymes [see Table 1 in Ovadi et al. (2004)]. Such binding increases HK activity to enhance the glycolytic flux in brain tissue (although this does not influence MT dynamics and structure). MT binding to PFK decreases the activity of the enzyme and results in a periodic cross-linking of the MTs. (cf. Fig. 7.1). MT binding to PK impedes MT assembly (but does not influence the activity of the enzyme). Moreover, the binding by MTs of individual enzymes is influenced by enzyme–enzyme interactions. For example, the assembly of PFK into an aldolase–PFK complex (where it is no longer subject to allosteric regulation) counters the association of PFK with MTs (Ovadi et al. 2004).

A two-way relationship exists between cytoskeleton-enzyme association and nucleoside triphosphate levels in which (1) the changing dynamics of the cytoskeletal filaments that result from enzyme association modifies nucleoside triphosphate levels and (2) changing nucleoside triphosphate levels modify the dynamics and hydrolytic activity of the cytoskeleton. For example, the efficiency of MT treadmilling depends on GTP concentration whilst MT dynamics depends on MT motor proteins (like Kin-I kinesins), which hydrolyze ATP (Moore and Wordeman 2004). Reciprocally, there is some evidence that cytoskeletal dynamics affects the levels of the nucleoside triphosphates. GTP hydrolysis, for example, depends on tubulin polymerization and is stimulated by MAPs (Sloboda et al. 1976). Moreover,

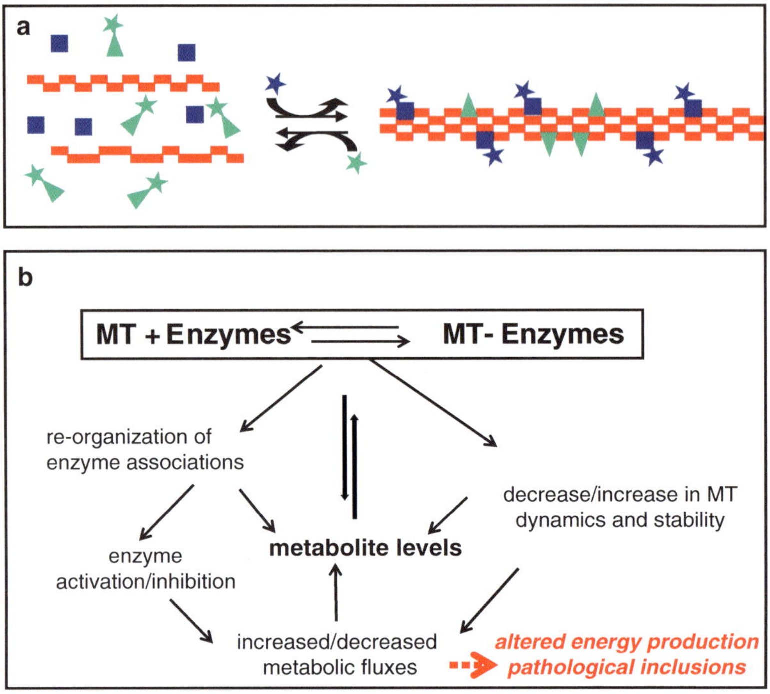

Fig. 7.4 (**a**) Microtubule (*red rectangles*) ultrastructures exist in dynamic equilibrium resulting from the association with enzymes (*green triangles* and *blue squares*). Microtubular systems are stabilized when these enzymes are either active in catalyzing their cognate reactions (*presence of star*) or when these enzymes are inactive (*absence of star*). (**b**) The network of interactions involved in sensing changes in levels of metabolites by the microtubular cytoskeleton. The *black arrows* represent the flow of information through the network following a change in metabolite levels. The *dotted red arrow* represents the possible pathological consequences of perturbations in the network

GTP impedes the association of the MAP TPPP/p25 with the MTs in the mitotic spindle to arrest mitosis (Tirian et al. 2003).

7.4 Enzyme Associations in Energy Production

Cells consist of a large number of nonlinearly interacting constituents exhibiting complex behavior during the course of their intra- and intermolecular communications. Systems Biology is based on a systemic approach to the analysis of the structural and functional interactions between components rather than on the analysis of the individual components themselves. Data gathered from the measured and computed fluxes through a pathway can be integrated and interpreted to obtain understanding at the system level. A typical example of a system-level concept is the metabolite channeling occurring in the microcompartments that comprise enzymes catalyzing sequential steps in a pathway. This type of analysis has been performed both theoretically and experimentally, mainly in vitro (Ovadi et al. 2004).

The supramolecular organization of metabolic enzymes and their interactions with one another and with subcellular structures constitute the basis for cellular microcompartmentation. Compartmentalized metabolic pathways or segment(s) of these pathways can overcome diffusive barriers within the crowded intracellular milieu since metabolism can successfully proceed and even be facilitated by metabolite channeling. In such channeling a direct transfer of intermediates from one enzyme to an adjacent enzyme happens without the need for free aqueous-phase diffusion. The enhanced probability of intermediates to be transferred from the active site of one enzyme to the active site of the following enzyme in the pathway requires stable or transient interactions between these enzymes. The structurally organized assembly of enzymes associated physically in a non-dissociable, static multienzyme complex would constitute a metabolon. Such metabolons, containing enzymes of a part or a whole metabolic pathway, might therefore be fundamental units in the control of these pathways. The formation of microcompartments is particularly important at metabolite crossroads where the association of the enzymes involved in one or other pathway determines the direction of the flux (Ovadi 1991). Thus, the supramolecular organization of the competing enzymes can control the behavior of metabolic systems by providing distributed control.

The control of the energy metabolism (e.g., ATP synthesis) is tightly coupled with several metabolic and signaling pathways [see, for example (Rostovtseva and Bezrukov 2012)], however, their exact relationship is unclear. Even less information is available on the interconnections between the pathological ultrastructures and energy (ATP) production at the molecular level.

Glucose is the major energy source in brain and is metabolized via glycolysis in the cytoplasm coupled with oxidation via the Krebs cycle in the mitochondrion synthesizing ATP as the key fuel for many metabolic and signaling pathways. The polarization state of the mitochondrial membrane is related to the energy state of the mitochondria which can be monitored in living cells by fluorescence microscopy using tetramethylrhodamine ethyl ester staining (Lehotzky et al. 2004). For example, K4 cells stably expressing EGFP-TPPP/p25 showed strikingly high fluorescence intensity as compared with control neuroblastoma (SK-N-MC) cells indicating that the expression of TPPP/p25 did not cause energy impairment but actually enhanced the membrane potential (Fig. 7.5). Consistent with this, the ATP concentration was found to be higher in the extract of the K4 cells as compared with that of control cells. Flux analysis of glucose metabolism in these cells revealed that the enhanced ATP concentration resulted in increased energy state in K4 cells (Orosz et al. 2004). The cellular, biochemical, and computation results suggest that the expression of the neomoonlighting TPPP/p25 protein in K4 cells is controlled by an unknown mechanism that maintains the amount of this protein below toxic levels (as revealed by the existence of a stable cell line); these results also suggest that its expression is coupled with increased energy production.

Mitochondrial impairment has been reported in the case of Huntington's disease (HD), a progressive neurodegenerative disorder caused by the insertion of a long

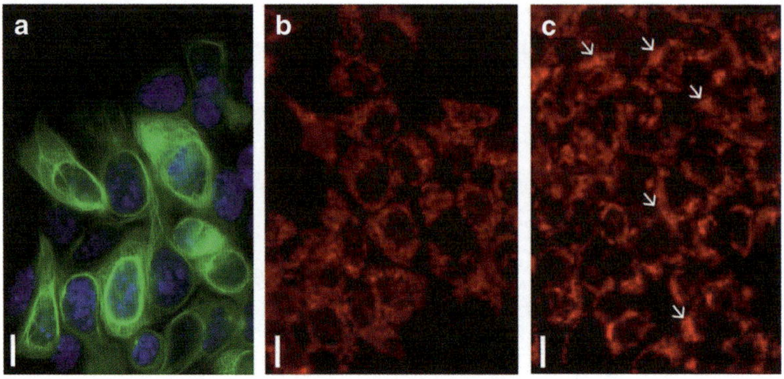

Fig. 7.5 Microscopic and kinetic studies of energy metabolism in K4 cells. (**a**) *Fluorescence* image of K4 cells expressing EGFP-TPPP/p25 fusion protein; nuclei are stained with DAPI. (**b** and **c**) Phase-contrast images of SK-N-MC (mother) cells (**b**) and K4 cells (*stable clone*) stained by tetramethylrhodamine ethyl ester for visualization of hyperpolarization of mitochondrial membrane (*red*). (Modified Fig.5 from Orosz et al. 2004)

CAG trinucleotide repeat into the gene encoding the N-terminal segment of the huntingtin protein. This polyglutamine tail is sticky and binds to GAPDH, a moonlighting glycolytic enzyme; this binding is probably responsible for the partial inactivation of this dehydrogenase as detected in the inclusion-containing region of transgene mice. In fact, energy metabolism in the disease produced by this mutation affects brain regions (as determined by immunohistochemistry) through changes in the ATP-producing systems of the cytosolic and mitochondrial compartments. In contrast to most of the speculation in the literature, our results showed that the neuronal damage in HD tissue was associated with increased energy metabolism at the tissue level leading to increased ATP concentration (Olah et al. 2008). Moreover, an increased conversion of glucose into lactate occurred in cytosolic extracts from the HD brain tissue. A mathematical model of the glycolytic pathway using the measured kinetic parameters of the individual enzymes and the well-established rate equations could simulate glycolytic fluxes in control and affected tissue of the transgene mice (Fig. 7.6). Data analysis from these comparative studies suggested a mechanism that might account for the observed effects. In the case of HD, GAPDH may be in closer proximity (perhaps because of binding of the enzyme to huntingtin) to aldolase thereby facilitating channeling between these enzymes. Hence, the association between these proteins would result in both an increased energy metabolism and the formation of pathological inclusions leading to neuronal damage.

The multifactorial character of molecular-conformational related diseases such as in the interrelationship between the etiology of metabolic and neurological disorders has been elucidated. Triosephosphate isomerase deficiency is an autosomal recessive multisystemic genetic disease coupled with hemolytic anemia and neurological disorder frequently leading to death in early childhood. Recent data

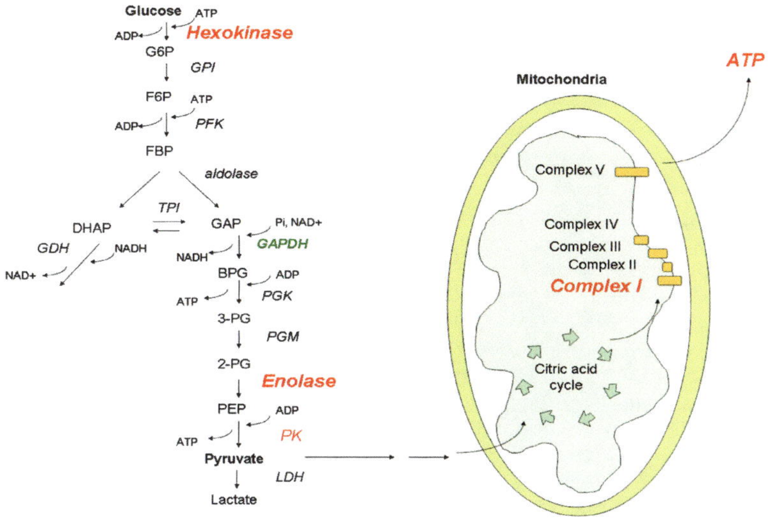

Fig. 7.6 Energy metabolism in HD (n171-82q) transgenic mouse. Activation of hexokinase, enolase, and pyruvate kinase (*PK*); inhibition of GAPDH; increased glycolytic flux; no alteration in mitochondrial complexes

suggest that it may not be a metabolic disease but, rather, a conformational one. In fact, various genetic mutations that alter this enzyme have been identified; these mutations result in decrease in the catalytic activity and likely in its oligomerization state. However, the mutations do not affect energy metabolism but result in accumulation of dihydroxyacetone phosphate followed by its chemical conversion into the toxic methylglyoxal. This results in the formation of advanced glycation end products and aberrant protein–protein interactions. Such interactions result in proteins sticking to microtubules and in the formation of aggregation-prone proteins, a typical trait of molecular conformational disorders.

7.5 Concluding Remark

In the cytoskeletal sensor hypothesis it has been proposed recently that the cytoskeleton senses and integrates the general metabolic activity of the cell (Norris et al. 2013). This potency of the cytoskeleton is related to the association of metabolic enzymes with the cytoskeleton which results in the mutual ultrastructural and functional changes presented in this chapter. These changes could be influenced by the nature of the nucleotides and intermediate metabolites. The evaluation by mathematical modeling of these complex processes at the molecular level, including MT dynamics coupled with ligand-mediated enzyme associations, will contribute enormously to our understanding the moonlighting functions of the enzyme-cytoskeleton ensemble. Work on this hypothesis is in progress in our laboratory.

Acknowledgments This work was supported by the European Commission (DCI-ALA/19.09.01/10/21526/245-297/ALFA 111(2010)29), the Hungarian National Scientific Research Fund Grants OTKA (T-101039), and Richter Gedeon Research Grant (RG-IPI-2011/TP5-001) to J. Ovádi.

References

Alexandre S, Cole G, Coutard S, Monnier C, Norris V, Margolin W, Yu X, Valleton J-M (2002) Interaction of FtsZ protein with a DPPE film. Colloids and Surfaces B: Biointerfaces 23:391–5

Amos LA, van den Ent F, Lowe J (2004) Structural/functional homology between the bacterial and eukaryotic cytoskeletons. Curr Opin Cell Biol 16:24–31

Balmer Y, Koller A, del Val G, Manieri W, Schurmann P, Buchanan BB (2003) Proteomics gives insight into the regulatory function of chloroplast thioredoxins. Proc Natl Acad Sci U S A 100(1): 370–375. doi:10.1073/pnas.232703799, 232703799 [pii]

Errington J, Daniel RA, Scheffers DJ (2003) Cytokinesis in bacteria. Microbiol Mol Biol Rev 67: 52–65

Ferguson PL, Shaw GS (2004) Human S100B protein interacts with the *Escherichia coli* division protein FtsZ in a calcium-sensitive manner. J Biol Chem 279:18806–13

Figge RM, Divakaruni AV, Gober JW (2004) MreB, the cell shape-determining bacterial actin homologue, coordinates cell wall morphogenesis in *Caulobacter crescentus*. Mol Microbiol 51:1321–32

Flores CL, Gancedo C (2011) Unraveling moonlighting functions with yeasts. IUBMB Life 63(7): 457–62. doi:10.1002/iub.454

Fridman K, Mader A, Zwerger M, Elia N, Medalia O (2012) Advances in tomography: probing the molecular architecture of cells. Nat Rev 13(11):736–42, nrm3453 [pii] 10.1038/nrm3453

Gonzalez JM, Velez M, Jimenez M, Alfonso C, Schuck P, Mingorance J, Vicente M, Minton AP, Rivas G (2005) Cooperative behavior of *Escherichia coli* cell-division protein FtsZ assembly involves the preferential cyclization of long single-stranded fibrils. Proc Natl Acad Sci USA 102:1895–900

Gundogdu ME, Kawai Y, Pavlendova N, Ogasawara N, Errington J, Scheffers DJ, Hamoen LW (2011) Large ring polymers align FtsZ polymers for normal septum formation. EMBO J 30 (3):617–26. doi:10.1038/emboj.2010.345, emboj2010345 [pii]

Hou S, Wieczorek SA, Kaminski TS, Ziebacz N, Tabaka M, Sorto NA, Foss MH, Shaw JT, Thanbichler M, Weibel DB, Nieznanski K, Holyst R, Garstecki P (2012) Characterization of *Caulobacter crescentus* FtsZ using dynamic light scattering. J Biol Chem. doi:10.1074/jbc. M111.309492, M111.309492 [pii]

Hsin J, Gopinathan A, Huang KC (2012) Nucleotide-dependent conformations of FtsZ dimers and force generation observed through molecular dynamics simulations. Proc Natl Acad Sci U S A 109(24):9432–7. doi:10.1073/pnas.1120761109, 1120761109 [pii]

Huberts DH, van der Klei IJ (2010) Moonlighting proteins: an intriguing mode of multitasking. Biochim Biophys Acta 1803(4):520–5. doi:10.1016/j.bbamcr.2010.01.022, S0167-4889(10) 00035-2 [pii]

Ingerson-Mahar M, Gitai Z (2012) A growing family: the expanding universe of the bacterial cytoskeleton. FEMS Microbiol Rev 36(1):256–66. doi:10.1111/j.1574-6976.2011.00316.x

Jeffery CJ (1999) Moonlighting proteins. Trends Biochem Sci 24(1):8–11, S0968-0004(98)01335-8 [pii]

Jeffery CJ (2003) Multifunctional proteins: examples of gene sharing. Ann Med 35(1):28–35

Jeffery CJ (2011) Proteins with neomorphic moonlighting functions in disease. IUBMB Life 63(7): 489–94. doi:10.1002/iub.504

Kamath K, Oroudjev E, Jordan MA (2010) Determination of microtubule dynamic instability in living cells. Methods Cell Biol 97:1–14. doi:10.1016/S0091-679X(10)97001-5, S0091-679X (10)97001-5 [pii]

Khersonsky O, Tawfik DS (2010) Enzyme promiscuity: a mechanistic and evolutionary perspective. Annu Rev Biochem 79:471–505. doi:10.1146/annurev-biochem-030409-143718

Kovacs GG, Laszlo L, Kovacs J, Jensen PH, Lindersson E, Botond G, Molnar T, Perczel A, Hudecz F, Mezo G, Erdei A, Tirian L, Lehotzky A, Gelpi E, Budka H, Ovadi J (2004) Natively unfolded tubulin polymerization promoting protein TPPP/p25 is a common marker of alpha-synucleinopathies. Neurobiol Dis 17(2):155–62. doi:10.1016/j.nbd.2004.06.006, S0969-9961 (04)00136-6 [pii]

Kumar JK, Tabor S, Richardson CC (2004) Proteomic analysis of thioredoxin-targeted proteins in *Escherichia coli*. Proc Natl Acad Sci USA 101:3759–64

Lehotzky A, Tirian L, Tokesi N, Lenart P, Szabo B, Kovacs J, Ovadi J (2004) Dynamic targeting of microtubules by TPPP/p25 affects cell survival. J Cell Sci 117(Pt 25):6249–59, 117/25/6249 [pii] 10.1242/jcs.01550

Lowe J, Amos LA (1998) Crystal structure of the bacterial cell-division protein FtsZ. Nature 391: 203–6

Martin-Garcia F, Salvarelli E, Mendieta-Moreno JI, Vicente M, Mingorance J, Mendieta J, Gomez-Puertas P (2012) Molecular dynamics simulation of GTPase activity in polymers of the cell division protein FtsZ. FEBS Lett 586(8):1236–9. doi:10.1016/j.febslet.2012.03.042, S0014-5793(12)00242-6 [pii]

Mazzola JL, Sirover MA (2003) Subcellular alteration of glyceraldehyde-3-phosphate dehydrogenase in Alzheimer's disease fibroblasts. J Neurosci Res 71(2):279–85

Meyer-Siegler K, Mauro DJ, Seal G, Wurzer J, deRiel JK, Sirover MA (1991) A human nuclear uracil DNA glycosylase is the 37-kDa subunit of glyceraldehyde-3-phosphate dehydrogenase. Proc Natl Acad Sci U S A 88(19):8460–4

Minton AP, Wilf J (1981) Effect of macromolecular crowding upon the structure and function of an enzyme: glyceraldehyde-3-phosphate dehydrogenase. Biochemistry 20(17):4821–6

Moore A, Wordeman L (2004) The mechanism, function and regulation of depolymerizing kinesins during mitosis. Trends Cell Biol 14(10):537–46. doi:10.1016/j.tcb.2004.09.001, S0962-8924(04)00233-8 [pii]

Norris V, den Blaauwen T, Cabin-Flaman A, Doi RH, Harshey R, Janniere L, Jimenez-Sanchez A, Jin DJ, Levin PA, Mileykovskaya E, Minsky A, Saier M Jr, Skarstad K (2007) Functional taxonomy of bacterial hyperstructures. Microbiol Mol Biol Rev 71(1):230–53

Norris V, Amar P, Legent G, Ripoll C, Thellier M, Ovadi J (2010) Hypothesis: the cytoskeleton is a metabolic sensor. In: Amar P, Képès F, Norris V (eds) Modelling complex biological systems in the context of genomics. EDP Sciences, Evry, pp 95–104

Norris V, Amar P, Legent G, Ripoll C, Thellier M, Ovádi J (2013) Sensor potency of the moonlighting enzyme-decorated cytoskeleton: the cytoskeleton as a metabolic sensor. BMC Biochem 14:3. doi:10.1186/1471-2091-14-3

Olah J, Klivenyi P, Gardian G, Vecsei L, Orosz F, Kovacs GG, Westerhoff HV, Ovadi J (2008) Increased glucose metabolism and ATP level in brain tissue of Huntington's disease transgenic mice. FEBS J 275(19):4740–55, EJB6612 [pii] 10.1111/j.1742-4658.2008.06612.x

Orosz F, Kovacs GG, Lehotzky A, Olah J, Vincze O, Ovadi J (2004) TPPP/p25: from unfolded protein to misfolding disease: prediction and experiments. Biol Cell 96(9):701–11, S0248-4900(04)00164-9 [pii] 10.1016/j.biolcel.2004.08.002

Ovadi J (1991) Physiological significance of metabolic channelling. J Theor Biol 152(1):1–22

Ovadi J (2011) Moonlighting proteins in neurological disorders. IUBMB Life 63(7):453–6. doi:10.1002/iub.491

Ovadi J, Orosz F (eds) (2009) Protein folding and misfolding: neurodegenerative diseases, vol 7, Focus on Structural Biology. Springer, Amsterdam

Ovadi J, Saks V (2004) On the origin of intracellular compartmentation and organized metabolic systems. Mol Cell Biochem 256–257:5–12

Ovadi J, Srere PA (2000) Macromolecular compartmentation and channeling. Int J Cytol 192: 255–80

Ovadi J, Orosz F, Hollan S (2004) Functional aspects of cellular microcompartmentation in the development of neurodegeneration: mutation induced aberrant protein-protein associations. Mol Cell Biochem 256–257(1–2):83–93

Partikian A, Olveczky B, Swaminathan R, Li Y, Verkman AS (1998) Rapid diffusion of green fluorescent protein in the mitochondrial matrix. J Cell Biol 140(4):821–9

Popp D, Iwasa M, Erickson HP, Narita A, Maeda Y, Robinson RC (2010) Suprastructures and dynamic properties of *Mycobacterium tuberculosis* FtsZ. J Biol Chem 285(15):11281–9, doi: M109.084079 [pii] 10.1074/jbc.M109.084079

Porter KR, Beckerle M, McNiven M (1983) The cytoplasmic matrix. In: McIntosh JR (ed) Spatial organization of eukaryotic cells–a symposium in honor of K. R. Porter. Alan R. Liss, New York, pp 259–302

Raamsdonk LM, Teusink B, Broadhurst D, Zhang N, Hayes A, Walsh MC, Berden JA, Brindle KM, Kell DB, Rowland JJ, Westerhoff HV, van Dam K, Oliver SG (2001) A functional genomics strategy that uses metabolome data to reveal the phenotype of silent mutations. Nat Biotechnol 19(1):45–50. doi:10.1038/83496

Romberg L, Levin PA (2003) Assembly dynamics of the bacterial cell division protein FtsZ: poised at the edge of stability. Annu Rev Microbiol 57:125–54

Rostovtseva TK, Bezrukov SM (2012) VDAC inhibition by tubulin and its physiological implications. Biochim Biophys Acta 1818(6):1526–35, doi:S0005-2736(11)00391-9 [pii] 10.1016/j.bbamem.2011.11.004

Sloboda RD, Dentler WL, Rosenbaum JL (1976) Microtubule-associated proteins and the stimulation of tubulin assembly in vitro. Biochemistry 15(20):4497–505

Sriram G, Martinez JA, McCabe ER, Liao JC, Dipple KM (2005) Single-gene disorders: what role could moonlighting enzymes play? Am J Hum Genet 76(6):911–924, doi:S0002-9297(07) 62890-0 [pii] 10.1086/430799

Swaminathan R, Hoang CP, Verkman AS (1997) Photobleaching recovery and anisotropy decay of green fluorescent protein GFP-S65T in solution and cells: cytoplasmic viscosity probed by green fluorescent protein translational and rotational diffusion. Biophys J 72(4):1900–7, doi: S0006-3495(97)78835-0 [pii] 10.1016/S0006-3495(97)78835-0

Thanedar S, Margolin W (2004) FtsZ exhibits rapid movement and oscillation waves in helix-like patterns in *Escherichia coli*. Curr Biol 14:1167–73

Tirian L, Hlavanda E, Olah J, Horvath I, Orosz F, Szabo B, Kovacs J, Szabad J, Ovadi J (2003) TPPP/p25 promotes tubulin assemblies and blocks mitotic spindle formation. Proc Natl Acad Sci U S A 100(24):13976–81, doi:10.1073/pnas.2436331100 2436331100 [pii]

Verkman AS (2002) Solute and macromolecular diffusion in cellular aqueous compartments. Trends Biochem Sci 27:27–32

Wágner G, Kovács J, Löw P, Orosz F, Ovádi J (2001) Tubulin and microtubule are potential targets for brain hexokinase binding. FEBS Lett 509(1):81–4

Weart RB, Lee AH, Chien AC, Haeusser DP, Hill NS, Levin PA (2007) A metabolic sensor governing cell size in bacteria. Cell 130(2):335–47

Chapter 8
Metabolic Dissipative Structures

Ildefonso Mtz. de la Fuente

Abstract The self-organization of metabolic processes, such as the spontaneous dissipative formation of macromolecular structures, the functional coordination between multienzymatic complexes, and the emergence of molecular rhythms, is one of the most relevant topics in the post-genomic era. Herein, I analyze some aspects of self-organization in metabolic processes utilizing information theory to quantifying biomolecular information flows in bits, an approach that enables the visualization of emergent effective connectivity structures. Specifically, I determined the emergent functional integrative processes arising from irreversible enzymatic steps in yeast glycolysis, and in the systemic metabolic structure. Experimental observations and numerical studies with dissipative metabolic networks have shown that enzymatic activity can spontaneously self-organize leading to the emergence of a systemic metabolic structure, characterized by a set of different enzymatic reactions always locked into active states, i.e., the metabolic core, while the remaining catalytic processes are intermittently active. This global metabolic structure was verified in *Escherichia coli*, *Helicobacter pylori*, and *Saccharomyces cerevisiae*, and it seems to be a common key feature for all cells. The observed effective connectivity in both the irreversible enzymatic steps of yeast glycolysis and the systemic metabolic structure is highly dynamic and characterized by significant variations of biomolecular causality flows. These effective connectivity flows reflect the integration of catalytic processes within multienzymatic systems, and their functional coordination. The resulting functional integrative structures appear to be fundamental motifs in the dissipative self-organization and self-regulation of cellular metabolism.

I.M. de la Fuente (✉)
BioCruces Health Research Institute, Barakaldo, Basque Country, Spain

Institute of Parasitology and Biomedicine "López Neyra", CSIC, Granada, Spain

Department of Mathematics, University of the Basque Country, EHU-UPV., Leioa, Spain

M.A. Aon et al. (eds.), *Systems Biology of Metabolic and Signaling Networks*,
Springer Series in Biophysics 16, DOI 10.1007/978-3-642-38505-6_8,

8.1 Introduction

Cells are complex metabolic systems characterized by continuous transformation and renewal of macromolecular structures, highly coordinated catalytic behavior, and emergent spatiotemporal metabolic rhythms.

Several studies about biochemical processes have shown that the concept of self-organization is central to understanding the formation of the cell's biomolecular architecture and its functional metabolic behavior (Glick 2007; Karsenti 2008; Kauffman 1993). In general, self-organization can be defined as the spontaneous emergence of macroscopic nonequilibrium dynamic structures, as a result of collective behavior of elements interacting nonlinearly with each other, to generate a system that increases its structural and functional complexity driven by energy dissipation (Halley and Winkler 2008; Misteli 2009). Another important ingredient of metabolic self-organization is given by nonlinear interaction mechanisms between processes, involving, e.g., autocatalysis, product activation, substrate inhibition.

It is well established that nonequilibrium states can be a source of order in the sense that the irreversible processes may lead to a new type of dynamic state in which the system becomes ordered in space and time. The emergent structures cannot be directly predicted from the individual properties of their elements, and this kind of self-organized process occurs only in association with energy dissipation.

Metabolic self-organization is based on the concept of dissipative structures, and its theoretical roots can be traced back to Ilya Prigogine (Nicolis and Prigogine 1977). According to this theory, an open system that operates far from thermodynamic equilibrium is capable of continuously importing matter and energy from the environment and, at the same time, exporting entropy. Consequently, the system's entropy can be either maintained at the same level or decreased, in contrast with the entropy of an isolated system which tends to increase towards a maximum at thermodynamic equilibrium. Therefore, the total entropy in an open system can decrease, and the negative entropy variation can be maintained over time by a continuous exchange of matter and energy with the environment avoiding the transition to thermodynamic equilibrium. A dissipative system works as an energy-transforming mechanism that uses some of the energy inflow to produce a new form of energy which has a higher thermodynamic value, i.e., lower entropy, and the negative entropy variation corresponds to a positive variation in the information of the system. This emergent information increases the complexity of molecular organization, producing highly ordered macrostructures and functional dynamic behavior (Ebeling et al. 1986; Klimontovich 1999; Prigogine et al. 1977).

In cells, dissipative self-organization is the main driving force of molecular order, involved in all fundamental processes, e.g., cell division (Tyson et al. 1996), mitosis (Bastiaens et al. 2006; Loose et al. 2008), genome organization (Misteli 2009), cell differentiation (Woodford and Zandstra 2012), bacterial chemotaxis

(Sourjik and Armitage 2010), cytoskeleton dynamics (Huber and Käs 2011; Mitchison 1992), and systemic metabolic processes (De la Fuente et al. 2008).

Self-assembly is another mechanism of molecular organization, but one which does not involve continuous energy dissipation, and exhibits a trend towards thermodynamic equilibrium. On a macroscopic level, self-assembly is driven by the interplay among local stereospecific interactions between molecular components, which remain unchanged throughout the assembly process (Kushner 1969; Misteli 2009; Whitesides and Grzybowski 2002). Although self-assembly is a spontaneous process because the energy of the unassembled elements is higher than in the self-assembled structure, it is non-dissipative and reversible, constituting an important natural mechanism for generating molecular order. Worth emphasizing is that self-assembly may also occur under nonequilibrium conditions, thus coexisting with dissipative self-organization (Halley and Winkler 2008). Examples of self-assembly are lipid bilayers (Tresset 2009), protein–protein interactions (Ceschini et al. 2000; Lee 2008), viral capsid formation (Cadena-Nava et al. 2012), actin polymerization in aster-like structures (Haviv et al. 2006), and the assembly of the 30S ribosomal subunit (Talkington et al. 2005).

Overall, self-organization is central to describing the complex molecular architecture of cells, and it together with self-assembly is the main source of biomolecular order, function, and complexity. Dynamic self-organization underlies spatiotemporal architecture of cellular processes including functional coordination between myriad of enzymatic reactions leading to the emergence of molecular rhythms.

8.1.1 Multienzymatic Complexes

Molecular crowding is prevalent in cells which results in self-organized and assembled processes. In particular, supramolecular organization of enzymatic complexes is of special relevance, for their essential catalytic role in cellular metabolism. Although most enzymes are proteins, a few RNA molecules called ribozymes or ribonucleic acid enzyme also exhibit catalytic activity (Cech 2000; Lilley 2005).

Proteomics studies have focused on individual proteins, but homologous or heterologous protein–protein interactions are the rule in the intracellular milieu (Pang et al. 2008). Analyses of the proteome of *Saccharomyces cerevisiae* have shown that at least 83 % of proteins form complexes comprised by two to eighty-three proteins. These complexes constitute macromolecular machines with enzymatic activity, while forming the structural basis of a modular network of biochemical reactions (Gavin et al. 2002). This kind of organization occurs in all sorts of cells, both eukaryotes and prokaryotes (Bobik 2006; Ho et al. 2002; Ito et al. 2001; Sutter et al. 2008; Uetz et al. 2000; Yeates et al. 2008).

Multienzyme complexes may allow for substrate channeling, i.e., direct transfer of intermediate metabolites from the active site of one enzyme to the next without prior diffusion into the bulk medium. Metabolic channeling decreases the transit time of reaction substrates, thus increasing the speed and efficiency of catalysis by preventing delays due to diffusion of reaction intermediates and their eventual loss (Clegg and Jackson 1990; Jovanović et al. 2007; Jorgensen et al. 2005; Negrutskii and Deutscher 1991; Ovádi and Srere 2000). Substrate channeling may also occur within channels or on the electrostatic surface of enzymes belonging to complexes (Milani et al. 2003; Ishikawa et al. 2004).

Additionally, reversible interactions between multienzyme complexes and structural proteins or membranes occur frequently in eukaryotic cells leading to the emergence of metabolic microcompartments (Lunn 2007; Monge et al. 2008, 2009; Saks et al. 2007, 2009. Ovádi and Saks 2004). Microcompartments also happen in prokaryotes, but in this case they consist of protein "shells" composed of thousands of protein subunits, some of them being enzymes that belong to specific metabolic pathways (Fan et al. 2010; Yeates et al. 2007). The dynamics of molecular processes that intervene in microcompartmentation and its maintenance remain unclear.

Multienzyme complexes, substrate channeling, and their integration into functional microcompartments seem to be a central feature of the spatiotemporal organization of cellular metabolism. This modular organization appears to be crucial for the regulation and efficiency of enzymatic processes and thus fundamental for understanding the molecular architecture of life.

8.1.2 Temporal Self-Organization of Multienzymatic Processes

Self-organization at the enzymatic level underlies the emergence of functional structures in the temporal organization of catalytic process. Besides forming complex catalytic associations, enzymes can exhibit molecular rhythms constituting a key self-organizational trait which allow functional coordination across multiple enzymatic complexes (De la Fuente et al. 2011). Cellular processes involving biosynthesis, turnover of molecular components, migration, and division require temporal organization across many simultaneous timescales (Chandrashekaran 2005; Lloyd and Murray 2005, 2006, 2007). The functional coordination implied by metabolic rhythms involves spatial and temporal aspects of localization and dynamics of enzymatic complexes (Hildebrandt 1982; Yates 1993; Aon and Cortassa 1997; Aon et al. 2000), including their synchronization (Wolf et al. 2000).

Dynamically, cells may exhibit quasi-stationary and oscillatory states. Quasi-stationarity arises from a slow drifting of metabolites' concentration over time. In cells, the proportion between oscillatory and quasi-stationary states is unknown, but existing evidence suggests that quasi-steady states are less frequent than oscillatory ones (Lloyd and Murray 2005).

During the last four decades, extensive studies of biochemical dynamics in both prokaryotic and eukaryotic cells have revealed that oscillations exist in most of the fundamental metabolic processes. For instance, specific ultradian oscillations were reported to occur in free fatty acids (Getty-Kaushik et al. 2005), NAD(P)H concentration (Rosenspire et al. 2001), biosynthesis of phospholipids (Marquez et al. 2004), cyclic AMP concentration (Holz et al. 2008), ATP (Ainscow et al. 2002) and other adenine nucleotide levels (Zhaojun et al. 2004), intracellular glutathione concentration (Lloyd and Murray 2005), actin polymerization (Rengan and Omann 1999), ERK/MAPK metabolism (Shankaran et al. 2009), mRNA levels (Zhaojun et al. 2004), intracellular free amino acid pools (Hans et al. 2003), cytokinins (Hartig and Beck 2005), cyclins (Hungerbuehler et al. 2007), transcription of cyclins (Shaul et al. 1996), gene expression (Chabot et al. 2007; Tian et al. 2005; Tonozuka et al. 2001; Klevecz et al. 2004), microtubule polymerization (Lange et al. 2004), membrane receptor activities (Placantonakis and Welsh 2001), membrane potential (De Forest and Wheeler 1999), intracellular pH (Sánchez-Armáss et al. 2006), respiratory metabolism (Lloyd et al. 2002), glycolysis (Dano et al. 1999), intracellular calcium concentration (Ishii et al. 2006), metabolism of carbohydrates (Jules et al. 2005), beta-oxidation of fatty acids (Getty et al. 2000), metabolism of mRNA (Klevecz and Murray 2001), tRNA (Brodsky et al. 1992), proteolysis (Kindzelskii et al. 1998), urea cycle (Fuentes et al. 1994), Krebs cycle (Wittmann et al. 2005), mitochondrial metabolic processes (Aon et al. 2008), nuclear translocation of the transcription factor (Garmendia-Torres et al. 2007), amino acid transports (Barril and Potter 1968), peroxidase-oxidase reactions (Møller et al. 1998), protein kinase activities (Chiam and Rajagopal 2007), and photosynthetic reactions (Smrcinová et al. 1998).

Experimental observations performed in *S. cerevisiae* during continuous culture have shown that most of transcriptome and the metabolome exhibit oscillatory dynamics (Klevecz et al. 2004; Murray et al. 2007). From transcriptome data, it has been inferred that at least 60 % of all gene expression oscillates with an approximate period of 300 min (Tu et al. 2005). Moreover, the entire transcriptome exhibits low-amplitude oscillatory behavior (Lloyd and Murray 2006) and this phenomenon has been described as a genome-wide oscillation (Klevecz et al. 2004; Lloyd and Murray 2005, 2006; Murray et al. 2007; Oliva et al. 2005; Tu et al. 2005). Evidence that cells exhibit multi-oscillatory metabolic processes with fractal properties has also been reported. This dynamic behavior appears to be consistent with scale-free dynamics spanning a wide range of frequencies of at least three orders of magnitude (Aon et al. 2008).

The temporal organization of the metabolic processes in terms of rhythmic phenomena spans periods ranging from milliseconds (Aon et al. 2006) to seconds (Roussel et al. 2006), minutes (Berridge and Galione 1988; Chance et al. 1973), and hours (Brodsky 2006). Transitions from simple periodic behavior to complex oscillations, including bursting (oscillations with one large spike and series of secondary oscillations) (Dekhuijzen and Bagust 1996) and chaotic phenomena (Olsen and Degn 1985), have often been observed.

A second kind of temporal metabolic structure is given by circadian rhythms. With a period close to 24 h (given by the Earth's rotation) circadian clocks enable cells to adapt their metabolism to the appropriate time of the day, synchronizing timing of metabolic reactions with cyclic changes in the external environment (Schibler and Sassone-Corsi 2002; Schibler and Naef 2005; Dunlap et al. 2004; Wijnen and Young 2006). Circadian rhythms govern a wide variety of metabolic and physiological processes in all organisms from prokaryotes to human cells (Wijnen and Young 2006). In some cells, at least 10 % of all cellular transcripts oscillate in a circadian manner (Nakahata et al. 2007), and in certain cells it has been observed that between 80 and 90 % of the transcripts seem to follow a pattern of circadian expression with a period of 24–26 h (Connor and Gracey 2011).

A third kind of molecular rhythms is given by spatial concentration waves. When spatial inhomogeneity elicits instabilities in the intracellular medium, propagating concentration waves can be triggered. This dynamic behavior is not only closely related to metabolic oscillations but also to synchronization. Biochemical waves are quite common and involve several cellular variables such as intracellular pH, membrane potential, flavoproteins, calcium, and NAD(P)H. They are linked to central metabolic processes and specific physiological functions, namely, signal transduction and intercellular communication (Petty 2006). There are several types of molecular waves and they vary in their chemical composition, velocity, shape, intensity, and location (Scemes and Giaume 2006; Galas et al. 2000; Guthrie et al. 1999). Examples of metabolic waves are Na^+ and Ca^{2+} (Bernardinelli et al. 2004), redox (Romashko et al. 1998), reactive oxygen species (Aon et al. 2004; Zhou et al. 2010), ATP (Ueda et al. 1990; Newman 2001), pH (Petty et al. 2000), NAD(P)H (Petty and Kindzelskii 2001), NAD(P)H coupled with calcium (Slaby and Lebiedz 2009), actin filament assembly during cell locomotion (Vicker 2002), and phosphatidylinositol (3,4,5)-trisphosphate (PIP_3) (Asano et al. 2008).

Metabolic rhythms constitute one of the most genuine properties of multienzymatic dynamics. The conditions required for the emergence and sustainability of these rhythms, and how they are regulated, represent a biological problem of the highest significance. However, in spite of its physiological importance many aspects of these spatiotemporal structures, such as their relationship to the cell cycle, are still poorly understood and thus deserve further attention.

8.1.3 Dissipative Multienzymatic Complexes: Metabolic Subsystems

Multienzymatic associations can be viewed as dissipative structures in which molecular rhythms and functional integrative processes can emerge increasing the efficiency and control of the catalytic reactions involved (De la Fuente 2010; De la Fuente and Cortes 2012). Self-organization and self-assembly processes allow for reversible interactions between multienzyme complexes and other molecular structures. This may lead to the formation of metabolic microcompartmentation and channeling resulting from a discrete reactive space where intermediate metabolites can be protected from being consumed by competing reactions (Milani et al. 2003). Elsewhere, we have referred as *metabolic subsystems* these self-organized multienzymatic complexes, which can be associated with other noncatalytic biomolecular structures, in which oscillations and quasi-steady-state patterns can spontaneously emerge (De la Fuente 2010). Thermodynamically, metabolic subsystems represent advantageous biochemical organizations, forming unique, well-defined autonomous dynamic systems (De la Fuente 2010).

Nonlinear kinetics, catalytic irreversibility, and the coupling between nonlinear reactions and diffusion are main source of spatiotemporal self-organization in metabolic subsystems (Goldbeter 2007; Nicolis and Prigogine 1977). A rich variety of dynamic patterns emerging from these metabolic subsystems can be associated with distinct activity regimes, independent of direct genomic control (De la Fuente et al. 2013).

Overall, a *metabolic subsystem* constitutes an elemental macromolecular machine, a catalytic module, which provides an efficient enzymatic activity. Associations between metabolic subsystems can form higher level complex molecular organizations, as for example it occurs with intracellular energetic units (ICEU) (Saks et al. 2006) and synaptosomes (Monge et al. 2008).

8.1.4 Dissipative Metabolic Networks

Structural observations have shown that, in cells, the overall enzymatic organization is given by modules of multienzyme complexes arrayed as a network (Gavin et al. 2002). Moreover, the cellular metabolic system behaves like a multi-oscillatory system (Lloyd 2005; Lloyd and Murray 2005, 2006; Murray et al. 2007; Roussel and Lloyd 2007; Vanin and Ivanov 2008) comprising mitochondrial, nuclear, transcriptional, and metabolic dynamics (Lloyd et al. 2006). In this context, redox rhythmicity has been suggested as a fundamental dynamic hub for intracellular temporal coherence (Lloyd and Murray 2007).

To address the function of cellular metabolism from the point of view of self-organization, the dissipative metabolic network (DMN) concept was proposed (De la Fuente et al. 1999a, 2008). Essentially, a DMN is a set of metabolic subsystems interconnected by substrate fluxes and three classes of regulatory signals: activatory (positive allosteric modulation), inhibitory (negative allosteric modulation), and all-none type, e.g., enzyme regulation by covalent modification. In a DMN, the output activity for each metabolic subsystem can be either steady state or predominantly oscillatory with different activity regimes.

The first model of a DMN put in evidence a singular systemic metabolic structure, characterized by a set of different metabolic subsystems always locked into active states (metabolic core) while the rest of the catalytic subsystems presented on–off dynamics. In this first numerical work it was also suggested that the systemic metabolic structure could be an intrinsic characteristic of metabolism, common to all living systems (De la Fuente et al. 1999a)

During 2004 and 2005, several studies applying flux balance analysis to experimental data contributed evidence that can be interpreted as a systemic functional structure (Almaas et al. 2004, 2005). Specifically, it was observed that a set of metabolic reactions belonging to different anabolic processes remained active under all investigated growth conditions. The remaining enzymatic reactions not belonging to this metabolic core stayed only intermittently active. These global catalytic processes were verified for *E. coli*, *H. pylori*, and *S. cerevisiae* (Almaas et al. 2004, 2005).

The systemic functional structure appears as a robust dynamical system (De la Fuente et al. 2009), in which self-organization, self-regulation, and long-term memory properties emerge (De la Fuente et al. 2010). Long-term correlations have been observed in different experimental studies, e.g., quantification of DNA patchiness (Viswanathan et al. 1997), physiological time series (Eke et al. 2002; Goldberger et al. 2002), NADPH series (Ramanujan et al. 2006), DNA sequences (Allegrini et al. 1988; Audit et al. 2004), K^+ channel activity (Kazachenko et al. 2007), mitochondrial processes (Aon et al. 2008), and neural electrical activity (De la Fuente et al. 2006; Mahasweta et al. 2003).

8.1.5 Introduction to the Effective Functional Connectivity

Understanding functional coordination and spontaneous synchronization between enzymatic processes needs a quantitative measure of effective connectivity.

In the field of information theory, transfer entropy (TE) has been proposed to be a rigorous, robust, and self-consistent method for the causal quantification of functional information flow between nonlinear processes (Schreiber 2000). TE quantifies the reduction in uncertainty that one variable has on its own future when adding another, allowing for a calculation of the functional influence between two variables in terms of effective connectivity.

The roots of TE are given by the Shannon entropy, which measures the average information required to determine a random variable X (Cover and Thomas 1991). The Shannon entropy is defined as

$$H(X) \equiv \sum_{x} -p(x) \log p(x), \tag{8.1}$$

where x is one of the possible states which characterize the dynamics of variable X. For instance, in tossing a coin, the two only possible states are head or tail. $p(x)$ defines the probability (or normalized occurrence) of measuring the variable X in the state x.

The amount of information shared between two variables can be quantified by mutual information:

$$I(X,Y) \equiv H(X) - H(X|Y), \tag{8.2}$$

which measures how much the uncertainty on variable X is reduced by conditioning it to another variable Y. Notice that because $H(X|Y) = H(X,Y) - H(Y)$, the mutual information is a symmetrical measure between the two variables X and Y, i.e., $I(X,Y) = I(Y,X)$

In contrast with mutual information, TE is not a symmetrical measure since the TE from X to Y is different from the one from Y to X. This asymmetry comes from the fact that to compute the TE one has to define the past and the future on one of the time series (say X) and only the past of the other variable (say Y). The TE is then defined as

$$\mathrm{TE}(Y \rightarrow X) \equiv H(X^{\mathrm{future}}|X^{\mathrm{past}}) - H(X^{\mathrm{future}}|X^{\mathrm{past}}, Y^{\mathrm{past}}), \tag{8.3}$$

which allows an easy interpretation of TE. The TE from Y to X measures how much the uncertainty of the X^{future} is reduced by knowing the Y^{past} and comparing this number with the situation in which only the X^{past} is reducing the uncertainty of X^{future}. If by adding Y^{past} the uncertainty of X^{future} is reduced more than adding only X^{past}, there is a nonzero TE from Y to X.

Another interpretation of TE comes from comparing (8.2) and (8.3); the TE is the mutual information between X^{future} and Y^{past} conditioning on Y^{past}, i.e., $\mathrm{TE}(Y \rightarrow X) = I(X^{\mathrm{future}}, Y^{\mathrm{past}}|X^{\mathrm{past}})$.

A recent application of TE to irreversible enzymatic steps from yeast glycolysis has shown that functional integration may emerge under certain experimental conditions (De la Fuente and Cortes 2012) and in the systemic metabolic structure (De la Fuente et al. 2011). Specifically, it was possible to quantify biomolecular information flows in bits between catalytic elements and determine the emergence of effective connectivity structures. These effective functional structures are involved both in the integration of catalytic processes belonging to a single metabolic subsystem and in the functional coordination between different metabolic subsystems.

The following section addresses these studies related with irreversible enzymes from glycolysis (Sect. 8.2) and in dissipative metabolic networks (Sect. 8.3).

8.2 Quantitative Analysis of the Effective Functional Structure in Yeast Glycolysis

Yeast glycolysis is considered one of the prototypical biochemical oscillators, and one of the most studied metabolic pathways. It was the first metabolic system in which spontaneous oscillations were observed (Duysens and Amesz 1957; Chance et al. 1964), and these studies led to the first models of this pathway based on enzyme kinetics (Goldbeter and Lefever 1972, Goldbeter 1973).

From a structural viewpoint, glycolytic enzymes can interact with structural proteins and membranes generating metabolic microcompartments. For instance, it has been observed that the entire glycolytic pathway is associated with the cytosolic face of the outer mitochondrial membrane (Graham et al. 2007), or forms a multienzyme complex on the inner surface of the plasma membrane (Campanella et al. 2008). The interaction of glycolytic enzymes with cytoskeletal proteins has also been suggested as another mechanism of compartmentation of this pathway (Clarke and Masters 1975; Waingeh et al. 2006).

Reported evidence shows substrate channeling in glycolytic enzymes (Malaisse et al. 2004; Shearer et al. 2005). For example, the three irreversible glycolytic enzymes hexokinase (Malaisse et al. 2004; Zhang et al. 2005), phosphofructokinase (Cascante et al. 2000; Commichau et al. 2009), and pyruvate kinase (Clarke and Masters 1975; Commichau et al. 2009; Waingeh et al. 2006) show substrate channeling or compartmentation. These experimental observations are consistent with the existence of a glycolytic multienzymatic complex. Mowbray and Moses (1976) proposed the existence of such a complex from studies performed using cell-free extracts (Mowbray and Moses 1976).

As in other dissipative structures, the metabolic dynamic patterns of yeast glycolysis find their roots in the nonlinear regulatory processes, e.g., stoichiometric autocatalysis, allosteric regulation and product activation (Cortassa et al. 1991; Goldbeter 2002, 2007), and other sources (Cortassa and Aon 1994; Olsen et al. 2009).

In yeast glycolysis, it has been shown that an instability-generating mechanism is given by the regulation of phosphofructokinase, specifically, the positive feed-back exerted by the reaction products, ADP and fructose-1,6-bisphosphate (Boiteux et al. 1975; Goldbeter and Lefever 1972; Goldbeter 2002). In yeast extracts, traveling waves of NADH and protons were observed associated with glycolysis (Mair et al. 2001). All this evidence shows that yeast glycolysis constitutes an example of a dissipatively structured enzymatic association that can display microcompartmentation, substrate channeling, and spatiotemporal dynamic behavior.

Over the last 30 years a large number of studies have been focusing on the molecular mechanisms underlying the emergence of self-organized glycolytic patterns (Dano et al. 1999; De la Fuente et al. 1995; Madsen et al. 2005; Olsen et al. 2009; Termonia and Ross 1981; Wolf et al. 2000). Nevertheless, and despite the significant advances made, we still lack a quantitative description of the effective functional structures involved in enzymatic activity coordination.

Functional connectivity quantifies how much the dynamics of one variable is statistically dependent on the dynamics of another. Therefore, although structural and functional connectivity are obviously related (i.e., structure shapes function), there is a striking difference between them: it is possible to have two variables which are functionally related (thus highly correlated) but structurally unconnected. The reason is because function can go beyond structure through neighbor–neighbor interactions. In order to quantify the functional structure arising from glycolytic enzymes catalyzing irreversible steps under dissipative conditions, the effective connectivity between enzyme and enzyme interactions, which account for the influence that the activity of one enzyme has on the future of another, was analyzed by transfer entropy (TE).

8.2.1 Determining the Rates of Irreversible Enzymatic Steps in Glycolysis

Although the kinetic behavior in vivo of most enzymes is unknown, in vitro studies can provide kinetic parameters and enzymatic rates. We used the latter strategy; for hexokinase we adopted a rate equation depending on glucose and ATP (Viola et al. 1982); for phosphofructokinase and pyruvate kinase a concerted transition model was applied (Goldbeter and Lefever 1972; Markus et al. 1980).

8.2.2 Modeling Glycolytic Processes Under Dissipative Conditions

In Fig. 8.1, the classical topological structure characterized by the specific location of enzymes, substrates, products, and feedback-regulatory metabolites is shown. When the metabolite S (glucose) feeds the glycolytic system, it is transformed by the first enzyme E_1 (hexokinase) into the product P_1 (glucose-6-phosphate). The enzymes E_2 (phosphofructokinase) and E_3 (pyruvatekinase) transform the substrates P'_1 (fructose 6-phosphate) and P'_2 (phosphoenolpyruvate) into the products P_2 (fructose 1-6-bisphosphate) and P_3 (pyruvate), respectively. The steps $P_1 \rightarrow P'_1$ and $P_2 \rightarrow P'$ are reversible. A part of P_1 is removed from the pathway with a constant rate of q_1 which is related to the activity of pentose phosphate pathway; likewise, q_2 is the constant rate for the sink of the product P_3 which is related to the activity of the pyruvate dehydrogenase complex.

Kinetic modeling with ordinary differential equations (ODE) is commonly used for studying metabolic systems. Consistently with the theory of dynamical systems, delays can be modeled by adding to the original variables other auxiliary functional

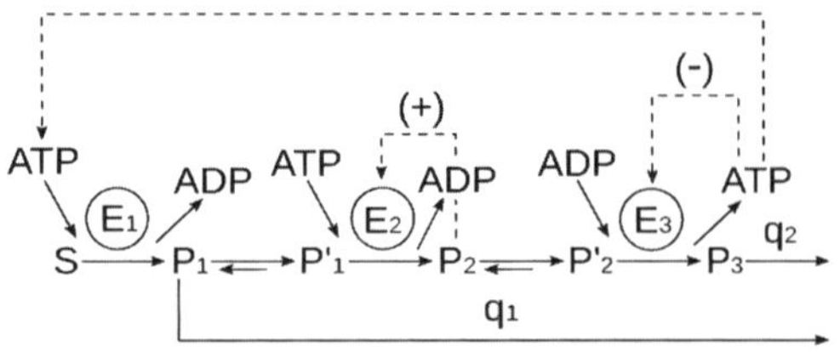

Fig. 8.1 *Multienzymatic instability-generating system in yeast glycolysis*. The irreversible stages correspond to the enzymes E_1 (hexokinase), E_2 (phosphofructokinase), and E_3 (pyruvate kinase). Metabolite S (glucose) is transformed by the first enzyme E_1 into the product P_1 (glucose-6-phosphate). P'_1, P_2, P'_2, and P_3 denote the concentrations of fructose 6-phospfate, fructose 1,6-bisphospfate, phosphoenolpyruvate, and pyruvate. q_1 is the rate first-order constant for the removal of P_1; q_2 is the rate constant for the sink of the product P_3. The figure includes the feedback activation of E_2 and the feedback inhibition of E_3. The ATP is consumed by E_1 and recycled by E_3. Adapted from Fig. 1 in (De la Fuente and Cortes 2012)

variables. By means of using differential equations with delay it is possible to consider initial functions (instead of the constant initial values of ODE systems) and to analyze the consequences of parametric variations (De la Fuente and Cortes 2012).

For spatially homogeneous conditions, the time evolution of the glycolytic system represented in Fig. 8.1 can be described by the following three delay differential equations:

$$\begin{aligned} \frac{\mathrm{d}\alpha}{\mathrm{d}t} &= z_1\sigma_1\Phi_1(\mu) - \sigma_2\Phi_2(\alpha,\beta) - q_1\alpha, \\ \frac{\mathrm{d}\beta}{\mathrm{d}t} &= z_2\sigma_2\Phi_2(\alpha,\beta) - \sigma_3\Phi_3(\beta,\beta',\mu), \\ \frac{\mathrm{d}\gamma}{\mathrm{d}t} &= z_3\sigma_3\Phi_3(\beta,\beta',\mu) - q_2\gamma, \end{aligned} \tag{8.4}$$

where the variables α, β, and γ denote the normalized concentrations of glucose-6-phosphate, fructose 1-6-bisphosphate, and pyruvate, respectively, with the following three enzymatic rate functions:

$$\Phi_1 = \mu S K_{D3}/(K_3K_2 + \mu K_{\mathrm{m}_1}K_{D3} + SK_2 + \mu S K_{D3}), \tag{8.5}$$

$$\Phi_2 = \frac{\alpha(1+\alpha)(1+d_1\beta)^2}{L_1(1+c\alpha)^2 + (1+\alpha)^2(1+d_1\beta)^2}, \tag{8.6}$$

$$\Phi_3 = \frac{d_2\beta'(1+d_2\beta')^3}{L_2(1+d_3\mu)^4 + (1+d_2\beta')^4} \tag{8.7}$$

and

$$\beta' = f(\beta(t-\lambda_1)),$$
$$\mu = h(\gamma(t-\lambda_2)).$$

The constants σ_1, σ_2, and σ_3 correspond to the maximum activity of the enzymes E_1, E_2, and E_3 (V_{m1}, V_{m2}, and V_{m3}) divided by the Michaelis constants K_{m1}, K_{m2}, and K_{m3}, respectively. The z constants are defined as $z_1 = K_{m1}/K_{m2}$, $z_2 = K_{m2}/K_{m3}$, and $z_3 = K_{m3}/K_{d3}$ with K_{d3} representing the dissociation constant of P_2 by E_3. L_1 and L_2 are the respective allosteric constant of E_2 and E_3. The constants d's are $d_1 = K_{m3}/K_{d2}$, $d_2 = K_{m3}/K_{d3}$, and $d_3 = K_{d3}/K_{d4}$ (K_{d3} and K_{d4} are the dissociation constant of P_2 by E_3 and the dissociation constant of MgATP, respectively). The c constant is the nonexclusive binding coefficient of the substrate P_1 (De la Fuente and Cortes 2012).

From a dissipative point of view, the essential enzymatic stages are those that correspond to the biochemical irreversible processes (Ebeling et al. 1986). To simplify the model, we did not consider the intermediate part of glycolysis belonging to the enzymatic reversible stages. In this way, the functions f and h are supposed to be the identity function:

$$f(\beta(t-\lambda_1)) = \beta(t-\lambda_1)$$
$$h(\gamma(t-\lambda_2)) = \gamma(t-\lambda_2)$$

The initial functions present a simple harmonic oscillation in the following form:

$$\alpha_0(t) = A + B\sin(2\pi t/P)$$
$$\beta_0(t) = C + D\sin(2\pi t/P)$$
$$\gamma_0(t) = E + F\sin(2\pi t/P)$$
$$P = 534\,s\,.$$

The dependent variables α, β, and γ are normalized by dividing them by K_{m2}, K_{m3}, and K_{d3}, the parameters λ_1 and λ_2 are time delays which correspond to phase shifts of the initial functions and P is the period (see for more details De la Fuente and Cortes 2012).

This glycolytic model has been exhaustively analyzed before, revealing a notable richness of temporal patterns which include the three main routes to chaos (De la Fuente et al. 1996a, b, 1999b), as well as a multiple coexisting stable states (De la Fuente et al. 1998a, 1999b), and persistent behavior (De la Fuente et al. 1998b, 1998c; 1999c).

8.2.3 Numerical Integration of the Glycolytic Model Under Sinusoidal Substrate Input Flux Values

In cell-free extracts, a sinusoidal input of glucose can produce periodic, quasiperiodic, or chaotic behavior (Markus et al. 1985a, b). Monitoring NADH fluorescence from glycolyzing baker's yeast cell-free extracts subjected to sinusoidal glucose input flux has shown that quasiperiodic dynamics is common at low input amplitudes whereas chaotic behavior happens at high amplitudes (Markus et al. 1985a, b). We adopted a similar approach, by analyzing the dynamic behavior of the glycolytic system considered under periodic input flux with a sinusoidal source of glucose $S = S_0 + A\sin(\omega t)$. Assuming the experimental value of $S_0 = 6\mathrm{mM/h}$ (Markus et al. 1984), after dividing by K_{m2}, the Michaelis constant of phosphofructokinase, a normalized input flux $S_0 = 0.033$ Hz was obtained (De la Fuente and Cortes 2012). Under these conditions, different types of dynamic patterns can be observed as a function of the amplitude A of the sinusoidal glucose input flux, as bifurcation parameter (De la Fuente et al. 1996b, 1999b).

Figure 8.2 shows different time series of the fructose 1,6-bisphospfate concentration after numerical integration of the glycolytic model (8.4) at different glucose input flux. A quasiperiodic route to chaos is observed (cf. left panel in Fig. 8.2). For $A = 0.001$ the biochemical oscillator exhibits a periodic pattern (Fig. 8.2a). After increasing the amplitude to $A = 0.005$ another Hopf bifurcation appears along with quasiperiodic behavior (Fig. 8.2b). Above $A = 0.021$, complex quasiperiodic oscillations take place (Fig. 8.2c), and after a new Hopf bifurcation the resulting dynamic behavior evolves into deterministic chaos ($A = 0.023$, Fig. 8.2d), as predicted (Ruelle and Takens 1971). This behavior corresponds to a typical quasiperiodic route to chaos, in agreement with experimental data (Markus et al. 1985b).

8.2.4 Measure of the Effective Functional Structure of Glycolysis

The time series of enzymatic activity obtained were analyzed using the nonlinear technique of transfer entropy in order to quantify the effective connectivity between glycolytic irreversible enzymes. We reasoned that the oscillatory patterns exhibited by metabolic intermediates might have information which can be captured by the TE measure. TE measures the influences between pairs of time series of catalytic activity, thus resulting in an asymmetric quantity that defines directionality in time and cause to effect, allowing to quantify the flow of functional information between processes behaving nonlinearly (Schreiber 2000).

For the calculation of TE, the different states of each dynamic variable were obtained simply by rounding the value of the variable at time t to the nearest integer. Thus, the state probabilities were computed by counting the number of times that

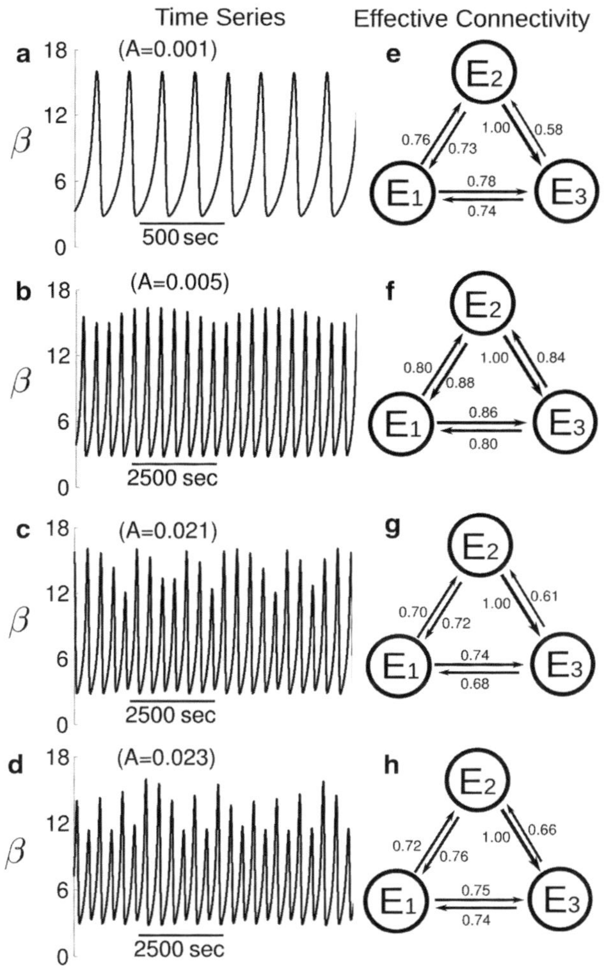

Fig. 8.2 *Dissipative functional structure involved in the integration of catalytic processes.* (**a**–**d**) Under dissipative conditions, a quasiperiodic route to chaos emerges in the glycolytic system when increasing the normalized amplitude of the periodic source of glucose substrate from A ~ 0:001 to A ~ 0:023. The figure shows the time evolution of the normalized concentration β, fructose 1,6-bisphospfate. (**a**) Periodic pattern. (**b**) Quasiperiodic oscillations. (**c**) Complex quasiperiodic motion indicating the transition towards chaos. (**d**) Deterministic chaos. (**e**–**h**) Biomolecular information flows measured in bits between the irreversible enzymes for the same conditions in the *left panel*. The strength of effective connectivity is plotted with *arrows* width proportional to the transfer entropy divided by its maximum value. The numerical integration of the system was performed with the package ODE Workbench, which is part of the Physics Academic Software. Internally this package uses a Dormand–Prince method of order 5 to integrate differential equations. Adapted from Fig. 2 in (De la Fuente and Cortes 2012)

each variable was in a given state. The set of all variables is given by α, β, and γ in (8.4).

The units of TE are in information bits, as the Log function appearing in the Shannon information is calculated on base 2. All the values were normalized to the maximum, thus the value of 1.00 corresponding in all cases to the TE between E_2 and E_3 which was the maximum in the system; the other values represent the ratio of information with respect to this maximum flow.

In Fig. 8.2 the biomolecular information flows between the irreversible enzymes of glycolysis are plotted. The values of functional influence obtained range from $0.58 \leq TE \leq 1.00$ (with mean $\pm$ SD $= 0.79 \pm 0.12$), which in general terms indicates a high effective connectivity in the multienzymatic subsystem.

8.2.5 Emergence of an Integrative Functional Structure in the Glycolytic Metabolic Subsystem

The data showed that the flows of functional connectivity can change significantly during the different metabolic transitions analyzed, exhibiting high TE values. The maximum source of information corresponds to the E_2 enzyme (phosphofructokinase) at the edge of chaos, when complex quasiperiodic oscillations emerge (cf. Fig. 8.2). This finding seems to be consistent with other studies showing that complexity is maximal when a dynamic system operates at the edge between order (e.g., periodic behavior) and chaos (Bertschinger and Natschlager 2004; Kauffman and Johnsen 1991).

The level of influence in terms of causal interactions between the enzymes is not always the same but varies depending on substrate fluxes and the particular dynamic regime. As a result of the dissipative self-organization, a biomolecular informative structure emerges in the metabolic subsystem, which is capable of modifying the catalytic activities of the glycolytic irreversible enzymes. The self-organization of the metabolic subsystem shapes a functional dynamic structure able to send biomolecular information between its catalytic elements, in such a way that the activity of each irreversible enzymatic step could be considered an information event. Each irreversible catalytic activity depends on the molecular information given by the substrate fluxes and regulatory signals, performing three simultaneous functions: as signal receptor, as signal integrator, and as a source of new molecular information. As a result of the overall process, the enzymatic activities are coordinated, and the metabolic subsystem operates as an information processing system which, at every moment, defines sets of biochemical instructions that make each irreversible enzyme evolve in a particular and precise catalytic pattern.

The simulations also show that for all cases analyzed the maximum effective connectivity corresponds to the transfer entropy from E Z to E Z, indicating the biggest information flow in this system. The total information flow was also analyzed as the difference between the TE output from an enzyme minus the total

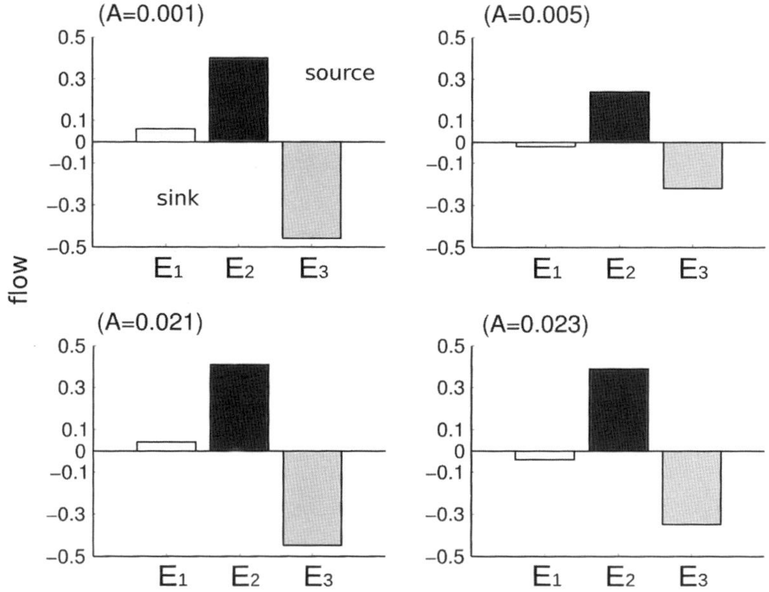

Fig. 8.3 *Total information flows Bars* represent the total information flow, defined per each enzyme as the total outward TE minus the total inward. The functionality attributed for each enzyme is an invariant and preserved along the route, i.e. E_2 is a source, E_3 is a sink, and E_1 has a quasi-zero flow. A is the normalized amplitude of the periodic source of glucose substrate. Adapted from Fig. 3 in (De la Fuente and Cortes 2012)

TE input (Fig. 8.3). Positive values from that difference mean that that enzyme is a source of causality flow, while negative numbers are interpreted as sinks or targets. The maximum source of total transfer information (0.41) corresponds to the E_2 enzyme (phosphofructokinase) for $A = 0.021$, when complex quasiperiodic oscillations appear in the glycolytic system. For all conditions the enzyme E_2 (phosphofructokinase) is the main source of effective influence and the enzyme E_3 (pyruvate kinase) a sink, which can be interpreted as a target from the point of view of its effective functionality; the enzyme E_1 (hexokinase) appears as less constrained with a flow close to zero. Accordingly, E_2 is the major source, E_3 is the major sink, and E_1 is a minor source with a value close to zero.

The results of our model revealed in a quantitative manner that the enzyme E Z (phosphofructokinase) is the major source of causal information and represents the key core of glycolysis in this model. Biochemically, phosphofructokinase has been considered a major checkpoint in the control of glycolysis (Gancedo and Serrano 1989; Heinisch et al. 1996). The main reason for this generalization is that phosphofructokinase exhibits a complex regulatory behavior that reflects its capacity to integrate many different signals (Stryer 1995). From a dissipative point of view, this enzyme catalyzes a reaction far from equilibrium and its regulation has been considered an important instability-generating mechanism for the emergence of oscillatory patterns in glycolysis (Goldbeter 2007). The TE studies give a quantification of the effective connectivity and confirm that E_2 (phosphofructokinase) is the key core of the pathway in this glycolytic model.

To summarize, the TE analysis shows the emergence of a new kind of dynamic functional structure, characterized by changing connectivity flows that reflect modulation of the kinetic behavior of the irreversible enzymes considered, and catalytic coordination.

8.3 Self-Organized Catalytic Behavior in Dissipative Metabolic Networks

Experimental observations (Almaas et al. 2004, 2005) and numerical studies with dissipative metabolic networks (De la Fuente et al. 1999a, 2008) have shown that cellular enzymatic activity self-organizes. This spontaneous organization leads to the emergence of a systemic metabolic structure characterized by a set of different enzymatic processes locked into active states (metabolic core) while others present on–off dynamics. Processes at the metabolic core and those exhibiting all-none dynamics are fundamental traits of the systemic metabolic structure which may be present in many different types of cells (Almaas et al. 2004, 2005).

Recently, in an attempt to provide a more accurate understanding of the functional coordination between several multienzymatic sets, the catalytic activities of a dissipative metabolic network were studied using transfer esntropy (De la Fuente et al. 2011). The results showed that a global functional structure of effective connectivity emerges, which is dynamic and characterized by significant variations of biomolecular information flows (De la Fuente et al. 2011).

In this study, a DMN of 18 metabolic subsystems, each one representing a set of self-organized enzymes (MSb), was first performed. Figure 8.4 illustrates the organization of substrate fluxes and substrate input fluxes of the DMN. Three types of biochemical signals were considered in the network: activating (positive allosteric modulation), inhibitory (negative allosteric modulation), and an all-or-none type.

For building the DMN, several factors were chosen at random: (1) the number of flux interactions, (2) the number of regulatory signals, (3) the parameters associated with the flux-integration functions, (4) the regulatory coefficients of the allosteric activities, and (4) the values of the initial conditions in the activities of all metabolic subsystems (De la Fuente et al. 2011).

Since metabolic networks are open systems, we considered substrate input fluxes from the environment. Here, MSb3 and MSb10 receive the constant substrate inputs of $S1 = 0.54$ and $S2 = 0.16$, arbitrarily fixed.

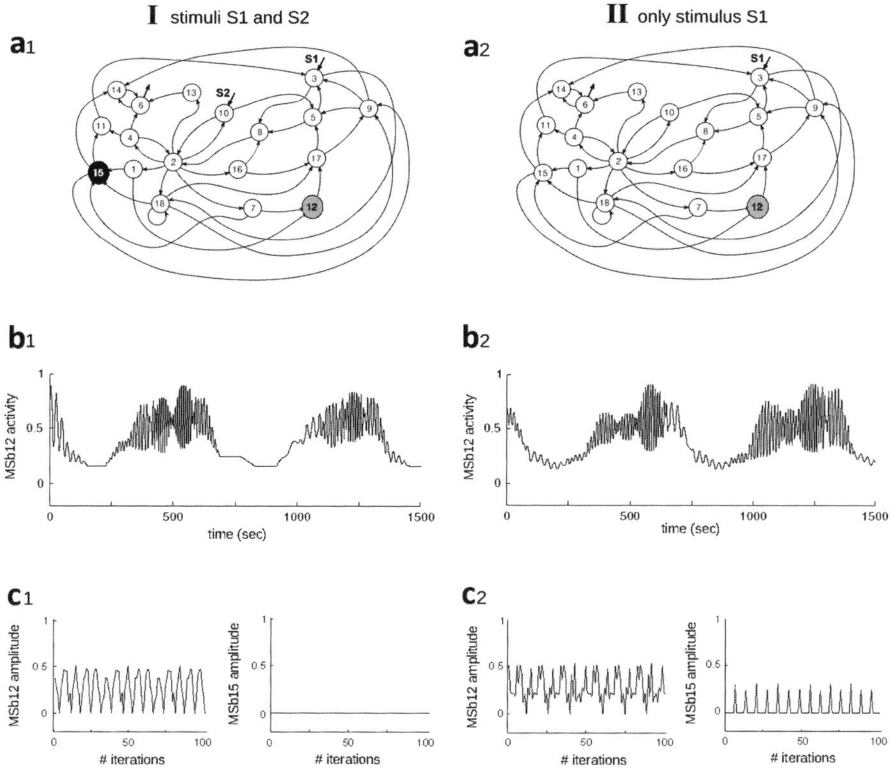

Fig. 8.4 *Dynamic catalytic behaviors of the dissipative metabolic network.* The network consisting of 18 self-organized multienzymatic complexes (metabolic subsystems) is depicted. The interconnection by fluxes through the network, and substrate input fluxes are shown. For simplification purposes, the topological architecture of regulatory signals considered in the network is not indicated. The network was perturbed by two different external conditions: (I) two substrate input fluxes S1 and S2 (*left column in panel*) and (II) only with the stationary stimulus S1 (*right column*) (*Panel* **a**1). Under condition I, a systemic metabolic structure spontaneously emerges in the network in which the enzymatic subsystem MSb12 is always active (metabolic core) (*gray circle*) and the catalytic subsystem MSb15 is inactive (*black circle*), whereas the rest of enzymatic sets exhibit on–off changing states (*white circle*) (*Panel* **a**1). Under condition II, the network preserves the systemic metabolic structure exhibiting flux plasticity which involves persistent changes in all the catalytic patterns (*Panels* **b** and **c**) and structural plasticity resulting in a persistent change in the dynamic state of the subsystem MSb15 which exhibits a transition from an off (*black circle*) to an on–off changing state (*white circle*) (*Panel* **a**2). Enzymatic activities of the MSb12 (metabolic core) exhibit a large number of different catalytic transitions between periodic oscillations and some steady states (*Panel* **b**1). Condition II produces persistent changes in the catalytic activity of subsystem MSb12 without steady states (flux plasticity) (*Panel* **b**2). Self-regulation the network under external conditions I and II implies changes in the amplitude of the activities from metabolic subsystem activities (flux plasticity) (*Panels* **c**1, **c**2). Adapted from Fig. 2 in (De la Fuente et al. 2011)

When the dynamics of this network was analyzed in the presence of two simultaneous external stimuli S1 and S2, a systemic metabolic structure emerged spontaneously. The MSb12 is always in an active state (metabolic core) and the MSb15 is always inactive whereas the remaining subsystems exhibit intermittent catalytic activities (on–off changing states) (Fig. 8.4). The active catalytic subsystems present complex output patterns with large periodic transitions between oscillatory and steady-state behaviors (105 transitions per period). Figure 8.4b1 displays a representative time series of activities from MSb12 showing 30 transitions between oscillatory and steady-state behavior.

When the external stimulus S2 was removed and only the stationary input flux of substrate S1 was considered, the network undergoes a drastic dynamic reorganization: flux (Almaas et al. 2005) and structural (Almaas et al. 2005; Almaas 2007) plasticity appear. The former involves persistent changes in all catalytic activities (see some examples in Fig. 8.4b2, c2), while the latter implies a persistent change in the state of the MSb15, i.e., under conditions of both stimuli S1 and S2 the MSb15 was in an off state, while the presence of only S1 stimulus MSb15 locks into an on–off changing dynamics. Despite the drastic catalytic changes observed in the temporal evolution of the subsystems dynamics, the network preserves the systemic metabolic structure, i.e., MSb12 is the metabolic core while the remaining subsystems exhibit intermittent dynamics.

Interestingly, the network adjusts the internal metabolic activities to the new environmental change (one or two stimuli) by means of flux and structural plasticities. This kind of behavior has been experimentally observed in several organisms as an adaptive response to external perturbations (Almaas et al. 2005). The complex dynamic behavior and transitions exhibited by the network studied are spontaneous and emerge from the regulatory structure, and nonlinear interactions.

8.3.1 Systemic Functional Structure of Biomolecular Information Flows

Next, the amplitude of the different catalytic patterns was used to study the effective connectivity based on transfer entropy.

For the condition corresponding to two simultaneous stimuli, the analyzed graph of effective connections shows that there are only nine metabolic subsystems with significant statistical values (Fig. 8.5d1). The arrows in the graph illustrate that the effective connectivity has directionality and the thickness is proportional to TE values. The maximum value of TE equal to 0.179 information bits corresponds to the link from MSb16 to MSb13.

Under conditions in which S2 was removed and only the stationary input flux of substrate S1 was considered, the TE values obtained are depicted in Fig. 8.5d2. It can be observed how the structure of the effective information flows is much more complex under one stimulus: (I) 10 of the 18 enzymatic sets have effective links,

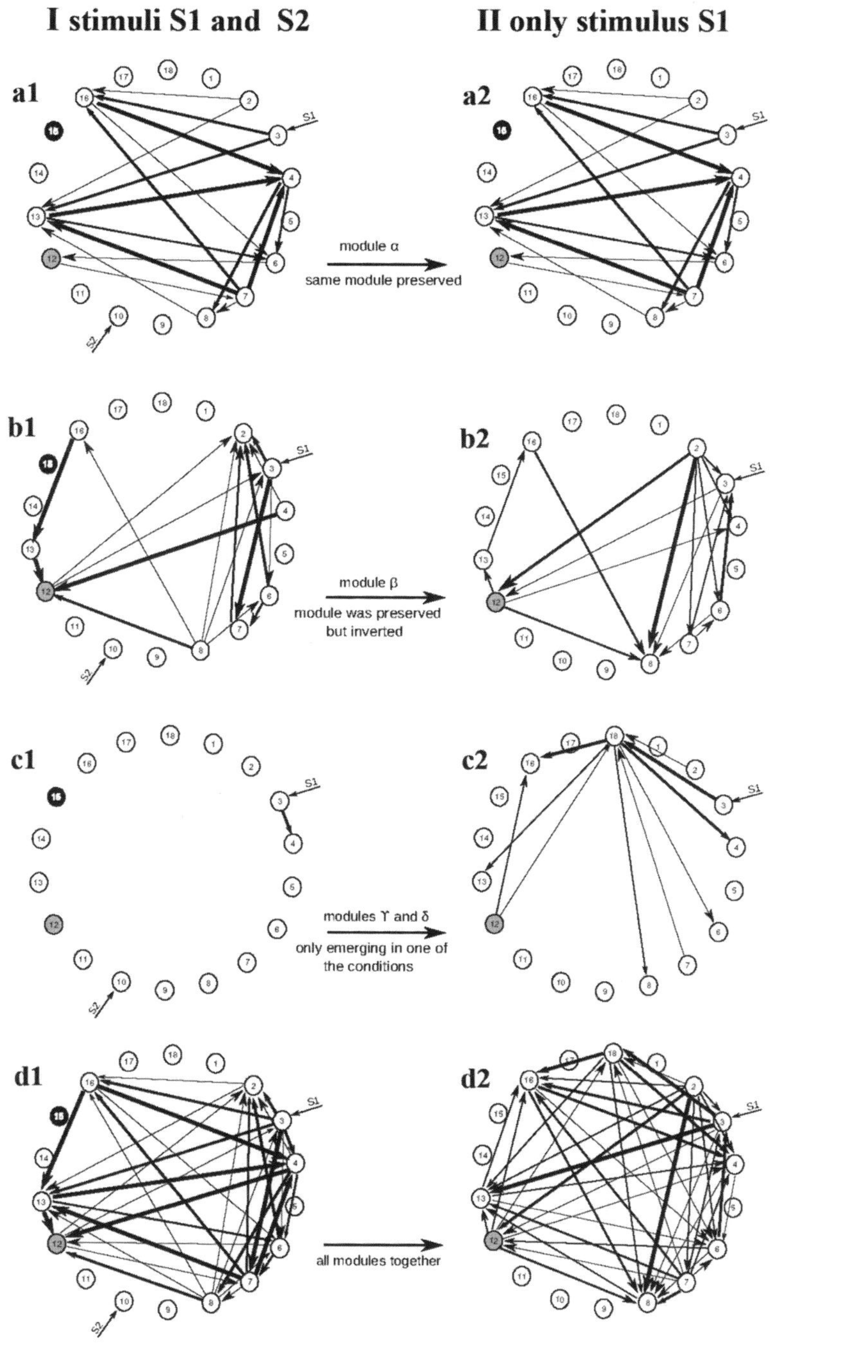

Fig. 8.5 *Systemic functional structure of biomolecular information flows*. The biomolecular information flows between the self-organized multienzymatic complexes (*circles*) are measured in bits, and they are responsible for their functional coordination. The arrow's thickness is proportional to the TE value except when TE was smaller than 0.1, for which the thickness does

(II) the density of effective links increases from 10.802 (two stimuli) to 13.580 (one stimulus), and (III) not only the density increased but the connections were made stronger (mean values increase from 0.079 to 0.133 and the maximum value of effective connectivity is 0.377 versus 0.179 molecular information bits).

The metabolic core also manifests qualitative and quantitative changes in the molecular information flows. Specifically, increasing the TE values and modifying the connectivity to several subsystems, for example, under stimuli S1 and S2, the core receives a causal information flow with a TE = 0.116 from MSb8 (Fig. 8.6). When the substrate input flux S2 is removed, the directionality of the signal reverses and the core sends an information flow of TE = 0.153 to MSb8 (see Fig. 8.6).

TE analysis reveals that a systemic functional structure of effective information flows emerges in the network, which is able to modify the catalytic activity of all the metabolic subsystems. Moreover, the level of information flows is highly responsive to environmental influences.

8.3.2 *Modular Organization of the Biomolecular Information Flows: Metabolic Switches*

Our analyses also showed a modular organization of the effective information flows in which some sets of catalytic subsystems are clustered forming functional metabolic sub-networks. A detailed study of TE data allows us to infer different modular organization: (I) a set of effective connections between certain subsystems is preserved under both external conditions (Module α) (Fig. 8.5a1–a2), (II) a second sub-network of effective information flows exhibit reverse directionality (Module β) (Fig. 8.5b1–b2), meaning that the TE connections are preserved but their direction is inversed, and (III) the third set of connections emerges only in one of the stimulation conditions (Modules γ and δ) (Fig. 8.5c1–c2).

Under both stimuli S1 and S2, the diversity of the enzymatic behavior is systemically self-regulated by means of modules α, β, and γ. However, when only a stationary input flux of substrate S1 is considered, the network undergoes a dramatic reorganization of all catalytic dynamics exhibiting flux and structural

---

Fig. 8.5 (continued) not scale with the TE value as it was plotted as thin as possible to be visualized. The values of TE are statistical significant (p value < 0.05, Bonferroni correction, $n = 50$ experiments). The metabolic subsystem activities shape four modules of effective connectivity. *Panels* **a**1–**a**2, the module α is preserved during both external stimuli; *Panels* **b**1–**b**2, a second sub-network of effective information flows are preserved but with inverted directionality (Module β); *Panels* **c**1–**c**2, a third class of modules emerges only in one of the external perturbations considered (Modules γ and δ); Panels **d**1–**d**2, all modules together. The transitions between modules (metabolic switches) provoke permanent changes in all catalytic activities of the metabolic subsystems, and these metabolic switches are triggered by changes in the external conditions (I: two stimuli S1 and S2, II: only stimulus S1). Adapted from Fig. 5 in (De la Fuente et al. 2011)

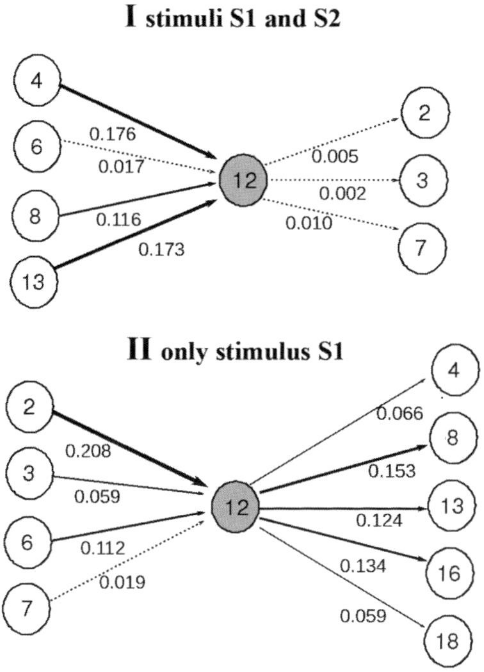

Fig. 8.6 *The dynamical structure of biomolecular information flows modify the catalytic activity of self-organized multienzymatic complexes*. The figure displays the in and out biomolecular information flows belonging to the metabolic core (*gray circle*). The white circles represent active multienzymatic complexes. The metabolic core (the metabolic subsystem 18) depends on the biomolecular information contained in substrate fluxes and regulatory signals; this core performs three functions simultaneously: signal reception and integration, and being a source for new biomolecular information. The dynamical functional structure depends strongly on the external stimuli (I and II)

plasticity. These drastic changes are self-regulated by means of Module α, an emergent Module δ, and Module β, but now in this module all the directions of effective connectivity reverse. Therefore, the transitions between the modules provoke permanent changes in all catalytic activities of the self-organized enzymatic subsystems, and these metabolic switches are triggered by changes in the external conditions.

Functionally, metabolic switches are discrete transitions between catalytic modules regulated by systemic dynamics of the biomolecular information flows. At a molecular level, a switch is a biochemical process in which determined enzymes can undergo a persistent change in their catalytic status. Experimentally, it has been observed that the molecular mechanisms involved in a switch depend on posttranslational modification of enzymatic activities (Miller et al. 2005) such as phosphorylation (Oesch-Bartlomowicz and Oesch 2003), acetylation (Pasini et al. 2010), and methylation (He et al. 2005). Molecular switches have been implicated in several metabolic processes including transcriptional regulation

(Lehman and Kelly 2002), Warburg effect (Levine and Puzio-Kuter 2010; Vasseur et al. 2010), cell cycle (Verkest et al. 2005), epigenetic processes (Lim and van Oudenaarden 2007), central carbon metabolism (Xu et al. 2008), DNA repair (He et al. 2005), growth cell metabolism (Tennessen et al. 2011), and T-cell activation and apoptosis (Perl et al. 2002).

In addition to the network topological structure, characterized by the specific location of enzymatic subsystems, molecular substrate fluxes, and regulatory signals, there is a global functional structure of biomolecular information flows which is dynamic and able to modify the catalytic activities of all participating enzymatic sets. Systemic metabolic structure not only involves a metabolic core and self-organized multienzymatic complexes in an on–off mode, but also determines a sophisticated structure of effective information flows which also provides integrative coordination and synchronization between all metabolic subsystems. The functional structure of biomolecular information flows is modular and the dynamic changes between modules correspond to metabolic switches which allow for critical transitions in the enzymatic activities. The modules of effective connectivity and the functional switches seem to be important elements of the systemic metabolic structure.

8.4 Concluding Remarks

In this work, we have addressed some aspects related with the self-organization of metabolic processes in terms of information theory. Specifically, we have quantified biomolecular information flows in bits between catalytic elements that put in evidence essential aspects of metabolic function. In particular, we have shown the emergence of effective connectivity structures and the functional coordination between catalytic elements.

As a continuation of these results, in another work we have performed analyses of different catalytic activities in a dissipative metabolic network based on statistical mechanics. We calculated the Shannon entropy and the energy function and found that enzymatic activities are systemically governed by Hopfield-like attractors with capacity to store functional catalytic patterns which can be correctly recovered from specific input stimuli. The metabolic attractors regulate the catalytic patterns, modify the efficiency in the connection between self-organized multienzymatic complexes, and stably store these modifications (De la Fuente et al. 2013). In the light of our results, the systemic metabolic structure appears to operate as a complex information processing system which continuously defines sets of biochemical instructions that make it to evolve into a particular and precise catalytic regime for each multienzymatic subsystem (De la Fuente et al. 2013).

At present, we are working on the molecular mechanisms that link the metabolic information emergent in the systemic metabolic structure with genetic information. Understanding the principles and quantitative laws that govern the systemic self-organization of enzymatic processes will be crucial to elucidate the structural and functional architecture of cellular dynamics.

References

Ainscow EK, Mirsham S, Tang T, Ashford MLJ, Rutter GA (2002) Dynamic imaging of free cytosolic ATP concentration during fuel sensing by rat hypothalamic neurones: evidence for ATP independent control of ATP-sensitive K^+ channels. J Physiol 544:429–45

Allegrini P, Buiatti M, Grigolini P, West BJ (1988) Fractal brownian motion as non stationary process: an alternative paradigm for DNA sequences. Phys Rev E 57:4558–62

Almaas E, Kovacs B, Vicsek T, Oltvai ZN, Barabási AL (2004) Global organization of metabolic fluxes in the bacterium Escherichia coli. Nature 427:839–43

Almaas E, Oltvai ZN, Barabasi AL (2005) The activity reaction core and plasticity of metabolic networks. PLoS Comput Biol 1(7):e68:0557–563

Almaas E (2007) Biological impacts and context of network theory. J Exp Biol 210:1548–58

Aon MA, Cortassa S (1997) Dynamic biological organization. Fundamentals as applied to cellular systems. Chapman & Hall, London, UK

Aon MA, Cortassa S, Lloyd D (2000) Chaotic dynamics and fractal space in biochemistry: simplicity underlies complexity. Cell Biol Int 24:581–87

Aon MA, Cortassa S, O'Rourke B (2004) Percolation and criticality in a mitochondrial network. Proc Natl Acad Sci USA 101:4447–52

Aon MA, Cortassa S, O'Rourke B (2006) The fundamental organization of cardiac mitochondria as a network of coupled oscillators. Biophys J 91:4317–27

Aon MA, Roussel MR, Cortassa S, O'Rourke B, Murray DB et al (2008) The scale-free dynamics of eukaryotic cells. PLoS One 3(e3624):1–12

Asano Y, Nagasaki A, Uyeda TQP (2008) Correlated waves of actin filaments and PIP3 in Dictyostelium cells. Cell Mot Cytosk 65:923–34

Audit B, Vaillant C, Arné A, D'Aubenton-Caraf Y, Thermes C (2004) Wavelet analysis of DNA bending profiles reveals structural constraints on the evolution of genomic sequences. J Biol Phys 30:33–81

Barril EF, Potter AR (1968) Systematic oscillations of amino acid transport in liver from rats adapted to controlled feeding schedules. J Nutrition 95:228–37

Bastiaens P, Caudron M, Niethammer P, Karsenti E (2006) Gradients in the self-organization of the mitotic spindle. Trends Cell Biol 16(3):125–34

Bernardinelli Y, Magistretti PJ, Chatton JY (2004) Astrocytes generate Na^+–mediated metabolic waves. Proc Natl Acad Sci USA 101:14937–42

Berridge MJ, Galione A (1988) Cytosolic calcium oscillators. FASEB J 2:3074–82

Bertschinger N, Natschlager T (2004) Real-time computation at the edge of Chaos in recurrent neural networks. Neural Comput 16:1413–36

Bobik TA (2006) Polyhedral organelles compartmenting bacterial metabolic processes. Appl Microbiol Biotechnol 70:517–25

Boiteux A, Goldbeter A, Hess B (1975) Control of oscillating glycolysis of yeast by stochastic, periodic, and steady source of substrate: a model and experimental study. Proc Natl Acad Sci USA 72:3829–33

Brodsky V, Boikov PY, Nechaeva NV, Yurovitsky YG, Novikova TE et al (1992) The rhythm of protein synthesis does not depend on oscillations of ATP level. J Cell Sci 103:363–70

Brodsky VY (2006) Direct cell-cell communication: a new approach derived from recent data on the nature and self-organisation of ultradian (circahoralian) intracellular rhythms. Biol Rev Camb Philos Soc 81:143–62

Cadena-Nava R, Comas-Garcia M, Garmann R, Rao ALN, Knobler C et al (2012) Self-assembly of viral capsid protein and RNA molecules of different sizes: requirement for a specific high protein/RNA mass ratio. J Virol 86(6):3318–26

Campanella ME, Chu H, Wandersee NJ, Peters LL, Mohandas N et al (2008) Characterization of glycolytic enzyme interactions with murine erythrocyte membranes in wild-type and membrane protein knockout mice. Blood 112(9):3900–6

Cascante M, Centelles JJ, Agius L (2000) Use of alpha-toxin from Staphylococcus aureus to test for channelling of intermediates of glycolysis between glucokinase and aldolase in hepatocytes. Biochem J 352(3):899–905
Cech T (2000) Structural biology. The ribosome is a ribozyme. Science 289(5481):878–9
Ceschini S, Lupidi G, Coletta M, Pon CL, Fioretti E et al (2000) Multimeric self-assembly equilibria involving the histone-like protein H-NS. A thermodynamic study. J Biol Chem 14(2):729–34
Chabot JR, Pedraza JM, Luitel P, van Oudenaarden A (2007) A Stochastic gene expression out-of-steady-state in the cyanobacterial circadian clock. Nature 450:1249–52
Chance B, Hess B, Betz A (1964) DPNH oscillations in a cell-free extract of S. carlsbergensis. Biochem Biophys Res Commun 16:182–7
Chance B, Pye EK, Ghosh AD, Hess B (1973) Biological and biochemical oscillations. Academic, New York, NY
Chandrashekaran MK (2005) Time in the living world. Universities Press, Hyderabad, India
Chiam HK, Rajagopal G (2007) Oscillations in intracellular signaling cascades. Phys Rev E 75: 061901
Clarke FM, Masters CJ (1975) On the association of glycolytic enzymes with structural proteins of skeletal muscle. Biochim Biophys Acta 381:37–46
Clegg JS, Jackson SA (1990) Glucose metabolism and the channelling of glycolytic intermediates in permeabilized L-929 cells. Arch Biochem Biophys 278:452–60
Commichau FM, Rothe FM, Herzberg C, Wagner E, Hellwig D, Lehnik-Habrink M, Hammer E, Völker U, Stülke J (2009) Novel activities of glycolytic enzymes in Bacillus subtilis: interactions with essential proteins involved in mRNA processing. Mol Cell Proteomics 8(6): 1350–60
Connor KM, Gracey AY (2011) Circadian cycles are the dominant transcriptional rhythm in the intertidal mussel Mytilus californianus. Proc Natl Acad Sci 108:16110–5
Cortassa S, Aon MA, Westerhoff HV (1991) Linear non equilibrium thermodynamics describes the dynamics of an autocatalytic system. Biophys J 60:794–803
Cortassa S, Aon MA (1994) Spatio-temporal regulation of glycolysis and oxidative phosphorylation in vivo in tumor and yeast cells. Cell Biol Int 18(7):687–713
Cover T, Thomas J (1991) Elements of information theory. John Wiley & Sons, Inc.
Dano S, Sorensen P, Hynne F (1999) Sustained oscillations in living cells. Nature 402:320–2
De Forest M, Wheeler CJ (1999) Coherent oscillations in membrane potential synchronize impulse bursts in central olfactory neurons of the crayfish. J Neurophysiol 81:1231–41
Dekhuijzen A, Bagust J (1996) Analysis of neural bursting: nonrhythmic and rhythmic activity in isolated spinal cord. J Neurosci Methods 67:141–7
De la Fuente IM, Martínez L, Veguillas J (1995) Dynamic behavior in glycolytic oscillations with phase shifts. Biosystems 35:1–13
De la Fuente IM, Martínez L, Veguillas J (1996a) Intermittency route to Chaos in a biochemical system. Biosystems 39:87–92
De la Fuente IM, Martínez L, Veguillas J, Aguirregabiria JM (1996b) Quasiperiodicity route to Chaos in a biochemical system. Biophys J 71:2375–79
De la Fuente IM, Martínez L, Veguillas J, Aguirregabiria JM (1998a) Coexistence of multiple periodic and chaotic regimes in biochemical oscillations. Acta Biotheor 46:37–51
De la Fuente IM, Martinez L, Aguirregabiria JM, Veguillas J (1998b) R/S analysis in strange attractors. Fractals 6(2):95–100
De la Fuente IM, Martınez L, Benitez N, Veguillas J, Aguirregabiria JM (1998c) Persistent behavior in a phase-shift sequence of periodical biochemical oscillations. Bull Math Biol 60: 689–702
De La Fuente IM, Benítez N, Santamaría A, Veguillas J, Aguirregabiria JM (1999a) Persistence in metabolic nets. Bull Mathemat Biol 61:573–95
De la Fuente IM (1999b) Diversity of temporal self-organized behaviors in a biochemical system. Biosystems 50:83–97

De la Fuente IM, Martínez L, Aguirregabiria JM, Veguillas J, Iriarte M (1999c) Long-range correlations in the phase-shifts of numerical simulations of biochemical oscillations and in experimental cardiac rhythms. J Biol Syst 7(2):113–30
De la Fuente IM, Pérez-Samartín A, Martínez L, García MA, Vera-López A (2006) Long-range correlations in rabbit brain neural activity. Ann Biomed Eng 34:295–9
De La Fuente IM, Martínez L, Pérez-Samartín AL, Ormaetxea L, Amezaga C et al (2008) Global Self-organization of the cellular metabolic structure. PLoS One 3:e3100
De La Fuente IM, Vadillo F, Pérez-Pinilla M-B, Vera-López A, Veguillas J (2009) The number of catalytic elements is crucial for the emergence of metabolic cores. PlosOne 4:e7510:1–e7510:11
De la Fuente IM (2010) Quantitative analysis of cellular metabolic dissipative. Self-organized structures. Int J Mol Sci 11(9):3540–99
De La Fuente IM, Vadillo F, Pérez-Samartín AL, Pérez-Pinilla M-B, Bidaurrazaga J et al (2010) Global self-regulations of the cellular metabolic structure. PlosOne 5:e9484:1–e9484:15
De la Fuente IM, Cortes J, Perez-Pinilla M, Ruiz-Rodriguez V, Veguillas J (2011) The metabolic core and catalytic switches are fundamental elements in the self-regulation of the systemic metabolic structure of cells. PLoS One 6:e27224
De la Fuente IM, Cortes JM (2012) Quantitative analysis of the effective functional structure in yeast glycolysis. PLoS One 7(2):e30162
De la Fuente IM, Cortes JM, Pelta D, Veguillas J (2013) Attractor metabolic networks. PLoS One 8:e58284
Dunlap JC, Loros JJ, DeCoursey P (2004) Chronobiology: biological timekeeping. Sinauer Associates, Sunderland, MA
Duysens L, Amesz J (1957) Fluorescence espectrophotometry of reduced phosphopyridine nucleotide in intact cells in the near-ultraviolet and visible region. Biochem Biophys Acta 24:19–26
Ebeling W, et al (1986) Selforganization by non-linear irreversible processes. In: Ebeling, W, Ulbricht H (eds) Thermodynamic aspects of selforganization. Springer, Berlin
Eke A, Herman P, Kocsis L, Kozak LR (2002) Fractal characterization of complexity in temporal physiological signals. Physiol Meas 23:R1–R38
Fan C, Cheng S, Liu Y, Escobar CM, Crowley CS (2010) Short N-terminal sequences package proteins into bacterial microcompartments. Proc Natl Acad Sci USA 107:7509–14
Fuentes JM, Pascual MR, Salido G, Soler JA (1994) Madrid oscillations in rat liver cytosolic enzyme activities of the urea cycle. Arch Physiol Biochem 102(5):237–41
Gancedo C, Serrano R (1989) The yeasts. In: AH Rose, Harrison JS (eds) The yeasts. Academic, London, UK.
Galas L, Garnier M, Lamacz M (2000) Calcium waves in frog melanotrophs are generated by intracellular inactivation of TTX-sensitive membrane Na_ channel. Mol Cell Endocrinol 170: 197–209
Garmendia-Torres C, Goldbeter A, Jacquet M (2007) Nucleocytoplasmic oscillations of the yeast transcription factor Msn2: evidence for periodic PKA activation. Curr Biol 17:1044–49
Gavin AC, Bosche M, Krause R, Grandi P, Marzioch M et al (2002) Functional organization of the yeast proteome by systematic analysis of protein complexes. Nature 415:141–7
Getty L, Panteleon AE, Mittelman SD, Dea MK, Bergman RN (2000) Rapid oscillations in omental lipolysis are independent of changing insulin levels in vivo. J Clin Invest 106:421–30
Getty-Kaushik L, Richard AM, Corkey BE (2005) Free fatty acid regulation of glucose-dependent intrinsic oscillatory lipolysis in perifused isolated rat adipocytes. Diabetes 54(3):629–37
Glick BS (2007) Let there be order. Nat Cell Biol 9:130–2
Goldberger AL, Amaral LA, Hausdorff JM, Ivanov PCH, Peng CK, Stanley HE (2002) Fractal dynamics in physiology: alterations with disease and aging. Proc Natl Acad Sci USA 99: 2466–72
Goldbeter A, Lefever R (1972) Dissipative structures for an allosteric model. Biophys J 12(10): 1302–1315
Goldbeter A (1973) Patterns of spatiotemporal organization in an allosteric enzyme model. Proc Natl Acad Sci USA 70:3255–59

Goldbeter A (2002) Computational approaches to cellular rhythms. Nature 420:238–45
Goldbeter A (2007) Biological rhythms as temporal dissipative structures. Adv Chem Phys 135: 253–95
Graham JW, Williams TC, Morgan M, Fernie AR, Ratcliffe RG et al (2007) Glycolytic enzymes associate dynamically with mitochondria in response to respiratory demand and support substrate channeling. Plant Cell 19(11):3723–38
Guthrie PB, Knappenberger J, Segal M, Bennett MVL, Charles AC et al (1999) ATP released from astrocytes mediates glial calcium waves. J Neurosci 19:520–8
Halley JD, Winkler DA (2008) Consistent concepts of self-organization and self-assembly. Complexity 14(2):10–17
Hans MA, Heinzle E, Wittmann C (2003) Free intracellular amino acid pools during autonomous oscillations in Saccharomyces cerevisiae. Biotechnol Bioeng 82(2):143–151
Hartig K, Beck E (2005) Endogenous cytokinin oscillations control cell cycle progression of tobacco BY-2 cells. Plant Biol 7(1):33–40
Haviv L, Brill-Karniely Y, Mahaffy R, Backouche F, Ben-Shaul A et al (2006) Reconstitution of the transition from lamellipodium to filopodium in a membrane-free system. Proc Natl Acad Sci USA 103:4906–11
He C, Hus JC, Sun LJ, Zhou P, Norman DP et al (2005) A methylation-dependent electrostatic switch controls DNA repair and transcriptional activation by E. coli ada. Mol Cell 20:117–29
Heinisch JJ, Boles E, Timpel C (1996) A yeast phosphofructokinase insensitive to the allosteric activator fructose-2,6-bisphosphate. J Biol Chem 271:15928–33
Hildebrandt G (1982) The time structure of adaptive processes. Georg Thieme Verlag, Stuttgart, Germany
Ho Y, Gruhler A, Heilbut A, Bader GD, Moore L et al (2002) Systematic identification of protein complexes in Saccharomyces cerevisiae by mass spectrometry. Nature 415:180–3
Holz GG, Emma Heart E, Leech CA (2008) Synchronizing Ca^{2+} and cAMP oscillations in pancreatic beta cells: a role for glucose metabolism and GLP-1 receptors? Am J Physiol Cell Physiol 294:c4–c6
Huber F, Käs J (2011) Self-regulative organization of the cytoskeleton. Cytoskeleton 68 (5):259–65
Hungerbuehler AK, Philippsen P, Gladfelter AS (2007) Limited functional redundancy and oscillation of cyclins in multinucleated Ashbya gossypii fungal cells. Eukaryot Cell 6(3): 473–86
Ishii K, Hirose K, Iino M (2006) Ca^{2+} shuttling between endoplasmic reticulum and mitochondria underlying Ca2+ oscillations. EMBO 7:390–6
Ito T, Chiba T, Ozawa R, Yoshida M, Hattori M et al (2001) Comprehensive two-hybrid analysis to explore the yeast protein interactome. Proc Natl Acad Sci USA 98:4569–74
Ishikawa M, Tsuchiya D, Oyama T, Tsunaka Y, Morikawa K et al (2004) Structural basis for channelling mechanism of a fatty acid bold beta-oxidation multienzyme complex. EMBO J 23:2745–54
Jorgensen K, Rasmussen AV, Morant M, Nielsen AH, Bjarnholt N et al (2005) Metabolon formation and metabolic channeling in the biosynthesis of plant natural products. Curr Opin Plant Biol 8:280–91
Jovanović S, Jovanović A, Crawford MR (2007) M-ldh serves as a regulatory subunit of the cytosolic substrate-channelling complex in vivo. J Mol Biol 371:349–61
Jules M, Francois J, Parrou JL (2005) Autonomous oscillations in Saccharomyces cerevisiae during batch cultures on trehalose. FEBS J 272:1490–1500
Karsenti E (2008) Self-organization in cell biology: a brief history. Nat Rev Mol Cell Biol 9(3): 255–62
Kauffman S, Johnsen S (1991) Coevolution to the edge of chaos: coupled fitness landscapes, poised states, and coevolutionary avalanches. J Theor Biol 149:467–505
Kauffman S (1993) Origins of order: self-organization and selection in evolution. Oxford University Press, Oxford

Kazachenko VN, Astashev ME, Grinevic AA (2007) Multifractal analysis of K^+ channel activity. Biochemistry (Moscow) 2:169–75
Kindzelskii AL, Zhou MJ, Haugland PR, Boxer AL, Petty RH (1998) Oscillatory pericellular proteolysis and oxidant deposition during neutrophil locomotion. Biophys J 74:90–7
Klevecz RR, Murray DB (2001) Genome wide oscillations in expression—wavelet analysis of time series data from yeast expression arrays uncovers the dynamic architecture of phenotype. Mol Biol Rep 28:73–82
Klevecz RR, Bolen J, Forrest G, Murray DB (2004) A genomewide oscillation in transcription gates DNA replication and cell cycle. Proc Natl Acad Sci 101(5):1200–05
Klimontovich YL (1999) Entropy and information of open systems. Phys-Uspekhi 42:375–84
Kushner DJ (1969) Self-assembly of biological structures. Bacteriol Rev 33:302–45
Lange G, Mandelkow EM, Jagla A, Mandelklow E (2004) Tubulin oligomers and microtubule oscillations antagonistic role of microtubule stabilizers and destabilizers. FEBS 178:61–9
Lehman JJ, Kelly DP (2002) Transcriptional activation of energy metabolic switches in the developing and hypertrophied heart. Clin Exp Pharmacol Physiol 29:339–45
Levine AJ, Puzio-Kuter AM (2010) The control of the metabolic switch in cancers by oncogenes and tumor suppressor genes. Science 330:1340–44
Lee CF (2008) Self-assembly of protein amyloids: a competition between amorphous and ordered aggregation. Phys Rev E 80(3):1–5
Lilley D (2005) Structure, folding and mechanisms of ribozymes. Curr Opin Struct Biol 15(3): 313–23
Lim HN, van Oudenaarden A (2007) A multistep epigenetic switch enables the stable inheritance of DNA methylation states. Nat Genet 39:269–75
Lloyd D, Eshantha L, Salgado J, Turner MP, Murray DB (2002) Respiratory oscillations in yeast: clock-driven mitochondrial cycles of energization. FEBS Lett 519(1–3):41–4
Lloyd D (2005) Systems dynamics of biology. J Appl Biomed 3:1.12
Lloyd D, Murray DB (2005) Ultradian metronome: timekeeper for orchestration of cellular coherence. Trends Biochem Sci 30(7):373–7
Lloyd D, Murray DB (2006) The temporal architecture of eukaryotic growth. FEBS Lett 580: 2830–35
Lloyd D, Murray DB (2007) Redox rhythmicity at the core of temporal coherence. Bioessays 29:465–73
Lloyd D, Douglas B, Murray DB (2006) The temporal architecture of eukaryotic growth. FEBS Lett 580:2830–5
Loose M, Fischer-Friedrich E, Ries J, Kruse K, Schwille P (2008) Spatial regulators for bacterial cell division self-organize into surface waves in vitro. Science 320(5877):789–92
Lunn EJ (2007) Compartmentation in plant metabolism. J Exp Bot 58:35–47
Mahasweta D, Gebber GL, Barman SM, Lewis CD (2003) Fractal properties of sympathetic nerve discharge. J Neurophysiol 89:833–40
Mair T, Warnke C, Muller SC (2001) Spatio-temporal dynamics in glycolysis. Faraday Discuss 120:249–59
Malaisse WJ, Zhang Y, Jijakli H, Courtois P, Sener A (2004) Enzyme-to-enzyme channelling in the early steps of glycolysis in rat pancreatic islets. Int J Biochem Cell Biol 36(8):1510–20
Madsen MF, Danø S, Sørensen PG (2005) On the mechanisms of glycolytic oscillations in yeast. FEBS J 272:2648–60
Markus M, Plesser T, Boiteux A, Hess B, Malcovati M (1980) Rate law of pyruvate kinase type I from Escherichia coli. Biochem J 189:421–33
Markus M, Kuschmitz D, Hess B (1984) Chaotic dynamics in yeast glycolysis under periodic substrate input flux. FEBS 172:235–38
Markus M, Kuschmitz D, Hess B (1985a) Properties of strange attractors in yeast glycolysis. Biophys Chem 22:95–105

Markus M, Muller S, Hess B (1985b) Observation of entrainment quasiperiodicity and chaos in glycolyzing yeast extracts under periodic glucose input. Ber Bunsen-Ges Phys Chem 89: 651–54
Marquez S, Crespo P, Carlini V, Garbarino-Pico E, Baler R et al (2004) The metabolism of phospholipids oscillates rhythmically in cultures of fibroblasts and is regulated by the clock protein PERIOD 1. FASEB J 18:519–21
Mitchison TJ (1992) Self-organization of polymer-motor systems in the cytoskeleton. Philos Trans R Soc Lond B Biol Sci 336:99–106
Milani M, Pesce A, Bolognesi M, Bocedi A, Ascenzi P (2003) Substrate channeling: molecular bases. Biochem Mol Biol 31:228–33
Miller P, Zhabotinsky AM, Lisman JE, Wang X-J (2005) The stability of a stochastic CaMKII switch: dependence on the number of enzyme molecules and protein turnover. PLoS Biol 3: e107
Misteli T (2009) Self-organization in the genome. PNAS 106(17):6885–6
Møller AC, Hauser MJB, Olsen LF (1998) Oscillations in peroxidase-catalyzed reactions and their potential function in vivo. Biophys Chem 72:63–72
Monge C, Beraud N, Kuznetsov AV, Rostovtseva T, Sackett D (2008) Regulation of respiration in brain mitochondria and synaptosomes: restrictions of ADP diffusion in situ, roles of tubulin, and mitochondrial creatine kinase. Mol Cell Biochem 318:147–65
Monge C, Grichine A, Rostovtseva T, Sackett P, Saks VA (2009) Compartmentation of ATP in cardiomyocytes and mitochondria. Kinetic studies and direct measurements. Biophys J 96:241
Mowbray J, Moses V (1976) The tentative identification in Escherichia coli of a multienzyme complex with glycolytic activity. J Biochem 66:25–36
Murray D, Beckmann M, Kitano H (2007) Regulation of yeast oscillatory dynamics. Proc Natl Acad Sci 104(7):2241–46
Nakahata Y, Grimaldi B, Sahar S, Hirayama J, Sassone-Corsi P (2007) Signaling to the circadian clock: plasticity by chromatin remodelling. Curr Opin Cell Biol 19:230–7
Negrutskii BS, Deutscher MP (1991) Channeling of aminoacyl tRNA for protein synthesis in vivo. Proc Natl Acad Sci USA 88:4991–5
Newman EA (2001) Propagation of intercellular calcium waves in retinal astrocytes and müller cells. J Neurosci 21:2215–23
Nicolis G, Prigogine I (1977) Self-organization in nonequilibrium systems. From dissipative structures to order through fluctuations. Wiley, New York
Oesch-Bartlomowicz B, Oesch F (2003) Cytochrome-P450 phosphorylation as a functional switch. Arch Biochem Biophys 409:228–34
Olsen LF, Degn H (1985) Chaos in biological systems. Q Rev Biophys 18:165–225
Olsen L, Andersen A, Lunding A, Brasen J, Poulsen A (2009) Regulation of glycolytic oscillations by mitochondrial and plasma membrane H^{+}–ATPases. Biophys J 96:38503861
Oliva A, Rosebrock A, Ferrezuelo F, Pyne S, Chen H et al (2005) The cell cycle-regulated genes of Schizosaccharomyces pombe. PLoS Biol 3:1239–60
Ovádi J, Srere PA (2000) Macromolecular compartmentation and channeling. Int Rev Cytol 192: 255–80
Ovádi J, Saks V (2004) On the origin of the ideas of intracellular compartmentation and organized metabolic systems. Mol Cell Biochem 256–257(1–2):5–12
Pang CN, Krycer JR, Lek A, Wilkins MR (2008) Are protein complexes made of cores, modules and attachments? Proteomics 8(3):425–34
Pasini D, Malatesta M, Jung HR, Walfridsson J, Willer A et al (2010) Characterization of an antagonistic switch between histone H3 lysine 27 methylation and acetylation in the transcriptional regulation of Polycomb group target genes. Nucleic Acids Res 38:4958–69
Perl A, Gergely P, Puskas F, Banki K (2002) Metabolic switches of T-cell activation and apoptosis. Antioxid Redox Signal 4:427–43
Petty HR, Worth RG, Kindzelskii AL (2000) Imaging sustained dissipative patterns in the metabolism of individual living cells. Phys Rev Lett 84:2754–57

Petty HR, Kindzelskii AL (2001) Dissipative metabolic patterns respond during neutrophil transmembrane signaling. Proc Natl Acad Sci USA 98:3145–49

Petty HR (2006) Spatiotemporal chemical dynamics in living cells: from information trafficking to cell physiology. Biosystems 83:217–24

Placantonakis DG, Welsh JP (2001) Two distinct oscillatory states determined by the NMDA receptor in rat inferior olive. J Physiol 534:123–40

Prigogine I, Mayné F, George C, De Haan M (1977) Microscopic theory of irreversible processes. Proc Nat Acad Sci USA 74:4152–56

Ramanujan VK, Biener G, Herman B (2006) Scaling behavior in mitochondrial redox fluctuations. Biophys J 90:L70–L72

Rengan R, Omann GM (1999) Regulation of oscillations in filamentous actin content in polymorphonuclear leukocytes stimulated with leukotriene B4 and platelet-activating factor. Biochem Biophys Res Commun 262:479–86

Romashko DN, Marban E, O'Rourke B (1998) Subcellular metabolic transients and mitochondrial redox waves in heart cells. Proc Natl Acad Sci USA 1998(95):1618–23

Rosenspire AJ, Kindzelskii AL, Petty H (2001) Pulsed DC electric fields couple to natural NAD(P) H oscillations in HT-1080 fibrosarcoma cells. J Cell Sci 114:1515–20

Roussel MR, Ivlev AA, Igamberdiev AU (2006) Oscillations of the internal CO_2 concentration in tobacco leaves transferred to low CO_2. J Plant Physiol 34:1188–96

Roussel MR, Lloyd D (2007) Observation of a chaotic multioscillatory metabolic attractor by real-time monitoring of a yeast continuous culture. FEBS J 274:1011–18

Ruelle D, Takens F (1971) On the nature of turbulence. Commun Math Phys 20:167–72

Saks V, Dzeja P, Schlattner U, Vendelin M, Terzic A et al (2006) Cardiac system bioenergetics: metabolic basis of the Frank-Starling law. J Physiol 571:253–73

Saks VA, Monge C, Anmann T, Dzeja P (2007) Integrated and organized cellular energetic systems: Theories of cell energetics, compartmentation and metabolic channeling. In: Saks VA (ed) Molecular system bioenergetics. Energy for life. Wiley-VCH, Weinheim, Germany

Saks V, Monge C, Guzun R (2009) Philosophical basis and some historical aspects of systems biology: from hegel to noble—applications for bioenergetic research. Int J Mol Sci 10:1161–92

Sánchez-Armáss S, Sennoune SR, Maiti D, Ortega F, Martínez-Zaguila R (2006) Spectral imaging microscopy demonstrates cytoplasmic pH oscillations in glial cells. Am J Physiol Cell Physiol 290:C524–C538

Scemes E, Giaume C (2006) Astrocyte calcium waves: what they are and what they do. Glia 54:716–25

Schibler U, Sassone-Corsi PA (2002) Web of circadian pacemakers. Cell 111:919–22

Schibler U, Naef F (2005) Cellular oscillators: rhythmic gene expression and metabolism. Curr Opin Cell Biol 17:223–9

Shaul O, Mironov V, Burssens S, Van Montagu M, Inze D (1996) Two Arabidopsis cyclin promoters mediate distinctive transcriptional oscillation in synchronized tobacco BY-2 cells. Proc Natl Acad Sci 93(10):4868–72

Shankaran H, Ippolito DL, Chrisler WB, Resat H, Bollinger K et al (2009) Rapid and sustained nuclear–cytoplasmic ERK oscillations induced by epidermal growth factor. Mol Syst Biol 332:1–13

Shearer G, Lee JC, Koo JA, Kohl DH (2005) Quantitative estimation of channeling from early glycolytic intermediates to CO in intact Escherichia coli. FEBS J 272(13):3260–9

Schreiber T (2000) Measuring information transfer. Phys Rev Lett 85:461–64

Slaby O, Lebiedz D (2009) Oscillatory NAD(P)H waves and calcium oscillations in neutrophils? A modeling study of feasibility. Biophys J 96:417–28

Smrcinová M, Sørensen PG, Krempasky J, Ballo P (1998) Chaotic oscillations in a chloroplast system under constant illumination. Int J Bifurcat Chaos 8:2467–70

Sourjik V, Armitage JP (2010) Spatial organization in bacterial chemotaxis. EMBO J 29(16): 2724–33

Stryer L (1995) Biochemistry. W.H. Freeman, New York
Sutter M, Boehringer D, Gutmann S, Günther S, Prangishvili D et al (2008) Structural basis of enzyme encapsulation into a bacterial nanocompartment. Nat Struct Mol Biol 15:939–47
Talkington MW, Siuzdak G, Williamson JR (2005) An assembly landscape for the 30S ribosomal subunit. Nature 438:628–32
Tennessen JM, Baker KD, Lam G, Evans J, Thummel CS (2011) The Drosophila estrogen-related receptor directs a metabolic switch that supports developmental growth. Cell Metab 13:139–48
Termonia Y, Ross J (1981) Oscillations and control features in glycolysis: Numerical analysis of a comprehensive models. Proc Natl Acad Sci USA 78:29522956
Tian B, Nowak DE, Brasier AR (2005) A TNF-induced gene expression program under oscillatory NF-κB control. BMC Genom 6:137
Tonozuka H, Wang J, Mitsui K, Saito T, Hamada Y et al (2001) Analysis of the upstream regulatory region of the GTS1 gene required for its oscillatory expression. J Biochem 130:589–95
Tresset G (2009) The multiple faces of self-assembled lipidic systems. PMC Biophys 17 2(1):3
Tu BP, Kudlicki A, Rowicka M, McKnight SL (2005) Logic of the yeast metabolic cycle: temporal compartmentalization of cellular processes. Science 310:1152–58
Tyson J, Novak B, Odell MG, Chen K, Thron CD (1996) Chemical kinetic theory: understanding cell-cycle regulation. TIBS 21:89–96
Ueda T, Nakagaki T, Yamada T (1990) Dynamic organization of ATP and birefringent fibrils during free locomotion and galvanotaxis in the plasmodium of Physarum polycephalum. J Cell Biol 110:1097–02
Uetz P, Giot L, Cagney G, Mansfield TA, Judson RS et al (2000) A comprehensive analysis of protein–protein interactions in *Saccharomyces cerevisiae*. Nature 403:623–27
Vanin AF, Ivanov VI (2008) Interaction of iron ions with oxygen or nitrogen monoxide in chromosomes triggers synchronous expression/suppression oscillations of compact gene groups ("genomewide oscillation"): Hypothesis. Nitric Oxide 18:147–52
Vasseur S, Tomasini R, Tournaire R, Iovanna JL (2010) Hypoxia induced tumor metabolic switch contributes to pancreatic cancer aggressiveness. Cancers 2:2138–52
Verkest A, Weinl C, Inzé D, De Veylder L, Schnittger A (2005) Switching the cell cycle. Kip-Related Proteins in Plant Cell Cycle Control Plant Physiol 139:1099–106
Vicker MG (2002) Eukaryotic cell locomotion depends on the propagation of self-organized reaction–diffusion waves and oscillations of actin filament assembly. Exp Cell Res 275:54–66
Viola E, Raushel M, Rendina R, Cleland W (1982) Substrate synergism and the kinetic mechanism of yeast hexokinase. Biochemistry 21:1295–302
Viswanathan GM, Buldyrev SV, Havlin S, Stanley HE (1997) Quantification of DNA patchiness using long-range correlation measures. Biophys J 72:866–75
Waingeh VF, Gustafson CD, Kozliak EI, Lowe SL, Knull HR, Thomasson KA (2006) Glycolytic enzyme interactions with yeast and skeletal muscle F-actin. Biophys J 90(4):1371–84
Whitesides GM, Grzybowski B (2002) Self-assembly at all scales. Science 295:2418–21
Wijnen H, Young MW (2006) Interplay of circadian clocks and metabolic rhythms. Annu Rev Genet 40:409–48
Wittmann C, Hans M, Van Winden AW, Ras C, Heijnen JJ (2005) Dynamics of intracellular metabolites of glycolysis and TCA cycle during cell-cycle-related oscillation in Saccharomyces cerevisiae. Biotechnol Bioeng 89:839–47
Wolf J, Passarge J, Somsen OJG, Snoep JL, Heinrich R, Westerhof HV (2000) Transduction of intracellular and intercellular dynamics in yeast glycolytic oscillations. Biophys J 78:1145–53
Woodford C, Zandstra PW (2012) Tissue engineering 2.0: guiding self-organization during pluripotent stem cell differentiation. Curr Opin Biotechnol 23:1–10
Xu SB, Li T, Deng ZY, Chong K, Xue Y et al (2008) Dynamic proteomic analysis reveals a switch between central carbon metabolism and alcoholic fermentation in rice filling grains. Plant Physiol 148:908–25

Yates FE (1993) Self-organizing systems. In: Boyd CAR, Noble D (eds) The logic of life. Oxford University Press, Oxford, UK

Yeates TO, Tsai Y, Tanaka S, Sawaya MR, Kerfeld CA (2007) Self-assembly in the carboxysome: a viral capsid-like protein shell in bacterial cells. Biochem Soc Trans 35:508–11

Yeates TO, Kerfeld CA, Heinhorst S, Cannon GC, Shively JM (2008) Protein-based organelles in bacteria: carboxysomes and related microcompartments. Nat Rev Microbiol 6:681–91

Zhang Y, Jijakli H, Courtois P, Sener A, Malaisse WJ (2005) Impaired enzyme-to-enzyme channelling between hexokinase isoenzyme(s) and phosphoglucoisomerase in rat pancreatic islets incubated at a low concentration of D-glucose. Cell Biochem Funct 23(1):15–21

Zhaojun XU, So-ichi Y, Kunio T (2004) Gts1p stabilizes oscillations in energy metabolism by activating the transcription of TPS1 encoding trehalose-6-phosphate synthase 1 in the yeast Saccharomyces cerevisiae. Biochem J 383:171–8

Zhou L, Aon MA, Almas T, Cortassa S, Winslow RL et al (2010) A reaction–diffusion model of ROS-induced ROS release in a mitochondrial network. PLoS Comput Biol 6(1):e1000657

Chapter 9
Systems Biology Approaches to Cancer Energy Metabolism

Alvaro Marín-Hernández, Sayra Y. López-Ramírez, Juan Carlos Gallardo-Pérez, Sara Rodríguez-Enríquez, Rafael Moreno-Sánchez, and Emma Saavedra

Abstract Application of Systems Biology approaches to energy metabolism of cancer cells help in the understanding of their controlling and regulatory mechanisms and identification of new drug targets. Our group built and validated a kinetic model of tumor glycolysis based on the experimental determination of all the enzyme/transporter kinetic parameters, metabolite concentrations, and fluxes in tumor cells. Model predictions enabled to understand how glycolysis is controlled and allowed identification of the main controlling steps which can be the most promising therapeutic targets. In this chapter, the model was extended to determine the contribution on the pathway function of the expression of different glycolytic isoforms displaying different catalytic properties, a feature commonly observed in tumor cells subjected to hypoxia. Model predictions now indicated that, by fully changing the glucose transporter (GLUT), hexokinase (HK), or both, from low- to high affinity isoforms, the glycolytic flux can be increased (GLUT + HK $>$ GLUT $>>$ HK); however, this concurred with a marked deregulation of the adenine nucleotides concentration. To gradually increase glycolytic flux with no alteration of adenine nucleotides homeostasis, which is closer to the physiological response of tumor cells, the model indicated that simultaneous expression in different ratios of GLUT and HK isoforms with different affinities should be accomplished. Mitochondrial metabolism is also active and essential for cancer cells. Therefore, a cancer energy metabolism model, including glycolysis and oxidative

A. Marín-Hernández • S.Y. López-Ramírez • J.C. Gallardo-Pérez • R. Moreno-Sánchez (✉) • E. Saavedra (✉)
Departamento de Bioquímica, Instituto Nacional de Cardiología Ignacio Chávez, Juan Badiano No. 1, Col. Sección XVI, Tlalpan, México D.F. 14080, Mexico
e-mail: rafael.moreno@cardiologia.org.mx; emma_saavedra2002@yahoo.com

S. Rodríguez-Enríquez
Departamento de Bioquímica, Instituto Nacional de Cardiología Ignacio Chávez, Juan Badiano No. 1, Col. Sección XVI, Tlalpan, México D.F. 14080, Mexico

Laboratorio de Medicina Traslacional, Instituto Nacional de Cancerología, México D.F. 14080, Mexico

M.A. Aon et al. (eds.), *Systems Biology of Metabolic and Signaling Networks*, Springer Series in Biophysics 16, DOI 10.1007/978-3-642-38505-6_9,

phosphorylation (Krebs cycle, respiratory chain, Pi/ADP transport, ATP synthase), should identify the most appropriate sites for successful multi-target therapies.

Abbreviations

ALDO	Aldolase
DHAP	Dihydroxyacetone phosphate
ENO	Enolase
Ery4P	Erythrose-4-phosphate
FBP	Fructose-1,6-bisphosphate
F6P	Fructose-6-phosphate
F2,6BP	Fructose-2,6-bisphosphate
C^{J}_{Ei} or FCC	Flux control coefficient
GAPDH	Glyceraldehyde-3-phosphate dehydrogenase
G3P	Glyceraldehyde-3-phosphate
G6P	Glucose-6-phosphate
GLUT	Glucose transporter
HK	Hexokinase
HPI	Hexosephosphate isomerase
KC	Krebs cycle
LDH	Lactate dehydrogenase
OxPhos	Oxidative phosphorylation
PEP	Phosphoenolpyruvate
Pyr	Pyruvate
PFK-1	Phosphofructokinase type 1
PFKFB3	Phosphofructokinase type 2 B3
6PG	6-phosphogluconate
PGK	Phosphoglycerate kinase
PGAM	3-phosphoglycerate mutase
PYK	Pyruvate kinase
TK	Transketolase
TPI	Triosephosphate isomerase.

9.1 Introduction

Cancer is a very complex disease that arises from the combination of multiple changes occurring at both genetic and biochemical levels. The existing treatments against this deadly disease are mostly based on the higher susceptibility of tumor cells to damage induced by radiation and chemotherapy compared to normal cells. However, very frequently the treatments have severe side effects on the patients and in many cases the tumors are refractory to its complete elimination because drug resistance emerges. For these reasons, it would be convenient to gain understanding of the differences between tumor and normal cells at the molecular level in the search for new therapeutic strategies that can be more selective and effective against cancer cells (Hornberg et al. 2007; Moreno-Sánchez et al. 2007, 2010; Kolodkin et al. 2012).

Due to its multifactorial nature, in cancer multiple metabolic and signal transduction pathways and cellular processes are affected, and hence application of systems biology in cancer research seems highly appropriate for the characterization and full understanding of the disease at the cellular, subcellular, and molecular levels as well as for providing new insights into the underlying biological process (Hornberg et al. 2006; Baker and Kramer 2011). Furthermore, a systems analysis of the disease may help in the identification of optimal drug targets at the time in which pharmaceutical industry has been facing a decline in the production of new drugs for successful cancer treatment (Soto et al. 2011). For the latter goal, the integration of Metabolic Control Analysis (MCA) and kinetic modeling, both "bottom-up" System Biology approaches (Westerhoff 2011), can help in the identification of promising drug targets by identifying the most controlling steps in essential metabolic pathways of cancer cells (Hornberg et al. 2007; Moreno-Sánchez et al. 2008). The advantage of inhibiting the main flux- (and/or metabolite concentration-) controlling steps of a metabolic pathway, or blocking the most controlling pathways in a cellular process, is that small changes in their activities will bring about a more pronounced effect in decreasing the pathway flux, or cell function, than inhibition of low or noncontrolling steps. This strategy may undoubtedly lead to the use of lower inhibitor concentrations, which may decrease undesirable side effects on healthy cells.

Most of systems biology studies on tumor cells have been focused on signaling pathways (e.g. Ca^{2+}, cAMPK/PKA, CaN-NFAT, I-kappaB/NF-kappa B, JAK-STAT, MAPK, p53, and Smad) due to their role in several essential cellular activities such as cell proliferation, apoptosis, angiogenesis, metastasis, or invasion (reviewed by Hübner et al. 2011). In one report, a kinetic model including 148 reactions was constructed for the study of the control of the epidermal growth factor-induced MAPK pathway. The authors concluded that amongst all the reactions, only the activity of the protein Raf exerted the highest control (Hornberg et al. 2005).

Other research groups have considered that the decrease in ATP synthesis in cancer cells might also be an alternative strategy to affect tumor cells proliferation; therefore, inhibition of both glycolysis and oxidative phosphorylation (OxPhos) in these cells emerges as another way to tackle the problem of cancer treatment (Cascante et al. 2002, 2010; Moreno-Sánchez et al. 2007, 2010; Sheng et al. 2009; Chen et al. 2012). Notwithstanding their relevance for tumor cell growth and survival, systems biology studies on glycolysis and/or OxPhos are scarce. In this chapter we have experimentally assessed and discussed the relevance of systems biology analysis of the energy metabolism of tumor cells for understanding the mechanisms that govern this function (i.e., cytosolic ATP supply), and how this approach may help in identifying drug targets against tumor energy metabolism.

9.2 Inhibiting Tumor Glycolysis

In most tumor cells the glycolytic flux is 2- to 17-fold higher compared to normal cells as a consequence of the over-expression of all the glycolytic enzymes and transporters induced by several oncogenes and the hypoxia-inducible factor-1 (HIF-1). This results in the excretion of large amounts of lactate even under aerobic conditions, a phenomenon known as the "Warburg effect" of cancer cells (reviewed by Moreno-Sánchez et al. 2007; Marín-Hernández et al. 2009). It is worth recalling that in addition to contributing to the ATP supply for cellular work, the increased rate of tumor glycolysis also provides various glycolytic intermediaries which are precursors for the synthesis of macromolecules (polysaccharides, nucleic acids, triglycerides, proteins) required for the constant and accelerated cell proliferation. Moreover, an increased glycolysis may predominate for ATP supply when mitochondria are damaged (Carew and Huang 2002) and an active mitochondrial degradation (mitophagy) occurs (Lock et al. 2011) or when tumor cells are under hypoxic conditions (Xu et al. 2005). In addition, a correlation between increased glycolysis and tumor resistance to chemo- and radiotherapy has been found (Fanciulli et al. 2000; Maschek et al. 2004; Xu et al. 2005; Lee et al. 2007). Therefore, it has been suggested that inhibition of this essential pathway might be a therapeutic option to increase the effectiveness of chemotherapy and radiotherapy (Pelicano et al. 2006).

The effect of individual inhibition of various glycolytic enzymes (hexokinase II, HKII; phosphofructokinase type 2 B3, PFKFB3; glyceraldehyde-3-phosphate dehydrogenase, GAPDH; lactate dehydrogenase A, LDH-A) or transporters (glucose transporter 1, GLUT1; monocarboxylate transporters, MCT) on tumor survival has been evaluated using a great variety of inhibitors (Pedersen et al. 2002; Pelicano et al. 2006; Kumagai et al. 2008; Bartrons and Caro 2007; Evans et al. 2008; Clem et al. 2008) with poor outcomes. In general, in such studies the targets have been chosen somewhat randomly and arbitrarily: most research groups focus on targeting the presumed rate-limiting steps reported on biochemistry textbooks (HK; phosphofructokinase type 1, PFK-1; pyruvate kinase, PYK), while others have used molecular biology tools (knock-down by RNAi, siRNA) to identify the essential or "key" enzyme (hexosephosphate isomerase, HPI; HKII, PYKM2, LDH-A)/transporter (GLUT1) for tumor growth (Funasaka et al. 2007; Amann et al. 2009; Zhou et al. 2010; Kim et al. 2011; Goldberg and Sharp 2012). However, drug target validation studies suggest that an almost complete pharmacological inhibition is required in order to obtain similar decreases in flux and pathway function to those reached by the genetic knockout of the pharmacological target; however, the use of elevated drug doses promotes toxic side effects in the host. Therefore, promising drug targets should be those enzymes/transporters for which a lower inhibitor dose suffices to produce a major effect on the pathway flux (flux-control) and/or the metabolite concentrations (homeostatic control). Otherwise stated, the steps that should be targeted are those controlling the pathway function (reviewed in Hornberg et al. 2007; Hellerstein 2008; Moreno-Sánchez et al. 2010).

9.3 Glycolysis Fingerprints in Tumor Cells Versus Normal Cells

Besides the higher glycolytic rate exhibited by tumor cells as compared with normal ones, the type of isoenzyme expression is another important difference. HK and PFK-1, which are the main controlling steps of glycolysis in normal cells, show changes in their regulatory mechanisms in tumor cells. In cancer cells, HK activity increases 5- to 500-fold (Nakashima et al. 1988; Smith 2000; Pedersen et al. 2002; Stubbs et al. 2003; Marín-Hernández et al. 2006), and it is preferentially bound to the outer mitochondrial membrane as compared with the tissue of origin or with other normal cells. Because of this localization, some authors have suggested that mitochondrial binding will enable HK to decrease its sensitivity to product inhibition by G6P (Nakashima et al. 1988; Widjojoatmodjo et al. 1990), but this could not be verified experimentally (Marín-Hernández et al. 2006).

In normal cells under aerobiosis, PFK-1 has a main role at the onset of the Pasteur effect through its allosteric inhibition, elicited by some mitochondrial intermediaries such as ATP, citrate, and H^+, which brings about a decrease of the glycolytic flux. In contrast, PFK-1 activity increases five times in some tumors versus normal cells (Vora et al. 1985; El-Bacha et al. 2003) and the expressed isoform exhibits 5- to 300-fold higher affinity for its allosteric activator F2,6BP (Oskam et al. 1985; Colomer et al. 1987; Staal et al. 1987), along with one to seven times lower affinity for ATP and citrate, respectively (Meldolesi et al. 1976; Oskam et al. 1985; Staal et al. 1987). Furthermore, in cancer cells there is a significant increase in the F2,6BP concentration due to the higher expression of PFKB-3, a PFK-2 isoform that maintains an elevated kinase/phosphatase activity ratio leading to predominating F2,6BP synthesis over its degradation (Yalcin et al. 2009). Consequently, elevated levels of F2,6BP as well as AMP and Pi (also PFK-1 allosteric activators) are found in cancer cells, which circumvents the inhibitory effect of ATP, citrate, and H^+ on PFK-1 causing the lack of Pasteur effect in tumor cells (Eigenbrodt et al. 1985; Moreno-Sánchez et al. 2012).

Isoenzyme expression changes in tumor cells indicate differences in the structure of control of glycolysis compared with normal cells. Hence, a Systems Biology approach appears to be appropriate for addressing the comprehensive genetic and biochemical remodeling underlying the well-documented increase in glycolysis shown by cancer cells. Indeed, several kinetic models of glycolysis in erythrocytes (du Preez et al. 2008), β-cells of pancreatic islets (Achs et al. 1991), and tumor cells (Marín-Hernández et al. 2011) have been developed. The analysis of this integrative approach as applied to tumor glycolysis is discussed in the next section.

9.4 Metabolic Control Analysis of Tumor Glycolysis

Metabolic Control Analysis (MCA) is a Systems Biology approach that analyzes metabolic networks with the goal of elucidating their underlying control and regulation mechanisms (reviewed by Fell 1997; Moreno-Sánchez et al. 2008, 2010; Westerhoff 2008). On this regard, MCA makes a clear distinction between control and regulation of metabolism. *"Control"* indicates the extent to which a flux through a pathway, or the concentration of an intermediary metabolite, is altered by changing the activity of one step, or group of steps, and is quantitatively represented by *flux-* and *concentration- control coefficients. "Regulation"* refers to how the flux of a pathway or a metabolite level is modified through the effect on the rate of an individual step by cellular factors, including metabolites different to substrates/products of that step, enzyme activity modulators, and ions, and is quantitatively represented by the *response coefficient* (Fell 1997). Initially used for metabolic pathways at steady state, the MCA principles have been recently applied to the analysis of time-dependent events and oscillations of metabolic and signal transduction pathways (Westerhoff 2008). MCA has also been extended to the regulation of cellular processes by gene expression and termed Hierarchical Control Analysis (Westerhoff 2008).

MCA studies have demonstrated that the flux control of a particular metabolic pathway is distributed, i.e., it is shared to different extents by all the participating steps; thus, the existence of a unique rate-limiting step can be ruled out. MCA allows to quantifying the degree of control exerted by each enzyme/transporter on the pathway flux (flux control coefficient; FCC or C^{J}_{ai}) and on the metabolite concentrations (concentration control coefficient; CCC or C^{X}_{ai}) where J is flux, X is the concentration of a pathway intermediary, and ai is the activity a in the cells of the pathway enzyme i. By applying this strategy using both "wet" and in silico experimentation, the main controlling steps can be identified, thereby becoming targets with the highest therapeutic potential.

Elasticity analysis, an in vivo experimental approach utilized in MCA to determine the control coefficients by groups of enzymes, was applied to glycolysis in AS-30D rat tumor cells (Marín-Hernández et al. 2006). The results indicated that the pathway flux is mainly controlled (71 %) by the glucose transporter (GLUT) and/or HK; PFK-1 controlled only by 6 %, whereas the rest of the control (25 %) resided in the enzymes from aldolase (ALDO) to lactate dehydrogenase (LDH). The low flux control exerted by PFK-1 is in agreement with the different control distribution exhibited by glycolysis in normal versus tumor cells. These findings also suggest that PFK-1 inhibition will probably be more harmful to normal than to tumor cells.

However, a limitation of the flux-control coefficients determined from elasticity coefficients is that their estimation requires reliable measurements of relatively small gradual changes in the pathway metabolites. In addition, this strategy did not allow us to elucidate the GLUT and HK flux control coefficients separately, neither to calculate the individual control coefficients from the downstream glycolytic steps

(Marín-Hernández et al. 2006). To circumvent these problems, we applied kinetic modeling combined with MCA to determine the control coefficients of tumor glycolysis.

9.5 Kinetic Modeling of Glycolysis in Tumor Cells

Kinetic modeling is a bottom-up systems biology approach that constructs detailed computational descriptions of metabolic/signal transduction pathways in order to understand how new properties emerge when the pathway components interact with each other (Westerhoff 2011). In kinetic models of metabolic pathways, each reaction is defined by a rate equation that includes the kinetic properties of the individual enzymes/transporters, namely, maximal velocity and the affinity constants for all the ligands. Kinetic models that also perform MCA have the advantage of enabling to understand the control and regulation mechanisms of a pathway which are not possible to elucidate from the kinetic properties of the individual enzymes (Moreno-Sánchez et al. 2008; van Gend and Snoep 2008; Cortassa et al. 2009). Recently, we built kinetic models of glycolysis in HeLa (human cervico-uterine) and AS-30D (rat ascitic hepatoma) tumor cell lines to identify the main controlling reactions in cells exposed to different environmental conditions (Marín-Hernández et al. 2011).

To build the models it was necessary to experimentally determine in cell-free extracts under physiological conditions of pH, temperature (pH 7.0 and 37 °C), and ion composition (high K^+), several kinetic parameters, and variables: (a) the affinity constants for ligands (substrates, products, activators, and inhibitors) of all enzymes; (b) the maximal enzyme activity (V_{max}) for the forward and reverse reactions within cells; (c) the intracellular intermediary metabolite concentrations; and (d) the glycolytic flux including its branches to glycogen synthesis/degradation, pentose phosphate pathway, and mitochondrial pyruvate oxidation. These determinations were performed under a defined metabolic steady state. The rate equations for each reaction were obtained from the literature in which parameters (a) and (b) were used to assign values to their constants. These equations and the initial concentration of substrates were assembled to construct the kinetic model using the metabolic simulators GEPASI v 3.3 (Mendes 1993) and COPASI (COmplex PAthway SImulator; Hoops et al. 2006). Comparisons of model simulations of metabolite concentrations and fluxes with the in vivo determinations (variables (*c*) and (*d*), respectively) were used to validate the kinetic model and predict the in vivo pathway behavior in two tumor cell lines.

For model refinement, a dynamic interplay between modeling and experimentation was carried out: model simulations helped to pinpoint what parameters/variables had to be experimentally reevaluated which were then fed to the model until simulation results converged with the in vivo pathway behavior. The three most important modifications performed in the model were: (1) the inhibition of hexose-6-phosphate isomerase (HPI) by some metabolites of glycolysis and pentose phosphate pathway; (2) the incorporation of a complete PFK-1 rate equation which includes the effect of allosteric modulators; and (3) the inclusion of phosphate in the GAPDH rate equation. These modifications are further elaborated below.

In a first attempt, the model rendered lower levels of G6P than those determined in vivo, and in silico titration of the activity of each enzyme indicated that the rate of HPI was too high. This prompted us to search for physiological inhibitors for this enzyme. We found that physiological concentrations of F1,6BP (glycolytic metabolite), and 6-phosphogluconate (6PG) and erythrose-4-phosphate (Ery4P) (metabolites of the PPP oxidative and non-oxidative sections, respectively) could be potential candidates. Although the inhibitory effect of these metabolites on HPI was described earlier (Zalitis and Oliver 1967; Chirgwin et al. 1975; Gaitonde et al. 1989), its relevance was only evident after the glycolytic flux-controlling property of HPI was elucidated.

Another difficulty found in our preliminary studies was that the simple rate equation used for PFK-1 (i.e., hyperbolic kinetics or Hill equation) was unable to describe its in vivo kinetic behavior, because it did not include the interaction with regulatory metabolites. Given the surprisingly few kinetic studies on PFK-1, it was necessary to thoroughly characterize its kinetic behavior to formulate a rate equation in both normal and tumor cells (Moreno-Sánchez et al. 2012). The general form of the new PFK-1 equation obeys the concerted transition model of Monod, Wyman, and Changeux (Segel 1975) for exclusive ligand binding (F6P, activators, and inhibitors) together with mixed-type hyperbolic activation by F2,6BP or AMP or Pi, and simple Michaelis–Menten terms for ATP and the reverse reaction (bi–bi random). The inhibitory allosteric effect of ATP (at high concentrations) and citrate could be reproduced with this equation. Due to the prevalence of the F2,6BP activating effect over the inhibitory effect exerted by ATP and citrate, a 50-fold decreased F6P level was obtained with the model.

Another finding of the model was that although GAPDH had a negligible control on the glycolytic flux, it exerted significant control on the concentration of F1,6BP and DHAP, depending on the cellular concentration of free Pi. The mechanistic rationale underlying this homeostatic GAPDH behavior is given by the low affinity of the enzyme for Pi. The relevance of this finding is readily apparent in the context of previous models (Bakker et al. 1999; Teusink et al. 2000; Saavedra et al. 2007) in which it was assumed that Pi was saturating thus irrelevant for regulating GAPDH, or any other enzyme activity. A recently published kinetic model indeed explored the important role of Pi in the regulation of glycolysis in bacteria (Levering et al. 2012). These examples are illustrative of the power of implementing iterative strategies of modeling-experimentation to generate validated kinetic models that accurately simulate available experimental data while predicting new behaviors that help understanding the control and regulatory mechanisms of metabolic pathways in vivo.

After incorporating into the model the three main changes mentioned above, it predicted that the main flux and ATP concentration control steps are HK $\left(C^{J}_{Ei} = 0.44\right)$,

HPI ($C^J_{Ei} = 0.4$) and GLUT ($C^J_{Ei} = 0.2$) in AS-30D hepatoma cells; and glycogen degradation ($C^J_{Ei} = 0.57$), GLUT ($C^J_{Ei} = 0.39$) and HK ($C^J_{Ei} = 0.08$) in HeLa cells. Another relevant finding was that in tumor cells exposed to hypoxia or hypoglycemia the main controlling steps remained the same with only slight quantitative changes in the values of the control coefficients (Marín-Hernández et al. 2011).

The model enabled identification of mechanisms by which these enzymes and transporters control the glycolytic flux in each cell type. In AS-30D cells, HK over-expression, as compared to healthy cells (up to 300-fold), and binding to the mitochondrial outer membrane could explain the diminished flux control displayed by this enzyme. Moreover, the HK over-expression and its ambiguous/ubiquitous character in cancer cells promote increased G6P levels that potently inhibit the enzyme, thus avoiding excessive ATP consumption that might compromise cell viability (Marín-Hernández et al. 2006). GLUT exerts high flux control as a consequence of its low in vivo catalytic efficiency (both low V_{max} and affinity for glucose). Despite being one of the fastest enzymes, HPI is also potently inhibited by physiological concentrations of several glycolytic and PPP intermediates, which is likely related with its strategic location at the G6P crossroad involving the glycolytic, glycogen, and PP pathways.

The standard concentration of 25 mM glucose present in the culture medium of HeLa cells promotes glycogen accumulation and over-expression of GLUT1, the isoform with low affinity for glucose ($K_m = 9$ mM). Consequently, the glycogen degradation branch exerts high control on the glycolytic flux. In contrast, AS-30D ascites cells that grow in the intraperitoneal cavity of rats where the glucose concentration is about 0.026 mM (Rodríguez-Enríquez et al. 2000) express the GLUT isoform with high affinity for glucose (GLUT3, $K_m = 0.5$ mM; Rodríguez-Enríquez et al. 2009) and maintain a low glycogen content. Under these conditions, the AS-30D cells rely more on extracellular glucose for glycolysis to proceed, leading to high HK control and negligible control by glycogen degradation (Marín-Hernández et al. 2011).

9.6 The Pattern of Enzyme Isoforms Expression and the Control Distribution of Glycolysis

Most of the glycolytic enzymes and transporters in mammalian cells, except for HPI and triose phosphate isomerase (TPI), have two to four different isoforms, each one with specific kinetic characteristics (Marín-Hernández et al. 2009). It is well documented that the HIF-1α transcription factor, which is stabilized by low oxygen conditions and found in high levels in tumor cells, upregulates tumor glycolysis. This stimulation happens through increasing transcription of a particular set of glycolytic protein isoforms that have higher affinity for substrates and catalytic capacity (V_{max}) in the forward reaction along with decreased sensitivity to their reaction products and physiological inhibitors (Marín-Hernández et al. 2009).

To explore the effect of changing isoforms on the glycolytic flux and metabolite concentrations, the kinetic model of HeLa cells previously reported (Fig. 9.1; Marín-Hernández et al. 2011) was used. The enzymes selected were those whose genes can be upregulated by HIF-1α under hypoxic conditions (Marín-Hernández et al. 2009). Among them, the catalytically more efficient isoforms of the glucose transporter (GLUT3; $K_{m_{Glu\ out}} = 0.52$ mM) and hexokinase (HKI $K_{m_{Glu\ in}} = 0.03$ mM), together with the phosphofructokinase-1 isoform exhibiting the higher affinity for the main activator F2,6BP (PFK-1-L; $Ka_{F2,6BP} = 0.53$ μM) (Rodríguez-Enríquez et al. 2009; Moreno-Sánchez et al. 2012), were manipulated. The in silico results were compared with those of the initial model built for the normoxic condition (Marín-Hernández et al. 2011) in which GLUT1 ($K_{m_{Glu\ out}} = 9.3$ mM), HKII ($K_{m_{Glu\ in}} = 0.1$ mM), and PFK-1-C ($Ka_{F2,6BP} = 1$ μM) are preferentially expressed (Rodríguez-Enríquez et al. 2009; Moreno-Sánchez et al. 2012) (Fig. 9.2).

First, only the K_m values for each enzyme were individually modified. The most important changes were attained when GLUT3 values were included, consisting of a 39 % increased glycolytic flux and 1.5–3 times increased metabolite concentrations (Table 9.1). When HK affinity was modified (replacing HKII by HKI), the flux increased only by 7 %, whereas the metabolites did not significantly change, except for intracellular glucose (Glu_{in}) and G6P, which as expected decreased and increased by 63 and 9 %, respectively. When the PFK-1 kinetic parameters were modified (by replacing PFK-1-C for PFK-1-L), no significant variation in the flux was observed, in agreement with the low control, determined by elasticity analysis, that this enzyme has on tumor glycolytic flux (Marín-Hernández et al. 2006). However, increases in the Glu_{in} (16 %), G6P (21 %), and F6P (44 %) concentrations were observed, indicating that PFK-1 indeed exerts homeostatic control of glycolysis.

Regarding the flux control coefficients (Table 9.2), significant changes were attained when GLUT3 or HKII replaced GLUT1 or HKII, respectively. In turn, changing PFK1-C by PFK1-L induced negligible variation in the control, i.e., C^J_{PFK1} values. The decrease in the flux control of glycogen degradation and increase in that of HK, elicited by GLUT3 "expression" (Table 9.2), both occur because the cells become more sensitive to changes in the Glu_{out} concentration due to the higher affinity of GLUT3. Remarkably, the main flux-controlling steps induced by changing to GLUT3 $K_{m_{Glu\ out}}$ or HKI $K_{m_{Glu\ in}}$ did not vary (glycogen degradation, GLUT, and HK), indicating the robust function of the glycolytic pathway under these conditions. These modeling results also indicated overcapacity of the remaining glycolytic enzymes, brought about by their over-expression, thus exerting negligible flux control (Table 9.2).

However, tumor cells exposed to hypoxia not only induce one isoform of one specific glycolytic enzyme/transporter; instead, they simultaneously express different isoforms of several enzymes and transporters. Therefore, the affinities were changed in the following two combinations (GLUT3 + HKI or GLUT3 + HKI + PFK-L). In both cases, the flux and metabolite concentrations increased by 65 %

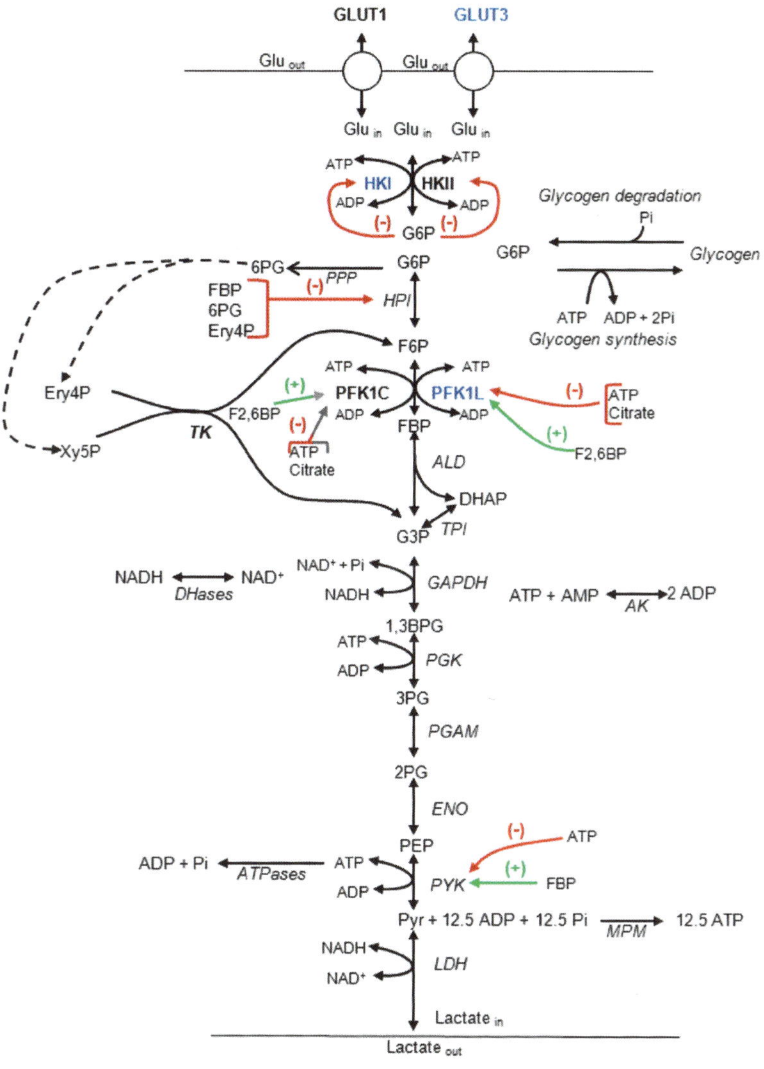

Fig. 9.1 Pathway reactions included in the kinetic model of glycolysis in HeLa tumor cells (modified from Marín-Hernández et al. 2011). *AK* adenylate kinase, *ALDO* fructose 1,6 bisphosphate aldolase, *ATPases* ATP-consuming processes, *DHAP* dihydroxyacetone phosphate, *DHases* NADH consuming reactions, *ENO* enolase, *Ery4P* erythrose-4-phosphate, *FBP* fructose-1,6-bisphosphate, *F6P* fructose-6-phosphate, *F2,6BP* fructose-2,6-bisphosphate, *GAPDH* glyceraldehyde-3-phosphate dehydrogenase, *G3P* glyceraldehyde-3-phosphate, *G6P* glucose-6-phosphate, *LDH* lactate dehydrogenase, *PEP* phosphoenolpyruvate, *PGAM* 3-phosphoglycerate mutase, *PGK* 3-phosphoglycerate kinase, *PGM* phosphoglucomutase, *PPP* pentose phosphate pathway, *Pyr* pyruvate, *PYK* pyruvate kinase, *Rib5P* ribose 5-phosphate, *TPI* triosephosphate isomerase, *TA* transaldolase, *TK* transketolase, *Xy5P* xylulose 5-phosphate, *2PG* 2-phosphoglycerate, *3PG* 3-phosphoglycerate, *6PG* 6-phosphogluconate

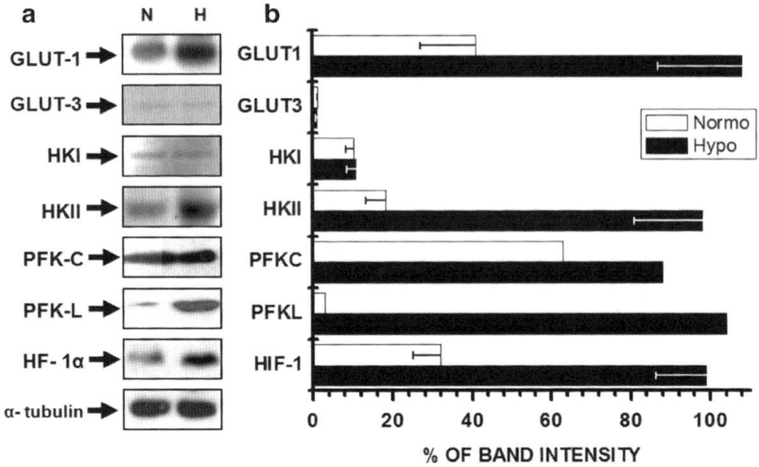

Fig. 9.2 Isoform expression of enzymes/transporters in HeLa cells subjected to normoxic or hypoxic conditions. (**a**) Western-blot analysis of isoform expression of GLUT, HK, and PFK-1 in monolayer cells cultured in the presence of 25 mM glucose and exposed to normoxia (N; 21 % O_2) or hypoxia (H; 0.1 % O_2) for 24 h (see Marín-Hernández et al. 2011, for further description of the experimental design). (**b**) Densitometric analysis with respect to α-tubulin. The GLUT1, PFK1-C, and PFK1-L data were taken from Marín-Hernández et al. (2011) and Moreno-Sánchez et al. (2012). The GLUT3, HKI, HKII, and HIF-1α data were determined in the present study. The Western-blot analyses were carried out according to published procedures (Rodríguez-Enríquez et al. 2009; Moreno-Sánchez et al. 2012)

and two to threefold, respectively (Table 9.1), whereas the flux control distribution was unchanged, emphasizing the role of GLUT as the main controlling step of glycolysis (Table 9.2).

These model results suggested that an effective way to increase the pathway flux is to increase the substrate affinity of the most controlling steps. However, these latter simulations were accompanied by decreased ADP (and AMP) pools (Table 9.1). To avoid this unrealistic response, the ATPase reaction, which accounts for the cellular ATP demand in the model, was increased 10 % to supply for the extra ADP required (Table 9.1). It is worth reminding that an undesired consequence of an unrestricted fast flux through the GLUT and HK reactions is ATP depletion, when such flux exceeds the one through the relatively slower PGK and PYK reactions (i.e., "turbo effect"; Teusink et al. 1998). The opposite response, i.e., faster ATP synthesis than consumption, would lead to ADP depletion, which would also be adverse for glycolysis and lethal for the cell. It is therefore expected that the cell has strict control on the expression of the different glycolytic enzyme isoforms.

In the simulations described above it was assumed that there was a complete shift in the isoform expression. However, cell physiology is not so simple, since tumor cells express different ratios of several isoforms for a particular enzyme. This means that for a specific reaction, the V_{max} experimentally determined in the cells

Table 9.1 Glycolytic flux and metabolite concentrations obtained from a model developed for HeLa cells

Metabolite	Control model[a]	Model with new isoforms +GLUT3	+HKI	+PFK-L	[b]GLUT3 + HKI	[b]GLUT3 + HKI + PFKL
Glu_{in}	0.7	2.1	0.26	0.81	1.3	1.43
G6P	0.78	1.3	0.85	0.94	1.9	2.2
F6P	0.018	0.031	0.02	0.026	0.046	0.07
FBP	0.14	0.46	0.17	0.13	0.94	0.79
DHAP	2.0	3.5	2.1	1.9	4.9	4.5
G3P	0.08	0.13	0.08	0.07	0.19	0.17
1,3BPG	0.0008	0.003	0.001	0.0008	0.005	0.004
3PG	0.006	0.008	0.006	0.006	0.01	0.009
2PG	0.002	0.004	0.003	0.002	0.004	0.004
PEP	0.0002	0.0003	0.0002	0.0002	0.0004	0.0004
Pyr	2.5	2.56	2.5	2.5	2.6	2.6
ATP	8.3	10.8	8.7	8.1	11.1	10.8
ADP	2.2	0.83	2	2.2	0.65	0.84
AMP	1.3	0.15	1.0	1.4	0.086	0.15
NADH	0.005	0.005	0.005	0.005	0.005	0.005
NAD^+	1.34	1.34	1.34	1.34	1.34	1.34
Glycolytic flux	19.6	27.2	20.9	19.2	32.3	31.5

Metabolite concentrations in mM; flux in nmol lactate min^{-1}(mg cellular protein)$^{-1}$. Metabolite concentration (mM) fixed values used for modeling were 0.0042 (F2,6BP), 1.7 (citrate), 7.2 (Pi), 33 (lactate), and 5 mM (glucose)

[a]The control model has GLUT 1, HKII, and PFK1-C. In addition, PFK-1 V_{max} value was determined in this study (24.7 nmol/min × mg cellular protein). The PFK1-C and PFK1-L kinetics parameters used in the model were determined at pH 7.5 in the presence of 140 mM K^+ and taken from Moreno-Sánchez et al. (2012)

[b]The k value for ATPase was 3.4×10^{-3} min^{-1}

results from the combination of distinct catalytic capacities from several isoforms. Figure 9.2 depicts the changes in the pattern of expression for GLUT, HK, and PFK isoforms in HeLa cells grown under normoxic and hypoxic conditions. Thereby, in order to account for isoform ratios and their respective kinetic properties, the rate equations have to be modified accordingly, to more accurately reproduce in silico the in vivo tumor glycolysis. To reach this goal, the rate equations of the initial model (Marín-Hernández et al. 2011) were modified as detailed below.

The expression for GLUT was modified according to a double mono-substrate reversible Michaelis–Menten equation (9.1)

$$v = V_{mf}\left(\left[\frac{f1\left([Glu_{out}] - \frac{[Glu_{in}]}{K_{eq}}\right)}{K_{Gluout1}\left(1 + \frac{[Glu_{in}]}{K_{Gluin1}}\right) + Glu_{out}}\right] + \left[\frac{f2\left([Glu_{out}] - \frac{[Glu_{in}]}{K_{eq}}\right)}{K_{Gluout2}\left(1 + \frac{[Glu_{in}]}{K_{Gluin2}}\right) + Glu_{out}}\right]\right) \quad (9.1)$$

Table 9.2 Flux control coefficients (CJEi) obtained with a kinetic model of glycolysis in HeLa cells

	C^J_{Ei}					
Enzyme	Control	GLUT3	HKI	PFK-L	GLUT3/HKI	GLUT3/HKI/PFKL
GLUT	0.37	0.23	0.47	0.35	0.33	0.32
HK	0.09	0.23	0.04	0.1	0.17	0.19
HPI	0.05	0.1	0.02	0.04	0.07	0.06
PPP	−0.009	−0.005	−0.009	−0.009	−0.004	−0.004
Glycogen synthesis	−0.1	−0.06	−0.1	−0.1	−0.05	−0.06
Glycogen degradation	0.57	0.33	0.56	0.58	0.27	0.29
PFK-1	0.05	0.16	0.02	0.07	0.14	0.2
ALDO + TPI + GAPDH + PGK + PGAM + ENO + PYK + LDH	0.006	0.05	0.003	0.006	0.07	0.05
MPM	−0.004	−0.001	−0.004	−0.003	−0.003	−0.0005
TK	0.014	0.009	0.014	0.015	0.007	0.008
ATPases	−0.03	−0.05	−0.013	−0.05	−0.001	−0.07

PPP pentose phosphate pathway, *MPM* mitochondrial pyruvate metabolism

Table 9.3 Ratios of GLUT, HK, and PFK isoforms in HeLa cells exposed to normoxia or hypoxia

	Normoxia		Hypoxia	
Isoform	% of band intensity	Fraction	% of band intensity	Fraction
GLUT1	41	0.97	108	0.99
GLUT3	1.3	0.03	1	0.01
$GLUT_{total}$	42.3		109	
HKI	10.3	0.36	10.8	0.1
HKII	18.3	0.64	98	0.9
HK_{total}	28.6		108.8	
PFK-C	63	0.95	88	0.46
PFK-L	3	0.05	104	0.54
PFK_{total}	66		192	

in which Glu_{out} and Glu_{in} are the extra- and intracellular glucose concentrations, $K_{Gluout1}$ and $K_{Gluout2}$ are the affinities for extracellular glucose of each GLUT isoform, K_{eq} is the equilibrium constant of the reaction, V_{mf} is the maximal velocity in the forward reaction determined in the cells, and $f1$ and $f2$ are the proportion of each isoform determined by western blot analysis (Fig. 9.2, Table 9.3).

The HK rate was formulated to represent a double random-bisubstrate Michaelis–Menten equation (9.2), in which A and B are Glu_{in} and ATP, and P and Q are G6P and ADP, respectively; $Ka1$ and $Ka2$ represent the affinity constants for Glu_{in} of each isoform whereas Kb is the affinity for ATP which did not change in both isoforms; Kp and Kq are the affinity constants for the corresponding products.

The rate equation for PFK-1 (9.3) obeys a double concerted transition model of Monod, Wyman, and Changeux for exclusive ligand binding (F6P, activators, inhibitors) together with mixed-type activation (F2,6BP or AMP or Pi) (Moreno-Sánchez et al. 2012) and simple Michaelis–Menten terms for ATP and the reverse reaction. ATP (at high concentrations) and citrate are the allosteric inhibitors. $L1$ and $L2$ are the allosteric transition constants; $Ka_{F2,6BP1}$ and $Ka_{F2,6BP2}$ are the activation constants for F2,6BP; Ki_{CIT1}, Ki_{CIT2} Ki_{ATP1}, and Ki_{ATP2} are the inhibition constants for citrate and ATP; $\alpha1$, $\alpha2$ and $\beta1$, $\beta2$ are the factors by which K_{F6P} (K_{F6P1} *and* K_{F6P2}) and V_{max} change, respectively, when an activator is bound to the active enzyme form (R conformation in the Monod model).

Each isoform ratio was determined (Fig. 9.2) considering that the antibodies have high specificity for their respective isoforms as previously determined (Rodríguez-Enríquez et al. 2009; Moreno-Sánchez et al. 2012) but assuming that they have similar specificity and detection sensitivity. The sum of the band intensities for both isoforms was considered the total protein content from which the corresponding fraction of each isoform was calculated (Table 9.3). These values are denoted by the coefficient f in the corresponding rate equations.

$$v = V_{\mathrm{mf}}\left(\left[\frac{\frac{f1}{Ka1Kb}\left([A][B]-\frac{[P][Q]}{K_{\mathrm{eq}}}\right)}{1+\frac{[A]}{Ka1}+\frac{[B]}{Kb}+\frac{[A][B]}{Ka1Kb}+\frac{[P]}{Kp}+\frac{[Q]}{Kq}+\frac{[P][Q]}{KpKq}+\frac{[A][Q]}{Ka1Kq}+\frac{[P][B]}{KpKb}}\right] + \left[\frac{\frac{f2}{Ka2Kb}\left([A][B]-\frac{[P][Q]}{K_{\mathrm{eq}}}\right)}{1+\frac{[A]}{Ka2}+\frac{[B]}{Kb}+\frac{[A][B]}{Ka2Kb}+\frac{[P]}{Kp}+\frac{[Q]}{Kq}+\frac{[P][Q]}{KpKq}+\frac{[A][Q]}{Ka2Kq}+\frac{[P][B]}{KpKb}}\right]\right) \tag{9.2}$$

$$\begin{aligned} v = V_{\mathrm{mf}} \Bigg[& \left(f1\left(\left(\frac{\frac{[\mathrm{ATP}]}{K_{\mathrm{ATP1}}}}{1+\frac{[\mathrm{ATP}]}{K_{\mathrm{ATP1}}}}\right)\left(\frac{1+\frac{\beta 1\mathrm{F2{,}6B}P}{\alpha 1 Ka_{\mathrm{F2{,}6BP1}}}}{1+\frac{\mathrm{F2{,}6B}P}{\alpha 1 Ka_{\mathrm{F2{,}6BP1}}}}\right)\right.\right. \\ & \times\left(\frac{\frac{\mathrm{F6P}\left(1+\frac{\mathrm{F2{,}6B}P}{\alpha 1 Ka_{\mathrm{F2{,}6BP1}}}\right)}{K_{\mathrm{F6P1}}\left(1+\frac{\mathrm{F2{,}6B}P}{Ka_{\mathrm{F2{,}6BP1}}}\right)}\left[1+\frac{\mathrm{F6P}\left(1+\frac{\mathrm{F2{,}6B}P}{\alpha 1 Ka_{\mathrm{F2{,}6BP1}}}\right)}{K_{F6P1}\left(1+\frac{\mathrm{F2{,}6BP}}{Ka_{\mathrm{F2{,}6BP1}}}\right)}\right]^3}{\frac{L1\left(1+\frac{[\mathrm{CIT}]}{Ki_{\mathrm{CIT1}}}\right)^4\left(1+\frac{[\mathrm{ATP}]}{Ki_{\mathrm{ATP1}}}\right)^4}{\left(1+\frac{\mathrm{F2{,}6B}P}{Ka_{\mathrm{F2{,}6BP1}}}\right)^4}+\left[1+\frac{\mathrm{F6P}\left(1+\frac{\mathrm{F2{,}6B}P}{\alpha 1 Ka_{\mathrm{F2{,}6BP1}}}\right)}{K_{\mathrm{F6P1}}\left(1+\frac{\mathrm{F2{,}6B}P}{Ka_{\mathrm{F2{,}6BP1}}}\right)}\right]^4}\right) \\ & \left.\left. -\left(\frac{\frac{[\mathrm{ADP}][\mathrm{FBP}]}{K_{\mathrm{ADP}}K_{\mathrm{FBP}}K_{\mathrm{eq}}}}{\frac{[\mathrm{ADP}]}{K_{\mathrm{ADP}}}+\frac{[\mathrm{FBP}]}{K_{\mathrm{FBP}}}+\frac{[\mathrm{ADP}][\mathrm{FBP}]}{K_{\mathrm{ADP}}K_{\mathrm{FBP}}}+1}\right)\right)\right)+ \\ & \left(f2\left(\left(\frac{\frac{[\mathrm{ATP}]}{K_{\mathrm{ATP2}}}}{1+\frac{[\mathrm{ATP}]}{K_{\mathrm{ATP2}}}}\right)\left(\frac{1+\frac{\beta 2\mathrm{F2{,}6B}P}{\alpha 2 Ka_{\mathrm{F2{,}6BP2}}}}{1+\frac{\mathrm{F2{,}6B}P}{\alpha 2 Ka_{\mathrm{F2{,}6BP2}}}}\right)\right.\right. \\ & \times\left(\frac{\frac{\mathrm{F6P}\left(1+\frac{\mathrm{F2{,}6B}P}{\alpha 2 Ka_{\mathrm{F2{,}6BP2}}}\right)}{K_{\mathrm{F6P2}}\left(1+\frac{\mathrm{F2{,}6B}P}{Ka_{\mathrm{F2{,}6BP2}}}\right)}\left[1+\frac{\mathrm{F6P}\left(1+\frac{\mathrm{F2{,}6B}P}{\alpha 2 Ka_{\mathrm{F2{,}6BP2}}}\right)}{K_{\mathrm{F6P2}}\left(1+\frac{\mathrm{F2{,}6B}P}{Ka_{\mathrm{F2{,}6BP2}}}\right)}\right]^3}{\frac{L2\left(1+\frac{[\mathrm{CIT}]}{Ki_{\mathrm{CIT2}}}\right)^4\left(1+\frac{[\mathrm{ATP}]}{Ki_{\mathrm{ATP2}}}\right)^4}{\left(1+\frac{\mathrm{F2{,}6B}P}{Ka_{\mathrm{F2{,}6BP2}}}\right)^4}+\left[1+\frac{\mathrm{F6P}\left(1+\frac{\mathrm{F2{,}6BP}}{\alpha 2 Ka_{\mathrm{F2{,}6BP2}}}\right)}{K_{\mathrm{F6P2}}\left(1+\frac{\mathrm{F2{,}6BP}}{Ka_{\mathrm{F2{,}6BP2}}}\right)}\right]^4}\right) \\ & \left.\left. -\left(\frac{\frac{[\mathrm{ADP}][\mathrm{FBP}]}{K_{\mathrm{ADP}}K_{\mathrm{FBP}}K_{\mathrm{eq}}}}{\frac{[\mathrm{ADP}]}{K_{\mathrm{ADP}}}+\frac{[\mathrm{FBP}]}{K_{\mathrm{FBP}}}+\frac{[\mathrm{ADP}][\mathrm{FBP}]}{K_{\mathrm{ADP}}K_{\mathrm{FBP}}}+1}\right)\right)\right)\Bigg] \end{aligned} \tag{9.3}$$

Model simulations indicated a marginal increase of 5–13 % in metabolite concentration, fluxes, and flux control coefficients under normoxic conditions compared with the model with no different isoforms (Tables 9.4 and 9.5). This result contrasted with the changes observed when GLUT3, alone or in combination with HKI, replaced GLUT1 (Tables 9.1 and 9.2). This behavior can be explained by the much lower expression of GLUT3 compared to GLUT1 (Fig. 9.2 and Table 9.3).

Table 9.4 Glycolytic flux and metabolite concentrations obtained with a kinetic model of glycolysis in HeLa cells accounting for enzyme isoforms expressed under normoxia or hypoxia

	Normoxia			Hypoxia		
Metabolite	In vivo[b]	Control model	Model *plus* isoform ratios	In vivo[b]	Control model[a]	Model *plus* isoforms ratios[a]
Glu $_{in}$	NM	0.7	0.49	NM	0.89	0.81
G6P	1.3	0.78	0.85	1.4	0.82	0.83
F6P	0.5	0.018	0.02	0.5	0.02	0.02
FBP	0.38	0.14	0.16	0.23	0.35	0.38
DHAP	0.93	2.0	2.1	0.54	3.0	3.1
G3P	ND	0.08	0.08	NM	0.12	0.12
1,3BPG	ND	0.0008	0.001	NM	0.002	0.001
3PG	ND	0.006	0.006	NM	0.008	0.008
2PG	ND	0.002	0.003	NM	0.003	0.003
PEP	0.32	0.0002	0.0002	NM	0.0002	0.0002
Pyr	8.5	2.5	2.5	4.2	2.6	2.6
ATP	8.7	8.3	8.7	7.9	8.7	8.9
ADP	2.7	2.2	2.0	1.8	1.6	1.9
AMP	0.4	1.3	1.1	NM	0.67	0.96
NADH	NM	0.005	0.005	NM	0.005	0.005
NAD^+	NM	1.34	1.34	NM	1.34	1.34
Glycolytic flux	16	19.6	20.7	21	26.1	26.8

Metabolite concentrations in mM; flux in nmol lactate min^{-1} (mg cellular protein)$^{-1}$
NM not measured, *ND* not detected (<1 nmol/15 mg cellular protein). The parameters used were 0.0042 (F2,6BP), 1.7 (citrate), 7.2 (Pi), 33 (lactate) mM, and 5 mM (glucose).
[a]ATPase *k* value was 3.6×10^{-3} min^{-1}. For the hypoxia model simulation, the V_{max} values of GLUT and PFK-1 were 30 and 33.8 nmol/min × mg cellular protein, respectively
[b]Values were taken from Marín-Hernández et al. (2011)

The modified model accounting for isoform ratios showed similar robust behavior, as the model with no different isoforms (Marín-Hernández et al. 2011). The model behavior did not significantly change after altering kinetic parameters (decreasing by half or increasing by two times) from most glycolytic steps and PPP flux. Exceptions were given by changes in GLUT, glycogen degradation, and GAPDH V_m values, which led to variations of 50 % or more in the flux-control distribution and/or metabolite concentrations. In addition, the FBP or DHAP concentrations varied between 9 and 100 % in response to changes in ALDO V_m and K_m values or TPI K_m values.

The model was sensitive to changes in the ATPase kinetics, revealing that this step is the weakest component in conferring stability to the pathway. The ATPase reaction in the present kinetic model accounts for the energy demand represented by a multitude of ATP-consuming cellular processes such as ion homeostasis (Na^+/K^+ ATPase, Ca^{2+} ATPases, H^+ ATPases, MDR ATPases), biosynthesis of proteins, nucleic acids, lipids, and polysaccharides, and specialized functions (signal transduction, secretion, proliferation). In addition, the rate equation used for the ATPase reaction has no affinity terms for substrates and products or K_{eq}, but it is rather a

Table 9.5 Flux control coefficients obtained with a kinetic model using the complete set of isoforms expressed in HeLa cells

	C^J_{Ei}			
	Normoxia		Hypoxia	
Enzyme	Control model	Model with total isoform ratios	Control model	Model with total isoform ratios
GLUT	**0.37**	**0.42**	**0.41**	**0.43**
HK	**0.09**	**0.07**	**0.13**	**0.13**
HPI	0.05	0.03	0.07	0.07
PPP	−0.009	−0.009	−0.006	−0.006
Glycogen synthesis	−0.1	−0.1	−0.08	−0.07
Glycogen degradation	**0.57**	**0.55**	**0.41**	**0.40**
PFK-1	0.05	0.04	0.07	0.06
ALDO + TPI + GAPDH + PGK + PGAM + ENO + PYK + LDH	0.006	0.006	0.03	0.03
MPM	−0.004	−0.004	−0.002	−0.001
TK	0.014	0.014	0.01	0.01
ATPases	−0.03	−0.02	−0.04	−0.05

PPP Pentose phosphate pathway, *MPM* Mitochondrial pyruvate metabolism. Underlined bold values indicate the main flux-controlling steps under each particular condition.

simple irreversible mass action law. Irrespective of this simplification, the majority of metabolite concentrations and fluxes simulated by the model were in quantitative agreement with the in vivo concentrations, except for F6P and PEP, which were 27 and 1,600 times lower, respectively. These remaining discrepancies indicated that further kinetic experimentation-based refinement of the model is still needed, mainly at the level of PFK-1, ALDO, PYK, and ATPases. In this regard, it could be useful to experimentally assess the effects of PFK-1, ALDO, and PYK binding to microtubules on their respective kinetic properties (reviewed by Cassimeris et al. 2012), which may account for the significantly lower modeled Fru6P and PEP concentrations.

An experimentally validated and robust kinetic model can be used to predict the control distribution under other metabolic steady states such as normoxia versus hypoxia and normoglycemia versus hypoglycemia. Under these conditions, not only a change in the isoform expression may occur but also changes in the enzyme amount (Table 9.3; Fig. 9.2) and V_{max} (Marín-Hernández et al. 2011) happen. Indeed, higher levels of metabolites and a 33 % increased flux were obtained under hypoxia compared to normoxia. Moreover, under hypoxia a higher HK flux control coefficient was determined at the expense of a decrease in the control of the glycogen degradation; this behavior is similar to what was observed when GLUT3 was modeled alone or in combination with HKI (Tables 9.1 and 9.2). Again, including the isoform ratios in the model did not modify metabolite concentrations, fluxes, and control coefficients, because GLUT3 expression is lower under hypoxia (Fig. 9.2).

The modeling results described above suggested that the initial assumption made, which considered that the enzyme activity in cells can be solely attributed to the predominant isoform expressed, is a convenient simplification that avoids the use of highly complex rate equations. In this regard, it is worth noting that cells grown in monolayer cultures maintain a high content of low affinity isoforms (GLUT1 and HKII) because they are always grown in the presence of excess (25 mM) glucose, even under hypoxia. Under such hyperglycemic conditions, cells do not require to express and use high affinity isoforms for glucose transport and phosphorylation (GLUT3 and HKI). Similarly, changes in GLUT isoform proportions can be achieved under other stressful conditions. For instance, pathological events such as ischemia, starvation, and mitochondria inhibition increase the content of GLUT3 in normal cells (Nagamatsu et al. 1994; Vannucci et al. 1996; Khayat et al. 1998). Same results have been observed in tumor cells, where hypoglycemia or hypoxia *plus* hypoglycemia increase the mRNA for GLUT3 (Natsuizaka et al. 2007).

In conclusion, the simplest and perhaps fastest way cancer cells have to modulate the glycolytic flux under cyclically varying conditions (e.g. hypoxia/normoxia; hypoglycemia/normoglycemia) is to change the ratio of enzyme isoform corresponding to the most controlling steps, without profoundly altering the installed, house-keeping over-expression pattern of all the pathway enzymes and transporters. In this regard, the kinetic model predicts that when GLUT3 is increased (Fig. 9.3), and hence the GLUT3/GLUT1 ratio, which simulates exposure

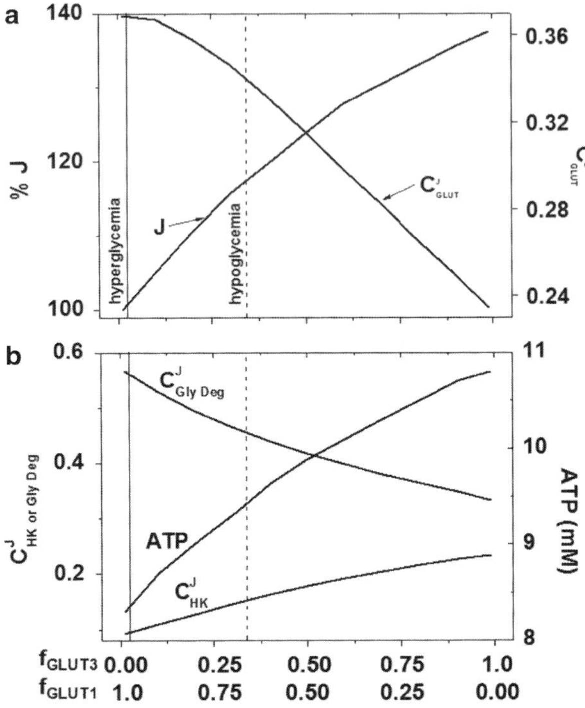

Fig. 9.3 Change in flux and control coefficients as a function of GLUT isoform proportions, as predicted by the model. (**a**) The glycolytic flux (*J*) and GLUT flux control coefficient (C^J_{GLUT}). (**b**) HK and glycogen degradation (*Gly Deg*) flux control coefficients (C^J_{HK} and $C^J_{Gly\ Deg}$) and ATP concentration. The *lines* indicate proportions of each GLUT isoform found in cells under hyperglycemia (Table 9.3) and hypoglycemia (Marín-Hernández et al. Unpublished results). f_{GLUT3} fraction of GLUT3, f_{GLUT1} fraction of GLUT1

of cancer cells to hypoglycemia, the glycolytic flux increases ≈ 20 % with respect to hyperglycemia, whereas the GLUT flux control coefficient and flux control distribution remain unchanged (Table 9.2; Fig. 9.3). Therefore, this strategy may seem useful for increasing the glucose uptake without shifting the entire pattern of enzyme and transporter isoforms, which may in turn lead to ADP and AMP pools depletion (Table 9.1; Fig. 9.3). Furthermore, the results presented enable us to gain understanding about why cancer cells differentially express multiple glucose transporters.

9.7 Advances in Modeling OxPhos in Tumor Cells

To decrease ATP levels to promote apoptosis in tumor cells, not only glycolysis has to be blocked but also OxPhos. In this regard, it has been profusely documented in recent years that ATP supply by mitochondria is as important as glycolysis for cancer cells (Moreno-Sánchez et al. 2007; Sheng et al. 2009; Chen et al. 2012). In fact, cells in solid tumors located far from the blood vessels in which a hypoxic environment prevails are predominantly glycolytic, whereas cells with a predominant oxidative (i.e. mitochondrial) metabolism localize closer to blood vessels or the O_2 supply (Sonveaux et al. 2008; Mandujano-Tinoco et al. 2013); fully blood-irrigated tumors are then dependent on OxPhos for ATP supply. Thus, the generalized belief, that mitochondrial function is absent or nonfunctional in cancer cells, has been challenged by abundant experimental evidence (see for reviews Moreno-Sánchez et al. 2007; Ralph et al. 2010a, b). Consequently, assessment of OxPhos in cancer cells should not be ignored. In addition, to improve understanding of the control of ATP supply in cancer cells, our glycolytic model should be extended to include OxPhos, which will enable us to have a more comprehensive assessment of the steps that exert the control of energy metabolism. In principle, this will allow to further substantiate the proposal of a multi-target versus mono-target therapy of the most flux- and metabolite concentration-controlling reactions.

So far modeling OxPhos in cancer cells has not been reported. However, advances have been carried out in modeling the Krebs cycle (KC) and OxPhos in normal cells. Kohn et al. (1979) built a kinetic model of the KC to explain the sequence of biochemical events that control metabolism of exogenous pyruvate in perfused rat hearts. This model made an important contribution in providing a useful repository data of kinetic mechanisms for the KC reactions (Wu et al. 2007). Another kinetic model of KC in rat liver mitochondrial was used to study the effect of salicylate on the energy metabolism and establish the mechanisms of the hepatotoxicity of this compound (Mogilevskaya et al. 2006).

Regarding OxPhos, several mathematical models have been published which predict the flux-control distribution and the ATP/ADP ratios, electrochemical H^+ gradient, and rates of O_2 consumption and ATP synthesis in isolated mitochondria under different metabolic states (reviewed by Mazat et al. 2010). The next step has been the integral modeling of both KC and OxPhos. With the use of simplified rate equations for OxPhos reactions, a mathematical model of cardiac mitochondria metabolism has been developed to predict how ATP supply may be governed by fluctuations in the matrix concentration of Ca^{2+}, a strong allosteric modulator of KC dehydrogenases (Cortassa et al. 2003), and the dynamics of other relevant ions such as Na^+, Pi, and H^+ (Wu et al. 2007; Wei et al. 2011). Although these models have been able to reproduce qualitatively and semiquantitatively the in vivo pathway behavior, several kinetic parameters/variables were adjusted or obtained under non-physiological experimental conditions, yielding some unrealistic responses and limiting their use for understanding the controlling/regulatory mechanisms of the pathway.

The kinetic database of KC and OxPhos in tumor cells is still incomplete hindering model development. Like in glycolysis, some cancer cells show increased activity of several KC/OxPhos enzymes (Dietzen and Davis 1993) and expression of specific isoforms with different kinetic values and regulatory properties to those of the isoforms expressed in normal cells/mitochondria (Siess et al. 1976). These changes dissuade the use of the kinetic parameters/variables reported for enzymes/transporters of normal cells in the modeling of cancer energy metabolism. Then, the immediate task at hand is to determine the kinetic parameters of each Krebs cycle and OxPhos enzyme/transporter in isolated mitochondria from cancer cells to build the corresponding kinetic model.

9.8 Conclusion

Kinetic modeling is a Systems Biology bottom-up approach that enables us to understand the controlling mechanisms underlying glycolysis in tumor cells. To build a reliable and robust kinetic model of a metabolic pathway, it is essential to use appropriate rate equations for each step or group of steps, in which the kinetic parameters for the forward and reverse reactions are adequately determined. For reactions with thermodynamic constraints (i.e., high negative ΔG° values > -5 kcal/mol), the kinetic parameters of the reverse reaction are difficult to determine, and hence $V_{\max_{\text{reverse}}}$ can be replaced by the equilibrium constant K_{eq} in the rate equation. The use of rate equations that accurately reproduce the enzyme/transporter behavior shields the kinetic modeling from arbitrarily introducing "adjustments" to the parameters to forcing correct simulation of the in vivo pathway behavior.

However, a drawback in constructing or extending kinetic models is the overwhelming amount of experimental data required for validation. That is the main reason why only few kinetic models have been developed (Hübner et al. 2011) and, in particular for cancer glycolysis (Marín-Hernández et al. 2011). The latter model has been used to construct a modular model of the most relevant metabolic processes regulated by the PI3K/Akt signaling pathway in the human embryonic kidney HEK293 cells (Mosca et al. 2012). Their model could reproduce the experimentally determined metabolic fluxes and allowed to predict the metabolic targets that may inhibit tumor cell growth under hyper activation of Akt kinase. Regarding glycolysis, models that include other levels of regulation such as gene expression and signal transduction are interesting aspects to be included in the future for a full understanding of the metabolic circuitry in these pathological cells.

Furthermore, it has been claimed that kinetic parameters determined in cellular extracts (i.e., in diluted cellular solutions) cannot reflect the kinetic properties of the enzymes in vivo because the cellular aqueous phases are crowded with macromolecules, altering the enzyme–substrate and enzyme–product interactions and in consequence the rate reaction (Agrawal et al. 2008). However, Vopel and Makhatadze (2012) have demonstrated that the presence of several synthetic crowding agents (ficoll, dextran 40,000, or albumin) does not significantly perturb the K_{m} and catalytic turnover number (k_{cat}) of several glycolytic enzymes such as HK, GAPDH, PGK, and LDH. Then, it seems that macromolecular crowding may only affect diffusion rates of metabolites (Agrawal et al. 2008). As these rates are orders of magnitude higher than the enzyme/transporter rate constants, modeling based on kinetic parameters determined in diluted solutions is expected not to be significantly modified by macromolecular crowding. This last conclusion is supported by the relatively elevated accuracy and robustness of the herein shown glycolytic model developed for tumor cells (Marín-Hernández et al. 2011).

Although our model of cancer glycolysis has only considered a few well-defined steady states, it may serve as a validated platform for constructing large-scale kinetic models, which can be applied to a wider variety of physiological conditions.

Moreover, integral kinetic modeling of glycolysis, KC, and OxPhos in cancer cells, using modular approaches for computational modeling (van Gend and Snoep 2008; Cortassa and Aon 2012; Mosca et al. 2012), is a future task that will most likely help to identifying suitable targets in the two exclusive energy provider pathways and that may lead to envision successful multi-target therapeutic strategies (Moreno-Sánchez et al. 2010; Mosca et al. 2012).

Acknowledgements The present work was partially supported by the following grants from CONACyT-México: Nos. 180322 (AM-H); 107183 (SR-E); 80534 and 123636 (RM-S); and 83084 and 178638 (ES); and from the Instituto de Ciencia y Tecnología del Distrito Federal No. PICS08-5 (RM-S).

References

Achs MJ, Garfinkel L, Garfinkel D (1991) A computer model of pancreatic islet glycolysis. J Theor Biol 150:109–35

Agrawal M, Santra SB, Anand R, Swaminathan R (2008) Effect of macromolecular crowding on the rate of diffusion-limited enzymatic reaction. Pramana 71:359–68

Amann T, Maegdefrau U, Hartmann A, Agaimy A, Marienhagen J, Weiss TS, Stoeltzing O, Warnecke C, Schölmerich J, Oefner PJ, Kreutz M, Bosserhoff AK, Hellerbrand C (2009) GLUT1 expression is increased in hepatocellular carcinoma and promotes tumorigenesis. Am J Pathol 174:1544–52

Baker SG, Kramer BS (2011) Systems biology and cancer: promises and perils. Prog Biophys Mol Biol 106:410–13

Bakker BM, Walsh MC, Ter Kuile BH, Mensonides FI, Michels PA, Opperdoes FR, Westerhoff HV (1999) Contribution of glucose transport to the control of the glycolytic flux in *Trypanosoma brucei*. Proc Natl Acad Sci USA 96:10098–103

Bartrons R, Caro J (2007) Hypoxia, glucose metabolism and the Warburg's effect. J Bioenerg Biomembr 39:223–9

Carew JS, Huang P (2002) Mitochondrial defects in cancer. Mol Cancer 1:9

Cascante M, Boros LG, Comin-Anduix B, de Atauri P, Centelles JJ, Lee PW (2002) Metabolic control analysis in drug discovery and disease. Nat Biotechnol 20:243–9

Cascante M, Benito A, Zanuy M, Vizán P, Marín S, de Atauri P (2010) Metabolic network adaptations in cancer as targets for novel therapies. Biochem Soc Trans 38:1302–06

Cassimeris L, Silva VC, Miller E, Ton Q, Molnar C, Fong J (2012) Fueled by microtubules: does tubulin dimer/polymer partitioning regulate intracellular metabolism? Cytoskeleton (Hoboken) 69:133–43

Chen V, Staub RE, Fong S, Tagliaferri M, Cohen I, Shtivelman E (2012) Bezielle selectively targets mitochondria of cancer cells to inhibit glycolysis and OXPHOS. PLoS One 7:e30300

Chirgwin JM, Parsons TF, Ernst AN (1975) Mechanistic implications of the pH independence of inhibition of phosphoglucose isomerase by neutral sugar phosphates. J Biol Chem 250:7277–79

Clem B, Telang S, Clem A et al (2008) Small-molecule inhibition of 6-phosphofructo-2-kinase activity suppresses glycolytic flux and tumor growth. Mol Cancer Ther 7:110–20

Colomer D, Vives-Corrons JL, Pujades A, Bartrons R (1987) Control of phosphofructokinase by fructose 2,6 bisphosphate in B-lymphocytes and B-chronic lymphocytic leukemia cells. Cancer Res 47:1859–62

Cortassa S, Aon MA (2012) Computational modeling of mitochondrial function. Methods Mol Biol 810:311–26

Cortassa S, Aon MA, Marbán E, Winslow RL, O'Rourke B (2003) An integrated model of cardiac mitochondrial energy metabolism and calcium dynamics. Biophys J 84:2734–55
Cortassa S, O'Rourke B, Winslow RL, Aon MA (2009) Control and regulation of integrated mitochondrial function in metabolic and transport networks. Int J Mol Sci 10:1500–13
Dietzen DJ, Davis EJ (1993) Oxidation of pyruvate, malate, citrate, and cytosolic reducing equivalents by AS-30D hepatoma mitochondria. Arch Biochem Biophys 305:91–102
du Preez FB, Conradie R, Penkler GP, Holm K, van Dooren FL, Snoep JL (2008) A comparative analysis of kinetic models of erythrocyte glycolysis. J Theor Biol 252:488–96
Eigenbrodt E, Fister P, Reinacher M (1985) New perspectives in carbohydrate metabolism in tumor cells. In: Reitner R (ed) Regulation of carbohydrate metabolism. CRC, Boca Raton, FL
El-Bacha T, de Freitas MS, Sola-Penna M (2003) Cellular distribution of phosphofructokinase activity and implications to metabolic regulation in human breast cancer. Mol Genet Metab 79:294–9
Evans A, Bates V, Troy H, Hewitt S, Holbeck S, Chung YL, Phillips R, Stubbs M, Griffiths J, Airley R (2008) GLUT1 as a therapeutic target: increased chemoresistance and HIF-1-independent link with cell turnover is revealed through compare analysis and metabolomic studies. Cancer Chemother Pharmacol 61:377–93
Fanciulli M, Bruno T, Giovannelli A, Gentile FP, Di Padova M, Rubiu O, Floridi A (2000) Energy metabolism of human LoVo colon carcinoma cells: correlation to drug resistance and influence of lonidamine. Clin Cancer Res 6:1590–97
Fell D (1997) Understanding the control of metabolism. Plenum, London
Funasaka T, Hu H, Hogan V, Raz A (2007) Down-regulation of phosphoglucose isomerase/autocrine motility factor expression sensitizes human fibrosarcoma cells to oxidative stress leading to cellular senescence. J Biol Chem 282:36362–69
Gaitonde MK, Murray E, Cunningham VJ (1989) Effect of 6-phosphogluconate on phosphoglucose isomerase in rat brain in vitro and in vivo. J Neurochem 52:1348–52
Goldberg MS, Sharp PA (2012) Pyruvate kinase M2-specific siRNA induces apoptosis and tumor regression. J Exp Med 209:217–24
Hellerstein MK (2008) A critique of the molecular target-based drug discovery paradigm based on principles of metabolic control: advantages of pathway-based discovery. Metab Eng 10:1–9
Hoops S, Sahle S, Gauges R, Lee C, Pahle J, Simus N, Singhal M, Xu L, Mendes P, Kummer U (2006) COPASI—a COmplex PAthway SImulator. Bioinformatics 22:3067–74
Hornberg JJ, Binder B, Bruggeman FJ, Schoeberl B, Heinrich R, Westerhoff HV (2005) Control of MAPK signalling: from complexity to what really matters. Oncogene 24:5533–42
Hornberg JJ, Bruggeman FJ, Westerhoff HV, Lankelma J (2006) Cancer: a systems biology disease. Biosystems 83:81–90
Hornberg JJ, Bruggeman FJ, Bakker BM, Westerhoff HV (2007) Metabolic control analysis to identify optimal drug targets. Prog Drug Res 64:173–89
Hübner K, Sahle S, Kummer U (2011) Applications and trends in systems biology in biochemistry. FEBS J 278:2767–857
Khayat ZA, McCall AL, Klip A (1998) Unique mechanism of GLUT3 glucose transporter regulation by prolonged energy demand: increased protein half-life. Biochem J 333:713–18
Kim JE, Ahn BC, Hwang MH, Jeon YH, Jeong SY, Lee SW, Lee J (2011) Combined RNA interference of hexokinase II and (131)I-sodium iodide symporter gene therapy for anaplastic thyroid carcinoma. J Nucl Med 52:1756–63
Kohn MC, Achs MJ, Garfinkel D (1979) Computer simulation of metabolism in pyruvate-perfused rat heart. II. Krebs cycle. Am J Physiol 237:R159–R166
Kolodkin A, Boogerd FC, Plant N et al (2012) Emergence of the silicon human and network targeting drugs. Eur J Pharm Sci 46:190–7
Kumagai S, Narasaki R, Hasumi K (2008) Glucose-dependent active ATP depletion by koningic acid kills high-glycolytic cells. Biochem Biophys Res Commun 365:362–8
Lee KA, Roth RA, LaPres JJ (2007) Hypoxia, drug therapy and toxicity. Pharmacol Ther 113:229–46

Levering J, Musters MW, Bekker M (2012) Role of phosphate in the central metabolism of two lactic acid bacteria: a comparative systems biology approach. FEBS J 279:1274–90

Lock R, Roy S, Kenific CM, Su JS, Salas E, Ronen SM, Debnath J (2011) Autophagy facilitates glycolysis during Ras-mediated oncogenic transformation. Mol Biol Cell 22:165–78

Mandujano-Tinoco EA, Gallardo-Pérez JC, Marín-Hernández A, Moreno-Sánchez R, Rodríguez-Enríquez S (2013) Anti-mitochondrial therapy in human breast cáncer multi-cellular spheroids. Biochim Biophys Acta 1833:541–51

Marín-Hernández A, Rodríguez-Enríquez S, Vital-González PA, Flores-Rodríguez FL, Macías-Silva M, Sosa-Garrocho M, Moreno-Sánchez R (2006) Determining and understanding the control of glycolysis in fast-growth tumor cells. Flux control by an over-expressed but strongly product-inhibited hexokinase. FEBS J 273:1975–88

Marín-Hernández A, Gallardo-Pérez JC, Ralph SJ, Rodríguez-Enríquez S, Moreno-Sánchez R (2009) HIF-1alpha modulates energy metabolism in cancer cells by inducing over-expression of specific glycolytic isoforms. Mini Rev Med Chem 9:1084–101

Marín-Hernández A, Gallardo-Pérez JC, Rodríguez-Enríquez S, Encalada R, Moreno-Sánchez R, Saavedra E (2011) Modeling cancer glycolysis. Biochim Biophys Acta 1807:755–67

Maschek G, Savaraj N, Priebe W, Braunschweiger P, Hamilton K, Tidmarsh GF, De Young LR, Lampidis TJ (2004) 2-deoxy-D-glucose increases the efficacy of adriamycin and paclitaxel in human osteosarcoma and non-small cell lung cancers in vivo. Cancer Res 64:31–4

Mazat JP, Fromentin J, Heiske M, Nazaret C, Ransac S (2010) Virtual mitochondrion: towards an integrated model of oxidative phosphorylation complexes and beyond. Biochem Soc Trans 38:1215–19

Meldolesi MF, Macchia V, Laccetti P (1976) Differences in phosphofructokinase regulation in normal and tumor rat thyroid cells. J Biol Chem 251:6244–51

Mendes P (1993) GEPASI: a software package for modelling the dynamics, steady states and control of biochemical and other systems. Comput Appl Biosci 9:563–71

Mogilevskaya E, Demin O, Goryanin I (2006) Kinetic model of mitochondrial krebs cycle: unraveling the mechanism of salicylate hepatotoxic effects. J Biol Phys 32:245–71

Moreno-Sánchez R, Rodríguez-Enríquez S, Marín-Hernández A, Saavedra E (2007) Energy metabolism in tumor cells. FEBS J 274:1393–418

Moreno-Sánchez R, Saavedra E, Rodríguez-Enríquez S, Olín-Sandoval V (2008) Metabolic control analysis: a tool for designing strategies to manipulate pathways. J Biomed Biotech 2008:597913

Moreno-Sánchez R, Saavedra E, Rodríguez-Enríquez S, Gallardo-Pérez JC, Quezada H, Westerhoff HV (2010) Metabolic control analysis indicates a change of strategy in the treatment of cancer. Mitochondrion 10:626–39

Moreno-Sánchez R, Marín-Hernández A, Gallardo-Pérez JC, Quezada H, Encalada R, Rodríguez-Enríquez S, Saavedra E (2012) Phosphofructokinase type 1 kinetics, isoform expression and gene polymorphisms in cancer cells. J Cell Biochem 113:1692–703

Mosca E, Barcella M, Alfieri R, Bevilacqua A, Canti G, Milanesi L (2012) Systems biology of the metabolic network regulated by the Akt pathway. Biotechnol Adv 30:131–41

Nagamatsu S, Sawa H, Inoue N, Nakamichi Y, Takeshima H, Hoshino T (1994) Gene expression of GLUT3 glucose transporter regulated by glucose in vivo in mouse brain and in vitro in neuronal cell cultures from rat embryos. Biochem J 300:125–31

Nakashima RA, Paggi MG, Scott LJ, Pedersen PL (1988) Purification and characterization of a bindable form of mitochondrial bound hexokinase from the highly glycolytic AS-30D rat hepatoma cell line. Cancer Res 48:913–19

Natsuizaka M, Ozasa M, Darmanin S, Miyamoto M, Kondo S, Kamada S, Shindoh M, Higashino F, Suhara W, Koide H, Aita K, Nakagawa K, Kondo T, Asaka M, Okada F, Kobayashi M (2007) Synergistic up-regulation of hexokinase-2, glucose transporters and angiogenic factors in pancreatic cancer cells by glucose deprivation and hypoxia. Exp Cell Res 313:3337–48

Oskam R, Rijksen G, Staal GEJ, Vora S (1985) Isozymic composition and regulatory properties of phosphofructokinase from well-differentiated and anaplastic medullary thyroid carcinomas of the rat. Cancer Res 45:135–42
Pedersen PL, Mathupala S, Rempel A, Geschwind JF, Ko YH (2002) Mitochondrial bound type II hexokinase: a key player in the growth and survival of many cancers and an ideal prospect for therapeutic intervention. Biochem Biophys Acta 1555:14–20
Pelicano H, Martin DS, Xu RH, Hung P (2006) Glycolysis inhibition for anticancer treatment. Oncogene 25:4633–46
Ralph SJ, Rodríguez-Enríquez S, Neuzil J, Moreno-Sánchez R (2010a) Bioenergetic pathways in tumor mitochondria as targets for cancer therapy and the importance of the ROS-induced apoptotic trigger. Mol Aspects Med 31:29–59
Ralph SJ, Rodríguez-Enríquez S, Neuzil J, Saavedra E, Moreno-Sánchez R (2010b) The causes of cancer revisited: "mitochondrial malignancy" and ROS-induced oncogenic transformation: why mitochondria are targets for cancer therapy. Mol Aspects Med 31:145–70
Rodríguez-Enríquez S, Torres-Márquez ME, Moreno-Sánchez R (2000) Substrate oxidation and ATP supply in AS-30D hepatoma cells. Arch Biochem Biophys 375:21–30
Rodríguez-Enríquez S, Marín-Hernández A, Gallardo-Pérez JC, Moreno-Sánchez R (2009) Kinetics of transport and phosphorylation of glucose in cancer cells. J Cell Physiol 221:552–9
Saavedra E, Marín-Hernández A, Encalada R, Olivos A, Mendoza-Hernández G, Moreno-Sánchez R (2007) Kinetic modeling can describe *in vivo* glycolysis in *Entamoeba histolytica*. FEBS J 274:4922–40
Segel IH (1975) Enzyme kinetics. Wiley, New York
Sheng H, Niu B, Sun H (2009) Metabolic targeting of cancers: from molecular mechanisms to therapeutic strategies. Curr Med Chem 16:1561–87
Siess EA, Nimmannit S, Wieland OH (1976) Kinetic and regulatory properties of pyruvate dehydrogenase from Ehrlich ascites tumor cells. Cancer Res 36:55–9
Smith TAD (2000) Mammalian hexokinase and their abnormal expression in cancer. Br J Biomed Sci 57:170–8
Sonveaux P, Végran F, Schroeder T, Wergin MC, Verrax J, Rabbani ZN, De Saedeleer CJ, Kennedy KM, Diepart C, Jordan BF, Kelley MJ, Gallez B, Wahl ML, Feron O, Dewhirst MW (2008) Targeting lactate-fueled respiration selectively kills hypoxic tumor cells in mice. J Clin Invest 118:3930–42
Soto AM, Sonnenschein C, Maini PK, Noble D (2011) Systems biology and cancer. Prog Biophys Mol Biol 106:337–9
Staal GE, Kalff A, Heesbeen EC, van Veelen CW, Rijksen G (1987) Subunit composition, regulatory properties, and phosphorylation of phosphofructokinase from human gliomas. Cancer Res 47:5047–51
Stubbs M, Bashford CL, Griffiths JR (2003) Understanding the tumor metabolic phenotype in the genomic era. Curr Mol Med 3:49–59
Teusink B, Walsh MC, van Dam K, Westerhoff HV (1998) The danger of metabolic pathways with turbo design. Trends Biochem Sci 23:162–9
Teusink B, Passarge J, Reijenga CA et al (2000) Can yeast glycolysis be understood in terms of in vitro kinetics of the constituent enzymes? Testing biochemistry. Eur J Biochem 267:5313–29
van Gend C, Snoep JL (2008) Systems biology model databases and resources. Essays Biochem 45:223–6
Vannucci SJ, Seaman LB, Vannucci RC (1996) Effects of hypoxia-ischemia on GLUT1 and GLUT3 glucose transporters in immature rat brain. J Cereb Blood Flow Metab 16:77–81
Vopel T, Makhatadze GI (2012) Enzyme activity in the crowded milieu. PLoS One 7:e39418
Vora S, Halper JP, Knowles DM (1985) Alterations in the activity and prolife of human phosphofructokinase during malignant transformation in vivo and in vitro: transformation and progression-linked discriminants of malignancy. Cancer Res 45:2993–3001

Wei AC, Aon MA, O'Rourke B, Winslow R, Cortassa S (2011) Mitochondrial energetic, pH regulation, and ion dynamics: a computational-experimental approach. Biophys J 100:2894–903

Westerhoff HV (2008) Signalling control strength. J Theor Biol 252:555–67

Westerhoff HV (2011) Systems biology left and right. Methods Enzymol 500:3–11

Widjojoatmodjo MN, Mancuso A, Blanch HW (1990) Mitochondrial hexokinase activity in a murine hybridoma. Biotech Lett 12:551–6

Wu F, Yang F, Vinnakota KC, Beard DA (2007) Computer modeling of mitocondrial tricarboxylic acid cycle, oxidative phosphorylation, metabolite transport, and electrophysiology. J Biol Chem 282:24525–37

Xu RH, Pelicano H, Zhou Y (2005) Inhibition of glycolysis in cancer cells: a novel strategy to overcome drug resistance associated with mitochondrial respiratory defect and hypoxia. Cancer Res 65:613–21

Yalcin A, Telang S, Clem B, Chesney J (2009) Regulation of glucose metabolism by 6-phosphofructo-2-kinase/fructose-2,6-bisphosphatases in cancer. Exp Mol Pathol 86:174–9

Zalitis J, Oliver IT (1967) Inhibition of glucose phosphate isomerase by metabolic intermediates of fructose. Biochem J 102:753–9

Zhou M, Zhao Y, Ding Y, Liu H, Liu Z, Fodstad O, Riker AI, Kamarajugadda S, Lu J, Owen LB, Ledoux SP, Tan M (2010) Warburg effect in chemosensitivity: targeting lactate dehydrogenase-A re-sensitizes taxol-resistant cancer cells to taxol. Mol Cancer 9:33

Part IV
Systems Biology of Organ Function

Chapter 10
Network Dynamics in Cardiac Electrophysiology

Zhilin Qu

Abstract In a network perspective, the heart is a network of cells that are composed of subnetworks of genes, proteins, metabolites, and organelles. In this chapter, we provide an overview of the networks in the heart and a current understanding of the network dynamics in the context of cardiac electrophysiology. We first review current knowledge of the genetic, signaling, and metabolic networks in the heart and their links to arrhythmias. We then review the emergent properties from the mitochondrial and calcium release unit networks, the cellular dynamics arising from integrated subnetworks, and the electrical dynamics arising from the cellular networks to manifest as normal rhythms and arrhythmias. Finally, we discuss future challenges and how systems biology approaches can overcome these challenges to uncover the mechanisms of normal heart rhythms and arrhythmias.

10.1 Introduction

Arrhythmias are irregular excitations in the heart, the major causes of sudden cardiac death (Zipes and Wellens 1998). Anti-arrhythmic drugs that block ion channels have been shown to be ineffective, and some even increase mortality (CAST 1989; Waldo et al. 1996). Defibrillators, merely based on the fact that the heart is electrically excitable, are the only effective therapeutics. This raises a serious question: what have been missed in our understanding of cardiac electrophysiology? Cardiac excitations are regulated by many factors from different scales of the heart, ranging from molecular interactions to tissue scale electrical wave dynamics (Fig. 10.1). At the molecular scale, proteins that form the ion channels behave randomly due to thermodynamic fluctuations and dynamics, resulting in

Z. Qu (✉)
Department of Medicine (Cardiology), David Geffen School of Medicine, University of California, A2-237 CHS, 650 Charles E. Young Drive South, Los Angeles, CA 90095, USA
e-mail: zqu@mednet.ucla.edu

M.A. Aon et al. (eds.), *Systems Biology of Metabolic and Signaling Networks*, Springer Series in Biophysics 16, DOI 10.1007/978-3-642-38505-6_10,

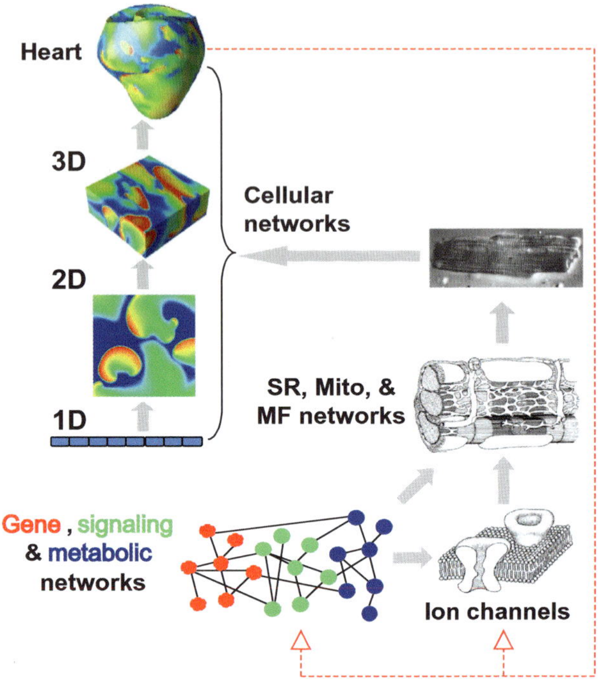

Fig. 10.1 Schematic plot of multi-scale and multi-network perspective of cardiac electrophysiology

random opening and closing of the ion channels. Genes, proteins, and metabolites form networks, which regulate intracellular calcium (Ca^{2+}) cycling, action potential properties, and many other cellular functions. At the subcellular scale, the dynamics of a single mitochondrion or a single Ca^{2+} release unit (CRU), such as Ca^{2+} sparks or mitochondrial flickers, are collective behaviors of many ion channels. Although a Ca^{2+} spark or a mitochondrial flicker exhibits certain randomness, it behaves very differently from random opening and closing of a single ion channel, which is a *collective* behavior arising from the synergistic interactions of many ion channels. The CRUs, mitochondria, and myofilaments form interacting networks, and novel dynamics, such as waves and oscillations, emerges from the networks to give rise to the *whole-cell* dynamics. A rich spectrum of dynamics have been observed at this scale in addition to the normal excitation–contraction–metabolism coupling, such as action potential duration alternans, early afterdepolarizations (EADs) and delayed afterdepolarizations (DADs), and automaticity. At the tissue scale, cardiac myocytes and other types of cells form cellular networks to generate and conduct electrical signals for contraction. Most of these wave dynamics are tissue-scale emergent properties that cannot be simply derived from single cell properties. In addition, *feedback loops* form between scales. For example, fast heart rate or arrhythmic wave dynamics of the heart may cause *ischemia*, intracellular Ca^{2+} and Na^{+} *accumulation*, and molecular and cellular *remodeling* (e.g., as indicated by the green arrows in Fig. 10.1), which then affect the wave dynamics in the heart. Therefore, in a network perspective, the heart is a network composed of many subnetworks. A defect (such as a gene mutation) may or may not cause erroneous electrical dynamics at the tissue scale to result in lethal arrhythmias, which depends not only on the defect itself but also on the network that the defect resides. To understand how such a complex system works and develop effective therapeutics, systems biology and multi-scale modeling approaches are needed to elucidate the underlying dynamics at each scale (or the dynamics of the subnetworks) and how the dynamics at smaller scales (subnetworks) integrate to result in complex dynamics at larger scales (whole networks).

In this chapter, we provide an overview of the networks in the heart and a current understanding of the network dynamics in the context of cardiac electrophysiology. We first review current knowledge of the genetic, signaling, and metabolic networks and their links to arrhythmias. We then review the emergent properties from the mitochondrial and CRU networks, the cellular dynamics, and the electrical dynamics arising from the cellular networks. Finally, we discuss future challenges and systems biology approaches to overcome these challenges.

10.2 Molecular and Organelle Networks and Network Dynamics in Cardiac Myocytes

A cell is a spatial entity, which contains not only networks of genes, proteins, and metabolites but also networks of spatially distributed organelles such as the CRU network, the mitochondrial network, and the myofilament network. Novel dynamics arise from these networks and from their interactions, which regulate cellular Ca^{2+} cycling and action potential dynamics for normal rhythms and arrhythmias of the heart.

10.2.1 Genetic, Signaling, and Metabolic Networks

Clinical, experimental, and computational studies have begun to reveal the gene, protein, and metabolic networks that link to cardiac arrhythmias.

Single gene mutations causing cardiac arrhythmias have been widely studied in the last two decades (Sanguinetti et al. 1995; Napolitano et al. 2012). These mutations, through altering ion channel conductance and kinetics to change the action potential and Ca^{2+} cycling dynamics, cause different diseases, such as long QT syndrome (Keating and Sanguinetti 2001; Sanguinetti and Tristani-Firouzi 2006; Moss and Kass 2005), Brugada syndrome (Hedley et al. 2009), and catecholaminergic polymorphic ventricular tachycardia (Cerrone et al. 2009). However, these monogenic diseases only account for a very small portion of the sudden cardiac death syndrome. Gene loci that are associated with common forms of cardiac diseases and arrhythmias have also been identified (Bezzina et al. 2010;

Arking et al. 2011; Jeyaraj et al. 2012). However, genes interact with each other and a genetic network perspective of cardiac electrophysiology is far from known and how to reveal and study the gene networks related to cardiac diseases is a great challenge (Weiss et al. 2012).

Many signaling pathways and their roles in cardiac excitation–contraction coupling and arrhythmias have been identified (Wang 2007; Swaminathan et al. 2012; Grimm et al. 2011), such as the β-adrenergic signaling pathways, the MAPK signaling pathways, the CaMKII signaling pathways, and the ROS activated signaling pathways. These signaling pathways are interlinked, causing complex effects that cannot be understood fully by experimental observations only. Mathematical models have been developed to quantitatively analyze the effects of these signaling pathways on cardiac diseases. Saucerman et al. (Saucerman et al. 2003, 2004) developed the first model to quantitatively study the effects of β-adrenergic signaling on cardiac contractility and excitation. In this model (Fig. 10.2), β-adrenergic stimulation activates cyclic AMP, which then activates protein kinase A (PKA). PKA phosphorylates L-type Ca^{2+} channel (LCC) and phospholamban, which increases the open probability of the LCC and the rate of sarcoplasmic reticulum (SR) Ca^{2+} uptake. The model was used to study the effects of isoproterenol on Ca^{2+} cycling and action potential dynamics (Saucerman et al. 2003, 2004). Other modeling studies (Hund and Rudy 2004; Hund et al. 2008; Saucerman and Bers 2008; Hashambhoy et al. 2009, 2010) have focused on the effects of CaMKII singling pathways on intracellular Ca^{2+} cycling and action potential dynamics. The synergy between β-adrenergic and CaMKII signaling has also been studied by computer modeling (Soltis and Saucerman 2010). These studies have provided important insights into cardiac signaling and diseases.

Cell metabolism is regulated by hundreds of metabolites that form a highly interconnected complex network (Feist et al. 2009). In cardiac myocytes, the metabolic network not only provides the energy needed for cardiac contraction but also affects cardiac electrophysiology. Mathematical models have been developed to study the dynamics of cardiac metabolism (Cortassa et al. 2003, 2004; Wu et al. 2007; Dash and Beard 2008; Zhou et al. 2005a, b; Jafri and Kotulska 2006). Emergent properties such as oscillations (Jafri and Kotulska 2006; Cortassa et al. 2004; Yang et al. 2008), occur due to the interactions of the metabolites and feedback loops in the metabolic networks. The metabolic oscillations result in ATP oscillations, which then causes oscillations in action potential duration due to opening of the ATP-sensitive potassium channels (O'Rourke et al. 1994; Yang et al. 2008). The oscillations may be responsible for arrhythmogenesis in postischemic hearts (Akar et al. 2005).

10.2.2 The CRU Network and Ca^{2+} Cycling Dynamics

Besides the molecular networks, organelles form spatial networks in cardiac cells (Fig. 10.3a). CRU network is the primary network generating intracellular Ca^{2+}

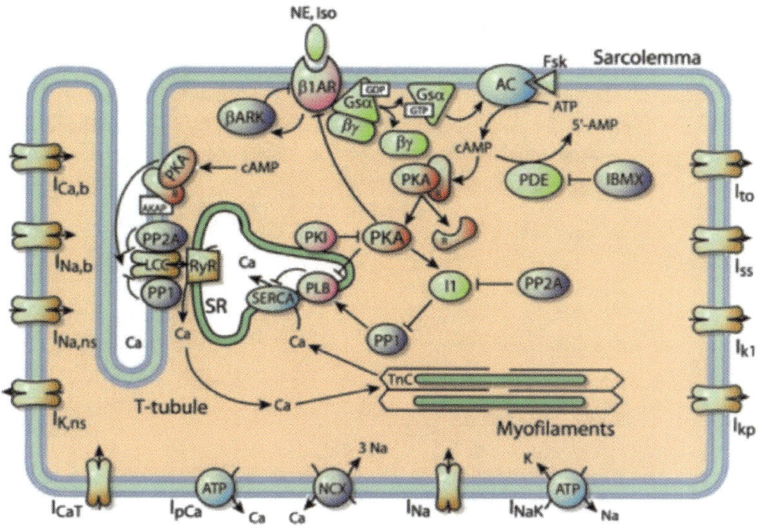

Fig. 10.2 Schematics of β-adrenergic singling network, calcium handling, and electrophysiology of a ventricular myocyte in a model developed by Saucerman et al. (2003)

cycling dynamics in cardiac myocytes. The Ca^{2+} release channel proteins, ryanodine receptors (RyRs), are clustered proximity to L-type Ca^{2+} channel clusters, forming CRUs. A unifying cardiac excitation–contraction coupling in terms of a CRU is illustrated in Fig. 10.3b (Bers 2002). A ventricular myocyte contains about 20,000 CRUs. CRUs are coupled via Ca^{2+} diffusion in the cytosolic and SR space, forming a three-dimensional (3D) CRU network inside a cell. A T-tubular system (Fig. 10.3c) facilitates effective communication of the 3D network with the extracellular space, resulting in synchronous Ca^{2+} release and thus synchronous contraction of the cell. Intracellular Ca^{2+} cycling dynamics emerging from the CRU network are responsible for the formation of Ca^{2+} clocks for sino-atrial nodal cells (Lakatta et al. 2010; Maltsev et al. 2011) as well as Ca^{2+} waves causing ventricular arrhythmias (Rovetti et al. 2010; Nivala et al. 2012b; ter Keurs and Boyden 2007).

Ca^{2+} release in a CRU exhibits a discrete and random behavior, called Ca^{2+} sparks (Cheng and Lederer 2008; Cheng et al. 1993). Many experimental studies (Cheng et al. 1996; Wier et al. 1997; Marchant and Parker 2001; Bootman et al. 1997) have demonstrated a Ca^{2+} signaling hierarchy, showing a transition from Ca^{2+} sparks to Ca^{2+} waves and whole-cell Ca^{2+} oscillations. A question is: How does the transition from sparks to waves and oscillations occur? Using a CRU network model we developed recently (Nivala et al. 2012a), we were able to recapitulate this Ca^{2+} signaling hierarchy (Fig. 10.4a). The amount of Ca^{2+} released from the SR due to random opening of one or two RyRs is small, which results in a quark (labeled as "q" in Fig. 10.4a). When several RyRs happen to randomly open at the same moment, the amount of Ca^{2+} released is large enough to initiate Ca^{2+}-induced

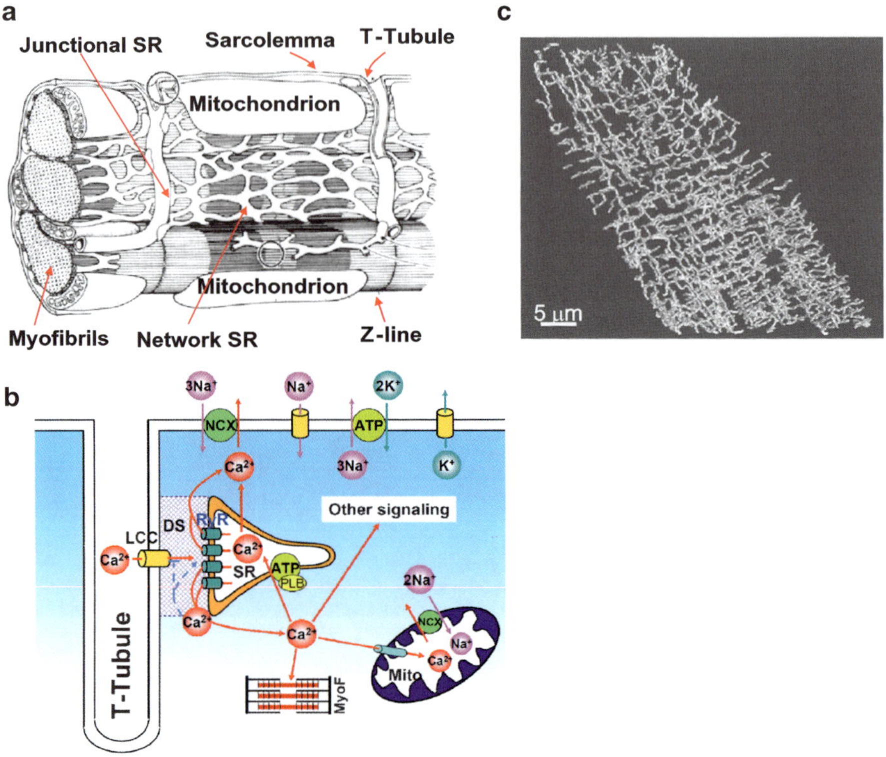

Fig. 10.3 (**a**) Schematic diagram of cellular structure of a ventricular myocyte (Katz 2011). (**b**). A T-tubule network image from a rat ventricular myocyte (Soeller and Cannell 1999). (**c**) Illustration of cardiac excitation–contraction system. In a normal action potential, voltage-dependent opening of the LCCs brings Ca^{2+} into the dyadic space (DS), a very small space between the LCC cluster and jSR (*shaded area*). Elevated Ca^{2+} concentration in the vicinity of the LCCs causes their inactivation. The RyR channels open stochastically and their open probability is sensitive to Ca^{2+} in the DS, a process called CICR. Therefore, the RyR channels can be triggered by Ca^{2+} entry from the LCCs, high myoplasmic and SR Ca^{2+}, and Ca^{2+} diffusing from neighboring CRUs. Ca^{2+} entered from the LCCs and released from the SR diffuses to the myofibrils to signal contraction and participates in many other signaling processes in the myocytes. Ca^{2+} is pumped back into the SR by the sarcoplasmic endoplasmic reticulum Ca^{2+} ATPase (SERCA) pump and extruded by Na^+–Ca^{2+} exchange (NCX). Ca^{2+} is also uptaken by mitochondria through the mitochondrial uniporter and released from mitochondria via NCX in the mitochondrial membrane and opening of other channels such as the mitochondrial permeation transition pore. LCC and NCX couple Ca^{2+} and voltage bidirectionally, but all other currents also affect this coupling either indirectly via their effects on voltage or directly via Ca^{2+} regulation of the ion channels

Ca^{2+} release, causing a large release event to result in a spark (labeled as "s" in Fig. 10.4a). The Ca^{2+} released in a spark may diffuse to cause its neighboring CRUs to fire, or neighboring CRUs may fire coincidentally together, forming spark clusters (labeled as "c" in Fig. 10.4a). When a cluster becomes large enough, it may propagate as a Ca^{2+} wave (labeled as "w" in Fig. 10.3a), depending on the

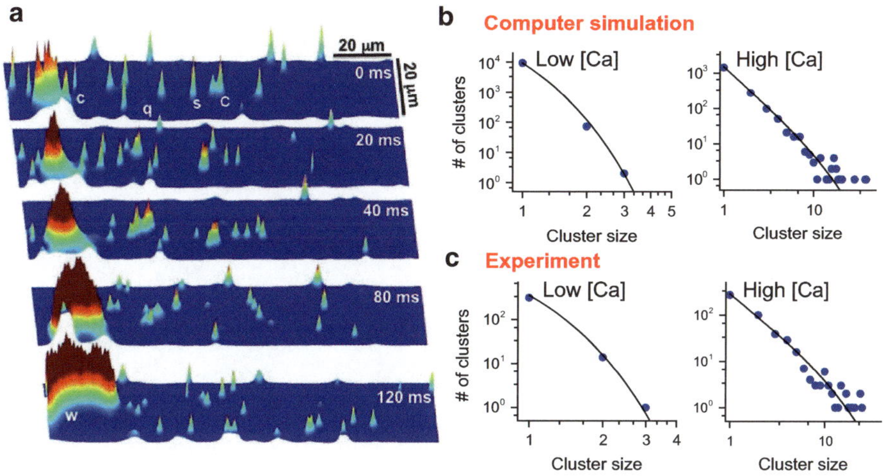

Fig. 10.4 *Transition from sparks to waves in a CRU network*. (**a**) 3D plots showing coexistence of Ca^{2+} quarks (q), sparks (s), spark clusters (c), and waves (w) in a slice of the 3D CRU network model with high Ca^{2+}. A large cluster (*left end*) eventually evolved in to a wave. (**b**) Spark cluster size distributions at low and high Ca^{2+} concentrations from the computer model. (**c**) Same as **b** but from a mouse ventricular myocyte

status of its surrounding CRUs. In another study (Nivala et al. 2012b), we combined computer simulation, theory, and experiments to show that *criticality* is responsible for the transition from local to global Ca^{2+} signaling, providing a general theoretical framework for understanding this transition. We calculated the distributions of the spark cluster sizes in both the computer model and in permeabilized mouse ventricular myocytes (Figs. 10.4b and c). At low Ca^{2+} loads, the cluster size distribution was exponential. As the Ca^{2+} load increased, the distribution changed toward a power law. We showed that the coupling between the CRUs plays a key role in the occurrence of power-law distribution. A power-law distribution is an indicator that a system is in a critical state, such as the critical phenomena of second-order phase transitions in thermodynamics and statistical physics (Stanley 1971, 1999) and self-organized criticality observed in many complex nonlinear systems in nature (Bak et al. 1988; Bak 1997; Turcotte and Rundle 2002). The fact that criticality is the governing mechanism of the Ca^{2+} signaling hierarchy has several implications: (1) Once a system is in a critical state, a tiny perturbation can grow into a macroscopic fluctuation due to the power-law distribution (Stanley 1971, 1999). This provides a general theoretical framework for understanding how single channel fluctuations may lead to macroscopic random oscillations; (2) Ca^{2+} oscillations are self-organized activities, which are emergent phenomena of the coupled CRU network, and do not require the preexistence of pacemaking sites. Due to the randomness in cluster formation in time and space, the whole-cell Ca^{2+} signal exhibits an irregular burst-like behavior (Skupin et al. 2008, 2010); and (3) in sino-atrial nodal cells of the heart, local Ca^{2+} release was shown to play a vital role in pacemaking activity (Lakatta et al. 2003). Local Ca^{2+} releases generating Ca^{2+} waves via criticality may provide a subcellular mechanism accounting for the fractal (i.e., power law) properties of heart rate variability (Ivanov et al. 1999; Ponard et al. 2007). (4) Why cardiac arrhythmias occur suddenly and unpredictably is a key clinical question (Zipes and Rubart 2006). The fact that a small random noise can result in a macroscopic fluctuation under criticality may provide some mechanistic insight into sudden cardiac death. In other words, the random and sudden occurrence of arrhythmias may originate for the random fluctuations at the single channel through the dynamics of criticality. This hypothesis needs to be validated in future studies.

10.2.3 The Mitochondrial Network and Spatiotemporal Depolarization Dynamics

A ventricular myocyte contains about 7,000–10,000 mitochondria. Similar to the CRU network, mitochondria form a network inside the cell coupled by Ca^{2+}, ATP, ROS, and many other metabolites. The mitochondrial network generates mitochondrial depolarization waves and oscillations (Brady et al. 2004; Aon et al. 2003, 2004; Kurz et al. 2010; Honda et al. 2005), which have also been modeled in computer simulations (Zhou et al. 2010; Zhou and O'Rourke 2012; Yang et al. 2010).

Similar to Ca^{2+} cycling dynamics, the transient single mitochondrial depolarizations, known as "flickers," tend to occur randomly in space and time. A question that needs to be answered is how the transition from random flicking to whole-cell oscillations occurs. In a recent study (Nivala et al. 2011), we developed a mathematical model to study how single mitochondrial flickering events self-organize to cause mitochondrial depolarization waves and whole-cell oscillations. We developed a Markov model of the inner membrane anion channel in which ROS-induced inner membrane anion channel opening causes transient mitochondrial depolarizations in a single mitochondrion, which occur in a nonperiodic manner, simulating flickering. We then coupled the individual mitochondria into a network, in which flickers occur randomly and sparsely (first panel in Fig. 10.5a) in the network when a small number of mitochondria are in the state of high superoxide production (determined by the parameter p). As the number of mitochondria in high superoxide production state increases, short lived or abortive waves due to ROS-induced ROS release coexist with flickers. When the number of mitochondria in high superoxide production state reaches a critical number, recurring propagating waves are observed. The origins of the waves occur randomly in space and are self-organized as a consequence of random flickering and local synchronization. At this critical state, the depolarization clusters exhibit a power-law distribution (Fig. 10.5b). In addition, the whole-cell mitochondrial membrane

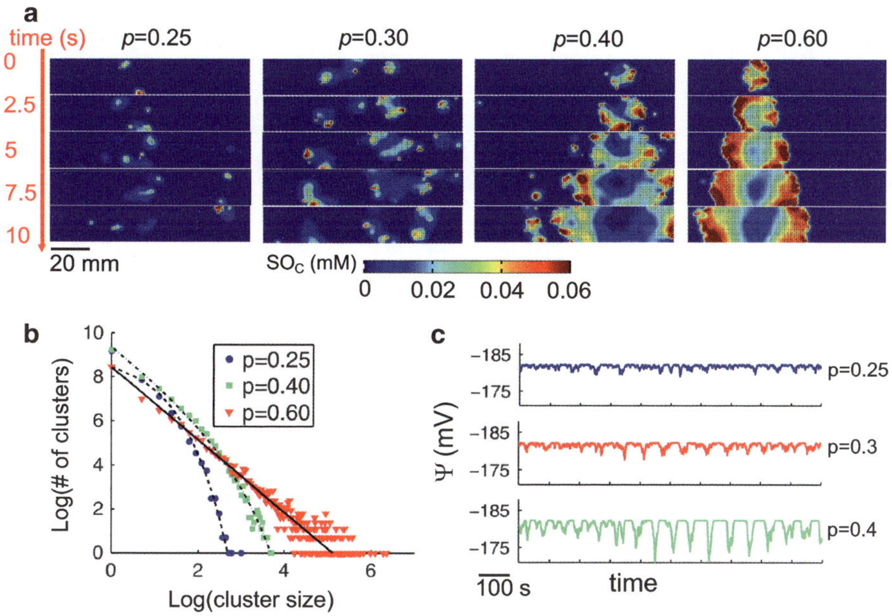

Fig. 10.5 *Transition from flickers to waves in a mitochondrial network.* (**a**) Snapshots of superoxide concentration in the cytoplasmic space at five time points and for different p. (**b**) Mitochondrial depolarization Cluster size distribution for different p. (**c**) Whole-cell mitochondrial potential versus time for different p

potential changes from exhibiting small random fluctuations to more periodic oscillations as superoxide production rate increases (Fig. 10.5c).

Since when criticality is reached, a random flicker may cause a cascade event, triggering a macroscopic mitochondrial depolarization event. The massive mitochondrial depolarization consumes ATP to a low level that opens the K_{ATP} channels, shortening the action potential. In other words, at criticality, a random single mitochondrial event may cause a macroscopic cellular event at the cellular level, which may trigger a sudden arrhythmia event at the tissue scale.

10.3 Cellular Electrophysiology: Dynamics from a Network of Networks

A cell is a network composed of molecular and organelle networks. The ionic currents generated by the ion channels and Ca^{2+} cycling regulate the excitation dynamics, Ca^{2+} cycling and the myofilament network generate the force for contraction, and the metabolic network provide the energy needed for contraction. Gene expression and signal transduction regulate excitation, contraction, and metabolism. However, the proper network dynamics and coupling between the networks are required for normal cardiac electrophysiology, and aberrations in one network may cause failure in another one. For example, local stretch affects Ca^{2+} release via ROS signaling to potentiate Ca^{2+} sparks (Prosser et al. 2011) or via Ca^{2+}-myofilament interaction to generate Ca^{2+} waves to cause DADs (ter Keurs et al. 2008); increased myofilament sensitivity to Ca^{2+} in cardiomyopathy potentiates arrhythmias (Huke and Knollmann 2010; Baudenbacher et al. 2008); metabolic stresses may potentiate Ca^{2+} alternans (Florea and Blatter 2010; Kockskamper et al. 2005; Huser et al. 2000; Belevych et al. 2009), Ca^{2+} sparks (Zhou et al. 2011), and EADs (Xie et al. 2009; Sato et al. 2009); and mitochondrial depolarization in ischemia causes membrane inexcitability and thus cardiac arrhythmias (Akar et al. 2005). Therefore, to understand the cellular dynamics of electrophysiology, a systems biology approach investigating the dynamics of the coupled networks is needed, and computational modeling is a key component of this approach.

Many mathematical models of cardiac action potential and excitation–contraction coupling have been developed (Noble and Rudy 2001; Greenstein and Winslow 2011). Most of the models included SR and cytosolic Ca^{2+}, and membrane voltage and ordinary differential equations were used to describe Ca^{2+} and voltage dynamics. Some of the models have integrated β-adrenergic and CaMKII signaling. Cortassa et al. (2006) and Matsuoka et al. (2004) developed multiple compartment models that integrated excitation, contraction, and metabolism to study the dynamics of excitation–contraction–metabolism (ECM) coupling. Hatano et al. (2011, 2012) developed an ECM coupling model which is a spatially distributed model taking into account diffusion of ions and metabolites in 3D spaces, described by partial differential equations.

However, a typical ventricular myocyte contains ~7,000–10,000 mitochondria and ~10,000–20,000 CRUs or couplons (Franzini-Armstrong et al. 1999). These organelles are intermingled in space, forming a complex ECM coupling network with local interactions to cause spatiotemporal dynamics (e.g., waves and oscillations) of intracellular Ca^{2+} cycling and metabolism (Cheng et al. 1996; Lukyanenko and Gyorke 1999; Aon et al. 2004; Brady et al. 2004). In addition, the elementary Ca^{2+} release events (e.g., Ca^{2+} sparks) tend to occur randomly due to random L-type Ca^{2+} channel and RyR openings (Cheng and Lederer 2008; Bridge et al. 1999). Similarly, the mitochondrial membrane potential flickering (De Giorgi et al. 2000; Thiffault and Bennett 2005) and ROS flashes (Wang et al. 2008; Pouvreau 2010) also tend to occur randomly at the single mitochondrion level. Therefore, both spatial distribution and random behaviors needed to be taken into account for ECM coupling dynamics. As shown in Figs. 10.4 and 10.5, complex emergent properties arise due to the coupling between the elements and the random firing of the elements. More importantly, as shown in Figs. 10.4 and 10.5, when the system is in criticality, a random event at the subcellular scale may trigger a cascade to result in macroscopic event at the cellular scale, which is the consequence of the coupled network of random elements. In recent studies (Cui et al. 2009; Rovetti et al. 2010; Nivala et al. 2011, 2012a, b), we started to develop models that integrate random opening of ion channels and the spatial distribution of CRUs and mitochondria to study the spatiotemporal dynamics. In future studies, models integrating excitation, Ca^{2+} cycling, signaling, metabolism, and contraction with random ion channel opening are needed to reveal the emergent dynamics of cellular electrophysiology.

10.4 Cellular Networks and Tissue-Scale Excitation Dynamics

The heart is a network of different types of cells which are electronically coupled via gap junctional conductance (Fig. 10.6a). Electrical impulses originating from the sino-atrial nodal region propagate to the atrium and the atrial-ventricular node, and then the Purkinje network and the ventricles to cause synchronous contraction of the heart. In normal ventricular tissue, a myocyte is coupled to about 11 myocytes, which is reduced to about six in ischemic tissue (Peters and Wit 1998). Recent studies have shown that fibroblasts may also be coupled to myocytes (Camelliti et al. 2004a, b).

In normal rhythm of the heart, the electrical impulse in heart is equivalent to a planar wave (Fig. 10.6b). Complex wave dynamics arise in cardiac tissue as results of cellular dynamics and coupling between cells (Qu and Weiss 2006; Qu 2011) to result in cardiac arrhythmias, which includes focal excitations, spiral reentry, spiral breakup (Fig. 10.6b), and mixture of focal and reentry excitations (Sato et al. 2009). These dynamics are emergent properties of the myocyte networks, which depend on the cellular properties and how the cells are coupled.

10.5 Conclusions

In a network perspective, a physiological system (such as the heart) is a network composed of subnetworks (Fig. 10.1). A biological function is an emergent property of the coupled networks, not of a single gene or of a single protein. While altering any of the elements may cause a change in biological functions, it is necessary to understand how an element affects other elements in the network, how properties emerge due to the interactions, and thus one can understand the underlying mechanisms. Systems biology approaches that combine experimental biology with computational biology are the likely solutions for dealing with such complex problems. Moreover, computational modeling becomes even more and more important since experimental tools are limited for revealing the complex dynamics. Since the first cardiac action potential model developed by Noble in

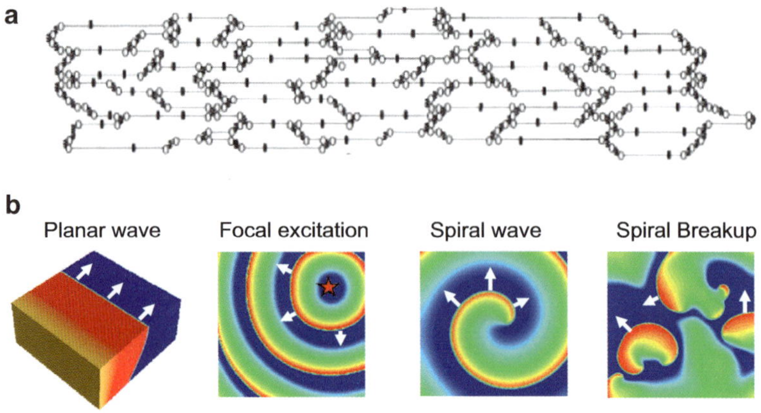

Fig. 10.6 (**a**) A schematic plot of a cell network reconstructed from real tissue (Spach and Heidlage 1995). (**b**) Excitation dynamics in cardiac tissue for normal rhythms (planar wave) and arrhythmias (focal excitation, spiral reentry, and spiral wave breakup). *Arrows* indicate directions of conduction. Voltage changes from low to high as the color changes from *blue* to *red*

1962 following the Hodgkin–Huxley model (Noble 1962), a great number of advanced models have been developed and used to study cardiac excitation and contraction dynamics in single cell, tissue, and whole heart models over the last four decades, setting the heart to be the most well-modeled organ. Moreover, many modeling studies are closely combined with experiments, which further enhance the applicability of these models.

With the experimental and computational studies, our current understanding of the mechanisms of cardiac arrhythmias has been greatly improved, however, effective therapeutics that can prevent cardiac arrhythmias are still lacking. Most of the antiarrhythmic drugs that were developed based on certain mechanistic insights of arrhythmias are not effective but rather cause more mortality (CAST 1989; Waldo et al. 1996). One can easily argue that we still do not understand the mechanisms due to the multi-scale complex regulations of the excitation and contraction dynamics in the heart. A full understanding of the system would be required to investigate the dynamics at each scale and how the dynamics at one scale affects the dynamics at another scale using both experimental and computational approaches. Therefore, systems biology approaches that combine computational biology and experimental biology are the likely means.

However, there are many challenges ahead. On the experimental side, one of the major problems is how to obtain the molecular and cellular information accurately that can be faithfully used in modeling. On the computational side, since one cannot develop a model of a heart or even a piece of tissue at the scale of molecules due to computational limitations and complexity, a multi-scale modeling strategy is needed. However, how to develop such a strategy and maintain the fidelity of information when changing from one scale to another of modeling is a big challenge (Qu et al. 2011). One possible solution is to closely combine experiments and modeling in which modeling generates hypothesis for experiments to obtain more accurate information for model improvement, i.e., an approach that continuous iterations between experiments and modeling may finally lead us to the true mechanisms of cardiac excitation and contraction dynamics for normal rhythms and arrhythmias. Finally, theoretical tools, such as nonlinear dynamics, are needed to draw general conclusions for the complex system, which is key to design effective experiments to unravel the complex dynamics, such as the excitation dynamics in the heart.

References

Akar FG, Aon MA, Tomaselli GF, O'Rourke B (2005) The mitochondrial origin of postischemic arrhythmias. J Clin Invest 115(12):3527–35

Aon MA, Cortassa S, Marban E, O'Rourke B (2003) Synchronized whole cell oscillations in mitochondrial metabolism triggered by a local release of reactive oxygen species in cardiac myocytes. J Biol Chem 278(45):44735–44

Aon MA, Cortassa S, O'Rourke B (2004) Percolation and criticality in a mitochondrial network. Proc Natl Acad Sci USA 101(13):4447–52

Arking DE, Junttila MJ, Goyette P, Huertas-Vazquez A, Eijgelsheim M, Blom MT, Newton-Cheh C, Reinier K, Teodorescu C, Uy-Evanado A, Carter-Monroe N, Kaikkonen KS, Kortelainen M-L, Boucher G, Lagace C, Moes A, Zhao X, Kolodgie F, Rivadeneira F, Hofman A, Witteman JCM, Uitterlinden AG, Marsman RF, Pazoki R, Bardai A, Koster RW, Dehghan A, Hwang S-J, Bhatnagar P, Post W, Hilton G, Prineas RJ, Li M, Kottgen A, Ehret G, Boerwinkle E, Coresh J, Kao WHL, Psaty BM, Tomaselli GF, Sotoodehnia N, Siscovick DS, Burke GL, Marban E, Spooner PM, Cupples LA, Jui J, Gunson K, Kesaniemi YA, Wilde AAM, Tardif J-C, O'Donnell CJ, Bezzina CR, Virmani R, Stricker BHC, Tan HL, Albert CM, Chakravarti A, Rioux JD, Huikuri HV, Chugh SS (2011) Identification of a sudden cardiac death susceptibility locus at 2q24.2 through Genome-Wide Association in European Ancestry Individuals. PLoS Genet 7(6):e1002158

Bak P (1997) How nature works: the science of self-organized criticality. Oxford University Press, New York

Bak P, Tang C, Wiesenfeld K (1988) Self-organized criticality. Phys Rev A 38(1):364–74

Baudenbacher F, Schober T, Pinto JR, Sidorov VY, Hilliard F, Solaro RJ, Potter JD, Knollmann BC (2008) Myofilament Ca^{2+} sensitization causes susceptibility to cardiac arrhythmia in mice. J Clin Invest 118(12):3893–903

Belevych AE, Terentyev D, Viatchenko-Karpinski S, Terentyeva R, Sridhar A, Nishijima Y, Wilson LD, Cardounel AJ, Laurita KR, Carnes CA, Billman GE, Gyorke S (2009) Redox modification of ryanodine receptors underlies calcium alternans in a canine model of sudden cardiac death. Cardiovasc Res 84(3):387–95

Bers DM (2002) Cardiac excitation–contraction coupling. Nature 415(6868):198–205

Bezzina CR, Pazoki R, Bardai A, Marsman RF, de Jong JS, Blom MT, Scicluna BP, Jukema JW, Bindraban NR, Lichtner P, Pfeufer A, Bishopric NH, Roden DM, Meitinger T, Chugh SS, Myerburg RJ, Jouven X, Kaab S, Dekker LR, Tan HL, Tanck MW, Wilde AA (2010) Genome-wide association study identifies a susceptibility locus at 21q21 for ventricular fibrillation in acute myocardial infarction. Nat Genet 42(8):688–91. doi:10.1038/ng.623

Bootman MD, Berridge MJ, Lipp P (1997) Cooking with calcium: the recipes for composing global signals from elementary events. Cell 91(3):367–73

Brady NR, Elmore SP, van Beek JJ, Krab K, Courtoy PJ, Hue L, Westerhoff HV (2004) Coordinated behavior of mitochondria in both space and time: a reactive oxygen species-activated wave of mitochondrial depolarization. Biophys J 87(3):2022–34

Bridge JH, Ershler PR, Cannell MB (1999) Properties of Ca^{2+} sparks evoked by action potentials in mouse ventricular myocytes. J Physiol 518:469–78

Camelliti P, Devlin GP, Matthews KG, Kohl P, Green CR (2004a) Spatially and temporally distinct expression of fibroblast connexins after sheep ventricular infarction. Cardiovasc Res 62(2):415–25

Camelliti P, Green CR, LeGrice I, Kohl P (2004b) Fibroblast network in rabbit sinoatrial node: structural and functional identification of homogeneous and heterogeneous cell coupling. Circ Res 94(6):828–35

Cast I (1989) Effect of encainide and flecainide on mortality in a random trial of arrhythmia suppression after myocardial infarction. N Engl J Med 321:406–12

Cerrone M, Napolitano C, Priori SG (2009) Catecholaminergic polymorphic ventricular tachycardia: a paradigm to understand mechanisms of arrhythmias associated to impaired Ca^{2+} regulation. Heart Rhythm 6(11):1652–59

Cheng H, Lederer WJ (2008) Calcium sparks. Physiol Rev 88(4):1491–1545

Cheng H, Lederer WJ, Cannell MB (1993) Calcium sparks: elementary events underlying excitation-contraction coupling in heart muscle. Science 262(5134):740–44

Cheng H, Lederer MR, Lederer WJ, Cannell MB (1996) Calcium sparks and $[Ca^{2+}]$i waves in cardiac myocytes. Am J Physiol 270(1 Pt 1):C148–59

Cortassa S, Aon MA, Marban E, Winslow RL, O'Rourke B (2003) An integrated model of cardiac mitochondrial energy metabolism and calcium dynamics. Biophys J 84(4):2734–55

Cortassa S, Aon MA, Winslow RL, O'Rourke B (2004) A mitochondrial oscillator dependent on reactive oxygen species. Biophys J 87(3):2060–73

Cortassa S, Aon MA, O'Rourke B, Jacques R, Tseng HJ, Marban E, Winslow RL (2006) A computational model integrating electrophysiology, contraction, and mitochondrial bioenergetics in the ventricular myocyte. Biophys J 91(4):1564–89

Cui X, Rovetti RJ, Yang L, Garfinkel A, Weiss JN, Qu Z (2009) Period-doubling bifurcation in an array of coupled stochastically excitable elements subjected to global periodic forcing. Phys Rev Lett 103(4):044102–104

Dash RK, Beard DA (2008) Analysis of cardiac mitochondrial Na^{+}–Ca^{2+} exchanger kinetics with a biophysical model of mitochondrial Ca^{2+} handling suggests a 3:1 stoichiometry. J Physiol 586(13):3267–85

De Giorgi F, Lartigue L, Ichas F (2000) Electrical coupling and plasticity of the mitochondrial network. Cell Calcium 28(5–6):365–70

Feist AM, Herrgard MJ, Thiele I, Reed JL, Palsson BO (2009) Reconstruction of biochemical networks in microorganisms. Nat Rev Microbiol 7(2):129–43. doi:10.1038/nrmicro1949

Florea SM, Blatter LA (2010) The role of mitochondria for the regulation of cardiac alternans. Front Physiol 1:141

Franzini-Armstrong C, Protasi F, Ramesh V (1999) Shape, size, and distribution of Ca(2+) release units and couplons in skeletal and cardiac muscles. Biophys J 77(3):1528–39

Greenstein JL, Winslow RL (2011) Integrative systems models of cardiac excitation-contraction coupling. Circ Res 108(1):70–84

Grimm M, Ling H, Brown JH (2011) Crossing signals: relationships between beta-adrenergic stimulation and CaMKII activation. Heart Rhythm 8(8):1296–98. doi:10.1016/j.hrthm.2011.02.027

Hashambhoy YL, Winslow RL, Greenstein JL (2009) CaMKII-induced shift in modal gating explains L-type Ca(2+) current facilitation: a modeling study. Biophys J 96(5):1770–85. doi:10.1016/j.bpj.2008.11.055

Hashambhoy YL, Greenstein JL, Winslow RL (2010) Role of CaMKII in RyR leak, EC coupling and action potential duration: a computational model. J Mol Cell Cardiol 49(4):617–24

Hatano A, Okada J, Washio T, Hisada T, Sugiura S (2011) A three-dimensional simulation model of cardiomyocyte integrating excitation-contraction coupling and metabolism. Biophys J 101(11):2601–10. doi:10.1016/j.bpj.2011.10.020

Hatano A, Okada J, Hisada T, Sugiura S (2012) Critical role of cardiac t-tubule system for the maintenance of contractile function revealed by a 3D integrated model of cardiomyocytes. J Biomech 45(5):815–23. doi:10.1016/j.jbiomech.2011.11.022

Hedley PL, Jørgensen P, Schlamowitz S, Moolman-Smook J, Kanters JK, Corfield VA, Christiansen M (2009) The genetic basis of Brugada syndrome: a mutation update. Hum Mutat 30(9):1256–66

Honda HM, Korge P, Weiss JN (2005) Mitochondria and ischemia/reperfusion injury. Ann N Y Acad Sci 1047:248–58

Huke S, Knollmann BC (2010) Increased myofilament Ca^{2+}–sensitivity and arrhythmia susceptibility. J Mol Cell Cardiol 48(5):824–33

Hund TJ, Rudy Y (2004) Rate dependence and regulation of action potential and calcium transient in a canine cardiac ventricular cell model. Circulation 110(20):3168–74

Hund TJ, Decker KF, Kanter E, Mohler PJ, Boyden PA, Schuessler RB, Yamada KA, Rudy Y (2008) Role of activated CaMKII in abnormal calcium homeostasis and I(Na) remodeling after myocardial infarction: insights from mathematical modeling. J Mol Cell Cardiol 45(3): 420–28

Huser J, Wang YG, Sheehan KA, Cifuentes F, Lipsius SL, Blatter LA (2000) Functional coupling between glycolysis and excitation–contraction coupling underlies alternans in cat heart cells. J Physiol 524(Pt 3):795–806

Ivanov PC, Amaral LA, Goldberger AL, Havlin S, Rosenblum MG, Struzik ZR, Stanley HE (1999) Multifractality in human heartbeat dynamics. Nature 399(6735):461–5

Jafri MS, Kotulska M (2006) Modeling the mechanism of metabolic oscillations in ischemic cardiac myocytes. J Theor Biol 242(4):801–17

Jeyaraj D, Haldar SM, Wan X, McCauley MD, Ripperger JA, Hu K, Lu Y, Eapen BL, Sharma N, Ficker E, Cutler MJ, Gulick J, Sanbe A, Robbins J, Demolombe S, Kondratov RV, Shea SA, Albrecht U, Wehrens XH, Rosenbaum DS, Jain MK (2012) Circadian rhythms govern cardiac repolarization and arrhythmogenesis. Nature 483(7387):96–9. doi:10.1038/nature10852

Katz AM (2011) Physiology of the heart, 5th edn. Lippincott Williams & Wilkins, Philadelphia, PA

Keating MT, Sanguinetti MC (2001) Molecular and cellular mechanisms of cardiac arrhythmias. Cell 104(4):569–80

Kockskamper J, Zima AV, Blatter LA (2005) Modulation of sarcoplasmic reticulum Ca^{2+} release by glycolysis in cat atrial myocytes. J Physiol (Lond) 564(3):697–714

Kurz FT, Aon MA, O'Rourke B, Armoundas AA (2010) Spatio-temporal oscillations of individual mitochondria in cardiac myocytes reveal modulation of synchronized mitochondrial clusters. Proc Natl Acad Sci USA 107(32):14315–20

Lakatta EG, Maltsev VA, Bogdanov KY, Stern MD, Vinogradova TM (2003) Cyclic variation of intracellular calcium: a critical factor for cardiac pacemaker cell dominance. Circ Res 92(3): e45–50

Lakatta EG, Maltsev VA, Vinogradova TM (2010) A coupled system of intracellular Ca^{2+} clocks and surface membrane voltage clocks controls the timekeeping mechanism of the heart's pacemaker. Circ Res 106(4):659–73

Lukyanenko V, Gyorke S (1999) Ca^{2+} sparks and Ca^{2+} waves in saponin-permeabilized rat ventricular myocytes. J Physiol 521(Pt 3):575–85

Maltsev AV, Maltsev VA, Mikheev M, Maltseva LA, Sirenko SG, Lakatta EG, Stern MD (2011) Synchronization of stochastic Ca(2)(+) release units creates a rhythmic Ca(2)(+) clock in cardiac pacemaker cells. Biophys J 100(2):271–83

Marchant JS, Parker I (2001) Role of elementary Ca(2+) puffs in generating repetitive Ca(2+) oscillations. EMBO J 20(1–2):65–76

Matsuoka S, Sarai N, Jo H, Noma A (2004) Simulation of ATP metabolism in cardiac excitation-contraction coupling. Prog Biophys Mol Biol 85(2–3):279–99

Moss AJ, Kass RS (2005) Long QT syndrome: from channels to cardiac arrhythmias. J Clin Invest 115(8):2018–24

Napolitano C, Bloise R, Monteforte N, Priori SG (2012) Sudden cardiac death and genetic ion channelopathies. Circulation 125(16):2027–34. doi:10.1161/circulationaha.111.055947

Nivala M, Korge P, Nivala M, Weiss JN, Qu Z (2011) Linking flickering to waves and whole-cell oscillations in a mitochondrial network model. Biophys J 101(9):2102–11. doi:10.1016/j.bpj.2011.09.038

Nivala M, de Lange E, Rovetti R, Qu Z (2012a) Computational modeling and numerical methods for spatiotemporal calcium cycling in ventricular myocytes. Front Physiol 3:114. doi:10.3389/fphys.2012.00114

Nivala M, Ko CY, Nivala M, Weiss JN, Qu Z (2012b) Criticality in intracellular calcium signaling in cardiac myocytes. Biophys J 102(11):2433–42. doi:10.1016/j.bpj.2012.05.001

Noble D (1962) A modification of the Hodgkin-Huxley equations applicable to Purkinje fibre action and pace-maker potentials. J Physiol 160:317–52

Noble D, Rudy Y (2001) Models of cardiac ventricular action potentials: iterative interaction between experiment and simulation. Philos Trans R Soc Lond Ser A 359(1783):1127–42. doi:10.1098/rsta.2001.0820

O'Rourke B, Ramza BM, Marban E (1994) Oscillations of membrane current and excitability driven by metabolic oscillations in heart cells. Science 265(5174):962–6

Peters NS, Wit AL (1998) Myocardial architecture and ventricular arrhythmogenesis. Circulation 97(17):1746–54

Ponard JG, Kondratyev AA, Kucera JP (2007) Mechanisms of intrinsic beating variability in cardiac cell cultures and model pacemaker networks. Biophys J 92(10):3734–52

Pouvreau S (2010) Superoxide flashes in mouse skeletal muscle are produced by discrete arrays of active mitochondria operating coherently. PLoS One 5(9)

Prosser BL, Ward CW, Lederer WJ (2011) X-ROS signaling: rapid mechano-chemo transduction in heart. Science 333(6048):1440–45. doi:10.1126/science.1202768

Qu Z (2011) Chaos in the genesis and maintenance of cardiac arrhythmias. Prog Biophys Mol Biol 105(3):247–57

Qu Z, Weiss JN (2006) Dynamics and cardiac arrhythmias. J Cardiovasc Electrophysiol 17:1042–49

Qu Z, Garfinkel A, Weiss JN, Nivala M (2011) Multi-scale modeling in biology: how to bridge the gaps between scales? Prog Biophys Mol Biol 107:21–31

Rovetti R, Cui X, Garfinkel A, Weiss JN, Qu Z (2010) Spark-induced sparks as a mechanism of intracellular calcium alternans in cardiac myocytes. Circ Res 106:1582–91. doi:10.1161/circresaha.109.213975

Sanguinetti MC, Tristani-Firouzi M (2006) hERG potassium channels and cardiac arrhythmia. Nature 440(7083):463–9

Sanguinetti MC, Jiang C, Curran ME, Keating MT (1995) A mechanistic link between an inherited and an acquired cardiac arrhythmia: HERG encodes the IKr potassium channel. Cell 81(2):299–307

Sato D, Xie LH, Sovari AA, Tran DX, Morita N, Xie F, Karagueuzian H, Garfinkel A, Weiss JN, Qu Z (2009) Synchronization of chaotic early afterdepolarizations in the genesis of cardiac arrhythmias. Proc Natl Acad Sci USA 106(9):2983–88

Saucerman JJ, Bers DM (2008) Calmodulin mediates differential sensitivity of CaMKII and calcineurin to local Ca^{2+} in cardiac myocytes. Biophys J 95(10):4597–612

Saucerman JJ, Brunton LL, Michailova AP, McCulloch AD (2003) Modeling beta-adrenergic control of cardiac myocyte contractility in silico. J Biol Chem 278(48):47997–8003

Saucerman JJ, Healy SN, Belik ME, Puglisi JL, McCulloch AD (2004) Proarrhythmic consequences of a KCNQ1 AKAP-binding domain mutation: computational models of whole cells and heterogeneous tissue. Circ Res 95(12):1216–24

Skupin A, Kettenmann H, Winkler U, Wartenberg M, Sauer H, Tovey SC, Taylor CW, Falcke M (2008) How does intracellular Ca^{2+} oscillate: by chance or by the clock? Biophys J 94(6):2404–11

Skupin A, Kettenmann H, Falcke M (2010) Calcium signals driven by single channel noise. PLoS Comput Biol 6(8):e1000870. doi:10.1371/journal.pcbi.1000870

Soeller C, Cannell MB (1999) Examination of the transverse tubular system in living cardiac rat myocytes by 2-photon microscopy and digital image–processing techniques. Circ Res 84(3):266–75

Soltis AR, Saucerman JJ (2010) Synergy between CaMKII substrates and beta-adrenergic signaling in regulation of cardiac myocyte Ca(2+) handling. Biophys J 99(7):2038–47. doi:10.1016/j.bpj.2010.08.016

Spach MS, Heidlage JF (1995) The stochastic nature of cardiac propagation at a microscopic level—electrical description of myocardial architecture and its application to conduction. Circ Res 76(3):366–80

Stanley HE (1971) Introduction to phase transitions and critical phenomena. Oxford University Press, London

Stanley HE (1999) Scaling, universality, and renormalization: three pillars of modern critical phenomena. Rev Mod Phys 71:S358–S366

Swaminathan PD, Purohit A, Hund TJ, Anderson ME (2012) Calmodulin-dependent protein kinase II: linking heart failure and arrhythmias. Circ Res 110(12):1661–77. doi:10.1161/CIRCRESAHA.111.243956

ter Keurs HEDJ, Boyden PA (2007) Calcium and arrhythmogenesis. Physiol Rev 87(2):457–506

ter Keurs HEDJ, Shinozaki T, Zhang YM, Zhang ML, Wakayama Y, Sugai Y, Kagaya Y, Miura M, Boyden PA, Stuyvers BDM, Landesberg A (2008) Sarcomere mechanics in uniform and non-uniform cardiac muscle: a link between pump function and arrhythmias. Prog Biophys Mol Biol 97(2–3):312–31

Thiffault C, Bennett JP Jr (2005) Cyclical mitochondrial deltapsiM fluctuations linked to electron transport, F0F1 ATP-synthase and mitochondrial Na^{+}/Ca^{2+} exchange are reduced in Alzheimer's disease cybrids. Mitochondrion 5(2):109–19

Turcotte DL, Rundle JB (2002) Self-organized complexity in the physical, biological, and social sciences. Proc Natl Acad Sci USA 99(Suppl 1):2463–65

Waldo AL, Camm AJ, deRuyter H, Friedman PL, Macneil DJ, Pauls JF, Pitt B, Pratt CM, Schwartz PJ, Veltri EP (1996) Effect of d-sotalol on mortality in patients with left ventricular dysfunction after recent and remote myocardial infarction. Lancet 348:7–12

Wang Y (2007) Mitogen-activated protein kinases in heart development and diseases. Circulation 116(12):1413–23

Wang W, Fang H, Groom L, Cheng A, Zhang W, Liu J, Wang X, Li K, Han P, Zheng M, Yin J, Wang W, Mattson MP, Kao JP, Lakatta EG, Sheu SS, Ouyang K, Chen J, Dirksen RT, Cheng H (2008) Superoxide flashes in single mitochondria. Cell 134(2):279–90

Weiss JN, Karma A, Maclellan WR, Deng M, Rau CD, Rees CM, Wang J, Wisniewski N, Eskin E, Horvath S, Qu Z, Wang Y, Lusis AJ (2012) "Good enough solutions" and the genetics of complex diseases. Circ Res 111(4):493–504. doi:10.1161/CIRCRESAHA.112.269084

Wier WG, ter Keurs HE, Marban E, Gao WD, Balke CW (1997) Ca^{2+} 'sparks' and waves in intact ventricular muscle resolved by confocal imaging. Circ Res 81(4):462–69

Wu F, Yang F, Vinnakota KC, Beard DA (2007) Computer modeling of mitochondrial tricarboxylic acid cycle, oxidative phosphorylation, metabolite transport, and electrophysiology. J Biol Chem 282(34):24525–37

Xie L-H, Chen F, Karagueuzian HS, Weiss JN (2009) Oxidative stress-induced afterdepolarizations and calmodulin kinase II signaling. Circ Res 104(1):79–86. doi:10.1161/circresaha.108.183475

Yang JH, Yang L, Qu Z, Weiss JN (2008) Glycolytic oscillations in isolated rabbit ventricular myocytes. J Biol Chem 283(52):36321–27

Yang L, Korge P, Weiss JN, Qu Z (2010) Mitochondrial oscillations and waves in cardiac myocytes: insights from computational models. Biophys J 98(8):1428–38
Zhou L, O'Rourke B (2012) Cardiac mitochondrial network excitability: insights from computational analysis. Am J Physiol Heart Circ Physiol 302(11):H2178–89. doi:10.1152/ajpheart.01073.2011
Zhou L, Salem JE, Saidel GM, Stanley WC, Cabrera ME (2005a) Mechanistic model of cardiac energy metabolism predicts localization of glycolysis to cytosolic subdomain during ischemia. Am J Physiol Heart Circ Physiol 288(5):H2400–2411
Zhou L, Stanley WC, Saidel GM, Yu X, Cabrera ME (2005b) Regulation of lactate production at the onset of ischaemia is independent of mitochondrial NADH/NAD+: insights from in silico studies. J Physiol 569(Pt 3):925–37
Zhou L, Aon MA, Almas T, Cortassa S, Winslow RL, O'Rourke B (2010) A reaction–diffusion model of ROS-induced ROS release in a mitochondrial network. PLoS Comput Biol 6(1): e1000657
Zhou L, Aon MA, Liu T, O'Rourke B (2011) Dynamic modulation of Ca^{2+} sparks by mitochondrial oscillations in isolated guinea pig cardiomyocytes under oxidative stress. J Mol Cell Cardiol 51(5):632–9. doi:10.1016/j.yjmcc.2011.05.007
Zipes DP, Rubart M (2006) Neural modulation of cardiac arrhythmias and sudden cardiac death. Heart Rhythm 3(1):108–13. doi:10.1016/j.hrthm.2005.09.021
Zipes DP, Wellens HJ (1998) Sudden cardiac death. Circulation 98(21):2334–51

Chapter 11
Systems Level Regulation of Cardiac Energy Fluxes Via Metabolic Cycles: Role of Creatine, Phosphotransfer Pathways, and AMPK Signaling

Valdur Saks, Uwe Schlattner, Malgorzata Tokarska-Schlattner, Theo Wallimann, Rafaela Bagur, Sarah Zorman, Martin Pelosse, Pierre Dos Santos, François Boucher, Tuuli Kaambre, and Rita Guzun

V. Saks (✉)
Laboratory of Fundamental and Applied Bioenergetics, Univ. Grenoble Alpes, Grenoble, France

INSERM, U1055 Grenoble, France

Laboratory of Bioenergetics, National Institute of Chemical Physics and Biophysics, Tallinn, Estonia
e-mail: valdur.saks@ujf-grenoble.fr

U. Schlattner (✉) • M. Tokarska-Schlattner • S. Zorman • M. Pelosse
Laboratory of Fundamental and Applied Bioenergetics, Univ. Grenoble Alpes, Grenoble, France

INSERM, U1055 Grenoble, France
e-mail: uwe.schlattner@ujf-grenoble.fr

T. Wallimann
ETH Zürich, Zürich, Switzerland

R. Bagur
Laboratory of Fundamental and Applied Bioenergetics, Univ. Grenoble Alpes, Grenoble, France

INSERM, U1055 Grenoble, France

TIMC-IMAG, Univ. Grenoble Alpes, Grenoble, France

CNRS, UMR5525, Grenoble, France

P.D. Santos
Univ. Bordeaux, Bordeaux, France

Centre de Recherche Cardiothoracique de Bordeaux, Inserm, U1045, Bordeaux, France

LIRYC, L'Institut de rythmologie et modélisation cardiaque, Université de Bordeaux, Bordeaux, France

F. Boucher
TIMC-IMAG, Univ. Grenoble Alpes, Grenoble, France

CNRS, UMR5525, Grenoble, France

M.A. Aon et al. (eds.), *Systems Biology of Metabolic and Signaling Networks*,
Springer Series in Biophysics 16, DOI 10.1007/978-3-642-38505-6_11,

Abstract Integrated mechanisms of regulation of energy metabolism at cellular, tissue, and organ levels are analyzed from a *systems biology* perspective. These integrated mechanisms comprise the coordinated function of three cycles of mass and energy transfer and conversion: (1) the Randle cycle of substrate supply, (2) the Krebs cycle coupled with energy transformation in mitochondrial oxidative phosphorylation, and (3) the kinase cycles of intracellular energy transfer and signal transduction for regulation of energy fluxes. These cycles are extended and partially governed by information transfer systems like those linked to protein kinase signaling. In the heart, these cycles are closely related to the Ca^{2+} cycle during excitation–contraction coupling. According to the view of integrated metabolic cycles, the phosphocreatine/creatine kinase system represents a most important subsystem determining the efficiency of regulation of metabolic and energy fluxes in heart, brain, and oxidative skeletal muscles. It carries about 80 % of the energy flux between mitochondria and cytoplasm in heart. The substrate uptake, respiration rate, and energy fluxes are regulated in response to workload via phosphotransfer pathways and Ca^{2+} cycling. We propose integrated network mechanisms to explain the linear relationship between myocardial oxygen consumption and heart work output under conditions of metabolic stability (metabolic aspect of Frank–Starling's law of the heart). The efficiency of energy transfer, force of contraction, and metabolic regulation of respiration and energy fluxes depend upon the intracellular concentration of total creatine, which is decreased in heart failure. The role of creatine, creatine kinase, and adenylate kinase phosphotransfer and AMP-activated protein kinase (AMPK) signaling systems and their interrelationship with substrate supply and Ca^{2+} cycles are analyzed. Finally, an introduction to the AMPK signaling network is provided with a particular emphasis on the heart in health and disease.

11.1 Introduction

In this chapter, we describe from a systems biology perspective the integration and regulation of substrate and energy supply in living organisms and the role of the creatine/creatine kinase (Cr/CK) system. Systems biology focuses on the mechanisms of interactions between system components at molecular, cellular,

T. Kaambre
Laboratory of Bioenergetics, National Institute of Chemical Physics and Biophysics, Tallinn, Estonia

R. Guzun (✉)
Laboratory of Fundamental and Applied Bioenergetics, Univ. Grenoble Alpes, Grenoble, France

INSERM, U1055 Grenoble, France

EFCR and Sleep Laboratory, Univ. Hospital of Grenoble (CHU), Grenoble, France
e-mail: rita.guzun@chu-grenoble.fr

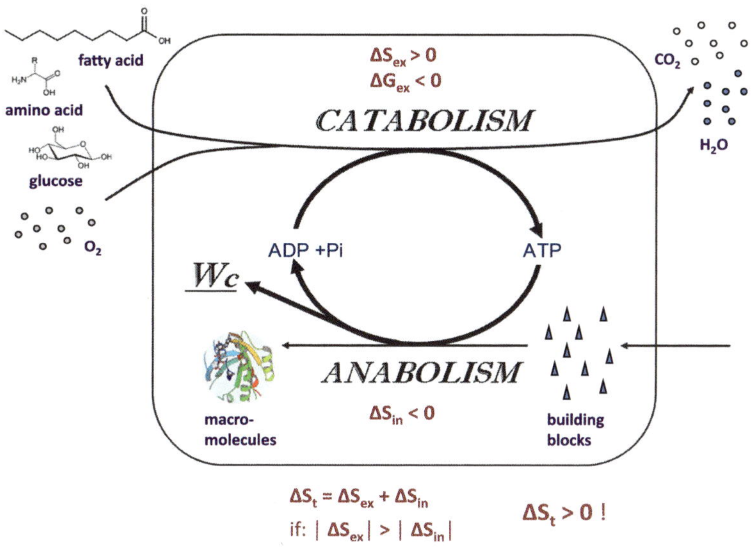

Fig. 11.1 General scheme of cellular metabolism. Catabolic reactions generating ATP (*top*), through coupling to anabolic reactions (biosynthesis, *bottom*) using ATP, maintain cell structural organization as an expression of the decrease of internal entropy ($\Delta S_{\text{in}} < 0$) and are also the source of energy for cellular work (W_c). Abbreviations: ΔS_{ex} external entropy, ΔS_{in} internal entropy, ΔS_t total entropy, ΔG_{ex} variatuion of Gibbs free energy. For further details, see text. Adapted from (Saks 2007) with permission

and organ levels, giving rise to biological function. As such, systems biology provides basic mechanistic insights about the principles that govern metabolic behavior in living systems. According to Schrödinger, the metabolic activity of living systems needs a continuous exchange of metabolites with the surroundings as a form of extracting free energy from the medium. This process enables cells and organisms to increase their internal organization such that they are able to perform biological work from anabolic reactions (Schrödinger 1944). An increase of internal order implies a decrease of entropy that should be compensated by an entropy increase in the environment. Catabolic and anabolic reactions are coupled to mediate biological work (e.g., muscle contraction) through processes of free energy conversion involving synthesis and utilization of ATP (Fig. 11.1). Coupling between cellular work, anabolism, and catabolism is achieved by cyclic processes involving mechanisms of feedback regulation. Herein, we introduce the theory of integrated metabolic cycles. Cycles of substrate supply (Randle cycle), intracellular energy conversion (Krebs cycle and mitochondrial oxidative phosphorylation), and phosphotransfer reactions (kinase cycles) constitute conspicuous examples of both substrate and energy provision and feedback regulation (Fig. 11.2). These cycles closely interact with calcium (Ca^{2+}) cycling (Fig. 11.2). Among the kinase cycles, a key role is played by the Cr/CK system, adenylate kinase, and AMPK in skeletal muscle, heart, brain, and other cell types (Wallimann et al. 1992, 2011; Schlattner et al. 2006a, b; Schlattner and Wallimann 2004; Wallimann 1996, 2007; Saks

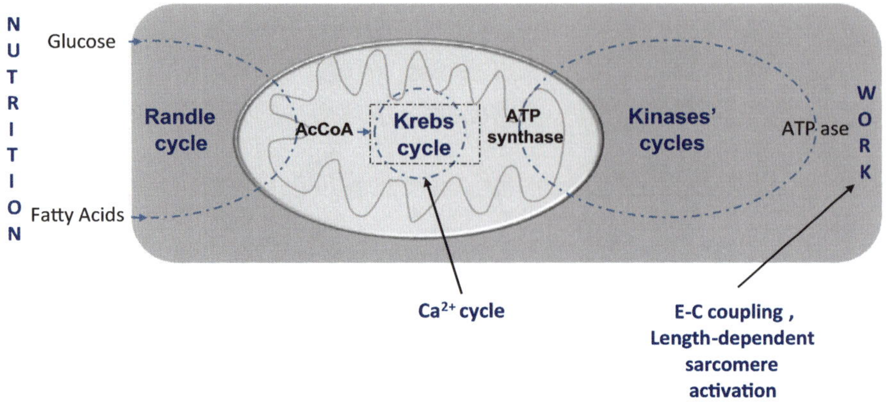

Fig. 11.2 General representation of regulation of energy fluxes via metabolic cycles at the cellular level. The regulatory action that energy transfer cycles, such as the creatine kinase (CK) and adenylate kinase systems (AK), exert on fuel supply is realized through the Randle cycle and energy transforming Krebs cycle, coupled to oxidative phosphorylation. Any decrease in the use of intracellular energy diminishes Krebs cycle activity and tends to favor the accumulation of substrates

et al. 1978, 2007a, 2010, 2012; Saks 2007; Dzeja and Terzic 2003, 2009). In the heart, contraction is initiated by excitation-contraction coupling that includes processes linked to intracellular Ca^{2+} cycling (Bers 2002; Bers and Despa 2006). Under physiological conditions, contractile force and cardiac work are regulated by ventricular filling and sarcomere length-dependent mechanism (Frank-Starling's law) at constant amplitude of Ca^{2+} transients. A main regulatory motif of cardiac energy fluxes is represented by metabolic feedback regulation through local changes in Pi, ADP, AMP, Cr, and phosphocreatine (PCr) ratios (Saks et al. 2006a, 2010, 2012; Bose et al. 2003; Dos Santos et al. 2000; Aliev et al. 2012). Under conditions of adrenergic stimulation, cardiac Ca^{2+} cycling in the cytoplasm and mitochondria becomes most important for energy flux regulation (Balaban 2002; Griffiths and Rutter 2009; Tarasov et al. 2012; Glancy and Balaban 2012). Control and regulation of mitochondrial respiration by both adenine nucleotides and Ca^{2+} have been analyzed in an integrated model of cardiomyocyte function (Cortassa et al. 2009).

In this work, we aim to analyze regulatory interactions involved in the modulation of energy supply and demand in the network comprised by Randle and Krebs cycles and phosphotransfer pathways in the heart. Contribution of calcium cycling to the regulation of energy supply–demand in the heart has been extensively reviewed elsewhere (Balaban 2002, 2009a, b, 2012; Tarasov et al. 2012; Glancy and Balaban 2012). The synchronization of the mitochondrial network in cardiac cells is treated by Cortassa and Aon in Chap. 5.

11.2 Structural Basis of Functional Organization of Cardiomyocyte Metabolism

In adult cardiac cells, mitochondria are localized at the A band level of sarcomeres between Z-lines close to T-tubular system and sarcoplasmic reticulum (SR). Estimation of the density distribution of mitochondria relative to their centers showed that neighboring mitochondria in cardiomyocytes are aligned according to a rectangle with distance between centers equal to 1.97 ± 0.43 μm and 1.43 ± 0.43 μm in the longitudinal and transverse direction, respectively (Vendelin et al. 2005). High temporal resolution analysis of mitochondrial dynamics in adult cardiomyocytes (one frame every 400 ms) revealed very rapid fluctuation of center positions that did not exceed the limit of the organelle (Beraud et al. 2009). These limited mitochondrial oscillations can be explained by inner membrane conformational changes likely elicited by changes in volume associated with energetic/redox states (Hackenbrock 1968; Mannella 2006). In vivo imaging of mitochondrial dynamics in cardiomyocytes showed separated individual organelles which do not fuse with each other (Gonzalez-Granillo et al. 2012). Figure 11.3 shows confocal images of mitochondria and α-actinin distribution in cardiomyocytes from adult rats. In this figure the fluorescence immunolabelling of α-actinin is used to mark sarcomeric Z-lines. Individual mitochondria regularly arranged between Z-lines can be visualized by flavoprotein autofluorescence (Fig. 11.3, green). The green fluorescence intensity profile shows the peaks distribution corresponding to mitochondrial fluorescence; the regions of "zero" intensity of α-actinin (Fig. 11.3 red) indicate intermyofibrillar localization of mitochondria between Z-lines without apparent fusion/fission (Gonzalez-Granillo et al. 2012). Possibly, fusion can happen in perinuclear mitochondrial clusters (Kuznetsov and Margreiter 2009).

Regular arrangement and limited morphodynamics of mitochondria in adult cardiomyocytes are determined by the cytoskeletal architecture, which includes myofilaments, inter-myofilaments, microtubules, and other structural proteins. Tubulin is one of the constituent cytoskeletal proteins with structural, transport, and metabolic functions (see also Chap. 7). Herein, we will focus on the structural role of β isotypes of tubulin. Tubulin is a heterodimeric complex formed by two globular and two C-terminal tails (CTT) of α and β proteins. Globular α and β proteins can be polymerized into microtubules, while α and β CTT can interact with other intracellular structures and proteins. Tubulin has additional binding sites that allow the filaments to join together laterally to form sheets of filaments. About 30 % of tubulins in adult cardiomyocytes are polymerized and 70 % are in the heterodimeric state (Tagawa et al. 1998). These two conformational states of protein are in a dynamic balance driven by polymerization–depolymerization processes (Sackett 2010). A study of the distribution of β tubulins by fluorescence confocal microscopy showed that βIV tubulin is polymerized creating a dense mesh of mainly longitudinally and obliquely oriented microtubules. βIII tubulin co-localizes with alpha-actinine in Z-lines while βI tubulin forms randomly dispersed short polymers

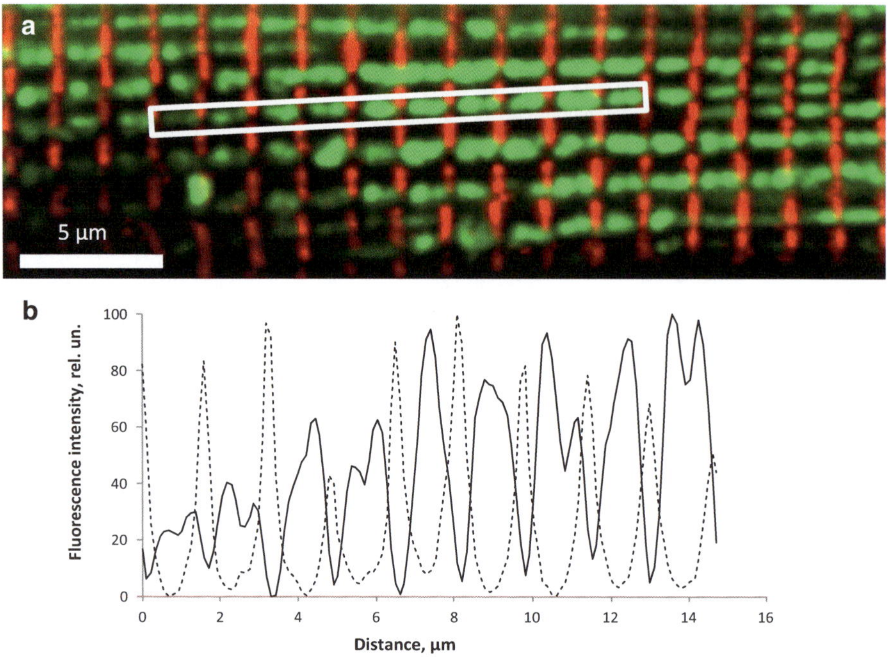

Fig. 11.3 Fluorescence confocal microscopy of mitochondria and alpha-actinin distribution in adult rat cardiomyocyte. (**a**) Regular distribution of individual mitochondria as visualized by autofluorescence of flavoproteines (*green color*) in between Z-lines that are labeled with rhodamine immunofluorescent for α-actinin (*red color*). (**b**) Analysis of fluorescence intensity along a selected line: *dotted* = α-actinin; *solid* = flavoproteins. Note that peaks of green fluorescence intensity corresponding to mitochondria are seen in the regions of "zero" intensity of red α-actinin fluorescence. Reproduced from (Gonzalez-Granillo et al. 2012) with permission

and dimers, and βII tubulin co-distributes with mitochondria (Saks et al. 2012; Gonzalez-Granillo et al. 2012; Guzun et al. 2011a, 2012). These findings are in agreement with data published first in 1990 by Saetersdal et al. regarding the link between β tubulin and mitochondria as revealed by immunogold labeling (Saetersdal et al. 1990). According to this study, β tubulin interacts with mitochondria through the outer membrane (MOM) creating links between the organelle and other cellular structures. The contribution of other cytoskeletal proteins to structural and functional interactions with mitochondria is under intensive investigation. Desmin and plectin are capable of interacting with voltage-dependent anion channel (VDAC) at MOM (Capetanaki et al. 2007; Capetenaki 2002; Liobikas et al. 2001; Schroder et al. 2002). The 1b isotype of plectin of cardiomyocytes co-localizes with mitochondria via direct interaction with VDAC, whereas plectin 1d isotype is specifically associated with sarcomeric Z-disks (Schroder et al. 2002).

Recently it has been proposed that the T-tubular system, which represents a network of tubular extensions from the sarcolemma, plays an important role in the structural organization of cardiac cell metabolism. The T-tubular system of rat ventricular cells creates a regular arrangement at the level of Z-line and along myofibrils (Fig. 11.4c) (Soeller and Cannell 1999). This system becomes disorganized with time in cardiac cells in culture. The functional role of T-tubules was described to provide a rapid inward spread of electrical excitation and Ca^{2+} influx that triggers Ca^{2+} release from the sarcoplasmic reticulum, as well as supply of each mitochondrion with oxygen and substrates. By using electron tomography (Hayashi et al. 2009) identified anatomical couplings between opposing membranes of T-tubules and sarcoplasmic reticulum (SR), these forming so-called Calcium Release Units (CRU). A close localization of mitochondria and CRU favors Ca^{2+} and metabolite microcompartmentation (Saks et al. 2012). Individual mitochondria localize at the level of the A-band of sarcomeres and at the Z-line they are in close contacts with jSR and the T-tubular system forming CRUs (Fig. 11.4b). This junctional cisterns of arrangement separates mitochondria from each other, also making their fusion unlikely. The 3D reconstruction of the T-tubular system in cardiac cells (Soeller and Cannell 1999) appears as an elaborated and effective system of Ca^{2+}, substrate, and oxygen supply from the extracellular medium. Its discovery about a decade ago profoundly changed our knowledge of the heart cell structure and the implications for metabolic regulation. As a matter of fact, according to this architecture no distinction is possible between intermyofibrillar and subsarcolemmal mitochondria, since both are in close contact with the T-tubular system. This is in agreement with results obtained from kinetic studies (Saks et al. 2012) and the fact that no electrical conduction occurs between individual mitochondria in cardiomyocytes (Beraud et al. 2009; Kuznetsov et al. 2009; Collins and Bootman 2003; Nivala et al. 2011; Zorov et al. 2000). Simultaneous measurements of sarcomere and mitochondrial dimensions in situ along the longitudinal axis of cardiomyocytes identified mitochondria as micron-sized spheres localized between sarcomeres and distributed throughout the cell in a crystal-like lattice without any visible fusion. In this organized lattice, transient mitochondrial depolarizations (flickers), elicited by ROS-induced opening of anion channels in the inner membrane, may propagate in cells as depolarization waves (Nivala et al. 2011; Yaniv et al. 2011). However, electron tomographic studies clearly revealed that there is no mitochondrial reticulum in cardiac cells; instead a regular lattice containing 5,000–10,000 single mitochondria seems to prevail (Nivala et al. 2011). In the heart, this forms the structural basis of the mitochondrial network described by Cortassa and Aon in Chap. 5. Taken together, all the data described above indicate that mitochondrial respiration depends upon localized events in their vicinity. These structurally organized functional domains—dubbed Intracellular Energetic Units (ICEUs) (Saks 2007; Saks et al. 2001, 2012) (Fig. 11.5)—comprise sites of ATP hydrolysis (myofibrillar ATPases, sarcoplasmic reticulum ATPase (SERCA), ion pumps) connected to ATP synthesis through phosphotransfer networks. Energy transduction within ICEUs involving the Randle and Krebs cycles of fuel supply and oxidative phosphorylation are governed by energy-demanding reactions. Next, we analyze cardiac energy metabolism from the perspective of regulatory interactions occurring in metabolic cycles.

11.3 Substrate Supply and Its Regulation (Randle and Krebs Cycles)

11.3.1 Mechanisms of Regulation of Fatty Acids Oxidation in Heart Muscle

Fatty acids are released from triacylglycerol (TAG) by activated lipoprotein lipase (LPL) and transferred in the cytoplasm bound to proteins. Free fatty acid transfer across mitochondrial membranes consumes ATP involving FFA conversion into an Acyl-CoA derivative and the transport-competent acyl-carnitine form by carnitine palmitoyl transferase (CPT). The MOM-localized CPT1 targeted by malonyl CoA inhibition constitutes an important regulatory step of β-oxidation of FAs (β-FAO) (Fig. 11.4) (Saks et al. 2006b). β-FAO is linked to the citric acid cycle and oxidative phosphorylation through NAD^+, FAD, and acyl-CoA. The NADH generated by the Krebs cycle and β-FAO is oxidized in the electron transport chain. Increased ATP utilization elicits ATP synthesis driven by the proton motive force, thus decreasing the NADH/NAD^+ ratio. Oxidation of the NADH pool increases the flux through the Krebs cycle through NAD^+-dependent isocitrate and α-ketoglutarate deshydrogenases, thus decreasing acetyl-CoA (AcCoA) levels. NAD^+ can also be reduced in β-FAO catalyzed by β-hydroxyacyl-CoA dehydrogenase and in the glycolytic pathway catalyzed by glyceraldehyde phosphate dehydrogenase (GAPDH). However, the transfer of NADH reduction potential from glycolysis towards the mitochondrial matrix via the malate–aspartate shuttle, being slower than direct NAD+ use by β-FAO, will prioritize the latter one (Kobayashi and Neely 1979). Thus, the GAPDH dependence on cytoplasmic NADH/NAD^+ ratio associated with the slow kinetics of malate–aspartate shuttle will rather slow down glycolysis. An increase in the rate of AcCoA utilization by the Krebs cycle will thus increase β-FAO. An accumulation of AcCoA does not influence significantly the rate of β-FAO due to the equilibrium constant of the reversible thiolase reaction which is in favor of AcCoA production (Neely and Morgan 1974).

At low ATP demand (decreased workload), the high NADH/NAD^+ ratio slows down the flux through NAD^+-dependent dehydrogenases, thus decreasing the rate of AcCoA oxidation through the Krebs cycle. An increased intra-mitochondrial AcCoA level is thought to favor its transfer towards the cytoplasm where it is converted into malonyl-CoA, an inhibitor of CPT-1-controlled FA transport into mitochondria. Malonyl-CoA levels are also controlled by acetyl-coA carboxylase (ACC), a cytosolic enzyme catalyzing conversion of AcCoA into malonyl-CoA, whose inactivation by AMPK during energy stress relieves CPT1 inhibition.

Preferential utilization of FAs involves inhibition of glucose transport, phosphofructokinase (PFK), and pyruvate dehydrogenase (PDH) reactions (Hue and Taegtmeyer 2009; Taegtmeyer 2010; Taegtmeyer et al. 2005). Glucose transport in muscle cells is realized through GLUT4, the expression of which in the sarcolemma is regulated by insulin and other signals. Increased NADH/NAD+ and

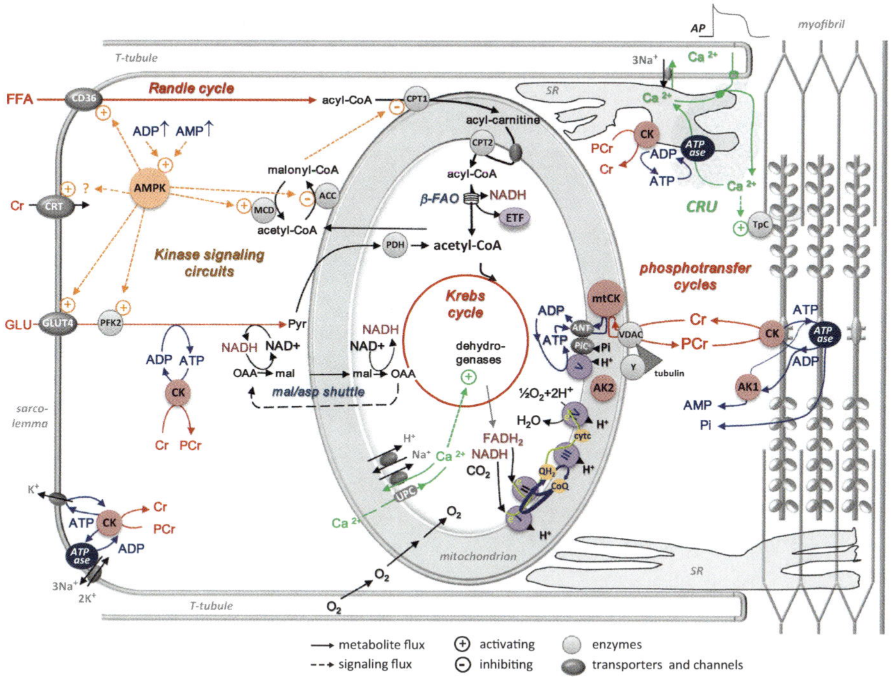

Fig. 11.4 Metabolic cycles and signaling networks in cardiomyocyte—Intracellular Energy Units (iEU). Free fatty acids (FFA, *upper left*) are taken up by a family of plasma membrane proteins (fatty acid transporter protein, FATP1, fatty acid translocase, CD36), and in the cytoplasm FAs are associated with fatty acid binding protein (FABP). FFAs are esterified to acyl-CoA via fatty acyl-CoA synthetase. The resulting acyl-CoA is then transported into mitochondria via carnitine palmitoyltransferase I (CPT and CPT II). Once inside, acyl-CoA becomes a substrate for the β-oxidation pathway, resulting in AcCoA production. Each round of β-oxidation produces 1 molecule of NADH, 1 molecule of $FADH_2$, and 1 molecule of AcCoA. AcCoA enters the Krebs cycle, where it is further oxidized to CO_2 with the concomitant generation of 3 molecules of NADH, 1 molecule of FADH2 and 1 molecule of ATP. Glucose (GLU) is taken up by glucose transporter-4 (GLUT-4, at the *left middle*) and enters the Embden–Meyerhof pathway, which converts glucose into 2 molecules of pyruvate (PYR). As a result of these reactions, 2 net ATP and 2 NADH are produced. NADH is transferred into mitochondria via the malate–aspartate shuttle. OAA, oxaloacetate; Glut, glutamate; αKG, α-ketoglutarate; ASP, aspartate; MAL, malate. Most of the metabolic energy derived from glucose can come from the entry of pyruvate into the Krebs cycle and oxidative phosphorylation via AcCoA. NADH and FADH2 issued from both metabolic pathways are oxidized in the respiratory chain. Mitochondrial creatine kinase (mtCK) catalyzes the direct transphosphorylation of intramitochondrial ATP and cytosolic creatine (Cr) into ADP and phosphocreatine (PCr). ADP enters the matrix space to stimulate oxidative phosphorylation, while PCr is transferred via the cytosolic Cr/PCr shuttle to be used in the functional coupling between CK and ATPases (acto-myosin ATPase and ion pumps, *black circles*). Feedback regulation of substrate supply occurs in the following way: the glucose–fatty acid (Randle) cycle: if glucose and FFAs are both present, FFAs inhibit the transport of glucose across the plasma membrane, and acyl-CoA oxidation increases the mitochondrial ratios of AcCoA/CoA and of NADH/NAD+ which inhibit the pyruvate dehydrogenase (PDH) complex. Citrate from increased production in the Krebs cycle can inhibit phosphofructokinase (PFK). These changes would slow down oxidation of glucose and pyruvate (PYR) and increase glucose-6-phosphate (G6P), which

AcCoA /CoA ratios inhibit PDH. Their inhibitory effect is realized through pyruvate dehydrogenase kinase (PDK) that phosphorylates and inhibits PDH (Randle 1998). Citrate that escapes oxidation in the Krebs cycle is transported to the cytosol where it inhibits PFK and glycolysis (Hue and Taegtmeyer 2009; Taegtmeyer 2010; Taegtmeyer et al. 2005).

Cell signaling via AMPK provides a parallel control of most of these processes, including substrate uptake via fatty acid and glucose transporters and flux via β-FAO and glycolysis (see Sect. 5.5). Activation of AMPK during energy stress situations stimulates all these activities.

Physiologically, the significance of the Randle cycle is to ensure the provision of FAs to high-energy demanding organs such as muscle and liver. Also, glucose is directed to organs such as brain, red blood cells, and other tissues dependent upon glucose oxidation and possessing relatively small stores of glycogen.

11.3.2 Which Substrate Is Better: Reductionism Versus Systems Biology

Living cells extract and transform energy from different sources distributing them between organs, as a function of their energy needs and metabolic potential. Unfortunately, there is not yet consensus on evaluating the amount of energy that may be extracted from different carbon sources. A reason for this is differences between reductionistic and systems biology type of approaches. The reductionist explanation of the competitive use of different energy sources by distinct organs is based on the oxygen needed to oxidize the different substrates and considerations of coupling of oxidative phosphorylation. All electrons from NADH produced in aerobic catabolism (i.e., from glycolysis and fatty acid oxidation) enter the respiratory chain via complex I, or electrons from $FADH_2$ formed in β-FAO are carried via electron transferring flavoprotein and complex III (Fig. 11.4), resulting in lower ATP/O ratio. In this way, the yield of 38 ATP for 12 atoms of oxygen consumed (P/O = 3.16) for glucose ($C_6H_{12}O_6$) oxidation and the yield of 129 ATP for 46 atoms of oxygen consumed (P/O = 2.8) for palmitic acid ($C_{16}H_{32}O_2$) oxidation are assumed to be sufficient to conclude that glucose is the preferential fuel for living organisms. This conclusion is further corroborated by measurements of oxygen consumption by direct calorimetry. When one liter of oxygen is used to burn substrates, the amount of energy obtained is 5.19 kcal/LO_2 for glucose and 4.81 kcal/LO_2 for palmitic acid (Leverve et al. 2006). However, these calculations

---

Fig. 11.4 (continued) would inhibit hexokinase (HK), and decrease glucose transport. *G6P* glucose 6-phosphate, *HK* hexokinase, *PFK* phosphofructokinase, *GLY* glycogen, *F1,6diP* fructose-1,6-bisphosphate, *GAPDH* glyceraldehyde 3 phosphate dehydrogenase, *1,3DPG* 1,3 diphosphoglycerate. AMPK signaling (orange) controls among others substrate uptake and flux via glycolysis and fatty acid oxidation under conditions of starvation, hypoxia and other triggers of energy stress. For details see text. Modified from (Saks et al. 2012) with permission

do not take into account that under aerobic physiological conditions oxygen is not a limiting factor for energy metabolism, but instead that there are many other factors to be taken into account in the whole system. And these factors were indeed taken into account by nature. Regarding the fuel supply to such a high-energy demanding organ as is the heart, Clark and collaborators were the first to show that glucose constituted less than 1/4 of the substrates oxidized by the isolated working frog heart (Clark et al. 1937). These authors were not able to figure out which substrate(s) were responsible for consuming the remnant oxygen. In 1954, Bing and collaborators showed that the respiratory quotient (RQ, VCO_2/VO_2) in post-absorptive state was about 0.7–0.75 while studying oxygen utilization during the aerobic metabolism of fats, ketones, and amino acids by human heart (Bing et al. 1954). This ratio was unchanged following overnight fasting but increased above 1 after ingestion of a high fat diet. The authors assumed that this increase could be due to utilization of intramuscular triacylglycerol (TAG) stores (Bing et al. 1954). Similar data were obtained in skeletal muscle. The average respiratory quotient (VCO_2/VO_2) of muscular tissue taken from de-pancreatized dogs was about 0.7 (Bing et al. 1954).

In the case of working heart, the preferential energy supply by FA can be understood from calculations specifying energy needs to realize work, energy content of different substrates per unit mass, and kinetics of reactions in Randle and Krebs cycles, rather than by oxygen consumed for oxidizing different fuels. A heart contracting with a frequency of 70 bpm exhibits a stroke volume of 0.07 L (i.e., cardiac output—5 L/min) that supports a pressure of 13 kPa (equivalent of 120/70 mm Hg) and realizes a work equal to 65 J/min or 93.6 kJ/day. ATP hydrolysis in the actomyosin reaction releases about 60 kJ/mol under physiological conditions. For the heart to accomplish a work equivalent to 100 kJ/day about 2.8 mol of ATP are needed ($n = W/\Delta G_{ATP}$ corrected for the reaction efficiency that in the case of actomyosin is about 60 %). This amount of ATP can be obtained from the oxidation of 0.074 mol glucose or 0.02 mol of palmitic acid. For glucose, supplemented with an equivalent molecular weight of 10 mol of water, 26.5 g glucose should be oxidized by the heart to perform work equivalent to 100 kJ/day. For palmitic acid only 5.5 g of this FA are necessary to perform a similar amount of work. Thus, the content of free energy per gram of mass that can be released during oxidation and converted into chemical energy in the form of ATP is much higher for FAs than for carbohydrates due to the much higher content of non-oxidized –C–C– and –C–H chemical bonds. Depending on the amount of bound water the difference in carbohydrates can range from three- to ninefold (Newsholme and Start 1973) (Fig. 11.5b). Thus, the kinetics of mass transfer in substrate supply is much more favorable when FAs, as compared to glucose, are used as substrates. And this fact explains the choice made by nature: heart and oxidative skeletal muscle clearly prefer FAs as substrates (Fig. 11.5a). Their preferred utilization by heart and oxidative muscle is achieved by multiple regulatory mechanisms involved in the Randle and Krebs cycles (Fig. 11.4).

Randle et al. (1963) were the first to propose the concept of selective supply of FAs over glucose for heart muscle (Randle et al. 1963). The glucose–FA cycle or

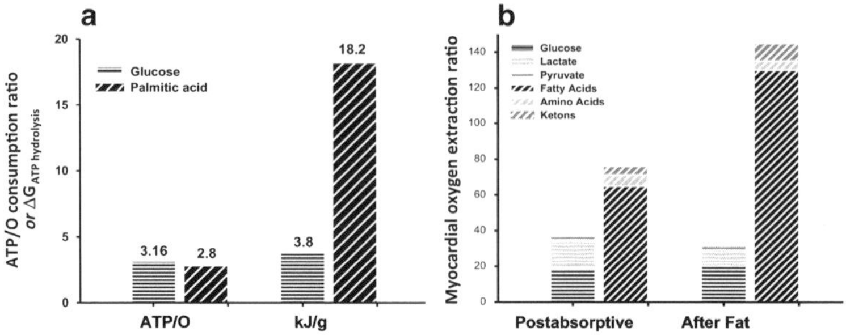

Fig. 11.5 The role of fatty acid oxidation in metabolism. (**a**) (i) ATP synthesis to oxygen consumption ratio in mitochondria for glucose and palmitate oxidation and (ii) the Gibbs free energy of ATP hydrolysis from the actin–myosin reaction obtained from the oxidation of one gram of glucose in comparison with the oxidation of one gram of palmitate. (**b**) Comparison of the myocardial oxygen extraction ratio of carbohydrates (glucose, pyruvate, and lactate) and non-carbohydrates (fatty acids, amino acids, ketones) in a post-absorptive state and after ingestion of FAs. In both states FAs oxidation is the prevalent source of energy for the heart [adapted from Bing et al. (1954) with permission]

Randle cycle outlined the restrictions imposed on muscle glucose metabolism by FA oxidation (Randle et al. 1963). Further mechanisms of regulation of the glucose–FA cycle in working heart were described by Neely and Morgan (1974) with new insights being revealed since then (Hue and Taegtmeyer 2009; Taegtmeyer 2010; Taegtmeyer et al. 2005). These mechanisms account for changes in the kinetics of fuels supply, mass transfer, and transformation including glucose transport and glycolysis, FA transport, β-FAO, and the Krebs cycle in response to variations in respiration rates and NADH oxidation.

11.4 Phosphotransfer Pathways (Kinase Cycles)

11.4.1 Creatine Biosynthesis and Transmembrane Transport

Creatine biosynthesis occurs in a two-step reaction; first, in the kidney and in pancreas, the amino acids arginine and glycine are combined to form guanidino acetic acid (GAA) by the enzyme AGAT (arginine-glycine amino-transferase), and second, in the liver, where GAA, taken up from blood serum via GABA-2 (gamma-aminobutyric acid transporte) (Tachikawa et al. 2012), is methylated to generate Cr by GAMT (guanidine-acetic acid methyltransferase) using SAM (S-adenosine-methionine) as a substrate (Wyss and Kaddurah-Daouk 2000). Creatine synthesized in the liver is released into the bloodstream by a still unknown mechanism. Since creatine is not produced in significant amounts in, e.g., heart, brain, skeletal, and smooth muscle, where it plays an important functional role, it has to be imported by these tissues from blood serum, using a specific creatine transporter (CRT) (Beard and Braissant 2010). In this way, creatine participates in the regulation of metabolism at the organ level. An increase in total Cr and PCr in cells also increases the PCr/ATP ratio and thus energy charge (Wallimann et al. 2011). Mutations in either of the genes coding for AGAT, GAMT (endogenous creatine synthesis), or CRT (creatine transport) in humans lead to the so-called creatine deficiency syndrome with a severe neuromuscular and neurological phenotype including developmental delay of expressive language and cognitive speech, mental retardation, autistic-like behavior, epilepsy, and brain atrophy (for review, see (Stockler et al. 2007)).

11.4.2 Direct Measurement of Energy Fluxes: Principal Role of the Phosphocreatine Pathway in Energy Transfer in the Heart

While Cr has been known for 175 years after its discovery by Michel Chevreul, the hypothesis of the PCr pathway was formulated by Samuel Bessman (Bessman and Carpenter 1985; Bessman and Fonyo 1966; Bessman and Geiger 1981) and independently by Martin Klingenberg (1970, 1976, 2008; Wallimann 1975; Turner et al. 1973; Saks et al. 1978) about 50 years ago. An important factual basis of this hypothesis is given by the observation made by Belitzer and Tsybakova (1939), who showed that Cr addition stimulated respiration in skeletal muscle homogenates, resulting in PCr production (Belitzer and Tsybakova 1939). A fundamental contribution to the existence of a PCr pathway of energy transfer in heart, muscle, brain, and other tissues was been made by Theo Wallimann's group. They showed that different CK isoenzymes belong to different compartments, with MtCK in mitochondria and cytosol and MM-CK in myofibrils and the membrane of sarcoplasmic reticulum. They also resolved the atomic structure of CKs and characterized interaction mechanisms with neighboring structures (Wallimann et al. 1992, 2007; Schlattner et al. 1998, 2006a, b; Schlattner and Wallimann 2004; Eder et al. 1999, 2000; Fritz-Wolf et al. 1996). MM-CK was also shown to localize in the sarcolemmal membrane (Saks et al. 1977). Such in vivo compartmentation of CK and ATP in muscle cells represents the cellular basis of the CK cycle, one of the phosphotransfer pathways of energy transport (Wallimann et al. 1992, 2007; Schlattner et al. 2006a, b; Schlattner and Wallimann 2004; Saks 2007, 2008, 2009; Aliev et al. 2012; Saks et al. 2007b). Detailed functional studies combining the use of mathematical modeling with experimental data have shown that within myofibrils, and in the subsarcolemmal area, the diffusion coefficient for ATP is decreased by factor of 10^5 as compared to water solution (Abraham et al. 2002; Alekseev et al. 2012; Selivanov et al. 2004). Diffusion limitations result in ATP compartmentation in cells, where the local ATP and ADP pools are connected by the phosphotransfer pathways. An equally important and fundamental contribution was been made by Dzeja and Terzic groups who measured quantitatively, using an isotope tracer method, energy fluxes between different cellular compartments involving kinase cycles (Dzeja and Terzic 2003, 2009; Dzeja et al. 1999; Nemutlu et al. 2012). Most effective and informative in bioenergetic studies of phosphoryl transfer has been the use of ^{18}O transfer (see the Chap. 6). This method is based on the following two reactions: ATP hydrolysis by water molecules containing ^{18}O and ATP resynthesis with formation of [^{18}O]γATP (Dzeja and Terzic 2009; Nemutlu et al. 2012):

$$\mathrm{ATP} + \left[{}^{18}\mathrm{O}\right]\mathrm{H_2O} \rightarrow \left[{}^{18}\mathrm{O}\right]\mathrm{Pi} + \mathrm{ADP} \tag{11.1}$$

$$\left[{}^{18}\mathrm{O}\right]\mathrm{Pi} + \mathrm{ADP} \rightarrow \left[{}^{18}\mathrm{O}\right]\gamma\mathrm{ATP} \tag{11.2}$$

Paul Boyer used this method for studying the ATP synthase reaction (Boyer 1997). Inclusion of [^{18}O]Pi into [^{18}O]γATP in the presence of uncouplers led him to the conclusion of the rotational binding change mechanism of mitochondrial ATP synthesis. Nelson Goldberg, Petras Dzeja, André Terzic, and coworkers have successfully applied this method for studying the kinetics of phosphoryl-transfer reactions and energy fluxes in vivo by measuring the rates of the following reactions (Dzeja and Terzic 2003, 2009; Nemutlu et al. 2012):

Creatine kinase phosphotransfer:

$$\left[{}^{18}\mathrm{O}\right]\gamma\mathrm{ATP} + \mathrm{Cr} \rightarrow \left[{}^{18}\mathrm{O}\right]\ \mathrm{PCr} + \mathrm{ADP} \tag{11.3}$$

Adenylate kinase phosphotransfer:

$$\left[{}^{18}\mathrm{O}\right]\gamma\mathrm{ATP} + \mathrm{AMP} \rightarrow \left[{}^{18}\mathrm{O}\right]\beta\mathrm{ADP} + \mathrm{ADP} \rightarrow \left[{}^{18}\mathrm{O}\right]\beta\mathrm{ATP} + \mathrm{AMP} \tag{11.4}$$

Glycolytic phosphotransfer:

$$\left[{}^{18}\mathrm{O}\right]\gamma\mathrm{ATP} + \mathrm{Glucose} \rightarrow \left[{}^{18}\mathrm{O}\right]\mathrm{G6P} + \mathrm{ADP} \tag{11.5}$$

If a direct transfer of ATP from mitochondria to MgATPases happens together with its immediate hydrolysis for contraction as sometimes proposed in the literature, only isotope transfer reactions 1 and 2 could be observed. In an excellent series of studies Dzeja's group showed that in normal cardiac cells about 80–85 % of phosphoryl groups are transferred out from mitochondria by the PCr flux, and about 10–15 % by adenylate kinase, with a minor contribution by glycolysis (Dzeja et al. 1999). In the heart, these fluxes increase linearly with workload energy demand under conditions of the Frank–Starling law (Saks et al. 2007c). Figure 11.6 shows that PCr fluxes measured experimentally can be quantitatively simulated with a mathematical model of compartmentalized energy transfer (Dos Santos et al. 2000; Aliev et al. 2012; Aliev and Saks 1997; Vendelin et al. 2000). This model was based on the experimental data obtained in studies of mitochondrial PCr synthesis in permeabilized cardiomyocytes. The role of the adenylate kinase system becomes important in hypoxia and pathological situations (Dzeja et al. 1999).

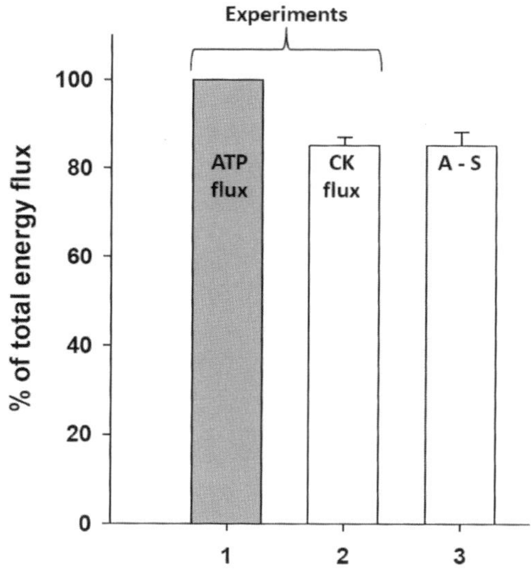

Fig. 11.6 Comparison of experimental data of energy flux measurements with results of simulations by mathematical models. ATP flux: the rate of ATP synthesis in mitochondria; CK flux: energy flux carried into cytoplasm by phosphocreatine measured experimentally by the ^{18}O transfer method [data summarized from Dzeja and Terzic (2003), Dzeja et al. (1996, 2001, 2007, 2011a), Pucar et al. (2001)]; A–S: Aliev and Saks models of compartmentalized energy transfer (Dos Santos et al. 2000; Aliev and Saks 1997). The mathematical model of the compartmentalized energy transfer system in cardiac myocytes includes mitochondrial synthesis of ATP by ATP synthase, PCr production in the coupled MtCK reaction, the myofibrillar and cytoplasmic CK reactions, ATP utilization by actomyosin ATPase during the contraction cycle, and diffusional exchange of metabolites between different compartments. The model gives a good fitting with the experimental data, showing that about 85 % of energy produced in mitochondria as ATP flux is transferred out of mitochondria as PCr flux, in agreement with the abundant experimental data reported by Dzeja and colleagues

Recently this method has been used in quantitative studies of metabolic cycles in human health and disease (Dzeja et al. 2011a).

11.4.3 Intracellular Energetic Units and Mitochondrial Interactosome: Local Signaling and Frank–Starling Law

In addition to the fundamental structural data from Wallimann and Schlattner and energy flux determinations by Dzeja and Terzic, another important question concerns the cellular mechanisms involved in the function of CKs and other phosphotransfer pathways. This question was addressed by the group of Valdur Saks utilizing permeabilized cells that enable the study of mitochondrial function in their natural environment (Saks et al. 1991, 1998, 2007a, d; Saks and Strumia

1993). A central bioenergetic question in muscle cells relates to the mechanism of PCr synthesis in mitochondria. This question arises because the equilibrium and kinetic constants of all CK isoforms would favor only the resynthesis of MgATP from PCr and MgADP (Saks et al. 2010; Guzun et al. 2009). Kinetic information available is in agreement with the role of MM-CK at the sites of local ATP regeneration in myofibrils and membranes of sarcolemmal and sarcoplasmic reticulum, but this is not the case for PCr synthesis in mitochondria. More insight can be obtained from the classical problem of cardiac physiology—the metabolic aspect of the basic Frank–Starling law of the heart (Saks et al. 2006c, 2012). Discovered in 1914–1926, the Frank–Starling law states that under physiological conditions contractile force, cardiac work, and the rate of oxygen consumption increase manifold with the filling of the left ventricle (Starling and Visscher 1927). Later it was found that this occurs without any changes in the ATP and PCr levels (metabolic stability) and Ca^{2+} transients (Neely et al. 1972; Balaban et al. 1986). The latter observation excludes any explanation involving a mechanism of control of mitochondrial respiration by changes in intracellular Ca^{2+}. A Ca^{2+}-mediated mechanism may be important only in the case of adrenergic activation of the heart (Tarasov et al. 2012; Balaban 2012). Assuming that ATP, ADP, PCr, and Cr are related through equilibrium relationships, the observation of metabolic stability was interpreted to exclude any other explanation of workload dependence of cardiac oxygen consumption than a mechanism involving the control of mitochondrial respiration by ADP or Pi only. The popular assumption of CK equilibrium, as in a mixed bag of enzymes (Wiseman and Kushmerick 1995), however, is in contradiction with the experimental evidence (Saks 2008; Guzun and Saks 2010). This includes recent high-resolution ^{31}P NMR experiments showing that the major part of adenine nucleotides, notably ATP in muscle cells, exists associated with macromolecules and that free ADP may be only transiently present in the cytoplasm (Nabuurs et al. 2010, 2013). We have shown that both high PCr fluxes in the heart detected by Dzeja and collaborators (Dzeja and Terzic 2003, 2009; Dzeja et al. 1999; Nemutlu et al. 2012) and the linear dependence of the rate of oxygen consumption on cardiac work may be explained by local signaling and metabolic channeling of adenine nucleotides in nonequilibrium CK reactions (Saks et al. 2012; Guzun et al. 2009; Timohhina et al. 2009). Actually, CK can catalyze within the same cell either the forward or the backward reaction depending on in which microcompartment the enzyme is located and where it functions as part of different multienzyme complexes.

Mechanisms involving the interaction of mitochondria and CKs with other cellular structures and multienzyme complexes are central for understanding metabolic stability in the heart. This implies a different perspective in the framework of systems biology. Figure 11.7a shows the localization of the tubulin isotype βII following the pattern of mitochondrial distribution in cardiac cells (Saks et al. 2012; Gonzalez-Granillo et al. 2012; Guzun et al. 2011b, 2012). Tubulin βII is part of the heterodimer tubulin that binds to VDAC in MOM, thus modulating the close probability of this channel specifically so that it is permeated by Cr or PCr but limited for ATP or ADP (Guzun et al. 2009; Timohhina et al. 2009). In cardiac

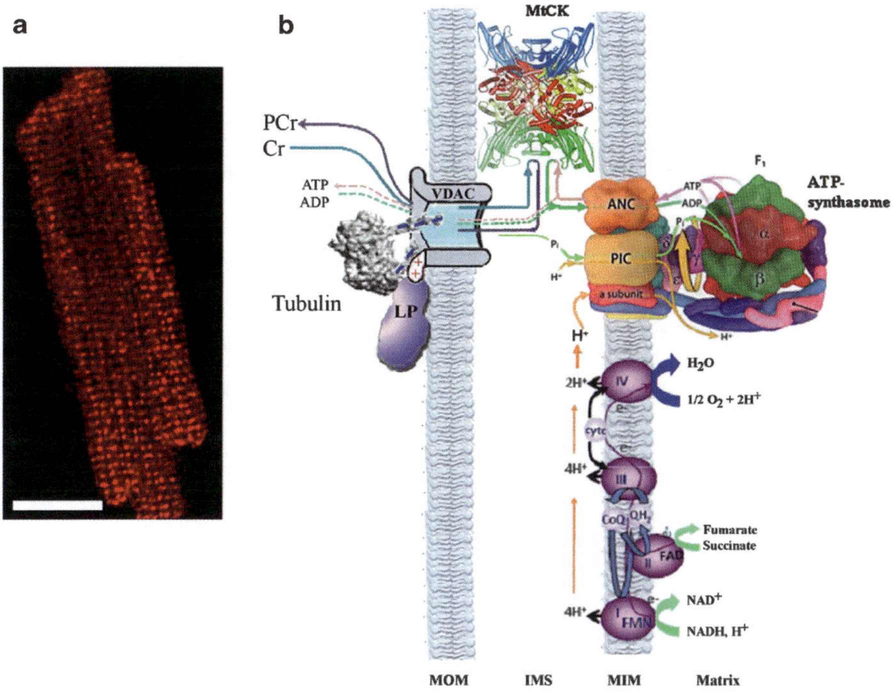

Fig. 11.7 Mitochondrial Interactosome. (**a**) Confocal image of a cardiomyocyte labeled with MitoTracker Red for mitochondria; scale bar 14 μm. (**b**) Scheme depicting the mitochondrial interactosome, a macromolecular complex formed by the ATP synthasome, in turn constituted by ATP synthase (subunits in different colors), adenine nucleotide translocase (ANC, *orange*), inorganic phosphate carrier (PIC, *yellow*), coupled to the respiratory complexes (I-IV, *purple circles*) in the mitochondrial inner membrane (MIM), octameric mitochondrial CK (MtCK, backbone structure with dimers in different color) in the intermembrane space (IMS) and the voltage-dependent anion channel (VDAC, *gray-blue*) in the mitochondrial outer membrane (MOM) interacting with cytoskeletal proteins tubulin (*gray*, surface structure representation) and putative linker protein (LP, *purple*). Metabolite fluxes are indicated by *arrows* in different colors. For ATP synthase, subunits of the F_1 part (*greek letters*) and F_0 part (*latin letters*) are indicated, as well as the rotation of the rotor (*yellow arrow*). For the respiratory chain, proton pumping (H^+, *yellow arrows*) and some redox centers (FMN, FAD) are indicated, as well as the two electron carriers coenzyme Q ($CoQ/CoQH_2$) and cytochrome c (cytc). Adapted from (Timohhina et al. 2009) and (Schlattner et al. 2009) with permission. Art work of the ATP synthasome in this figure was reproduced with kind permission from P.L. Pedersen and is the result of the combined efforts of Drs. Young H. Ko and David J. Blum; MtCK structure and membrane topology is reproduced from (Schlattner et al. 2006b) with permission

cells, the heterodimeric tubulin αβII and VDAC form a supercomplex with MtCK and the ATP synthasome—the mitochondrial interactosome (MI) (Fig. 11.7b) (Timohhina et al. 2009). Within this supramolecular structure, ATP and ADP cycle between ATP synthasome and MtCK maintaining oxidative phosphorylation effectively coupled to the synthesis of PCr. In the MI, MtCK functions unidirectionally toward PCr synthesis utilizing mitochondrial ATP supplied by ANT (direct channeling). This process moves ADP back into mitochondria, because of the differential permeability of VDAC in interaction with tubulin that impedes ADP release from mitochondria. These coupled reactions of oxidative phosphorylation and PCr synthesis in MI are effectively regulated by Cr (Fig. 11.8). In the presence of an extra-mitochondrial ADP trapping system (pyruvate kinase, PK; phosphoenolpyruvate, PEP), Cr addition rapidly increases the respiration rate to its maximal value, revealing a preferential accessibility of the ADP produced by MtCK to matrix ATPase, not to the cytosolic trapping system. Metabolic control analysis of mitochondrial respiration in permeabilized cardiac cells showed high flux control coefficients (FCC) for reactions involving ADP recycling coupled to MtCK and PCr production (Fig. 11.9a). Actually, the sum of control coefficients exceeds the theoretical value for linear systems by a factor of 4 (Tepp et al. 2011). This can be interpreted in terms of MtCK-controlled reactions in MI acting as very effective amplifiers of metabolic signals from cytoplasm (Tepp et al. 2011; Aon and Cortassa 2012). According to Kholodenko, Westerhoff, and their coworkers, the sum of the FCC of the metabolic pathway components exceeding one indicates a direct channeling in the pathway (Moreno-Sanchez et al. 2008). On the contrary, in isolated heart mitochondria and permeabilized cardiac fibers the sum of FCC of respiratory chain complexes, ATP synthase, and metabolite carriers, estimated under conditions of respiration activated by ADP, is close to 1, corresponding to a linear metabolic pathway (Moreno-Sanchez et al. 2008; Kuznetsov et al. 1996; Doussiere et al. 1984; Fell and Thomas 1995; Groen et al. 1982). The high efficiency of energy flux control in MI makes this supercomplex a key site for the feedback of metabolic regulation of mitochondrial respiration in cardiac cells (Saks et al. 2012; Tepp et al. 2011).

Figure 11.9b depicts the possible role of both Cr and ADP in the control of respiration in situ. Extra- and intra-mitochondrial ADP in the regulation of respiration was studied by MgATP titration in the absence or presence of Cr, i.e., activated MtCK (Saks et al. 2012; Guzun et al. 2009; Guzun and Saks 2010; Timohhina et al. 2009). The influence of mitochondrial ADP alone on respiration was estimated by removing extra-mitochondrial ADP through the PEP-PK trapping system mimicking glycolytic ADP consumption. From Fig. 11.9b we can see that stimulation of the extra-mitochondrial ADP producing system by MgATP alone cannot effectively activate respiration. The high apparent K_m for exogenous MgATP (157.8 ± 40.1 μM) corresponds to the apparent K_m of myofibrillar ATPase reaction for MgATP. However, when oxidative phosphorylation is stimulated by both extra- and intra-mitochondrial ADP (in the presence of Cr to activate MtCK and MM-CK in myofibrils), the respiration rate increases rapidly up to maximal values and the apparent K_m for ATP decreases from 157.8 ± 40.1 μM to 24.9 ± 0.8 μM. Removal of extra-mitochondrial ADP by PEP-PK provokes an increase of K_m for MgATP up to 2.04 ± 0.10 mM. These results show that local endogenous ADP in ICEUs is an important regulatory factor of respiration but only in the presence of Cr and activated MtCK. The stimulatory effect of respiration by endogenous ADP is strongly amplified by functional coupling of MtCK with ANT

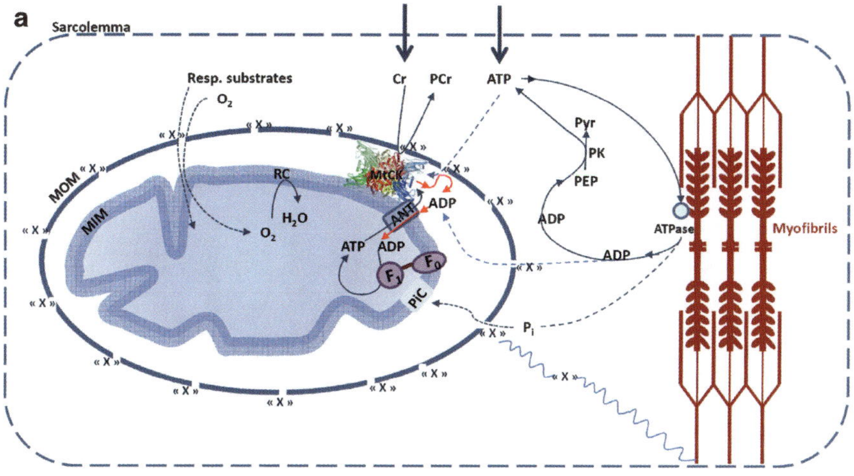

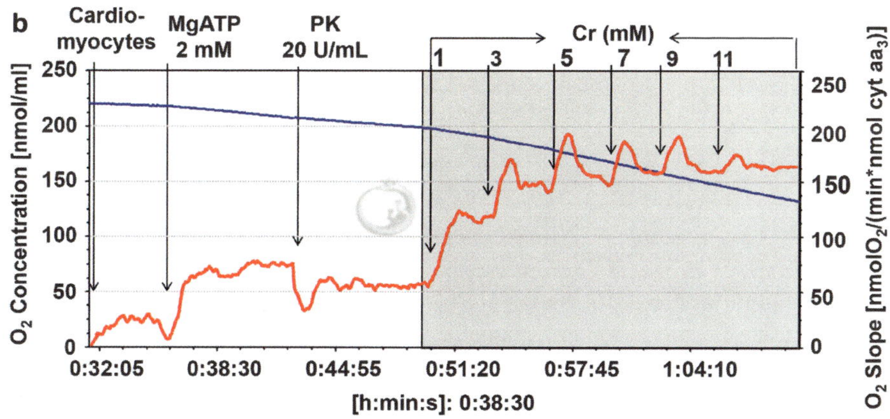

Fig. 11.8 Control of mitochondrial respiration by creatine in permeabilized cardiomyocytes. (**a**) Schematic representations of an oxygraph experiment and of a mitochondrion in a permeabilized cardiac cell, surrounded by cytoskeletal proteins and myofibrils. First, added ATP is hydrolyzed by cellular ATPases and the ADP produced stimulates respiration. Phosphoenolpyruvate (PEP) and pyruvate kinase (PK) continuously trap extra-mitochondrial ADP to regenerate ATP. Stepwise addition of Cr in the presence of ATP stimulates mitochondrial creatine kinase (MtCK) that controls respiration through continuous intra-mitochondrial re-cycling of ADP from ATP. (**b**) Oxygraph recording of Cr stimulated respiration. This experiment enables the estimation of the apparent affinity of MtCK for Cr. The *left scale* and the *blue* trace indicate the oxygen concentration (nmol O_2 ml^{-1}) in the experimental milieu. The *right scale* and the *red* trace denote the rate of oxygen uptake (in nmol O_2 min^{-1} $nmol^{-1}$ cyt. aa3). Adapted from (Guzun et al. 2009) with permission

that increases adenine nucleotides recycling within the MI (Saks et al. 2012; Guzun et al. 2009; Timohhina et al. 2009; Jacobus and Saks 1982). The loss of Cr-stimulated respiration in transgenic MtCK-knockout mice confirms the central role of MtCK in respiration regulation (Kay et al. 2000).

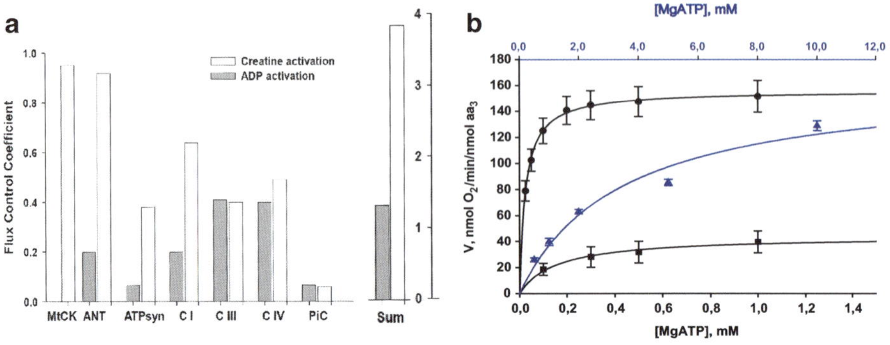

Fig. 11.9 *The energy flux control in permeabilized cardiomyocytes: creatine stimulation of mitochondrial respiration*. (**a**) Flux control coefficients for MtCK, adenine nucleotide translocase (ANT), ATP synthasome (ATPsyn), respiratory complexes I (C I), III (C III), IV (C IV), and inorganic phosphate carrier (PiC). The *right panel* shows the sum of flux control coefficients. Reproduced from (Tepp et al. 2011) with permission. (**b**) The role of endogenous ADP produced in MgATPase reactions at different concentrations of MgATP in the regulation of mitochondrial respiration in permeabilized cardiomyocytes under different conditions: (*square*)—without ADP trapping system (PEP-PK) and in the absence of Cr; (*filled circle*)—without PEP-PK system but in the presence of 20 mM Cr (i.e., activated MtCK); (*triangle*)—in the presence of both trapping system for free ADP and 20 mM Cr. Reproduced from (Timohhina et al. 2009) with permission

Taken together this information allows explaining the linear relationship existing between oxygen consumption and cardiac work by local metabolic feedback signaling within ICUEs (Saks et al. 2010, 2012; Aliev et al. 2012) (Fig. 11.10). Direct flux determination and mathematical modeling show that not more than 10 % of free energy is transported out of mitochondria by ATP flux needed to equilibrate the information-carrying flux of ADP into mitochondria. According to this model, ADP released from actomyosin cross-bridges stimulates the local MM-CK reaction in the myofibrillar space within ICEUs while at the same time forms a concentration gradient towards mitochondria (Fig. 11.10a–c) (Dos Santos et al. 2000; Aliev et al. 2012; Aliev and Saks 1997; Vendelin et al. 2000). The amplitude of displacement of MM-CK from equilibrium, as well as cyclic changes in ADP, is proportionally increased with workload (Fig. 11.10b, c). The rephosphorylation of ADP in the MM-CK reaction increases locally the Cr/PCr ratio that is transferred towards MtCK via the CK/PCr shuttle. Regulation of VDAC permeability by βII tubulin is a key element mediating the linear response of mitochondrial respiration to local signaling within ICEUs. When MOM is permeable, as in isolated mitochondria, modulation of respiration is impossible because of saturating ADP concentrations used under these conditions. The latter exceeds manifold the apparent affinity of oxidative phosphorylation for free ADP (K_m^{app}ADP = 7.9 ± 1.6 μM), even in diastolic phase (about 40 μM) (Fig. 11.11a). On the contrary, when ADP diffusion is restricted at the level of

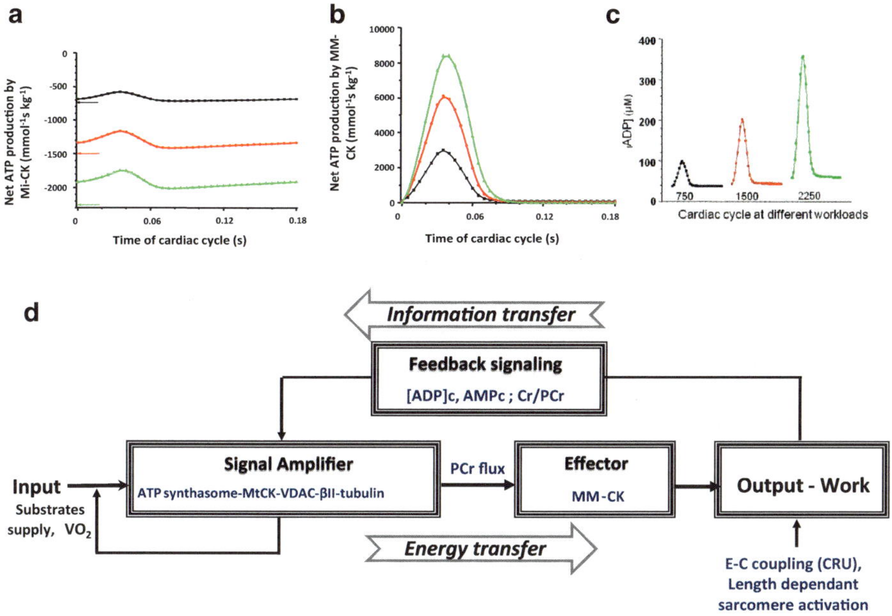

Fig. 11.10 *Mechanisms of regulation of mitochondrial respiration controlled by MtCK and of energy fluxes in cardiac muscle cells.* (**a–c**) Results from a mathematical model of cardiac energy metabolism (Vendelin-Aliev-Saks-Dos Santos model). (**a,b**) Calculated net PCr production rates in nonequilibrium steady state MtCK reaction (**a**) and cyclic changes in rates of ATP regeneration in nonequilibrium myofibrillar MM-CK reaction (**b**) during contraction cycles at different workloads corresponding to oscillations of [ADP]c indicated in Fig. 11.11. (**c**) Mathematically modeled oscillations of ADP concentrations in the core of myofibrils over cardiac cycle at workloads equivalent to 750 (*black*), 1,500 (*red*) and 2,250 (*green*) μmol ATP s^{-1} kg^{-1}. According to this model, the ATP cyclically produced during contractions (**b**) is associated with cyclical oscillations of ADP and Pi concentrations in myofibrils (**c**) and subsequent PCr production in the MtCK reaction (**a**). Reproduced from (Dos Santos et al. 2000; Aliev et al. 2012; Aliev and Saks 1997; Vendelin et al. 2000) with permission. (**d**) Schematic representation of feedback metabolic signaling in regulation of energy metabolism within ICEUs in cardiac cells. Due to the nonequilibrium MtCK and cyclic MM-CK reactions, intracellular ATP utilization (output) and mitochondrial ATP regeneration (input) are linked via cyclic fluctuations of cytosolic ADP and Cr/PCr. See the text for explanation

MOM, as in mitochondria in situ, the apparent K_m for free ADP increases to about 370.75 ± 30.57 μM and the respiration rate becomes almost linearly dependent on local ADP concentration. Under these conditions, the initial respiratory rate can be approximated by its linear dependence on ADP within the range of values corresponding to the increase in workload (Fig. 11.11b) (Guzun et al. 2009; Timohhina et al. 2009). Thus, cyclic changes in local ADP concentrations within the myofibrillar space of ICEUs become an effective regulatory signal due to (1) the nonequilibrium state of CK reactions, (2) the restricted VDAC permeability to metabolites elicited by association with βII tubulin, and (3) the presence of

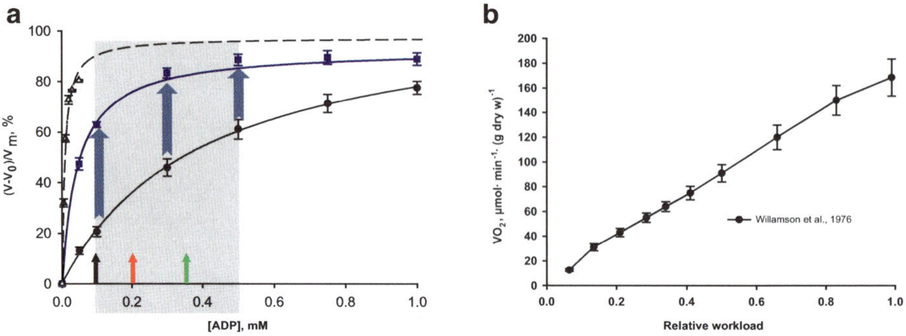

Fig. 11.11 *The role of restriction of ADP diffusion in the regulation of mitochondrial respiration.* (**a**) Kinetic analysis of ADP-activated respiration. The ADP concentrations corresponding to mathematically modeled fluctuations of ADP by Michaelis–Menten graph representation with colored small arrows (*black, red and green*), contained in the area of physiological cytosolic ADP concentration (indicated by a *gray box*). When MOM is permeable, as in isolated mitochondria (Δ, K_m^{app}ADP—7.9 ± 1.6 μM), the regulation of respiration is impossible because of a saturated ADP concentration for the minimal workload. When the ADP diffusion is restricted at the level of MOM, as in mitochondria in permeabilized cardiomyocytes (*circle*, K_m^{app}ADP—370.75 ± 30.57 μM), the respiration rates become linearly dependant on ADP concentrations, in fact also on heart workloads in accordance with the Frank–Starling law (**b**). This linear dependence under physiological conditions can be amplified by creatine (see large *blue arrows* in **a**) in the presence of activated MtCK (*Square*, K_m^{app}ADP—50.24 ± 7.98 μM). Reproduced from (Guzun et al. 2009) with permission. (**b**) The metabolic aspect of the Frank-Starling's law of the heart is expressed by linear dependence between the increase of left ventricular end-diastolic volume and the increase of respiration rates in the absence of measurable changes in the intracellular ATP and PCr content. Reproduced from (Saks et al. 2006c) with permission

Cr. When these conditions are fulfilled, activation of the coupled MtCK within MI by Cr induces ADP/ATP recycling and increases respiration rate, thus amplifying the effect of cytoplasmic ADP; under these conditions, the apparent K_m for ADP becomes equal to 50.24 ± 7.98 μM (Fig. 11.11a). These data suggest that modulation of respiration by local changes in ADP concentration, under condition of restriction of adenine nucleotide diffusion across mitochondrial membranes, is mediated by the structural organization of the MI. The MtCK reaction amplifies the ADP signal due to its functional coupling with ATP Synthasome (Fig. 11.7), thus increasing the steady-state rate of adenine nucleotides cycling in mitochondria and the rate of respiration. The coupled reactions of muscle type MM-CK in myofibrils and MtCK in mitochondria perform under nonequilibrium conditions and proceed in opposite directions (Fig. 11.10a–c) (Saks et al. 2012; Guzun et al. 2009; Guzun and Saks 2010; Timohhina et al. 2009). This mode of function results in separation of energy fluxes (mass and energy transfer by PCr) and signaling (information transfer by oscillations of cytosolic ADP concentrations, Pi and PCr/Cr ratio) that is amplified within the MI. As a result, reactions catalyzed by different isoforms of compartmentalized CK tend to maintain the intracellular metabolic stability. The separation of energy and information transfer is illustrated by the scheme depicted in Fig. 11.10d. This scheme shows feedback regulation of respiration in vivo according to Norbert Wiener's cybernetic principles (Saks et al. 2012; Guzun and Saks 2010): the usage of ATP (or release of free energy of ATP hydrolysis, ΔG_{ATP}, to perform work, marked as output) and ATP regeneration (or extraction of ΔG_{ATP} from substrates by oxidative phosphorylation, denoted as input) are interconnected via the feedback signaling through oscillations of cytoplasmic concentrations of ADP, AMP, Pi, and Cr/PCr amplified within MI. In this framework, the role of βII tubulin in association with MOM in cardiomyocytes would be to induce the linear response of mitochondrial respiration to workload-dependent metabolic signals. This elegant feedback mechanism of regulation of respiration on a beat-to-beat basis ensures metabolic stability necessary for normal heart function and explains well the metabolic aspect of the Frank–Starling's law of the heart (Saks 2007; Saks et al. 2006a, 2012). Importantly, recycling of adenine nucleotides within MI when coupled to PCr production significantly decreases ROS levels ensuring maximal efficiency of free energy transduction in mitochondria while inhibiting permeability transition pore opening, thus protecting the heart under stress conditions (Schlattner et al. 2006b; Meyer et al. 2006).

While the mechanisms described above represent local signaling within ICEUs, important mechanisms of synchronization of mitochondrial activity between ICEUs and their integration into structurally and functionally organized cellular systems are described by Cortassa and Aon in Chap. 5. The role of Ca^{2+} cycle in maintaining high respiratory activity of mitochondria within ICEUs has been described by Balaban's group and studied by mathematical modeling by Cortassaet al. (2009).

11.4.4 Intracellular Creatine Concentration as a Regulatory Factor in Heart Energetics

Many experimental and clinical studies have shown that intracellular Cr concentration is an important factor, determining the efficiency of intracellular energy transfer in heart cells (Saks et al. 1978, 2012; Wyss and Kaddurah-Daouk 2000; Nascimben et al. 1996). The results of an earlier work of ours published more than 30 years ago are reproduced in Fig. 11.12. This experiment shows that removal of Cr from the frog heart cells results in decreased PCr content and diminished contractile force; all parameters return to their initial value after restoration of Cr content (Saks et al. 1978). Similar results were recently reported by (Nabuurs et al. 2013) by assessing morphological, metabolic, and functional consequences of systemic Cr depletion in skeletal muscle. These data were obtained in a mouse model deficient in *L*-arginine:glycine amidino transferase ($AGAT^{-/-}$) which catalyzes the first step of Cr biosynthesis. In this work, systemic Cr depletion resulted in mitochondrial dysfunction and intracellular energy deficiency, as well as structural and physiological abnormalities. In vivo magnetic resonance

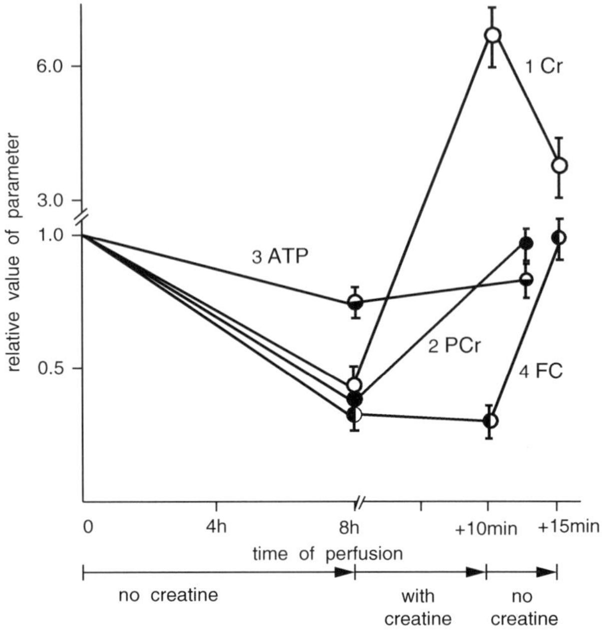

Fig. 11.12 *The role of Creatine in the regulation of contraction in frog heart.* After 8 h of perfusion without creatine, frog heart strips assume a hypodynamic state with decreased contractile force (FC) and lowered Cr and PCr levels. Addition of 20 mM Cr to the perfusate restored to normal the values of all these variables. Reproduced with permission from (Saks et al. 1978)

spectroscopy showed a near-complete absence of Cr and PCr in resting hind limb muscle of AGAT$^{-/-}$ mice. Compared to wild type, the inorganic phosphate/β-ATP ratio was increased fourfold, while ATP levels were reduced to nearly half and overall mitochondrial content was increased. The Cr-deficient AGAT$^{-/-}$ mice presented with significantly reduced grip strength and suffered from severe muscle atrophy. Oral Cr administration led to rapid accumulation in skeletal muscle (faster than in brain) and reversed all muscle abnormalities revealing that the condition of the AGAT$^{-/-}$ mice can be switched between Cr-deficient and normal simply by dietary manipulation. The consequences of AGAT deficiency were more pronounced than those of muscle-specific CK deficiency (Nabuurs et al. 2013), which suggests a multifaceted involvement of Cr in addition to its role in the PCr–CK system and in muscle energy homeostasis, as, e.g., by direct effects on biomembranes (Tokarska-Schlattner et al. 2012). It was also shown by the group of Stefan Neubauer in Oxford that a moderate elevation of total Cr levels in the heart by approximately 50 % in transgenic mice overexpressing the Cr transporter (CRT) conveyed significant protection and improved recovery of the hearts upon experimental induction of ischemia/reperfusion (Lygate et al. 2012). In one of their most important work the Neubauer's group has shown that a decrease of PCr content in the heart of patients with dilated cardiomyopathy is accompanied with significantly increased mortality rates (Neubauer 2007).

The role of altered phosphotransfer pathways in heart pathology of animal models, as well as human patients, is well documented and has been described in a number of reviews (Ingwall and Weiss 2004; Ingwall 2006; Ventura-Clapier et al. 2002, 2004). Most recently, two younger Chinese patients with acute myocardial infarction and presenting with muscle MM-CK deficiency have been diagnosed with somatic mutations in the M-CK gene. These mutants at amino acid E79 prevent correct folding and dimerization of M-CK. In parallel, correct targeting of the enzyme to subcellular structures is hampered and enzymatic CK activity dramatically lowered (Wu et al. 2013). These data with human cardiac infarction patients have shown that active dimeric MM-CK together with its substrates Cr and PCr are important for normal heart function.

Thus, the current opinion, supported by a host of data derived from different experimental approaches and provided by a number of different independent laboratories, is that Cr and PCr together with microcompartmentalized CK isoforms are physiologically essential for normal body function, specifically for optimal performance of skeletal and heart muscle, brain, neuronal cells, skin, retina and auditory cell, spermatozoa, and other cells of intermittant high-energy requirements (Wallimann et al. 1992, 2011). This fundamental hypothesis is strongly supported by the more or less severe phenotypes observed in double and single CK isoenzyme knockout mice, respectively (Steeghs et al. 1997; Streijger et al. 2005; Heerschap et al. 2007), as well as by the phenotypes of AGAT and GAMT knockout mice, presenting with disturbed energy metabolism body weight control, hampered fertility, muscular dystrophy, and cognitive and behavioral impairment, etc. (Nabuurs et al. 2013; Schmidt et al. 2004; ten Hove et al. 2005; Torremans et al. 2005).

In a most recent, provocative publication, entitled "Life without creatine", Lygate and colleagues purport that the phenotype of the GAMT knockout mouse was basically "normal". Specifically, they do not find a skeletal or cardiac muscle phenotype (Lygate et al. 2013). This contradicts the phenotype of the same transgenic mouse described earlier (Schmidt et al. 2004; ten Hove et al. 2005; Kan et al. 2005). Most importantly it is in contrast to the AGAT knockout creatine deficiency mouse (Nabuurs et al. 2013). This latter AGAT knockout mouse, in contrast to the GAMT knockout mouse, does not synthesize guanidine acetate (GAA), which in the GAMT knockout skeletal muscle was shown to be phosphorylated by CK to form an alternative energy-rich phosphagen, phospho-GAA, which still can be utilized as high-energy phosphagen, albeit at lower efficiency (Heerschap et al. 2007). Thus, it will be most important to reevaluate cardiac energy metabolism and heart phenotype in the GAMT knockout mouse to completely rule out any compensatory effects of the high concentrations of phospho-GAA accumulated in these knockout mice and to rule out a still possible contribution of phospho-GAA as an still alternative energy source. Corresponding experiments with the pure creatine deficiency AGAT knockout, presenting with a rather severe phenotype that is reversible by simple creatine supplementation (Nabuurs et al. 2013), are warranted and should provide some answers to these pending questions. To get a physiologically more meaningful answer to the true function of CK in heart it will be paramount to stress the heart to maximal performance and work output, where one would expect to see the true effects of creatine deficiency also in this organ.

In conclusion, until the enigma of the results provided by (Lygate et al. 2013) (see above) is solved, all available data still indicate that the CK system together with PCr and Cr is central to the regulatory mechanisms of metabolic and energy fluxes in those cells under intermitantly fluctuating high-energy requirements, including the heart (Taegtmeyer and Ingwall 2013).

11.5 The Signaling Network of AMP-Activated Protein Kinase (AMPK) in the Heart

11.5.1 Protein Kinase Signaling Networks in Metabolic Control of Cardiac Function

Metabolic cycles as described before provide immediate metabolic feedback for changes in energy input (nutrient supply) and energy output (workload). They are particularly important in the heart, an organ that maintains a high degree of metabolic stability and a particularly well-controlled energy homeostasis. An additional layer of regulation, which ascertains this metabolic stability, is achieved by information transfer via protein kinase signaling. All major protein kinase pathways were shown to play important roles in the heart, controlling contraction force, contractility, and heart rate in particular during cardiac development, under prolonged strong stimulation, and under emerging pathological conditions.

The possibly best studied example is the cyclic adenosine nucleotide (cAMP)-dependent protein kinase A (PKA) (Taylor et al. 2008), together with its homologous cGMP-dependent protein kinase G (PKG) (Takimoto 2012). Their control of cardiac contraction strength, ion fluxes, and hypertrophy relies on a precise spatiotemporal regulation of substrate phosphorylation. In case of PKA, A-kinase anchoring proteins (AKAPs) and cyclic nucleotide phosphodiesterases (PDEs) play a major role in this spatiotemporal organization and the occurrence of cAMP microdomains (Edwards et al. 2012; Mika et al. 2012; Diviani et al. 2013). This emphasizes the importance of cellular localization and organization for protein kinase-mediated information fluxes, as already outlined above for cardiac CK isoenzymes.

Also some other protein kinase signaling pathways have to be considered as relevant for cardiac metabolism. Protein kinase B (PKB or Akt) is an essential component of the growth response of an organism to nutritional input. In the heart, it participates in the regulation of myocyte growth under physiological conditions (Walsh 2006; Hers et al. 2011). PKC isoforms regulate cardiac contraction and hypertrophic responses, as well as other signaling pathways in more pathological situations such as ischemia and reperfusion injuries (Steinberg 2012). While calcium-regulated PKD is a more recent addition to the kinome, less well studied in respect to the cardiovascular system (Avkiran et al. 2008), members of the mitogen-activated protein kinase (MAPK) family are prominent regulators of cardiac function with both protective and detrimental effects (Rose et al. 2010).

The protein kinase most relevant in the context of metabolic stability and energy homeostasis, however, is AMP-activated protein kinase (AMPK). It has often been described as a major "signaling hub," "fuel gauge," "metabolic sensor," or "metabolic master switch" because it plays a central role in sensing and regulating energy homeostasis at the cellular, organ, and whole-body level (Winder and Hardie 1999; Hardie and Carling 1997). Activation of AMPK is triggered by a diverse array of signals linked to limited energy availability in physiological and pathological situations, including extracellular (e.g., hormones, cytokines, nutrients) and intracellular stimuli (e.g., AMP, ADP) (Hardie et al. 2012a). Activation involves covalent phosphorylations and allosteric binding of AMP or ADP that cooperate in a complex manner. In general, these regulations are coordinated to activate AMPK in situations of energy deficits and aim at compensating ATP loss via accelerated catabolism and inhibited anabolism. However, pleiotropic control exerted by AMPK affects not only metabolic pathways but also other physiological functions like cell growth and proliferation, cell polarity and motility, apoptosis, autophagy, and central appetite control by regulating enzyme activities and gene transcription. This has made the kinase also a prime pharmacological target for treating metabolic disorders or possibly also cancer (Hardie 2007; Zhang et al. 2009; Finckenberg and Mervaala 2010; Inoki et al. 2012; Srivastava et al. 2012).

Although earlier work on AMPK mainly focused on tissues like liver and skeletal muscle, more recent research has revealed novel molecular mechanisms of AMPK regulation and downstream action that are relevant also to cardiovascular function. Activation of the AMPK pathway plays a particularly important role in the myocardial response to pathological stimuli like ischemia–reperfusion (Kudo et al. 1995; Russell et al. 2004), pressure overload (Tian et al. 2001; Kim et al. 2012), or heart failure (Sasaki et al. 2009). Pharmacological activation of AMPK also holds promise as a therapeutic strategy for treating cardiovascular diseases (e.g., Sasaki et al. 2009; Calvert et al. 2008; Shinmura et al. 2007). The following paragraphs will briefly summarize the key elements of AMPK signaling with an emphasis on metabolic regulation in the heart. For more complete reviews of this issue, the reader is referred to some excellent recent overviews dedicated to general AMPK signaling (Hardie et al. 2012a; Inoki et al. 2012; Carling et al. 2012; Oakhill et al. 2012) or AMPK functions in the heart (Zaha and Young 2012; Harada et al. 2012; Ahn et al. 2012; Horman et al. 2012) and other organs [see e.g., a recent review series in *Mol Cell Endocrinol* (Steinberg 2013)]. Thus, AMPK signaling may constitute a fourth module for a systems description of the cardiac metabolic network.

11.5.2 Network Elements: AMPK Isoforms and Their Distribution in Cells and Tissues

AMPK is an evolutionary conserved and ubiquitously expressed serine/threonine kinase that presents complex structural and functional features. Structurally, it occurs in vertebrates as an obligatory heterotrimeric complex composed of one catalytic subunit (α) and two regulatory subunits (β and γ). As a first layer of complexity, all subunits exist in form of different isoforms ($\alpha 1$, $\alpha 2$, $\beta 1$, $\beta 2$, $\gamma 1$, $\gamma 2$, and $\gamma 3$) and splice variants (of $\gamma 2$ and $\gamma 3$), generating multiple heterotrimeric complexes. The precise physiological significance of these isoforms is not yet entirely clear. However, there is some evidence that they determine intracellular distribution, protein recognition, or tissue-specific functions of AMPK, all of which could provide selectivity for specific subsets of substrates within the ever increasing list of AMPK substrates (Hardie et al. 2012a, b; Carling et al. 2012).

11.5.2.1 Subcellular Localization

The subcellular distribution and recruitment of AMPK to specific sites are likely an important factor for its signaling function, but so far only few details are known, in particular in heart. AMPK is generally observed as a soluble complex with diffuse cytosolic localization, but at least $\alpha 2$-containing complexes in their activated form can translocate into the nucleus to phosphorylate important substrates (e.g., transcription factors, histones, histone deacetylases) as seen, e.g., after exercise in skeletal muscle (McGee et al. 2003, 2008; Suzuki et al. 2007; McGee and Hargreaves 2008). Minor but important portions of AMPK may associate with cellular structures like specific membranes, where processes are regulated by AMPK (e.g., ion channel activity, cell polarity, or cell junction formation) (Forcet and Billaud 2007; Andersen and Rasmussen 2012; Nakano and Takashima 2012). Myristoylation of the AMPK β-subunit can localize the kinase complex to membranes and increases its activability, possibly favoring activation of membrane-bound complexes (Suzuki et al. 2007; Oakhill et al. 2010).

AMPK may also be recruited into specific complexes via interaction with its upstream kinases, downstream substrates, or more general scaffolding proteins. However, the AMPK interactome is only partially known so far from some targeted and non-biased interaction studies conducted by us and others (e.g., Ewing et al. 2007; Moreno et al. 2009; Behrends et al. 2010; Klaus et al. 2012), and more research is needed on this issue, in particular in the heart. AMPK interaction with LKB1, its major upstream kinase in the heart, could localize AMPK to places of LKB1 localization, including the mitochondrial surface or E-cadherin in adherens junctions (Sebbagh et al. 2009). Scaffolding proteins can provide specificity in cell signaling by isolating activated kinases from bulk signaling and directing the information flow into specific pathways. In heart, for example, AMPK competes with p38 MAPK for binding to the scaffolding protein

TAK-1-binding protein-1, thus blunting p38 activation during ischemia (Li et al. 2005). Mitochondrial VDAC may represent yet another anchor protein recruiting AMPK to this organelle (Strogolova et al. 2012). There is also some evidence that AMPK subunit isoforms determine specific protein interactions. The β-subunit may in some cases confer substrate specificity, as seen with the yeast and plant orthologues (Vincent and Carlson 1999; Polge et al. 2008), but also putative mammalian AMPK interactors (IntAct database, (Kerrien et al. 2012)). We recently found the β2-isoform interacting with Mu- and Pi-type glutathione transferases (GSTs) to favor glutathionylation of the α-subunit (Klaus et al. 2013). However, in case of fumarate hydratase (FH), we identified a specific interaction with α2-containing AMPK complexes (Klaus et al. 2012).

11.5.2.2 Expression Patterns

AMPK isoforms also show some differences in their tissue- and developmental-specific expression patterns, also in heart, although the physiological significance is still uncertain. While the α1β1γ1 complex is probably the most abundant in most cell types, differences seem to occur in the amount of additional isoforms like α2- and γ3 in skeletal muscle or β2 and a specific intermediate length γ2 splice variant (γ2–3B) in the heart (Stapleton et al. 1996; Thornton et al. 1998; Li et al. 2006; Pinter et al. 2012). There are also pathological and developmental changes in cardiac AMPK expression. The α2, β2, and γ2 isoforms are all upregulated by pressure overload or heart failure in rodents, although in patients rather the content of α1, β1, and γ2 (an intermediate form) increases with different forms of cardiomyopathy (Tian et al. 2001; Kim et al. 2012). During embryonic development in rodents, γ1 increases while high levels of γ3 disappear, and the embryonically predominant full-length γ2 form is replaced by γ2–3B in heart, but by short γ2b in other tissues (Pinter et al. 2012). These developmental and tissue particularities may also explain why γ2 gene mutations in the CBS domains cause heriditary hypertrophic cardiomyopathy (see below) but no other pathological symptoms. Full-length γ2 and γ2–3B share an N-terminal domain with unknown function that could localize the AMPK complex to specific compartments or signaling pathways (Pinter et al. 2012). Total AMPK activity increases after birth, contributing to the switch to predominant use of fatty acids (Makinde et al. 1997). AMPK levels may also be determined by ubiquitin-dependent mechanisms (Qi et al. 2008; Moreno et al. 2010), but its role in the heart is not known.

11.5.3 Network Elements: Molecular Structure of AMPK

Given the interest in AMPK as a putative drug target in metabolic diseases, recent years have seen intense efforts to elucidate the molecular structure of AMPK.

By solving several X-ray structures for AMPK domains and truncated core complexes, the topology of the heterotrimer, the conserved global fold of large parts of the subunits, and the putative activation mechanisms could be deduced (Townley and Shapiro 2007; Amodeo et al. 2007; Xiao et al. 2007, 2011; Chen et al. 2009; Oakhill et al. 2011). However, a high-resolution structure of full-length heterotrimeric complex in both active and inactive states is still lacking. The so far most complete X-ray structure covers most of the α1-subunit except a C-terminal linker region (although not all of the sequence present is well resolved in the electron density map), the C-terminal domain of the β2-subunit, and the entire γ1-subunit (Xiao et al. 2007, 2011).

11.5.3.1 α-Subunit

The catalytic α-subunit contains an N-terminal serine-threonine kinase domain with an activation loop carrying the critical Thr172 residue. Phosphorylation of this residue by upstream kinases is essential for AMPK activation and often used as an indicator for AMPK activity (Hawley et al. 1996). The C-terminal α-domain carries the motif interacting with the β-subunit and further structural determinants that are involved in AMPK activation. These include an autoinhibitory domain (AID) and loop(s) contacting the γ-subunit (called α-hook or α-RIM1 and α-RIM2) (Xiao et al. 2007; Chen et al. 2009, 2013a; Pang et al. 2007). Their exact roles are, however, disputed and further structural studies will be necessary to delineate their function in autoinhibition and α–γ communication. Earlier in vitro studies suggested that the α2 subunit has a higher sensitivity to allosteric activation by AMP (Salt et al. 1998).

11.5.3.2 β-Subunit

The regulatory β-subunit is often described as a scaffold for α- and γ-subunits, a function that indeed is provided by the C-terminal domain. The N-terminal domain whose structure is not entirely resolved at molecular resolution carries additional regulatory elements. A glycogen-binding domain (GBD) seems to provide regulation of AMPK by glucose α1–6-branched glycogen that inhibits AMPK activation (Polekhina et al. 2003; McBride et al. 2009). The N-terminal β-domain may also be involved in protein interactions of AMPK (see above).

11.5.3.3 γ-Subunit

While α- and β-isoforms are very homologous, γ-subunits and their splice variants differ in length. They all share the C-terminal part that consists of four cystathionine β synthetase (CBS) domains that are arranged in tandem in so-called Bateman domains (Bateman 1997). The symmetrical structure of this domain theoretically provides four binding sites for adenylates [referred to as sites 1–4 (Kemp et al. 2007; Hardie et al. 2011)], but only sites 1, 3, and 4 can be occupied in the mammalian enzyme, while site 2 is nonfunctional. The precise role of these sites is still unclear. Initial evidence suggested that site 4 binds AMP tightly in a non-exchangeable manner, while site 1 is a high-affinity site for AMP involved in allosteric activation (see below) and site 3 represents a lower affinity site (binding AMP, ADP, and ATP) more involved in regulating α-Thr172 phosphorylation (Xiao et al. 2007). A more recent study suggests that also site 4 can bind ATP (and causes site 3 to be empty) and that both sites 3 and 4 may play a role in allosteric activation (Chen et al. 2012). The γ2- and γ3-isoforms contain N-terminal extensions that are subject to truncation by RNA splicing and whose molecular structure and function are currently unknown. Mutations in the CBS domains of the AMPK γ2 subunit, expressed at particularly high levels in heart, cause the Wolff–Parkinson–White (WPW) syndrome, a hereditary cardiomyopathy of varying severity, involving cardiac hypertrophy, contractile dysfunction, and arrhythmias. Mutations impair adenylate binding and thus AMPK activation (Scott et al. 2004; Burwinkel et al. 2005), but the major cause for the cardiomyopathy is the increased AMP-independent basal AMPK activity. This leads to higher glucose uptake, accumulation of glycogen in cardiac myocytes, and finally impairment of normal heart muscle development (Burwinkel et al. 2005; Davies et al. 2006).

11.5.4 Network Connectivity: AMPK Input Signals and Upstream Regulation

AMPK integrates various intra- and extracellular signals and maintains cross talk with other signaling pathways, all related to the cellular energy and nutrient state. This makes the kinase a central signaling hub in sensing and regulating cellular energetics and ATP-dependent functions. Indeed, most recent research revealed that AMPK activation is much more complex than initially anticipated and depends on multiple covalent modifications and allosteric effectors (Fig. 11.13). AMPK regulation also evolved from a more simple state as, e.g., in the yeast AMPK homologues that lack allosteric activation by AMP (see below) to the more complex regulation present in mammals.

11.5.4.1 Covalent Regulation by Phosphorylation

The phosphorylation state at α-Thr172 defines the primary activation of AMPK. This is determined by the balance of different upstream kinases and phosphatases. There are potentially three mammalian AMPK upstream kinases, with the major one in most cell types, including heart, being Liver Kinase B1 (LKB1, also called

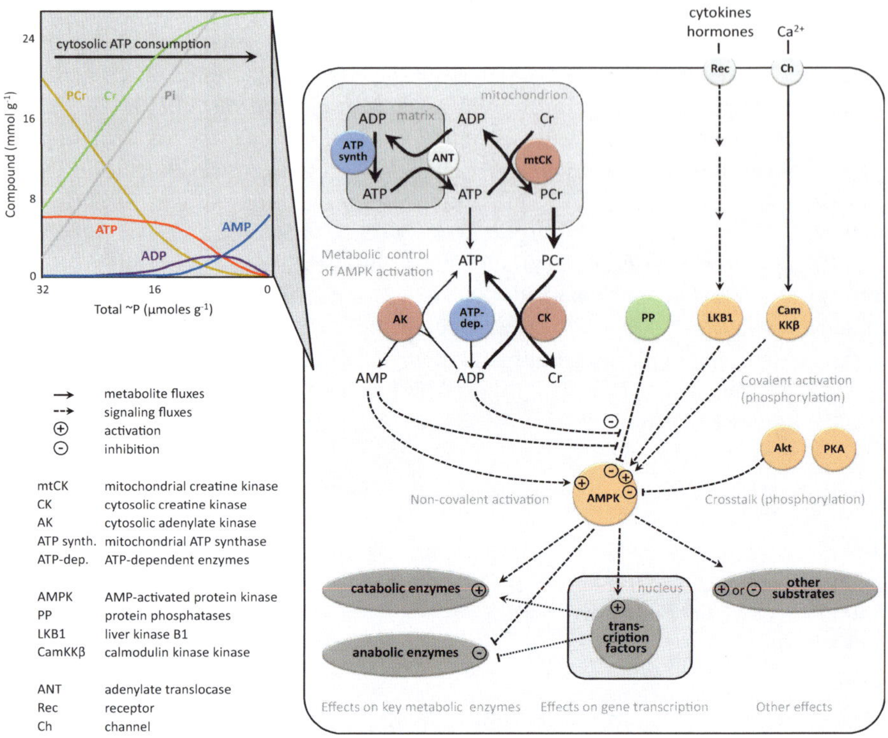

Fig. 11.13 *AMPK signaling. Large scheme: Activation of AMPK by intra- and extracellular metabolic and endocrine signals and major fields of downstream signaling*. Activation of AMPK is determined by upstream kinases (covalent activation by LKB1, CamKKβ, inhibition by Akt and PKA) and phosphatases. They mediate mainly extracellular signals carrying, e.g., information on the energy and nutrient state of the cellular environment and the entire organism (endocrine signals). Covalent activation also depends on some intracellular parameters (Ca^{2+}, possibly also ROS/RNS). As a second layer of regulation, AMPK is activated by ADP and in particular AMP (allosteric regulation), both acting as second messengers of cellular energy stress. This signaling is linked to conversion of nucleotides via the adenylate kinase (AK) and creatine kinase (CK) reactions. Activated AMPK compensates for ATP loss by accelerating catabolism, inhibiting anabolism, and further effects on cell motility, growth, proliferation, and others, via regulation of key enzymes and transcription factors. For further details see text. *Insert: Connection of AMPK signaling and phosphotransfer reactions (CK and AK)* (Neumann et al. 2003). Global cellular concentration changes of phosphocreatine [PCr] and adenine nucleotides ([ATP], [ADP] and [AMP]), inorganic phosphate [Pi], and creatine [Cr], calculated from the reactions of CK, AK, and a generalized ATPase at decreasing "high-energy" phosphates (corresponding to a transition from rest to high work-load). Note that with "high-energy" phosphate consumption, [ATP] remains constant until more than 80 % of the PCr pool is consumed. Only then, there is a transient increase in [ADP] and finally [AMP] starts to rise dramatically. The exponential rise in [AMP] makes this nucleotide an ideal second messenger for a lowered cellular energy state. This simplified model assumes that the CK and AK reactions work at equilibrium (which is unlikely to be true in vivo) and does not account for specific subcellular localizations of CK and AK

STK11). LKB1 is upstream of an entire family of 12 other AMPK-related kinases in the human kinome, and like AMPK forms heterotrimers with two accessory subunits, STRADα/β and MO25α/β (Hawley et al. 2003; Woods et al. 2003). LKB1 has originally been identified as a tumor suppressor whose inactivating mutations lead to the Peutz–Jeghers syndrome, an inherited susceptibility to different human cancers (Alessi et al. 2006). However, LKB1 seems to mostly exhibit constitutive activity and may thus not be the limiting step in AMPK activation. An alternative upstream kinase much less expressed in heart is Ca^{2+}/calmodulin-dependent protein kinase kinase β (CamKKβ) that mediates Ca^{2+}-dependent AMPK activation (Hawley et al. 2005; Hurley et al. 2005; Woods et al. 2005). Although such CamKKβ-mediated AMPK activation might anticipate an increasing energy turnover that accompanies a rise in cytosolic Ca^{2+} during muscle contraction, its role in the heart is not well understood. More recently, the transforming growth factor-β-activated kinase-1 (TAK1) that phosphorylates the yeast AMPK homologue Snf1 was proposed as an AMPK upstream kinase (Momcilovic et al. 2006; Xie et al. 2006a). Although TAK-1 is present in heart, it is not activated during ischemia and it is unclear whether it acts via direct AMPK phosphorylation (Xie et al. 2006a).

Protein phosphatases are possibly the more critical parameter governing the α-Thr172 phosphorylation state, and this may also apply to the heart. However, their identity and regulatory role in vivo remain to be confirmed. Both seem to depend on cell type and/or stimulus. Different phosphatases can act on AMPK, including PP1, PP2A, and PP2C in vitro (Davies et al. 1995), as well as PP1-R6 in MIN6 beta cells (Garcia-Haro et al. 2010) and metal-dependent phosphatase PPM1E/F in HEK-293 cells in vivo (Voss et al. 2011). In heart and endothelial cells, expression levels PP2C and 2A, respectively, correlate with AMPK activation (Wang and Unger 2005; Wu et al. 2007).

The α-Thr172 phosphorylation state is further negatively controlled by hierarchical phosphorylation at other sites in the AMPK heterotrimer. Phosphorylation at α1-Ser485 (α2-Ser491) by PKA or at α-Ser173 by PKB/Akt reduces α-Thr172 phosphorylation (Hurley et al. 2006; Horman et al. 2006; Djouder et al. 2010). Further phosphorylation sites were identified in both α- and β-subunits, many of them targeted by autophosphorylation, but their functional role remains uncertain.

11.5.4.2 Non-Covalent Allosteric Regulation by AMP and ADP

The activation of AMPK by low cellular energy state is triggered by increased concentrations of AMP and, as discovered more recently, also of ADP, since the kinase can sense AMP/ATP and ADP/ATP concentration ratios (Xiao et al. 2011; Oakhill et al. 2011). In many cell types and in particular in heart, breakdown of ATP to ADP at the onset of high workload or cellular stress has only minor immediate effects on ATP levels. Due to the energy buffer and transfer function of the Cr/CK system (see above), global and local ATP is rapidly replenished (Wallimann et al. 2011; Neumann et al. 2003). Thus, ATP is not a very suitable signal for indicating developing energy deficits. However, minor decreases in ATP levels lead to more pronounced relative increases in free ADP and even more in AMP due to the adenylate kinase (AK) reaction (Fig. 11.13). Under these conditions, AK uses two ADP to regenerate ATP and AMP, thus increasing AMP concentrations from the sub-micromolar range under resting conditions to the lower micromolar range (Frederich and Balschi 2002). To lesser extent, AMP levels also depend on pyrophosphates (cleaving the β-phosphate bond of ATP) and the activity of AMP degradation pathways [AMP-deaminase and 5′-nucleotidase, whose inhibition may be useful to activate AMPK (Kulkarni et al. 2011)]. As a consequence, a decrease in ATP levels by only 10 % translates into a 10- to 100-fold increase in AMP, making AMP an ideal second messenger of energy stress (Fig. 11.13, upper left). Regulation of AMPK activation by the balance between ATP, ADP, and AMP concentrations resembles to what was put forward by Atkinson 40 years ago as "energy charge" regulation (Xiao et al. 2011; Oakhill et al. 2011; Atkinson 1968; Hardie and Hawley 2001).

The molecular basis of allosteric AMPK activation is not yet fully understood, but certainly involves multiple interconnected mechanisms. The nucleotide ratios are sensed at the γ-subunit binding sites (sites 1, 3, and 4), which possess high affinity for AMP and ADP, but less for ATP in its major, Mg^{2+}-complexed form. AMP or ADP binding to AMPK has several consequences: (1) it makes α-Thr172 a better substrate for phosphorylation, (2) it protects P-α-Thr172 from dephosphorylation, and (3), only in case of AMP, it exerts direct allosteric activation of AMPK (Xiao et al. 2011; Oakhill et al. 2011; Davies et al. 1995; Suter et al. 2006). All these effects require close communication between the AMP-binding γ- and the catalytic α-subunit. The three adenylate binding sites participate differentially in these mechanisms. Diverging models have been proposed that involve different structural elements of the α-subunit (Xiao et al. 2011; Chen et al. 2013a). We and our collaborators have proposed that all these mechanisms involve an AMP- (or ADP)- induced conformational switch within the full-length AMPK complex that is not seen in the X-ray structures of AMPK core complexes solved so far (Chen et al. 2009, 2012; Riek et al. 2008; Zhu et al. 2011).

11.5.4.3 Other Covalent and Non-Covalent Regulations

An increasing number of additional secondary protein modifications adds to the complex scheme of AMPK activation. Myristoylation at Gly2 in the β-subunit increases the sensitivity of AMPK for allosteric activation and promotes Thr172 phosphorylation (Oakhill et al. 2010). Acetylation of α-subunits is determined by the reciprocal actions of the acetylase p300 and the histone deacetylase 1. AMPK deacetylation promotes its activation via LKB1 interaction (Lin et al. 2012). LKB1 itself is also regulated by acetylation, since deacetylated LKB1 shifts from nucleus to the cytosol, where it forms active complexes with STRAD (Lan et al. 2008). Thus, acetylation is a potentially important factor for activating the LKB1–AMPK pathway (Ruderman et al. 2010), but its role in the heart has not been examined so far. Glutathionylation at Cys299 and Cys304 in the α-subunit activates the kinase under oxidative conditions in cellular models and is favored by binding to certain GST isoforms (Klaus et al. 2013; Zmijewski et al. 2010). This latter mechanism may be part of a more general redox regulation of the kinase (Han et al. 2010; Jeon et al. 2012). ROS and RNS activate AMPK, but it is unclear whether this happens via increases in [ADP] and [AMP], or whether noncanonical mechanisms at the level of AMPK (like glutathionylation) or upstream kinases play a role. Vice versa, AMPK regulates NADPH homeostasis and an entire battery of ROS-detoxifying enzymes. Another non-covalent allosteric regulator is glycogen as well as other synthetic branched oligosaccharides that inhibit AMPK activity by binding to the β-GBD domain (McBride et al. 2009).

11.5.4.4 Upstream Regulation in Cardiac (Patho) Physiology

In the heart, AMPK activity is increased by a wide array of signals acting via upstream kinases and modulation of adenylate levels under both pathological and physiological stress and involving various hormones and cytokines (Zaha and Young 2012). Classical physiological stimuli of AMPK are exercise or hypoxia. Both also occur in the heart (Coven et al. 2003; Musi et al. 2005; Frederich et al. 2005) and promote the metabolism of glucose and fatty acids via different downstream targets (see below). However, it is unclear whether this activation is due to altered energy state as in skeletal muscle or rather relies on alternative upstream signaling. AMPK is also involved in the adaptive response of the heart to caloric restriction (Chen et al. 2013b), but nutrient effects in the heart may be more complex (Clark et al. 2004). Possibly, within the physiological range, the role of cardiomyocyte AMPK is different from other cell types, because of the remarkable metabolic stability of this organ maintained by multiple other mechanisms, including the metabolic cycles outlined before.

As pathological stimulus, ischemia is the best studied in form of both no-flow and partial ischemia in isolated perfused animal hearts, as well as regional ischemia due to coronary ligation in vivo (Russell et al. 2004; Kudo et al. 1996; Wang et al. 2009; Paiva et al. 2011; Kim et al. 2011a), for a review, see (Young 2008). They both lead to rapid and lasting AMPK activation. As already mentioned, besides energetic stress, oxidative stress may be a determinant of such activation, acting through different forms of ROS (Sartoretto et al. 2011; Zou et al. 2002). In endothelial cells, it is rather peroxynitrite formation that affects AMPK via the protein kinase Cζ–LKB1 axis (Zou et al. 2004; Xie et al. 2006b), while in other non-excitable cells it may be rather a ROS-induced Ca^{2+} release that triggers the CamKKβ axis (Mungai et al. 2011). ROS-facilitated glutathionylation of AMPK (see above) as observed in cellular systems represents yet another direct activation mechanism, but still has to be verified in cardiomyocytes (Klaus et al. 2013; Zmijewski et al. 2010). However, the signaling function of ROS may be lost at more intense oxidative stress that simply inactivates AMPK. In models of cardiotoxicity induced by the anticancer drug doxorubicine, AMPK has been found inactivated in most cases, despite pronounced oxidative, energetic, and genotoxic stress (Tokarska-Schlattner et al. 2005; Gratia et al. 2012). This is probably due to activation of PKB/Akt via DNA-damage signaling kinases that induce the inhibitory cross talk via AMPKα1-Ser485 phosphorylation. In other situations, also LKB1 may become inactivated (Dolinsky et al. 2009). Stress resulting from many but not all forms of pressure overload also results in AMPK activation, mainly increasing glucose uptake and glycolysis (Tian et al. 2001; Li et al. 2007; Allard et al. 2007; Zhang et al. 2008), as well as changing the gene expression profile (Hu et al. 2011).

Information about the cellular environment and whole-body energy and nutrient state is connected to AMPK signaling via endocrine, paracrine, and autocrine mechanisms. These include a diverse array of hormones and cytokines identified in noncardiac cells that act via largely unknown cellular signaling cascades on AMPK upstream kinases, including adiponectin (Shibata et al. 2004), leptin (Minokoshi et al. 2004), resistin (Kang et al. 2011), ghrelin (Kola et al. 2005), IL6 (Kelly et al. 2004), and CNTF (Watt et al. 2006). Regulation of AMPK by these factors partially depends on the tissue. While in peripheral tissues leptin activates and ghrelin inhibits AMPK in the regulation of fatty acid oxidation and glucose uptake, their effects in hypothalamus are different, since they inhibit (leptin) or stimulate (ghrelin) AMPK-controlled food intake [for reviews see (Kahn et al. 2005; Steinberg and Kemp 2009)]. In the heart, AMPK seems to be involved in the positive effects of adiponectin for cardioprotection during ischemia and for reduced cardiac hypertrophy (Shibata et al. 2004, 2005). For example, AMPK limits accumulation and densification of microtubules that occur in response to hypertrophic stress (Fassett et al. 2013). Also leptin may modulate AMPK in the heart, since impaired leptin signaling correlates with reduced AMPK activation and metabolic defects or reduced postconditioning after ischemia (McGaffin et al. 2009; Bouhidel et al. 2008). Proinflammatory cytokines like IL-6 rather reduce AMPK protein and activation (Ko et al. 2009), although there may be opposite effects in specific tissues like skeletal muscle due to a specific autocrine–paracrine effect (Kelly et al. 2004). Other cytokines with functions in the heart include macrophage migration inhibitory factor (MIF), which is involved in AMPK activation during ischemia and hypoxia and its decrease with age in mice seems to reduce AMPK activation during ischemia (Miller et al. 2008; Ma et al. 2010).

11.5.4.5 Evolution of Cellular Homeostasis Signaling Circuits

From a phylogenetic perspective, it is interesting to note that AMPK homologues evolved early with eukaryotic life. However, yeast homologues of AMPK lack the direct allosteric AMP-activation, although they already possess the ADP-regulation of the α-Thr172 phosphorylation state (Mayer et al. 2011). Since such lower eukaryotes neither express a CK/PCr system, it can be concluded that they still tolerate larger fluctuations in energy state. It seems that those more sophisticated regulatory circuits evolved only with multicellular life. It will be interesting to examine when during metazoan evolution AMP has been established as a second messenger for energy stress and activation of AMPK. Creatine kinase and other closely homologous phosphagen kinases have emerged quite early at the dawn of the radiation of metazoans (Ellington 2001; Ellington and Suzuki 2007). Recently, besides identifying arginine kinase in unicellular organisms, a novel taurocyamine phosphagen kinase has been identified even in a unicellular protist (Uda et al. 2013).

In addition, at least in vertebrates a crosstalk has evolved between AMPK signaling and the Cr/CK system (Neumann et al. 2003; Ju et al. 2012). Although AMPK is not directly activated by Cr as postulated earlier (Ponticos et al. 1998; Ingwall 2002; Taylor et al. 2006), the PCr/Cr ratio will also determine cellular ATP/ADP ratios via the CK reaction and thus indirectly AMPK activation, as well. Knockdown of cytosolic CK activates AMPK (Li et al. 2013), and similar control of AMPK signaling is observed when manipulating the cellular levels of adenylate kinase isoenzymes (Dzeja et al. 2011b). Such indirect mechanisms may also cause the additional AMPK activation observed after Cr supplementation in cellular models of skeletal muscle (Ceddia and Sweeney 2004), in the muscles of patients undergoing exercise programs in different pathological settings (Alves et al. 2012), and in Huntington disease models (Mochel et al. 2012), although these findings need further investigation.

Vice versa, AMPK complexes interact with cytosolic CKs and are able to phosphorylate them (Ponticos et al. 1998; Dieni and Storey 2009). Since this does not affect CK enzyme activity, at least in rodents (Ingwall 2002; Taylor et al. 2006) this phosphosphorylation remained enigmatic. Our most recent unpublished data indicate that BB-CK phosphorylation by AMPK may determine subcellular localization of this enzyme which is known to partially associate with ATP-requiring cellular structures and ATPases. In myocytes, active AMPK may also increase cellular Cr uptake by positively acting on Cr transporter (Alves et al. 2012; Darrabie et al. 2011), while an inverse effect was found in kidney epithelial cells (Li et al. 2010). If the latter cells are under energy stress, either physiological or pathological, this mechanism would prevent them to spend additional energy required for Cr uptake from the glomerular filtrate.

11.5.5 Network Connectivity: AMPK Output Signals and Downstream Regulation

AMPK integrates a large number of signals from inside and outside the cell that carry information on the nutrient and energy state from the cellular to organism level with the aim to mount a coordinated response (Fig. 11.13). This response includes compensation for ATP loss by stimulating catabolic and inhibiting several anabolic pathways, but also control of many other energy-related biological checkpoints in cell growth and proliferation, cell motility and polarity, apoptosis, autophagy, and central functions like appetite control. To date, about 50 AMPK substrates have been described in different tissues, including metabolic enzymes, transcription (co)factors, and other cellular signaling elements. They all are activated or inactivated by phosphorylation at Thr or Ser residues within a more or less conserved AMPK recognition motif. We will give here only some examples pertinent to heart; more complete descriptions can be found in recent reviews (Hardie et al. 2012a, b; Carling et al. 2012; Steinberg and Kemp 2009).

11.5.5.1 Metabolic Pathways

AMPK control of cellular substrate uptake, transport, and metabolism is the historically best described and possibly most important function of AMPK, also in the heart, since it is critical for ATP generation (Fig. 11.5). Activated AMPK stimulates cellular glucose and fatty acid uptake via translocation of GLUT4 (Kurth-Kraczek et al. 1999; Yamaguchi et al. 2005) and FAT/CD36 (van Oort et al. 2009), respectively, to the plasma membrane, involving among others phosphorylation of the Rab-GTPase activating protein TBC1D1 (Frosig et al. 2010). The subsequent substrate flux via glycolysis is increased by phosphorylation and activation of 6-phosphofructosekinase-2 (PFK2), whose product fructose-2,6-bisphosphate is an allosteric activator of the glycolytic enzyme 6-phosphofructokinase-1 (Marsin et al. 2000) and in long term by stimulation of hexokinase II (HKII) transcription (Stoppani et al. 2002). Substrate flux via fatty acid β-oxidation is increased by inhibition of mitochondria-associated acetyl-CoA carboxylase (ACC2), whose product malonyl-CoA is an allosteric inhibitor of carnitine palmitoyltransferase 1 (CPT1), the rate-limiting enzyme for of mitochondrial fatty acid import and oxidation (Merrill et al. 1997). At the same time, inhibition of cytosolic ACC1 will repress ATP-consuming fatty acid synthesis for which malonyl-CoA is the precursor (Davies et al. 1992). In other organs with multiple anabolic functions like liver, several other anabolic pathways like gluconeogenesis or triglyceride and cholesterol synthesis are inhibited (Bultot et al. 2012; Muoio et al. 1999; Clarke and Hardie 1990).

Active AMPK also affects gene expression of many of these metabolic enzymes by phosphorylation of transcription (co)factors and histone deacetylases (HDACs). Activation of peroxisome proliferator-activated receptor gamma co-activator-1 alpha (PGC-1α) increases the expression of nuclear-encoded mitochondrial genes that favor mitochondrial biogenesis (Irrcher et al. 2003; Jager et al. 2007), and further catabolic genes including substrate transporters (e.g., GLUT4). Mainly in the liver, expression of several genes in anabolic lipogenesis (e.g., ACC1) and gluconeogenesis is reduced via inhibition of ChREBP or SREBP (Kawaguchi et al. 2002; Li et al. 2011) and CRTC2 or class II HDACs (Koo et al. 2005; Mihaylova et al. 2011), respectively. Cellular redox regulation by AMPK also occurs mainly at the transcriptional level. AMPK directly phosphorylates transcription factor FOXO3, which increases transcription of many genes, mainly in oxidative stress defense (Greer et al. 2007) and activates, possibly more indirectly, class

III deacetylase SIRT1, which deacetylates and thus activates FOXO1/3 and PGC-1α (Canto et al. 2009).

11.5.5.2 Protein Metabolism, Cell Growth, and Proliferation

AMPK also acts via cross talk with other major cellular signaling hubs. The most important may be the mammalian target of rapamycin complex 1 (mTORC1) which is inhibited by activated AMPK via multiple mechanisms, including phosphorylation of tuberous sclerosis complex protein-2 (TSC2) (Inoki et al. 2003) upstream of mTORC1, or direct phosphorylating the mTORC1 subunit Raptor (Gwinn et al. 2008). This reduces the multiple TORC1 functions in stimulation of protein biosynthesis and cell cycle (Kwiatkowski and Manning 2005) and inhibition of autophagy (Meijer and Codogno 2007). Autophagy is also directly stimulated by AMPK-induced phosphorylation of the protein kinase ULK1 (Kim et al. 2011b; Egan et al. 2011). AMPK further reduces protein synthesis more indirectly by inhibiting eukaryotic elongation factor 2 kinase (eEF2K) (Browne et al. 2004) and downregulating ribosomal RNA (Hoppe et al. 2009) and several cyclins (Wang et al. 2002). Phosphorylation of the tumor suppressor p53 and the cyclin-dependent kinase inhibitor $p27^{KIP1}$ will both contribute to cell cycle arrest and eventual autophagy (Imamura et al. 2001; Jones et al. 2005; Liang et al. 2007). AMPK also stimulates protein-ubiquitination and proteasome-dependent degradation (Viana et al. 2008; Solaz-Fuster et al. 2008).

11.5.5.3 Cell Contractility, Dynamics, and Shape

AMPK phosphorylates cardiac troponin I (cTnI) during ischemia and thus increases its calcium sensitivity, suggesting that AMPK activation improves myocyte contraction (Oliveira et al. 2012a). Cellular models also suggest that AMPK controls microtubule dynamics through phosphorylation of the microtubule plus end protein CLIP-170 (Nakano et al. 2010) and dynamics of cells and in particular of the mitotic spindle via different indirect mechanisms that increase phosphorylation of the non-muscle myosin regulatory light chain (MRLC) (Lee et al. 2007; Banko et al. 2011).

11.5.5.4 Cellular Ion Homeostasis

The maintenance of ion gradients across cell membranes and intracellular ion homeostasis are further highly energy-demanding processes. Thus it is little surprising that AMPK, like CK, might also regulate these processes. Indeed, different ion transporters are inhibited by AMPK, directly or indirectly, including cystic fibrosis transmembrane conductance regulator Cl-channel (CFTR) (Hallows et al. 2003) and ATP-sensitive potassium (KATP) channel (Kir6.2) (Chang et al. 2009), and several more cell-type specific ion transporters, like epithelial Na^+ channel (eNaC) (Carattino et al. 2005), renal Na^+-K^+-$2Cl^-$ cotransporter (NKCC2) (Fraser et al. 2007), and neuronal voltage-gated, delayed rectifier K^+ channel (Kv2.1), whose phosphorylation reduces the frequency of highly energy-consuming action potentials (Ikematsu et al. 2011).

11.5.6 Modeling Signaling Networks in Heart and Beyond

Given the complexity and interconnectivity of cell signaling networks, only recently emerging systems biology approaches hold promises for understanding and predicting the higher order properties and behavior of such networks (Arkin and Schaffer (2011) and following papers of this Cell *Leading Edge Review* series). However, the mathematical modeling necessary for such systems approaches is still in its infancy. Modeling needs a solid base of both quantitative and qualitative data, including the spatiotemporal component as outlined above. So far, both bottom-up and top-down systems approaches have been applied to obtain comprehensive databases of protein kinase signaling. Typically, bottom-up, hypothesis-, or model-driven approaches were used to study the role of individual components, also facilitating first studies of dynamic systems properties. More recently, with the broad availability of "omics" approaches, more top-down so-called hypothesis-free studies have mapped interactomes and phosphoproteomes of protein kinases, mainly in yeast (Breitkreutz et al. 2010; Bensimon et al. 2012; Oliveira et al. 2012b). Such large-scale data are necessary to construct first network models needed to advance mathematical modeling in this field. Indeed, progress is also made in the computational methodology and the mathematical description of signaling pathways (Frey et al. 2008; Ideker et al. 2011; Telesco and Radhakrishnan 2012). However, similar modeling approaches with mammalian cells, in particular under physiological and pathological conditions relevant to humans, are still scarce (Benedict et al. 2011; Basak et al. 2012; Rogne and Tasken 2013), except a strong history of modeling in cardiac electrophysiology (Amanfu and Saucerman 2011). Such models would be highly valuable for in silico drug target identification, drug screening, and development (Benedict et al. 2011). Important steps in such approaches are (1) to establish a network structure, (2) to obtain quantitative dynamic datasets for basic systems properties, (3) to generate dynamic mathematical models, and (4) to test and iteratively improve the models by prediction and experimental verification of systems perturbations (Frey et al. 2008). First models also including AMPK signaling are currently emerging (Marcus 2008; Sonntag et al. 2012), but sustained interdisciplinary efforts in the field will be necessary to obtain models that allow meaningful predictions of AMPK systems behavior.

Acknowledgments The authors would like to thank all group members involved in work cited in this review. Work of the contributing groups was funded among others by EU FP6 and FP7 programs (contract LSHM-CT-2004-005272 EXGENESIS to U.S.; by reintegration grants

ANTHRAWES, 041870; ANTHRAPLUS, 249202 to M.T.S.), the French Agence Nationale de Recherche ("chaire d'excellence" to U.S., project ANR N RA0000C407 to V.S. and R.G.), and by the National Council of Science and Technology of Mexico (CONACYT).

References

Abraham MR, Selivanov VA, Hodgson DM, Pucar D, Zingman LV, Wieringa B et al. (2002) Coupling of cell energetics with membrane metabolic sensing. Integrative signaling through creatine kinase phosphotransfer disrupted by M-CK gene knock-out. J Biol Chem 277 (27):24427–34. doi:10.1074/jbc.M201777200

Ahn YJ, Kim H, Lim H, Lee M, Kang Y, Moon S et al. (2012) AMP-activated protein kinase: implications on ischemic diseases. BMB Rep 45(9):489–95

Alekseev AE, Reyes S, Selivanov VA, Dzeja PP, Terzic A (2012) Compartmentation of membrane processes and nucleotide dynamics in diffusion-restricted cardiac cell microenvironment. J Mol Cell Cardiol 52(2):401–9. doi:10.1016/j.yjmcc.2011.06.007

Alessi DR, Sakamoto K, Bayascas JR (2006) LKB1-dependent signaling pathways. Annu Rev Biochem 75:137–63. doi:10.1146/annurev.biochem.75.103004.142702

Aliev MK, Saks VA (1997) Compartmentalized energy transfer in cardiomyocytes: use of mathematical modeling for analysis of in vivo regulation of respiration. Biophys J 73 (1):428–45

Aliev M, Guzun R, Karu-Varikmaa M, Kaambre T, Wallimann T, Saks V (2012) Molecular system bioenergics of the heart: experimental studies of metabolic compartmentation and energy fluxes versus computer modeling. Int J Mol Sci 12(12):9296–331. doi:10.3390/ijms12129296

Allard MF, Parsons HL, Saeedi R, Wambolt RB, Brownsey R (2007) AMPK and metabolic adaptation by the heart to pressure overload. Am J Physiol Heart Circ Physiol 292(1): H140–8. doi:10.1152/ajpheart.00424.2006

Alves CR, Ferreira JC, de Siqueira-Filho MA, Carvalho CR, Lancha AH Jr, Gualano B (2012) Creatine-induced glucose uptake in type 2 diabetes: a role for AMPK-alpha? Amino Acids 43 (4):1803–7. doi:10.1007/s00726-012-1246-6

Amanfu RK, Saucerman JJ (2011) Cardiac models in drug discovery and development: a review. Crit Rev Biomed Eng 39(5):379–95

Amodeo GA, Rudolph MJ, Tong L (2007) Crystal structure of the heterotrimer core of Saccharomyces cerevisiae AMPK homologue SNF1. Nature 449(7161):492–5. doi:10.1038/nature06127

Andersen MN, Rasmussen HB (2012) AMPK: A regulator of ion channels. Commun Integr Biol 5 (5):480–4. doi:10.4161/cib.21200

Aon MA, Cortassa S (2012) Mitochondrial network energetics in the heart. Wiley Interdiscip Rev Syst Biol Med 4(6):599–613. doi:10.1002/wsbm.1188

Arkin AP, Schaffer DV (2011) Network news: innovations in 21st century systems biology. Cell 144(6):844–9. doi:10.1016/j.cell.2011.03.008

Atkinson DE (1968) The energy charge of the adenylate pool as a regulatory parameter. Interaction with feedback modifiers. Biochemistry 7(11):4030–4

Avkiran M, Rowland AJ, Cuello F, Haworth RS (2008) Protein kinase d in the cardiovascular system: emerging roles in health and disease. Circ Res 102(2):157–63. doi:10.1161/CIRCRESAHA.107.168211

Balaban RS (2002) Cardiac energy metabolism homeostasis: role of cytosolic calcium. J Mol Cell Cardiol 34(10):1259–71

Balaban RS (2009a) Domestication of the cardiac mitochondrion for energy conversion. J Mol Cell Cardiol 46:832–41. doi:10.1016/yjmcc.2009.02.018

Balaban RS (2009b) The role of Ca^{2+} signaling in the coordination of mitochondrial ATP production with cardiac work. Biochim Biophys Acta 1787(11):1334–41. doi:10.1016/j.bbabio.2009.05.011

Balaban RS (2012) Perspectives on: SGP symposium on mitochondrial physiology and medicine: metabolic homeostasis of the heart. J Gen Physiol 139(6):407–14. doi:10.1085/jgp.201210783

Balaban RS, Kantor HL, Katz LA, Briggs RW (1986) Relation between work and phosphate metabolite in the in vivo paced mammalian heart. Science (New York) 232(4754):1121–3

Banko MR, Allen JJ, Schaffer BE, Wilker EW, Tsou P, White JL et al. (2011) Chemical genetic screen for AMPKalpha2 substrates uncovers a network of proteins involved in mitosis. Mol Cell 44(6):878–92. doi:10.1016/j.molcel.2011.11.005

Basak S, Behar M, Hoffmann A (2012) Lessons from mathematically modeling the NF-kappaB pathway. Immunol Rev 246(1):221–38. doi:10.1111/j.1600-065X.2011.01092.x

Bateman A (1997) The structure of a domain common to archaebacteria and the homocystinuria disease protein. Trends Biochem Sci 22(1):12–3

Beard E, Braissant O (2010) Synthesis and transport of creatine in the CNS: importance for cerebral functions. J Neurochem 115(2):297–313. doi:10.1111/j.1471-4159.2010.06935.x

Behrends C, Sowa ME, Gygi SP, Harper JW (2010) Network organization of the human autophagy system. Nature 466(7302):68–76. doi:10.1038/nature09204

Belitzer VA, Tsybakova ET (1939) About mechanism of phosphorylation, respiratory coupling. Biokhimiya 4:516–34

Benedict KF, Mac Gabhann F, Amanfu RK, Chavali AK, Gianchandani EP, Glaw LS et al. (2011) Systems analysis of small signaling modules relevant to eight human diseases. Ann Biomed Eng 39(2):621–35. doi:10.1007/s10439-010-0208-y

Bensimon A, Heck AJ, Aebersold R (2012) Mass spectrometry-based proteomics and network biology. Annu Rev Biochem 81:379–405. doi:10.1146/annurev-biochem-072909-100424

Beraud N, Pelloux S, Usson Y, Kuznetsov AV, Ronot X, Tourneur Y et al. (2009) Mitochondrial dynamics in heart cells: very low amplitude high frequency fluctuations in adult cardiomyocytes and flow motion in non beating Hl-1 cells. J Bioenerg Biomembr 41 (2):195–214

Bers DM (2002) Cardiac excitation-contraction coupling. Nature 415(6868):198–205

Bers DM, Despa S (2006) Cardiac myocytes Ca^{2+} and Na^{+} regulation in normal and failing hearts. J Pharmacol Sci 100(5):315–22

Bessman SP, Carpenter CL (1985) The creatine-creatine phosphate energy shuttle. Annu Rev Biochem 54:831–62

Bessman SP, Fonyo A (1966) The possible role of the mitochondrial bound creatine kinase in regulation of mitochondrial respiration. Biochem Biophys Res Commun 22(5):597–602

Bessman SP, Geiger PJ (1981) Transport of energy in muscle: the phosphorylcreatine shuttle. Science 211(4481):448–52

Bing RJ, Siegel A, Ungar I, Gilbert M (1954) Metabolism of the human heart. II. Studies on fat, ketone and amino acid metabolism. Am J Med 16(4):504–15

Bose S, French S, Evans FJ, Joubert F, Balaban RS (2003) Metabolic network control of oxidative phosphorylation: multiple roles of inorganic phosphate. J Biol Chem 278(40):39155–65

Bouhidel O, Pons S, Souktani R, Zini R, Berdeaux A, Ghaleh B (2008) Myocardial ischemic postconditioning against ischemia-reperfusion is impaired in ob/ob mice. Am J Physiol Heart Circ Physiol 295(4):H1580–6. doi:10.1152/ajpheart.00379.2008

Boyer PD (1997) The ATP, synthase—a splendid molecular machine. Annu Rev Biochem 66:717–49

Breitkreutz A, Choi H, Sharom JR, Boucher L, Neduva V, Larsen B et al. (2010) A global protein kinase and phosphatase interaction network in yeast. Science 328(5981):1043–6. doi:10.1126/science.1176495

Browne GJ, Finn SG, Proud CG (2004) Stimulation of the AMP-activated protein kinase leads to activation of eukaryotic elongation factor 2 kinase and to its phosphorylation at a novel site, serine 398. J Biol Chem 279(13):12220–31. doi:10.1074/jbc.M309773200

Bultot L, Guigas B, Von Wilamowitz-Moellendorff A, Maisin L, Vertommen D, Hussain N et al. (2012) AMP-activated protein kinase phosphorylates and inactivates liver glycogen synthase. Biochem J 443(1):193–203. doi:10.1042/BJ20112026

Burwinkel B, Scott JW, Buhrer C, van Landeghem FK, Cox GF, Wilson CJ et al. (2005) Fatal congenital heart glycogenosis caused by a recurrent activating R531Q mutation in the gamma 2-subunit of AMP-activated protein kinase (PRKAG2), not by phosphorylase kinase deficiency. Am J Hum Genet 76(6):1034–49. doi:10.1086/430840

Calvert JW, Gundewar S, Jha S, Greer JJ, Bestermann WH, Tian R et al. (2008) Acute metformin therapy confers cardioprotection against myocardial infarction via AMPK-eNOS-mediated signaling. Diabetes 57(3):696–705. doi:10.2337/db07-1098

Canto C, Gerhart-Hines Z, Feige JN, Lagouge M, Noriega L, Milne JC et al. (2009) AMPK regulates energy expenditure by modulating NAD+ metabolism and SIRT1 activity. Nature 458(7241):1056–60. doi:10.1038/nature07813

Capetanaki Y, Bloch RJ, Kouloumenta A, Mavroidis M, Psarras S (2007) Muscle intermediate filaments and their links to membranes and membranous organelles. Exp Cell Res 313 (10):2063–76

Capetenaki Y (2002) Desmin cytoskeleton: a potential regulator of muscle mitochondrial behaviour and function. Trends Cardiovasc Med 12(8):339–48

Carattino MD, Edinger RS, Grieser HJ, Wise R, Neumann D, Schlattner U et al. (2005) Epithelial sodium channel inhibition by AMP-activated protein kinase in oocytes and polarized renal epithelial cells. J Biol Chem 280(18):17608–16. doi:10.1074/jbc.M501770200

Carling D, Thornton C, Woods A, Sanders MJ (2012) AMP-activated protein kinase: new regulation, new roles? Biochem J 445(1):11–27. doi:10.1042/BJ20120546

Ceddia RB, Sweeney G (2004) Creatine supplementation increases glucose oxidation and AMPK phosphorylation and reduces lactate production in L6 rat skeletal muscle cells. J Physiol 555 (Pt 2):409–21. doi:10.1113/jphysiol.2003.056291

Chang TJ, Chen WP, Yang C, Lu PH, Liang YC, Su MJ et al. (2009) Serine-385 phosphorylation of inwardly rectifying K^+ channel subunit (Kir6.2) by AMP-dependent protein kinase plays a key role in rosiglitazone-induced closure of the K(ATP) channel and insulin secretion in rats. Diabetologia 52(6):1112–21

Chen L, Jiao ZH, Zheng LS, Zhang YY, Xie ST, Wang ZX et al. (2009) Structural insight into the autoinhibition mechanism of AMP-activated protein kinase. Nature 459(7250):1146–9. doi:10.1038/nature08075

Chen L, Wang J, Zhang YY, Yan SF, Neumann D, Schlattner U et al. (2012) AMP-activated protein kinase undergoes nucleotide-dependent conformational changes. Nat Struct Mol Biol 19(7):716–8. doi:10.1038/nsmb.2319

Chen L, Xin FJ, Wang J, Hu J, Zhang YY, Wan S, Cao LS, Lu C, Li P, Yan SF, Neumann D, Schlattner U, Xia B, Wang ZX, Wu JW (2013a) Conserved regulatory elements in AMPK. Nature 498:E8–E10

Chen K, Kobayashi S, Xu X, Viollet B, Liang Q (2013b) AMP activated protein kinase is indispensable for myocardial adaptation to caloric restriction in mice. PLoS One 8(3): e59682. doi:10.1371/journal.pone.0059682

Clark AJ, Gaddie R, Stewart CP (1937) The aerobic metabolism of the isolated frog's heart poisoned by iodoacetic acid. J Physiol 90(3):335–46

Clark H, Carling D, Saggerson D (2004) Covalent activation of heart AMP-activated protein kinase in response to physiological concentrations of long-chain fatty acids. Eur J Biochem 271 (11):2215–24. doi:10.1111/j.1432-1033

Clarke PR, Hardie DG (1990) Regulation of HMG-CoA reductase: identification of the site phosphorylated by the AMP-activated protein kinase in vitro and in intact rat liver. EMBO J 9(8):2439–46

Collins TJ, Bootman MD (2003) Mitochondria are morphologically heterogeneous within cells. J Exp Biol 206(Pt 12):1993–2000

Cortassa S, O'Rourke B, Winslow RL, Aon MA (2009) Control and regulation of mitochondrial energetics in an integrated model of cardiomyocyte function. Biophys J 96(6):2466–78
Coven DL, Hu X, Cong L, Bergeron R, Shulman GI, Hardie DG et al. (2003) Physiological role of AMP-activated protein kinase in the heart: graded activation during exercise. Am J Physiol Endocrinol Metab 285(3):E629–36. doi:10.1152/ajpendo.00171.2003
Darrabie MD, Arciniegas AJ, Mishra R, Bowles DE, Jacobs DO, Santacruz L (2011) AMPK and substrate availability regulate creatine transport in cultured cardiomyocytes. Am J Physiol Endocrinol Metab 300(5):E870–6. doi:10.1152/ajpendo.00554.2010
Davies SP, Carling D, Munday MR, Hardie DG (1992) Diurnal rhythm of phosphorylation of rat liver acetyl-CoA carboxylase by the AMP-activated protein kinase, demonstrated using freeze-clamping. Effects of high fat diets. Eur J Biochem 203(3):615–23
Davies SP, Helps NR, Cohen PT, Hardie DG (1995) 5′-AMP inhibits dephosphorylation, as well as promoting phosphorylation, of the AMP-activated protein kinase. Studies using bacterially expressed human protein phosphatase-2C alpha and native bovine protein phosphatase-2AC. FEBS Lett 377(3):421–5
Davies JK, Wells DJ, Liu K, Whitrow HR, Daniel TD, Grignani R et al. (2006) Characterization of the role of gamma2 R531G mutation in AMP-activated protein kinase in cardiac hypertrophy and Wolff-Parkinson-White syndrome. Am J Physiol Heart Circ Physiol 290(5):H1942–51. doi:10.1152/ajpheart.01020.2005
Dieni CA, Storey KB (2009) Creatine kinase regulation by reversible phosphorylation in frog muscle. Comp Biochem Physiol B Biochem Mol Biol 152(4):405–12. doi:10.1016/j.cbpb.2009.01.012
Diviani D, Maric D, Perez Lopez I, Cavin S, Del Vescovo CD (2013) A-kinase anchoring proteins: Molecular regulators of the cardiac stress response. Biochim Biophys Acta 1833(4):901–8. doi:10.1016/j.bbamcr.2012.07.014
Djouder N, Tuerk RD, Suter M, Salvioni P, Thali RF, Scholz R et al. (2010) PKA phosphorylates and inactivates AMPKalpha to promote efficient lipolysis. EMBO J 29(2):469–81. doi:10.1038/emboj.2009.339
Dolinsky VW, Chan AY, Robillard Frayne I, Light PE, Des Rosiers C, Dyck JR (2009) Resveratrol prevents the prohypertrophic effects of oxidative stress on LKB1. Circulation 119 (12):1643–52. doi:10.1161/CIRCULATIONAHA.108.787440
Dos Santos P, Aliev MK, Diolez P, Duclos F, Besse P, Bonoron-Adele S et al. (2000) Metabolic control of contractile performance in isolated perfused rat heart. Analysis of experimental data by reaction:diffusion mathematical model. J Mol Cell Cardiol 32(9):1703–34
Doussiere J, Ligeti E, Brandolin G, Vignais PV (1984) Control of oxidative phosphorylation in rat heart mitochondria. The role of the adenine nucleotide carrier. Biochim Biophys Acta 766 (2):492–500
Dzeja PP, Terzic A (2003) Phosphotransfer networks and cellular energetics. J Exp Biol 206 (Pt 12):2039–47
Dzeja P, Terzic A (2009) Adenylate kinase and AMP signaling networks: metabolic monitoring, signal communication and body energy sensing. Int J Mol Sci 10(4):1729–72
Dzeja PP, Zeleznikar RJ, Goldberg ND (1996) Suppression of creatine kinase-catalyzed phosphotransfer results in increased phosphoryl transfer by adenylate kinase in intact skeletal muscle. J Biol Chem 271(22):12847–51
Dzeja PP, Vitkevicius KT, Redfield MM, Burnett JC, Terzic A (1999) Adenylate kinase-catalyzed phosphotransfer in the myocardium: increased contribution in heart failure. Circ Res 84 (10):1137–43
Dzeja PP, Holmuhamedov EL, Ozcan C, Pucar D, Jahangir A, Terzic A (2001) Mitochondria: gateway for cytoprotection. Circ Res 89(9):744–6
Dzeja PP, Bast P, Pucar D, Wieringa B, Terzic A (2007) Defective metabolic signaling in adenylate kinase AK1 gene knock-out hearts compromises post-ischemic coronary reflow. J Biol Chem 282(43):31366–72. doi:10.1074/jbc.M705268200

Dzeja PP, Hoyer K, Tian R, Zhang S, Nemutlu E, Spindler M et al. (2011a) Rearrangement of energetic and substrate utilization networks compensate for chronic myocardial creatine kinase deficiency. J Physiol 589(Pt 21):5193–211. doi:10.1113/jphysiol.2011.212829

Dzeja PP, Chung S, Faustino RS, Behfar A, Terzic A (2011b) Developmental enhancement of adenylate kinase-AMPK metabolic signaling axis supports stem cell cardiac differentiation. PLoS One 6(4):e19300. doi:10.1371/journal.pone.0019300

Eder M, Schlattner U, Becker A, Wallimann T, Kabsch W, Fritz-Wolf K (1999) Crystal structure of brain-type creatine kinase at 1.41 A resolution. Protein Sci 8(11):2258–69

Eder M, Fritz-Wolf K, Kabsch W, Wallimann T, Schlattner U (2000) Crystal structure of human ubiquitous mitochondrial creatine kinase. Proteins 39(3):216–25. doi:10.1002/(SICI)1097-0134(20000515)39

Edwards HV, Christian F, Baillie GS (2012) cAMP: novel concepts in compartmentalised signalling. Semin Cell Dev Biol 23(2):181–90. doi:10.1016/j.semcdb.2011.09.005

Egan DF, Shackelford DB, Mihaylova MM, Gelino S, Kohnz RA, Mair W et al. (2011) Phosphorylation of ULK1 (hATG1) by AMP-activated protein kinase connects energy sensing to mitophagy. Science 331(6016):456–61. doi:10.1126/science.1196371

Ellington WR (2001) Evolution and physiological roles of phosphagen systems. Annu Rev Physiol 63:289–325. doi:10.1146/annurev.physiol.63.1.289

Ellington WR, Suzuki T (2007) Early evolution of the creatine kinase gene family and the capacity for creatine biosynthesis and membrane transport. Subcell Biochem 46:17–26

Ewing RM, Chu P, Elisma F, Li H, Taylor P, Climie S et al. (2007) Large-scale mapping of human protein-protein interactions by mass spectrometry. Mol Syst Biol 3:89. doi:10.1038/msb4100134

Fassett JT, Hu X, Xu X, Lu Z, Zhang P, Chen Y et al. (2013) AMPK attenuates microtubule proliferation in cardiac hypertrophy. Am J Physiol Heart Circ Physiol 304(5):H749–58. doi:10.1152/ajpheart.00935.2011

Fell DA, Thomas S (1995) Physiological control of metabolic flux: the requirement for multisite modulation. Biochem J 311(Pt 1):35–9

Finckenberg P, Mervaala E (2010) Novel regulators and drug targets of cardiac hypertrophy. J Hypertens 28(Suppl 1):S33–8. doi:10.1097/01.hjh.0000388492.73954.0b

Forcet C, Billaud M (2007) Dialogue between LKB1 and AMPK: a hot topic at the cellular pole. Sci STKE 2007(404):pe51. doi:10.1126/stke.4042007pe51

Fraser SA, Gimenez I, Cook N, Jennings I, Katerelos M, Katsis F et al. (2007) Regulation of the renal-specific Na^{+}-K^{+}-$2Cl^{-}$ co-transporter NKCC2 by AMP-activated protein kinase (AMPK). Biochem J 405(1):85–93. doi:10.1042/BJ20061850

Frederich M, Balschi JA (2002) The relationship between AMP-activated protein kinase activity and AMP concentration in the isolated perfused rat heart. J Biol Chem 277(3):1928–32. doi:10.1074/jbc.M107128200

Frederich M, Zhang L, Balschi JA (2005) Hypoxia and AMP independently regulate AMP-activated protein kinase activity in heart. Am J Physiol Heart Circ Physiol 288(5):H2412–21. doi:10.1152/ajpheart.00558.2004

Frey S, Millat T, Hohmann S, Wolkenhauer O (2008) How quantitative measures unravel design principles in multi-stage phosphorylation cascades. J Theor Biol 254(1):27–36. doi:10.1016/j.jtbi.2008.04.037

Fritz-Wolf K, Schnyder T, Wallimann T, Kabsch W (1996) Structure of mitochondrial creatine kinase. Nature 381(6580):341–5. doi:10.1038/381341a0

Frosig C, Pehmoller C, Birk JB, Richter EA, Wojtaszewski JF (2010) Exercise-induced TBC1D1 Ser237 phosphorylation and 14-3-3 protein binding capacity in human skeletal muscle. J Physiol 588(Pt 22):4539–48. doi:10.1113/jphysiol.2010.194811

Garcia-Haro L, Garcia-Gimeno MA, Neumann D, Beullens M, Bollen M, Sanz P (2010) The PP1-R6 protein phosphatase holoenzyme is involved in the glucose-induced dephosphorylation and inactivation of AMP-activated protein kinase, a key regulator of insulin secretion, in MIN6 beta cells. FASEB J 24(12):5080–91. doi:10.1096/fj.10-166306

Glancy B, Balaban RS (2012) Role of mitochondrial Ca^{2+} in the regulation of cellular energetics. Biochemistry 51(14):2959–73. doi:10.1021/bi2018909
Gonzalez-Granillo M, Grichine A, Guzun R, Usson Y, Tepp K, Chekulayev V, Shevchuk I, Karu-Varikmaa M, Kuznetsov AV, Grimm M, Saks V, Kaambre T (2012) Studies of the role of tubulin beta II isotype in regulation of mitochondrial respiration in intracellular energetic units in cardiac cells. J Mol Cell Cardiol 52:437–47
Gratia S, Kay L, Potenza L, Seffouh A, Novel-Chate V, Schnebelen C et al. (2012) Inhibition of AMPK signalling by doxorubicin: at the crossroads of the cardiac responses to energetic, oxidative, and genotoxic stress. Cardiovasc Res 95(3):290–9. doi:10.1093/cvr/cvs134
Greer EL, Oskoui PR, Banko MR, Maniar JM, Gygi MP, Gygi SP et al. (2007) The energy sensor AMP-activated protein kinase directly regulates the mammalian FOXO3 transcription factor. J Biol Chem 282(41):30107–19. doi:10.1074/jbc.M705325200
Griffiths EJ, Rutter GA (2009) Mitochondrial calcium as a key regulator of mitochondrial ATP production in mammalian cells. Biochim Biophys Acta 1787:1324–33
Groen AK, Wanders RJ, Westerhoff HV, van der Meer R, Tager JM (1982) Quantification of the contribution of various steps to the control of mitochondrial respiration. J Biol Chem 257 (6):2754–7
Guzun R, Saks V (2010) Application of the principles of systems biology and Wiener's cybernetics for analysis of regulation of energy fluxes in muscle cells in vivo. Int J Mol Sci 11 (3):982–1019
Guzun R, Timohhina N, Tepp K, Monge C, Kaambre T, Sikk P et al. (2009) Regulation of respiration controlled by mitochondrial creatine kinase in permeabilized cardiac cells in situ. Importance of system level properties. Biochim Biophys Acta 1787(9):1089–105. doi:10.1016/j.bbabio.2009.03.024
Guzun R, Karu-Varikmaa M, Gonzalez-Granillo M, Kuznetsov AV, Michel L, Cottet-Rousselle C et al. (2011a) Mitochondria-cytoskeleton interaction: distribution of beta-tubulins in cardiomyocytes and HL-1 cells. Biochim Biophys Acta 1807(4):458–69
Guzun R, Timohhina N, Tepp K, Gonzalez-Granillo M, Shevchuk I, Chekulayev V, Kuznetsov AV, Kaambre T, Saks VA (2011b) Systems bioenergetics of creatine kinase networks: physiological roles of creatine and phosphocreatine in regulation of cardiac cell function. Amino Acids 40:1333–48
Guzun R, Gonzalez-Granillo M, Karu-Varikmaa M, Grichine A, Usson Y, Kaambre T et al. (2012) Regulation of respiration in muscle cells in vivo by VDAC through interaction with the cytoskeleton and MtCK within mitochondrial interactosome. Biochim Biophys Acta 1818 (6):1545–54. doi:10.1016/j.bbamem.2011.12.034
Gwinn DM, Shackelford DB, Egan DF, Mihaylova MM, Mery A, Vasquez DS et al. (2008) AMPK phosphorylation of raptor mediates a metabolic checkpoint. Mol Cell 30(2):214–26. doi:10.1016/j.molcel.2008.03.003
Hackenbrock CR (1968) Chemical and physical fixation of isolated mitochondria in low-energy and high-energy states. Proc Natl Acad Sci USA 61(2):598–605
Hallows KR, McCane JE, Kemp BE, Witters LA, Foskett JK (2003) Regulation of channel gating by AMP-activated protein kinase modulates cystic fibrosis transmembrane conductance regulator activity in lung submucosal cells. J Biol Chem 278(2):998–1004. doi:10.1074/jbc.M210621200
Han Y, Wang Q, Song P, Zhu Y, Zou MH (2010) Redox regulation of the AMP-activated protein kinase. PLoS One 5(11):e15420. doi:10.1371/journal.pone.0015420
Harada M, Nattel SN, Nattel S (2012) AMP-activated protein kinase: potential role in cardiac electrophysiology and arrhythmias. Circ Arrhythm Electrophysiol 5(4):860–7. doi:10.1161/CIRCEP.112.972265
Hardie DG (2007) AMP-activated protein kinase as a drug target. Annu Rev Pharmacol Toxicol 47:185–210. doi:10.1146/annurev.pharmtox.47.120505.105304
Hardie DG, Carling D (1997) The AMP-activated protein kinase–fuel gauge of the mammalian cell? Eur J Biochem 246(2):259–73

Hardie DG, Hawley SA (2001) AMP-activated protein kinase: the energy charge hypothesis revisited. Bioessays 23(12):1112–9. doi:10.1002/bies.10009

Hardie DG, Carling D, Gamblin SJ (2011) AMP-activated protein kinase: also regulated by ADP? Trends Biochem Sci 36(9):470–7. doi:10.1016/j.tibs.2011.06.004

Hardie DG, Ross FA, Hawley SA (2012a) AMPK: a nutrient and energy sensor that maintains energy homeostasis. Nat Rev Mol Cell Biol 13(4):251–62. doi:10.1038/nrm3311

Hardie DG, Ross FA, Hawley SA (2012b) AMP-activated protein kinase: a target for drugs both ancient and modern. Chem Biol 19(10):1222–36. doi:10.1016/j.chembiol.2012.08.019

Hawley SA, Davison M, Woods A, Davies SP, Beri RK, Carling D et al. (1996) Characterization of the AMP-activated protein kinase kinase from rat liver and identification of threonine 172 as the major site at which it phosphorylates AMP-activated protein kinase. J Biol Chem 271 (44):27879–87

Hawley SA, Boudeau J, Reid JL, Mustard KJ, Udd L, Makela TP et al. (2003) Complexes between the LKB1 tumor suppressor, STRAD alpha/beta and MO25 alpha/beta are upstream kinases in the AMP-activated protein kinase cascade. J Biol 2(4):28. doi:10.1186/1475-4924-2-28

Hawley SA, Pan DA, Mustard KJ, Ross L, Bain J, Edelman AM et al. (2005) Calmodulin-dependent protein kinase kinase-beta is an alternative upstream kinase for AMP-activated protein kinase. Cell Metab 2(1):9–19. doi:10.1016/j.cmet.2005.05.009

Hayashi T, Martone ME, Yu Z, Thor A, Doi M, Holst MJ et al. (2009) Three-dimensional electron microscopy reveals new details of membrane systems for Ca^{2+} signaling in the heart. J Cell Sci 122(Pt 7):1005–13. doi:10.1242/jcs.028175

Heerschap A, Kan HE, Nabuurs CI, Renema WK, Isbrandt D, Wieringa B (2007) In vivo magnetic resonance spectroscopy of transgenic mice with altered expression of guanidinoacetate methyltransferase and creatine kinase isoenzymes. Subcell Biochem 46:119–48

Hers I, Vincent EE, Tavare JM (2011) Akt signalling in health and disease. Cell Signal 23 (10):1515–27. doi:10.1016/j.cellsig.2011.05.004

Hoppe S, Bierhoff H, Cado I, Weber A, Tiebe M, Grummt I et al. (2009) AMP-activated protein kinase adapts rRNA synthesis to cellular energy supply. Proc Natl Acad Sci USA 106 (42):17781–6. doi:10.1073/pnas.0909873106

Horman S, Vertommen D, Heath R, Neumann D, Mouton V, Woods A et al. (2006) Insulin antagonizes ischemia-induced Thr172 phosphorylation of AMP-activated protein kinase alpha-subunits in heart via hierarchical phosphorylation of Ser485/491. J Biol Chem 281 (9):5335–40. doi:10.1074/jbc.M506850200

Horman S, Beauloye C, Vanoverschelde JL, Bertrand L (2012) AMP-activated protein kinase in the control of cardiac metabolism and remodeling. Curr Heart Fail Rep 9(3):164–73. doi:10.1007/s11897-012-0102-z

Hu X, Xu X, Lu Z, Zhang P, Fassett J, Zhang Y et al. (2011) AMP activated protein kinase-alpha2 regulates expression of estrogen-related receptor-alpha, a metabolic transcription factor related to heart failure development. Hypertension 58(4):696–703. doi:10.1161/HYPERTENSIONAHA.111.174128

Hue L, Taegtmeyer H (2009) The Randle cycle revisited: a new head for an old hat. Am J Physiol Endocrinol Metab 297(3):E578–91

Hurley RL, Anderson KA, Franzone JM, Kemp BE, Means AR, Witters LA (2005) The Ca^{2+}/calmodulin-dependent protein kinase kinases are AMP-activated protein kinase kinases. J Biol Chem 280(32):29060–6. doi:10.1074/jbc.M503824200

Hurley RL, Barre LK, Wood SD, Anderson KA, Kemp BE, Means AR et al. (2006) Regulation of AMP-activated protein kinase by multisite phosphorylation in response to agents that elevate cellular cAMP. J Biol Chem 281(48):36662–72. doi:10.1074/jbc.M606676200

Ideker T, Dutkowski J, Hood L (2011) Boosting signal-to-noise in complex biology: prior knowledge is power. Cell 144(6):860–3. doi:10.1016/j.cell.2011.03.007

Ikematsu N, Dallas ML, Ross FA, Lewis RW, Rafferty JN, David JA et al. (2011) Phosphorylation of the voltage-gated potassium channel Kv2.1 by AMP-activated protein kinase regulates membrane excitability. Proc Natl Acad Sci USA 108(44):18132–7

Imamura K, Ogura T, Kishimoto A, Kaminishi M, Esumi H (2001) Cell cycle regulation via p53 phosphorylation by a 5'-AMP activated protein kinase activator, 5-aminoimidazole-4-carboxamide-1-beta-D-ribofuranoside, in a human hepatocellular carcinoma cell line. Biochem Biophys Res Commun 287(2):562–7. doi:10.1006/bbrc.2001.5627

Ingwall JS (2002) Is creatine kinase a target for AMP-activated protein kinase in the heart? J Mol Cell Cardiol 34(9):1111–20

Ingwall JS (2006) On the hypothesis that the failing heart is energy starved: lessons learned from the metabolism of ATP and creatine. Curr Hypertens Rep 8(6):457–64

Ingwall JS, Weiss RG (2004) Is the failing heart energy starved? On using chemical energy to support cardiac function. Circ Res 95(2):135–45

Inoki K, Zhu T, Guan KL (2003) TSC2 mediates cellular energy response to control cell growth and survival. Cell 115(5):577–90

Inoki K, Kim J, Guan KL (2012) AMPK and mTOR in cellular energy homeostasis and drug targets. Annu Rev Pharmacol Toxicol 52:381–400. doi:10.1146/annurev-pharmtox-010611-134537

Irrcher I, Adhihetty PJ, Sheehan T, Joseph AM, Hood DA (2003) PPARgamma coactivator-1alpha expression during thyroid hormone- and contractile activity-induced mitochondrial adaptations. Am J Physiol Cell Physiol 284(6):C1669–77. doi:10.1152/ajpcell.00409.2002

Jacobus WE, Saks VA (1982) Creatine kinase of heart mitochondria: changes in its kinetic properties induced by coupling to oxidative phosphorylation. Arch Biochem Biophys 219 (1):167–78

Jager S, Handschin C, St-Pierre J, Spiegelman BM (2007) AMP-activated protein kinase (AMPK) action in skeletal muscle via direct phosphorylation of PGC-1alpha. Proc Natl Acad Sci USA 104(29):12017–22. doi:10.1073/pnas.0705070104

Jeon SM, Chandel NS, Hay N (2012) AMPK regulates NADPH homeostasis to promote tumour cell survival during energy stress. Nature 485(7400):661–5. doi:10.1038/nature11066

Jones RG, Plas DR, Kubek S, Buzzai M, Mu J, Xu Y et al. (2005) AMP-activated protein kinase induces a p53-dependent metabolic checkpoint. Mol Cell 18(3):283–93. doi:10.1016/j.molcel.2005.03.027

Ju TC, Lin YS, Chern Y (2012) Energy dysfunction in Huntington's disease: insights from PGC-1alpha, AMPK, and CKB. Cell Mol Life Sci 69(24):4107–20. doi:10.1007/s00018-012-1025-2

Kahn BB, Alquier T, Carling D, Hardie DG (2005) AMP-activated protein kinase: ancient energy gauge provides clues to modern understanding of metabolism. Cell Metab 1(1):15–25

Kan HE, Buse-Pot TE, Peco R, Isbrandt D, Heerschap A, de Haan A (2005) Lower force and impaired performance during high-intensity electrical stimulation in skeletal muscle of GAMT-deficient knockout mice. Am J Physiol Cell Physiol 289(1):C113–9. doi:10.1152/ajpcell.00040.2005

Kang S, Chemaly ER, Hajjar RJ, Lebeche D (2011) Resistin promotes cardiac hypertrophy via the AMP-activated protein kinase/mammalian target of rapamycin (AMPK/mTOR) and c-Jun N-terminal kinase/insulin receptor substrate 1 (JNK/IRS1) pathways. J Biol Chem 286 (21):18465–73. doi:10.1074/jbc.M110.200022

Kawaguchi T, Osatomi K, Yamashita H, Kabashima T, Uyeda K (2002) Mechanism for fatty acid "sparing" effect on glucose-induced transcription: regulation of carbohydrate-responsive element-binding protein by AMP-activated protein kinase. J Biol Chem 277(6):3829–35. doi:10.1074/jbc.M107895200

Kay L, Nicolay K, Wieringa B, Saks V, Wallimann T (2000) Direct evidence for the control of mitochondrial respiration by mitochondrial creatine kinase in oxidative muscle cells in situ. J Biol Chem 275(10):6937–44

Kelly M, Keller C, Avilucea PR, Keller P, Luo Z, Xiang X et al. (2004) AMPK activity is diminished in tissues of IL-6 knockout mice: the effect of exercise. Biochem Biophys Res Commun 320(2):449–54. doi:10.1016/j.bbrc.2004.05.188

Kemp BE, Oakhill JS, Scott JW (2007) AMPK structure and regulation from three angles. Structure 15(10):1161–3. doi:10.1016/j.str.2007.09.006

Kerrien S, Aranda B, Breuza L, Bridge A, Broackes-Carter F, Chen C et al. (2012) The IntAct molecular interaction database in 2012. Nucleic Acids Res 40(Database issue):D841–6. doi:10.1093/nar/gkr1088

Kim AS, Miller EJ, Wright TM, Li J, Qi D, Atsina K et al. (2011a) A small molecule AMPK activator protects the heart against ischemia-reperfusion injury. J Mol Cell Cardiol 51 (1):24–32. doi:10.1016/j.yjmcc.2011.03.003

Kim J, Kundu M, Viollet B, Guan KL (2011b) AMPK and mTOR regulate autophagy through direct phosphorylation of Ulk1. Nat Cell Biol 13(2):132–41. doi:10.1038/ncb2152

Kim M, Shen M, Ngoy S, Karamanlidis G, Liao R, Tian R (2012) AMPK isoform expression in the normal and failing hearts. J Mol Cell Cardiol 52(5):1066–73. doi:10.1016/j.yjmcc.2012.01.016

Klaus A, Polge C, Zorman S, Auchli Y, Brunisholz R, Schlattner U (2012) A two-dimensional screen for AMPK substrates identifies tumor suppressor fumarate hydratase as a preferential AMPKalpha2 substrate. J Proteomics 75(11):3304–13. doi:10.1016/j.jprot.2012.03.040

Klaus A, Zorman S, Berthier A, Polge C, Ramirez-Rios S, Michelland S et al. (2013) Glutathione S-transferases interact with AMP-activated protein kinase: evidence for S-glutathionylation and activation in vitro. PLoS One 8:e62497

Klingenberg M (1970) Mitochondria metabolite transport. FEBS Lett 6(3):145–54

Klingenberg M (1976) The state of ADP or ATP fixed to the mitochondria by bongkrekate. Eur J Biochem/FEBS 65(2):601–5

Klingenberg M (2008) The ADP, and ATP transport in mitochondria and its carrier. Biochim Biophys Acta 1778(10):1978–2021

Ko HJ, Zhang Z, Jung DY, Jun JY, Ma Z, Jones KE et al. (2009) Nutrient stress activates inflammation and reduces glucose metabolism by suppressing AMP-activated protein kinase in the heart. Diabetes 58(11):2536–46. doi:10.2337/db08-1361

Kobayashi K, Neely JR (1979) Control of maximum rates of glycolysis in rat cardiac muscle. Circ Res 44(2):166–75

Kola B, Hubina E, Tucci SA, Kirkham TC, Garcia EA, Mitchell SE et al. (2005) Cannabinoids and ghrelin have both central and peripheral metabolic and cardiac effects via AMP-activated protein kinase. J Biol Chem 280(26):25196–201. doi:10.1074/jbc.C500175200

Koo SH, Flechner L, Qi L, Zhang X, Screaton RA, Jeffries S et al. (2005) The CREB coactivator TORC2 is a key regulator of fasting glucose metabolism. Nature 437(7062):1109–11. doi:10.1038/nature03967

Kudo N, Barr AJ, Barr RL, Desai S, Lopaschuk GD (1995) High rates of fatty acid oxidation during reperfusion of ischemic hearts are associated with a decrease in malonyl-CoA levels due to an increase in 5'-AMP-activated protein kinase inhibition of acetyl-CoA carboxylase. J Biol Chem 270(29):17513–20

Kudo N, Gillespie JG, Kung L, Witters LA, Schulz R, Clanachan AS et al. (1996) Characterization of 5′AMP-activated protein kinase activity in the heart and its role in inhibiting acetyl-CoA carboxylase during reperfusion following ischemia. Biochim Biophys Acta 1301(1–2):67–75

Kulkarni SS, Karlsson HK, Szekeres F, Chibalin AV, Krook A, Zierath JR (2011) Suppression of 5′-nucleotidase enzymes promotes AMP-activated protein kinase (AMPK) phosphorylation and metabolism in human and mouse skeletal muscle. J Biol Chem 286(40):34567–74. doi:10.1074/jbc.M111.268292

Kurth-Kraczek EJ, Hirshman MF, Goodyear LJ, Winder WW (1999) 5′ AMP-activated protein kinase activation causes GLUT4 translocation in skeletal muscle. Diabetes 48(8):1667–71

Kuznetsov AV, Margreiter R (2009) Heterogeneity of mitochondria and mitochondrial function within cells as another level of mitochondrial complexity. Int J Mol Sci 10(4):1911–29

Kuznetsov AV, Clark JF, Winkler K, Kunz WS (1996) Increase of flux control of cytochrome c oxidase in copper-deficient mottled brindled mice. J Biol Chem 271(1):283–8

Kuznetsov AV, Hermann M, Troppmair J, Margreiter R, Hengster P (2009) Complex patterns of mitochondrial dynamics in human pancreatic cells revealed by fluorescent confocal imaging. J Cell Mol Med 14:417–25

Kwiatkowski DJ, Manning BD (2005) Tuberous sclerosis: a GAP at the crossroads of multiple signaling pathways. Hum Mol Genet 14(Spec No. 2):R251–8. doi:10.1093/hmg/ddi260

Lan F, Cacicedo JM, Ruderman N, Ido Y (2008) SIRT1 modulation of the acetylation status, cytosolic localization, and activity of LKB1. Possible role in AMP-activated protein kinase activation. J Biol Chem 283(41):27628–27635. doi:10.1074/jbc.M805711200

Lee JH, Koh H, Kim M, Kim Y, Lee SY, Karess RE et al. (2007) Energy-dependent regulation of cell structure by AMP-activated protein kinase. Nature 447(7147):1017–20. doi:10.1038/nature05828

Leverve X, Fontaine E, Peronnet F, (eds) (2006) Bioénergétique. Traité de nutrition artificielle de l'adulte: Springer

Li J, Miller EJ, Ninomiya-Tsuji J, Russell RR 3rd, Young LH (2005) AMP-activated protein kinase activates p38 mitogen-activated protein kinase by increasing recruitment of p38 MAPK to TAB1 in the ischemic heart. Circ Res 97(9):872–9. doi:10.1161/01.RES.0000187458.77026.10

Li J, Coven DL, Miller EJ, Hu X, Young ME, Carling D et al. (2006) Activation of AMPK alpha- and gamma-isoform complexes in the intact ischemic rat heart. Am J Physiol Heart Circ Physiol 291(4):H1927–34. doi:10.1152/ajpheart.00251.2006

Li HL, Yin R, Chen D, Liu D, Wang D, Yang Q et al. (2007) Long-term activation of adenosine monophosphate-activated protein kinase attenuates pressure-overload-induced cardiac hypertrophy. J Cell Biochem 100(5):1086–99. doi:10.1002/jcb.21197

Li H, Thali RF, Smolak C, Gong F, Alzamora R, Wallimann T et al. (2010) Regulation of the creatine transporter by AMP-activated protein kinase in kidney epithelial cells. Am J Physiol Renal Physiol 299(1):F167–77. doi:10.1152/ajprenal.00162.2010

Li Y, Xu S, Mihaylova MM, Zheng B, Hou X, Jiang B et al. (2011) AMPK phosphorylates and inhibits SREBP activity to attenuate hepatic steatosis and atherosclerosis in diet-induced insulin-resistant mice. Cell Metab 13(4):376–88. doi:10.1016/j.cmet.2011.03.009

Li XH, Chen XJ, Ou WB, Zhang Q, Lv ZR, Zhan Y et al. (2013) Knockdown of creatine kinase B inhibits ovarian cancer progression by decreasing glycolysis. Int J Biochem Cell Biol 45(5):979–86. doi:10.1016/j.biocel.2013.02.003

Liang J, Shao SH, Xu ZX, Hennessy B, Ding Z, Larrea M et al. (2007) The energy sensing LKB1-AMPK pathway regulates p27(kip1) phosphorylation mediating the decision to enter autophagy or apoptosis. Nat Cell Biol 9(2):218–24. doi:10.1038/ncb1537

Lin YY, Kiihl S, Suhail Y, Liu SY, Chou YH, Kuang Z, Lu JY, Khor CN, Lin CL, Bader JS, Irizarry R, Boeke JD (2012) Functional dissection of lysine deacetylases reveals that HDAC1 and p300 regulate AMPK. Nature 482(7384):251–255. doi:10.1038/nature10804

Liobikas J, Kopustinskiene DM, Toleikis A (2001) What controls the outer mitochondrial membrane permeability for ADP: facts for and against the role of oncotic pressure. Biochim Biophys Acta 1505(2–3):220–5

Lygate CA, Bohl S, ten Hove M, Faller KM, Ostrowski PJ, Zervou S et al. (2012) Moderate elevation of intracellular creatine by targeting the creatine transporter protects mice from acute myocardial infarction. Cardiovasc Res 96(3):466–75. doi:10.1093/cvr/cvs272

Lygate CA, Aksentijevic D, Dawson D, Ten Hove M, Phillips D, de Bono JP et al. (2013) Living without creatine: unchanged exercise capacity and response to chronic myocardial infarction in creatine-deficient mice. Circ Res 112(6):945–55. doi:10.1161/CIRCRESAHA.112.300725

Ma H, Wang J, Thomas DP, Tong C, Leng L, Wang W et al. (2010) Impaired macrophage migration inhibitory factor-AMP-activated protein kinase activation and ischemic recovery in the senescent heart. Circulation 122(3):282–92. doi:10.1161/CIRCULATIONAHA.110.953208

Makinde AO, Gamble J, Lopaschuk GD (1997) Upregulation of 5′-AMP-activated protein kinase is responsible for the increase in myocardial fatty acid oxidation rates following birth in the newborn rabbit. Circ Res 80(4):482–9

Mannella CA (2006) Structure and dynamics of the mitochondrial inner membrane cristae. Biochim Biophys Acta 1763(5–6):542–8. doi:10.1016/j.bbamcr.2006.04.006
Marcus FB (2008) Bioinformatics and systems biology - collaborative research and resources. Springer, Berlin, Heidelberg
Marsin AS, Bertrand L, Rider MH, Deprez J, Beauloye C, Vincent MF et al. (2000) Phosphorylation and activation of heart PFK-2 by AMPK has a role in the stimulation of glycolysis during ischaemia. Curr Biol 10(20):1247–55
Mayer FV, Heath R, Underwood E, Sanders MJ, Carmena D, McCartney RR et al. (2011) ADP regulates SNF1, the Saccharomyces cerevisiae homolog of AMP-activated protein kinase. Cell Metab 14(5):707–14. doi:10.1016/j.cmet.2011.09.009
McBride A, Ghilagaber S, Nikolaev A, Hardie DG (2009) The glycogen-binding domain on the AMPK beta subunit allows the kinase to act as a glycogen sensor. Cell Metab 9(1):23–34. doi:10.1016/j.cmet.2008.11.008
McGaffin KR, Moravec CS, McTiernan CF (2009) Leptin signaling in the failing and mechanically unloaded human heart. Circ Heart Fail 2(6):676–83. doi:10.1161/CIRCHEARTFAILURE.109.869909
McGee SL, Hargreaves M (2008) AMPK and transcriptional regulation. Front Biosci 13:3022–33
McGee SL, Howlett KF, Starkie RL, Cameron-Smith D, Kemp BE, Hargreaves M (2003) Exercise increases nuclear AMPK alpha2 in human skeletal muscle. Diabetes 52(4):926–8
McGee SL, van Denderen BJ, Howlett KF, Mollica J, Schertzer JD, Kemp BE et al. (2008) AMP-activated protein kinase regulates GLUT4 transcription by phosphorylating histone deacetylase 5. Diabetes 57(4):860–7. doi:10.2337/db07-0843
Meijer AJ, Codogno P (2007) AMP-activated protein kinase and autophagy. Autophagy 3 (3):238–40
Merrill GF, Kurth EJ, Hardie DG, Winder WW (1997) AICA riboside increases AMP-activated protein kinase, fatty acid oxidation, and glucose uptake in rat muscle. Am J Physiol 273(6 Pt 1): E1107–12
Meyer LE, Machado LB, Santiago AP, Da-Silva WS, De Felice FG, Holub O et al. (2006) Mitochondrial creatine kinase activity prevents reactive oxygen species generation: antioxidant role of mitochondrial kinase-dependent ADP re-cycling activity. J Biol Chem 281 (49):37361–71. doi:10.1074/jbc.M604123200
Mihaylova MM, Vasquez DS, Ravnskjaer K, Denechaud PD, Yu RT, Alvarez JG et al. (2011) Class IIa histone deacetylases are hormone-activated regulators of FOXO and mammalian glucose homeostasis. Cell 145(4):607–21. doi:10.1016/j.cell.2011.03.043
Mika D, Leroy J, Vandecasteele G, Fischmeister R (2012) PDEs create local domains of cAMP signaling. J Mol Cell Cardiol 52(2):323–9. doi:10.1016/j.yjmcc.2011.08.016
Miller EJ, Li J, Leng L, McDonald C, Atsumi T, Bucala R et al. (2008) Macrophage migration inhibitory factor stimulates AMP-activated protein kinase in the ischaemic heart. Nature 451 (7178):578–82. doi:10.1038/nature06504
Minokoshi Y, Alquier T, Furukawa N, Kim YB, Lee A, Xue B et al. (2004) AMP-kinase regulates food intake by responding to hormonal and nutrient signals in the hypothalamus. Nature 428 (6982):569–74. doi:10.1038/nature02440
Mochel F, Durant B, Meng X, O'Callaghan J, Yu H, Brouillet E et al. (2012) Early alterations of brain cellular energy homeostasis in Huntington disease models. J Biol Chem 287(2):1361–70. doi:10.1074/jbc.M111.309849
Momcilovic M, Hong SP, Carlson M (2006) Mammalian TAK1 activates Snf1 protein kinase in yeast and phosphorylates AMP-activated protein kinase in vitro. J Biol Chem 281 (35):25336–43. doi:10.1074/jbc.M604399200
Moreno D, Viana R, Sanz P (2009) Two-hybrid analysis identifies PSMD11, a non-ATPase subunit of the proteasome, as a novel interaction partner of AMP-activated protein kinase. Int J Biochem Cell Biol 41(12):2431–9. doi:10.1016/j.biocel.2009.07.002
Moreno D, Towler MC, Hardie DG, Knecht E, Sanz P (2010) The laforin-malin complex, involved in Lafora disease, promotes the incorporation of K63-linked ubiquitin chains into

AMP-activated protein kinase beta subunits. Mol Biol Cell 21(15):2578–88. doi:10.1091/mbc.E10-03-0227
Moreno-Sanchez R, Saavedra E, Rodriguez-Enriquez S, Olin-Sandoval V (2008) Metabolic control analysis: a tool for designing strategies to manipulate metabolic pathways. J Biomed Biotechnol 2008:597913
Mungai PT, Waypa GB, Jairaman A, Prakriya M, Dokic D, Ball MK et al. (2011) Hypoxia triggers AMPK activation through reactive oxygen species-mediated activation of calcium release-activated calcium channels. Mol Cell Biol 31(17):3531–45. doi:10.1128/MCB.05124-11
Muoio DM, Seefeld K, Witters LA, Coleman RA (1999) AMP-activated kinase reciprocally regulates triacylglycerol synthesis and fatty acid oxidation in liver and muscle: evidence that sn-glycerol-3-phosphate acyltransferase is a novel target. Biochem J 338(Pt 3):783–91
Musi N, Hirshman MF, Arad M, Xing Y, Fujii N, Pomerleau J et al. (2005) Functional role of AMP-activated protein kinase in the heart during exercise. FEBS Lett 579(10):2045–50. doi:10.1016/j.febslet.2005.02.052
Nabuurs C, Huijbregts B, Wieringa B, Hilbers CW, Heerschap A (2010) 31P saturation transfer spectroscopy predicts differential intracellular macromolecular association of ATP and ADP in skeletal muscle. J Biol Chem 285(51):39588–96. doi:10.1074/jbc.M110.164665
Nabuurs CI, Choe CU, Veltien A, Kan HE, van Loon LJ, Rodenburg RJ et al. (2013) Disturbed energy metabolism and muscular dystrophy caused by pure creatine deficiency are reversible by creatine intake. J Physiol 591(Pt 2):571–92. doi:10.1113/jphysiol.2012.241760
Nakano A, Takashima S (2012) LKB1 and AMP-activated protein kinase: regulators of cell polarity. Genes Cells 17(9):737–47. doi:10.1111/j.1365-2443.2012.01629.x
Nakano A, Kato H, Watanabe T, Min KD, Yamazaki S, Asano Y et al. (2010) AMPK controls the speed of microtubule polymerization and directional cell migration through CLIP-170 phosphorylation. Nat Cell Biol 12(6):583–90. doi:10.1038/ncb2060
Nascimben L, Ingwall JS, Pauletto P, Friedrich J, Gwathmey JK, Saks V et al. (1996) Creatine kinase system in failing and nonfailing human myocardium. Circulation 94(8):1894–901
Neely J, Morgan H (1974) Relationship between carbohydrate and lipid metabolism and the energy balance of heart muscle. Annu Rev Physiol 63:413–59
Neely JR, Denton RM, England PJ, Randle PJ (1972) The effects of increased heart work on the tricarboxylate cycle and its interactions with glycolysis in the perfused rat heart. Biochem J 128 (1):147–59
Nemutlu E, Zhang S, Gupta A, Juranic NO, Macura SI, Terzic A et al. (2012) Dynamic phosphometabolomic profiling of human tissues and transgenic models by 18O-assisted (3)(1)P NMR and mass spectrometry. Physiol Genomics 44(7):386–402. doi:10.1152/physiolgenomics.00152.2011
Neubauer S (2007) The failing heart–an engine out of fuel. N Engl J Med 356(11):1140–51
Neumann D, Schlattner U, Wallimann T (2003) A molecular approach to the concerted action of kinases involved in energy homoeostasis. Biochem Soc Trans 31(Pt 1):169–74. doi:10.1042/
Newsholme EA, Start C (1973) Regulation in metabolism. Wiley, London
Nivala M, Korge P, Weiss JN, Qu Z (2011) Linking flickering to waves and whole-cell oscillations in a mitochondrial network model. Biophys J 101(9):2102–11. doi:10.1016/j.bpj.2011.09.038
Oakhill JS, Chen ZP, Scott JW, Steel R, Castelli LA, Ling N et al. (2010) beta-Subunit myristoylation is the gatekeeper for initiating metabolic stress sensing by AMP-activated protein kinase (AMPK). Proc Natl Acad Sci USA 107(45):19237–41. doi:10.1073/pnas.1009705107
Oakhill JS, Steel R, Chen ZP, Scott JW, Ling N, Tam S et al. (2011) AMPK is a direct adenylate charge-regulated protein kinase. Science 332(6036):1433–5. doi:10.1126/science.1200094
Oakhill JS, Scott JW, Kemp BE (2012) AMPK functions as an adenylate charge-regulated protein kinase. Trends Endocrinol Metab 23(3):125–32. doi:10.1016/j.tem.2011.12.006
Oliveira SM, Zhang YH, Solis RS, Isackson H, Bellahcene M, Yavari A et al. (2012a) AMP-activated protein kinase phosphorylates cardiac troponin I and alters contractility of murine ventricular myocytes. Circ Res 110(9):1192–201. doi:10.1161/CIRCRESAHA.111.259952

Oliveira AP, Ludwig C, Picotti P, Kogadeeva M, Aebersold R, Sauer U (2012b) Regulation of yeast central metabolism by enzyme phosphorylation. Mol Syst Biol 8:623. doi:10.1038/msb.2012.55

Paiva MA, Rutter-Locher Z, Goncalves LM, Providencia LA, Davidson SM, Yellon DM et al. (2011) Enhancing AMPK activation during ischemia protects the diabetic heart against reperfusion injury. Am J Physiol Heart Circ Physiol 300(6):H2123–34. doi:10.1152/ajpheart.00707.2010

Pang T, Xiong B, Li JY, Qiu BY, Jin GZ, Shen JK et al. (2007) Conserved alpha-helix acts as autoinhibitory sequence in AMP-activated protein kinase alpha subunits. J Biol Chem 282 (1):495–506. doi:10.1074/jbc.M605790200

Pinter K, Grignani RT, Czibik G, Farza H, Watkins H, Redwood C (2012) Embryonic expression of AMPK gamma subunits and the identification of a novel gamma2 transcript variant in adult heart. J Mol Cell Cardiol 53(3):342–9. doi:10.1016/j.yjmcc.2012.05.017

Polekhina G, Gupta A, Michell BJ, van Denderen B, Murthy S, Feil SC et al. (2003) AMPK beta subunit targets metabolic stress sensing to glycogen. Curr Biol 13(10):867–71

Polge C, Jossier M, Crozet P, Gissot L, Thomas M (2008) Beta-subunits of the SnRK1 complexes share a common ancestral function together with expression and function specificities; physical interaction with nitrate reductase specifically occurs via AKINbeta1-subunit. Plant Physiol 148 (3):1570–82. doi:10.1104/pp. 108.123026

Ponticos M, Lu QL, Morgan JE, Hardie DG, Partridge TA, Carling D (1998) Dual regulation of the AMP-activated protein kinase provides a novel mechanism for the control of creatine kinase in skeletal muscle. EMBO J 17(6):1688–99. doi:10.1093/emboj/17.6.1688

Pucar D, Dzeja PP, Bast P, Juranic N, Macura S, Terzic A (2001) Cellular energetics in the preconditioned state: protective role for phosphotransfer reactions captured by 18O-assisted 31P NMR. J Biol Chem 276(48):44812–9. doi:10.1074/jbc.M104425200

Qi J, Gong J, Zhao T, Zhao J, Lam P, Ye J et al. (2008) Downregulation of AMP-activated protein kinase by Cidea-mediated ubiquitination and degradation in brown adipose tissue. EMBO J 27 (11):1537–48. doi:10.1038/emboj.2008.92

Randle PJ (1998) Regulatory interactions between lipids and carbohydrates: the glucose fatty acid cycle after 35 years. Diabetes Metab Rev 14(4):263–83

Randle PJ, Garland PB, Hales CN, Newsholme EA (1963) The glucose fatty-acid cycle. Its role in insulin sensitivity and the metabolic disturbances of diabetes mellitus. Lancet 1(7285):785–9

Riek U, Scholz R, Konarev P, Rufer A, Suter M, Nazabal A et al. (2008) Structural properties of AMP-activated protein kinase: dimerization, molecular shape, and changes upon ligand binding. J Biol Chem 283(26):18331–43. doi:10.1074/jbc.M708379200

Rogne M, Tasken K (2013) Cell signalling analyses in the functional genomics era. N Biotechnol 30(3):333–8. doi:10.1016/j.nbt.2013.01.003

Rose BA, Force T, Wang Y (2010) Mitogen-activated protein kinase signaling in the heart: angels versus demons in a heart-breaking tale. Physiol Rev 90(4):1507–46. doi:10.1152/physrev.00054.2009

Ruderman NB, Xu XJ, Nelson L, Cacicedo JM, Saha AK, Lan F, Ido Y (2010) AMPK and SIRT1: a long-standing partnership? Am J Physiol Endocrinol Metab 298(4):E751–E760. doi:10.1152/ajpendo.00745.2009

Russell RR 3rd, Li J, Coven DL, Pypaert M, Zechner C, Palmeri M et al. (2004) AMP-activated protein kinase mediates ischemic glucose uptake and prevents postischemic cardiac dysfunction, apoptosis, and injury. J Clin Invest 114(4):495–503. doi:10.1172/JCI19297

Sackett D (2010) Evolution and coevolution of tubulin's carboxy-terminal tails and mitochondria. In: Svensson OL (ed) Mitochondria: structure, function and dysfunction. Nova Biomedical Books, New York, pp 789–810

Saetersdal T, Greve G, Dalen H (1990) Associations between beta-tubulin and mitochondria in adult isolated heart myocytes as shown by immunofluorescence and immunoelectron microscopy. Histochemistry 95(1):1–10

Saks V (ed) (2007) Molecular system bioenergetics—energy for life, basic principles, organization and dynamics of cellular energetics. Wiley-VCH, Weinheim

Saks V (2008) The phosphocreatine-creatine kinase system helps to shape muscle cells and keep them healthy and alive. J Physiol 586(Pt 12):2817–8

Saks V (2009) Molecular system bioenergetics-new aspects of metabolic research. Int J Mol Sci 10 (8):3655–7

Saks V, Strumia E (1993) Phosphocreatine: molecular and cellular aspects of the mechanism of cardioprotective action. Curr Ther Res 53(5):565–98

Saks VA, Lipina NV, Sharov VG, Smirnov VN, Chazov E, Grosse R (1977) The localization of the MM isozyme of creatine phosphokinase on the surface membrane of myocardial cells and its functional coupling to ouabain-inhibited (Na+, K+)-ATPase. Biochim Biophys Acta 465 (3):550–8

Saks VA, Rosenshtraukh LV, Smirnov VN, Chazov EI (1978) Role of creatine phosphokinase in cellular function and metabolism. Can J Physiol Pharmacol 56(5):691–706

Saks VA, Belikova YO, Kuznetsov AV (1991) In vivo regulation of mitochondrial respiration in cardiomyocytes: specific restrictions for intracellular diffusion of ADP. Biochim Biophys Acta 1074(2):302–11

Saks V, Dos Santos P, Gellerich FN, Diolez P (1998) Quantitative studies of enzyme-substrate compartmentation, functional coupling and metabolic channelling in muscle cells. Mol Cell Biochem 184(1–2):291–307

Saks VA, Kaambre T, Sikk P, Eimre M, Orlova E, Paju K et al. (2001) Intracellular energetic units in red muscle cells. Biochem J 356(Pt 2):643–57

Saks V, Dzeja P, Schlattner U, Vendelin M, Terzic A, Wallimann T (2006a) Cardiac system bioenergetics: metabolic basis of the Frank-Starling law. J Physiol 571(Pt 2):253–73

Saks V, Favier R, Guzun R, Schlattner U, Wallimann T (2006b) Molecular system bioenergetics: regulation of substrate supply in response to heart energy demands. J Physiol 577(Pt 3):769–77

Saks V, Dzeja P, Schlattner U, Vendelin M, Terzic A, Wallimann T (2006c) Cardiac system bioenergetics: metabolic basis of the Frank-Starling law. J Physiol 571(Pt 2):253–73. doi:10.1113/jphysiol.2005.101444

Saks V, Kaambre T, Guzun R, Anmann T, Sikk P, Schlattner U et al. (2007a) The creatine kinase phosphotransfer network: thermodynamic and kinetic considerations, the impact of the mitochondrial outer membrane and modelling approaches. Subcell Biochem 46:27–65

Saks V, Anmann T, Guzun R, Kaambre T, Sikk P, Schlattner U et al. (2007b) The creatine kinase phosphotransfer network: thermodynamic and kinetic considerations, the impact of the mitochondrial outer membrane and modelling approaches. In: Wyss M, Salomons G (eds) Creatine and creatine kinase in health and disease. Springer, Dordrecht, pp 27–66

Saks VA, Dzeja P, Guzun R, Aliev MK, Vendelin M, Terzic A, Wallimann T (2007c) System analysis of cardiac energetics—excitation–contraction coupling: integration of mitochondrial respiration, phosphotransfer pathways, metabolic pacing and substrate supply in the heart. In: Saks V (ed) Molecular system bioenergetics. Energy for Life. Wiley, Weinheim, GmbH, pp 367–405

Saks V, Monge C, Anmann T, Dzeja P (2007d) Integrated and organized cellular energetic systems: theories of cell energetics, compartmentation and metabolic channeling. In: Saks V (ed) Molecular system bioenergetics. Energy for life. Wiley, Weinheim, GmbH, pp 59–110

Saks V, Guzun R, Timohhina N, Tepp K, Varikmaa M, Monge C et al. (2010) Structure-function relationships in feedback regulation of energy fluxes in vivo in health and disease: mitochondrial interactosome. Biochim Biophys Acta 1797(6–7):678–97

Saks V, Kuznetsov AV, Gonzalez-Granillo M, Tepp K, Timohhina N, Karu-Varikmaa M et al. (2012) Intracellular energetic units regulate metabolism in cardiac cells. J Mol Cell Cardiol 52:419–36

Salt I, Celler JW, Hawley SA, Prescott A, Woods A, Carling D et al. (1998) AMP-activated protein kinase: greater AMP dependence, and preferential nuclear localization, of complexes containing the alpha2 isoform. Biochem J 334(Pt 1):177–87

Sartoretto JL, Kalwa H, Pluth MD, Lippard SJ, Michel T (2011) Hydrogen peroxide differentially modulates cardiac myocyte nitric oxide synthesis. Proc Natl Acad Sci USA 108(38):15792–7. doi:10.1073/pnas.1111331108

Sasaki H, Asanuma H, Fujita M, Takahama H, Wakeno M, Ito S et al. (2009) Metformin prevents progression of heart failure in dogs: role of AMP-activated protein kinase. Circulation 119 (19):2568–77. doi:10.1161/CIRCULATIONAHA.108.798561

Schlattner U, Wallimann T (2004) Metabolite channeling: creatine kinase microcompartments. In: Lennarz WJ, Lane MD (eds) In encyclopedia of biological chemistry. Academic, New York, USA, pp 646–51

Schlattner U, Forstner M, Eder M, Stachowiak O, Fritz-Wolf K, Wallimann T (1998) Functional aspects of the X-ray structure of mitochondrial creatine kinase: a molecular physiology approach. Mol Cell Biochem 184(1–2):125–40

Schlattner U, Tokarska-Schlattner M, Wallimann T (2006a) Molecular structure and function of mitochondrial creatine kinases. In: Vial C (ed) Creatine kinase. Nova, New York, pp 123–70

Schlattner U, Tokarska-Schlattner M, Wallimann T (2006b) Mitochondrial creatine kinase in human health and disease. Biochim Biophys Acta 1762(2):164–80

Schlattner U, Tokarska-Schlattner M, Ramirez S, Bruckner A, Kay L, Polge C et al. (2009) Mitochondrial kinases and their molecular interaction with cardiolipin. Biochim Biophys Acta 1788(10):2032–47. doi:10.1016/j.bbamem.2009.04.018

Schmidt A, Marescau B, Boehm EA, Renema WK, Peco R, Das A et al. (2004) Severely altered guanidino compound levels, disturbed body weight homeostasis and impaired fertility in a mouse model of guanidinoacetate N-methyltransferase (GAMT) deficiency. Hum Mol Genet 13(9):905–21. doi:10.1093/hmg/ddh112

Schroder R, Kunz WS, Rouan F, Pfendner E, Tolksdorf K, Kappes-Horn K et al. (2002) Disorganization of the desmin cytoskeleton and mitochondrial dysfunction in plectin-related epidermolysis bullosa simplex with muscular dystrophy. J Neuropathol Exp Neurol 61 (6):520–30

Schrödinger E (ed) (1944) What is life? The physical aspect of the living cell. Cambridge University Press, Cambridge, UK

Scott JW, Hawley SA, Green KA, Anis M, Stewart G, Scullion GA et al. (2004) CBS domains form energy-sensing modules whose binding of adenosine ligands is disrupted by disease mutations. J Clin Invest 113(2):274–84. doi:10.1172/JCI19874

Sebbagh M, Santoni MJ, Hall B, Borg JP, Schwartz MA (2009) Regulation of LKB1/STRAD localization and function by E-cadherin. Curr Biol 19(1):37–42. doi:10.1016/j.cub.2008.11.033

Selivanov VA, Alekseev AE, Hodgson DM, Dzeja PP, Terzic A (2004) Nucleotide-gated KATP channels integrated with creatine and adenylate kinases: amplification, tuning and sensing of energetic signals in the compartmentalized cellular environment. Mol Cell Biochem 256–257 (1–2):243–56

Shibata R, Ouchi N, Ito M, Kihara S, Shiojima I, Pimentel DR et al. (2004) Adiponectin-mediated modulation of hypertrophic signals in the heart. Nat Med 10(12):1384–9. doi:10.1038/nm1137

Shibata R, Sato K, Pimentel DR, Takemura Y, Kihara S, Ohashi K et al. (2005) Adiponectin protects against myocardial ischemia-reperfusion injury through AMPK- and COX-2-dependent mechanisms. Nat Med 11(10):1096–103. doi:10.1038/nm1295

Shinmura K, Tamaki K, Saito K, Nakano Y, Tobe T, Bolli R (2007) Cardioprotective effects of short-term caloric restriction are mediated by adiponectin via activation of AMP-activated protein kinase. Circulation 116(24):2809–17. doi:10.1161/CIRCULATIONAHA.107.725697

Soeller C, Cannell MB (1999) Examination of the transverse tubular system in living cardiac rat myocytes by 2-photon microscopy and digital image-processing techniques. Circ Res 84 (3):266–75

Solaz-Fuster MC, Gimeno-Alcaniz JV, Ros S, Fernandez-Sanchez ME, Garcia-Fojeda B, Criado Garcia O et al. (2008) Regulation of glycogen synthesis by the laforin-malin complex is modulated by the AMP-activated protein kinase pathway. Hum Mol Genet 17(5):667–78. doi:10.1093/hmg/ddm339

Sonntag AG, Dalle Pezze P, Shanley DP, Thedieck K (2012) A modelling-experimental approach reveals insulin receptor substrate (IRS)-dependent regulation of adenosine monosphosphate-dependent kinase (AMPK) by insulin. FEBS J 279(18):3314–28. doi:10.1111/j.1742-4658.2012.08582.x

Srivastava RA, Pinkosky SL, Filippov S, Hanselman JC, Cramer CT, Newton RS (2012) AMP-activated protein kinase: an emerging drug target to regulate imbalances in lipid and carbohydrate metabolism to treat cardio-metabolic diseases. J Lipid Res 53(12):2490–514. doi:10.1194/jlr.R025882

Stapleton D, Mitchelhill KI, Gao G, Widmer J, Michell BJ, Teh T et al. (1996) Mammalian AMP-activated protein kinase subfamily. J Biol Chem 271(2):611–4

Starling EH, Visscher MB (1927) The regulation of the energy output of the heart. J Physiol 62 (3):243–61

Steeghs K, Benders A, Oerlemans F, de Haan A, Heerschap A, Ruitenbeek W et al. (1997) Altered Ca^{2+} responses in muscles with combined mitochondrial and cytosolic creatine kinase deficiencies. Cell 89(1):93–103

Steinberg SF (2012) Cardiac actions of protein kinase C isoforms. Physiology (Bethesda) 27 (3):130–9. doi:10.1152/physiol.00009.2012

Steinberg GR (2013) AMPK and the endocrine control of energy metabolism. Mol Cell Endocrinol 366(2):125–6. doi:10.1016/j.mce.2013.01.003

Steinberg GR, Kemp BE (2009) AMPK in health and disease. Physiol Rev 89(3):1025–78. doi:10.1152/physrev.00011.2008

Stockler S, Schutz PW, Salomons GS (2007) Cerebral creatine deficiency syndromes: clinical aspects, treatment and pathophysiology. Subcell Biochem 46:149–66

Stoppani J, Hildebrandt AL, Sakamoto K, Cameron-Smith D, Goodyear LJ, Neufer PD (2002) AMP-activated protein kinase activates transcription of the UCP3 and HKII genes in rat skeletal muscle. Am J Physiol Endocrinol Metab 283(6):E1239–48. doi:10.1152/ajpendo.00278.2002

Streijger F, Oerlemans F, Ellenbroek BA, Jost CR, Wieringa B, Van der Zee CE (2005) Structural and behavioural consequences of double deficiency for creatine kinases BCK and UbCKmit. Behav Brain Res 157(2):219–34. doi:10.1016/j.bbr.2004.07.002

Strogolova V, Orlova M, Shevade A, Kuchin S (2012) Mitochondrial porin Por1 and its homolog Por2 contribute to the positive control of Snf1 protein kinase in Saccharomyces cerevisiae. Eukaryot Cell 11(12):1568–72. doi:10.1128/EC.00127-12

Suter M, Riek U, Tuerk R, Schlattner U, Wallimann T, Neumann D (2006) Dissecting the role of 5′-AMP for allosteric stimulation, activation, and deactivation of AMP-activated protein kinase. J Biol Chem 281(43):32207–16. doi:10.1074/jbc.M606357200

Suzuki A, Okamoto S, Lee S, Saito K, Shiuchi T, Minokoshi Y (2007) Leptin stimulates fatty acid oxidation and peroxisome proliferator-activated receptor alpha gene expression in mouse C2C12 myoblasts by changing the subcellular localization of the alpha2 form of AMP-activated protein kinase. Mol Cell Biol 27(12):4317–27. doi:10.1128/MCB.02222-06

Tachikawa M, Ikeda S, Fujinawa J, Hirose S, Akanuma S, Hosoya K (2012) Gamma-Aminobutyric acid transporter 2 mediates the hepatic uptake of guanidinoacetate, the creatine biosynthetic precursor, in rats. PLoS One 7(2):e32557. doi:10.1371/journal.pone.0032557

Taegtmeyer H (2010) Tracing cardiac metabolism in vivo: one substrate at a time. J Nucl Med 51 (Suppl 1):80S–7S

Taegtmeyer H, Ingwall JS (2013) Creatine—a dispensable metabolite? Circ Res 112(6):878–80. doi:10.1161/CIRCRESAHA.113.300974

Taegtmeyer H, Wilson CR, Razeghi P, Sharma S (2005) Metabolic energetics and genetics in the heart. Ann N Y Acad Sci 1047:208–18

Tagawa H, Koide M, Sato H, Zile MR, Carabello BA, Cooper GT (1998) Cytoskeletal role in the transition from compensated to decompensated hypertrophy during adult canine left ventricular pressure overloading. Circ Res 82(7):751–61

Takimoto E (2012) Cyclic GMP-dependent signaling in cardiac myocytes. Circ J 76(8):1819–25

Tarasov AI, Griffiths EJ, Rutter GA (2012) Regulation of ATP production by mitochondrial Ca^{2+}. Cell Calcium 52(1):28–35. doi:10.1016/j.ceca.2012.03.003

Taylor EB, Ellingson WJ, Lamb JD, Chesser DG, Compton CL, Winder WW (2006) Evidence against regulation of AMP-activated protein kinase and LKB1/STRAD/MO25 activity by creatine phosphate. Am J Physiol Endocrinol Metab 290(4):E661–9. doi:10.1152/ajpendo.00313.2005

Taylor SS, Kim C, Cheng CY, Brown SH, Wu J, Kannan N (2008) Signaling through cAMP and cAMP-dependent protein kinase: diverse strategies for drug design. Biochim Biophys Acta 1784(1):16–26. doi:10.1016/j.bbapap.2007.10.002

Telesco SE, Radhakrishnan R (2012) Structural systems biology and multiscale signaling models. Ann Biomed Eng 40(11):2295–306. doi:10.1007/s10439-012-0576-6

ten Hove M, Lygate CA, Fischer A, Schneider JE, Sang AE, Hulbert K et al. (2005) Reduced inotropic reserve and increased susceptibility to cardiac ischemia/reperfusion injury in phosphocreatine-deficient guanidinoacetate-N-methyltransferase-knockout mice. Circulation 111(19):2477–85. doi:10.1161/01.CIR.0000165147.99592.01

Tepp K, Timohhina N, Chekulayev V, Shevchuk I, Kaambre T, Saks V (2011) Metabolic control analysis of integrated energy metabolism in permeabilized cardiomyocytes—experimental study. Acta Biochim Pol 57(4):421–30

Thornton C, Snowden MA, Carling D (1998) Identification of a novel AMP-activated protein kinase beta subunit isoform that is highly expressed in skeletal muscle. J Biol Chem 273 (20):12443–50

Tian R, Musi N, D'Agostino J, Hirshman MF, Goodyear LJ (2001) Increased adenosine monophosphate-activated protein kinase activity in rat hearts with pressure-overload hypertrophy. Circulation 104(14):1664–9

Timohhina N, Guzun R, Tepp K, Monge C, Varikmaa M, Vija H et al. (2009) Direct measurement of energy fluxes from mitochondria into cytoplasm in permeabilized cardiac cells in situ: some evidence for Mitochondrial Interactosome. J Bioenerg Biomembr 41(3):259–75. doi:10.1007/s10863-009-9224-8

Tokarska-Schlattner M, Zaugg M, da Silva R, Lucchinetti E, Schaub MC, Wallimann T et al. (2005) Acute toxicity of doxorubicin on isolated perfused heart: response of kinases regulating energy supply. Am J Physiol Heart Circ Physiol 289(1):H37–47. doi:10.1152/ajpheart.01057.2004

Tokarska-Schlattner M, Epand RF, Meiler F, Zandomeneghi G, Neumann D, Widmer HR et al. (2012) Phosphocreatine interacts with phospholipids, affects membrane properties and exerts membrane-protective effects. PLoS One 7(8):e43178. doi:10.1371/journal.pone.0043178

Torremans A, Marescau B, Possemiers I, Van Dam D, D'Hooge R, Isbrandt D et al. (2005) Biochemical and behavioural phenotyping of a mouse model for GAMT deficiency. J Neurol Sci 231(1–2):49–55. doi:10.1016/j.jns.2004.12.014

Townley R, Shapiro L (2007) Crystal structures of the adenylate sensor from fission yeast AMP-activated protein kinase. Science 315(5819):1726–9. doi:10.1126/science.1137503

Turner DC, Wallimann T, Eppenberger HM (1973) A protein that binds specifically to the M-line of skeletal muscle is identified as the muscle form of creatine kinase. Proc Natl Acad Sci USA 70(3):702–5

Uda K, Hoshijima M, Suzuki T (2013) A novel taurocyamine kinase found in the protist Phytophthora infestans. Comp Biochem Physiol B Biochem Mol Biol 165(1):42–8. doi:10.1016/j.cbpb.2013.03.003

van Oort MM, van Doorn JM, Hasnaoui ME, Glatz JF, Bonen A, van der Horst DJ et al. (2009) Effects of AMPK activators on the sub-cellular distribution of fatty acid transporters CD36 and FABPpm. Arch Physiol Biochem 115(3):137–46. doi:10.1080/13813450902975090

Vendelin M, Kongas O, Saks V (2000) Regulation of mitochondrial respiration in heart cells analyzed by reaction-diffusion model of energy transfer. Am J Physiol Cell Physiol 278(4): C747–64

Vendelin M, Beraud N, Guerrero K, Andrienko T, Kuznetsov AV, Olivares J et al. (2005) Mitochondrial regular arrangement in muscle cells: a "crystal-like" pattern. Am J Physiol Cell Physiol 288(3):C757–67

Ventura-Clapier R, De Sousa E, Veksler V (2002) Metabolic myopathy in heart failure. News Physiol Sci 17:191–6

Ventura-Clapier R, Garnier A, Veksler V (2004) Energy metabolism in heart failure. J Physiol 555 (Pt 1):1–13

Viana R, Aguado C, Esteban I, Moreno D, Viollet B, Knecht E et al. (2008) Role of AMP-activated protein kinase in autophagy and proteasome function. Biochem Biophys Res Commun 369 (3):964–8. doi:10.1016/j.bbrc.2008.02.126

Vincent O, Carlson M (1999) Gal83 mediates the interaction of the Snf1 kinase complex with the transcription activator Sip4. EMBO J 18(23):6672–81. doi:10.1093/emboj/18.23.6672

Voss M, Paterson J, Kelsall IR, Martin-Granados C, Hastie CJ, Peggie MW et al. (2011) Ppm1E is an in cellulo AMP-activated protein kinase phosphatase. Cell Signal 23(1):114–24. doi:10.1016/j.cellsig.2010.08.010

Wallimann T (1975) Creatine kinase isoenzymes and myofibrillar structure [Nr. 5437]. ETH Zürich, Switzerland

Wallimann T (1996) 31P-NMR-measured creatine kinase reaction flux in muscle: a caveat! J Muscle Res Cell Motil 17(2):177–81

Wallimann T (2007) Introduction–creatine: cheap ergogenic supplement with great potential for health and disease. Sub-Cell Biochem 46:1–16

Wallimann T, Wyss M, Brdiczka D, Nicolay K, Eppenberger HM (1992) Intracellular compartmentation, structure and function of creatine kinase isoenzymes in tissues with high and fluctuating energy demands: the 'phosphocreatine circuit' for cellular energy homeostasis. Biochem J 281(Pt 1):21–40

Wallimann T, Tokarska-Schlattner M, Neumann D, Epand RF, Andres RH, Widmer HR, Hornemann T, Saks V, Agarkova I, Schlattner U (2007) The phosphocreatine circuit: molecular and cellular physiology of creatine kinases, sensitivity to free radicals, and enhancement by creatine supplementation. In: Saks V (ed) Molecular system bioenergetics. Energy for life. Wiley, Weinheim, GmbH, pp 195–264

Wallimann T, Tokarska-Schlattner M, Schlattner U (2011) The creatine kinase system and pleiotropic effects of creatine. Amino Acids 40(5):1271–96

Walsh K (2006) Akt signaling and growth of the heart. Circulation 113(17):2032–4. doi:10.1161/CIRCULATIONAHA.106.615138

Wang MY, Unger RH (2005) Role of PP2C in cardiac lipid accumulation in obese rodents and its prevention by troglitazone. Am J Physiol Endocrinol Metab 288(1):E216–21. doi:10.1152/ajpendo

Wang W, Fan J, Yang X, Furer-Galban S, Lopez de Silanes I, von Kobbe C et al. (2002) AMP-activated kinase regulates cytoplasmic HuR. Mol Cell Biol 22(10):3425–36

Wang Y, Gao E, Tao L, Lau WB, Yuan Y, Goldstein BJ et al. (2009) AMP-activated protein kinase deficiency enhances myocardial ischemia/reperfusion injury but has minimal effect on the antioxidant/antinitrative protection of adiponectin. Circulation 119(6):835–44. doi:10.1161/CIRCULATIONAHA.108.815043

Watt MJ, Dzamko N, Thomas WG, Rose-John S, Ernst M, Carling D et al. (2006) CNTF reverses obesity-induced insulin resistance by activating skeletal muscle AMPK. Nat Med 12(5):541–8. doi:10.1038/nm1383

Winder WW, Hardie DG (1999) AMP-activated protein kinase, a metabolic master switch: possible roles in type 2 diabetes. Am J Physiol 277(1 Pt 1):E1–10

Wiseman RW, Kushmerick MJ (1995) Creatine kinase equilibration follows solution thermodynamics in skeletal muscle. 31P NMR studies using creatine analogs. J Biol Chem 270 (21):12428–38

Woods A, Johnstone SR, Dickerson K, Leiper FC, Fryer LG, Neumann D et al. (2003) LKB1 is the upstream kinase in the AMP-activated protein kinase cascade. Curr Biol 13(22):2004–8

Woods A, Dickerson K, Heath R, Hong SP, Momcilovic M, Johnstone SR et al. (2005) Ca^{2+}/calmodulin-dependent protein kinase kinase-beta acts upstream of AMP-activated protein kinase in mammalian cells. Cell Metab 2(1):21–33. doi:10.1016/j.cmet.2005.06.005

Wu Y, Song P, Xu J, Zhang M, Zou MH (2007) Activation of protein phosphatase 2A by palmitate inhibits AMP-activated protein kinase. J Biol Chem 282(13):9777–88. doi:10.1074/jbc.M608310200

Wu QY, Li F, Guo HY, Cao J, Chen C, Chen W et al. (2013) Disrupting of E79 and K138 interaction is responsible for human muscle creatine kinase deficiency diseases. Int J Biol Macromol 54:216–24. doi:10.1016/j.ijbiomac.2012.12.034

Wyss M, Kaddurah-Daouk R (2000) Creatine and creatinine metabolism. Physiol Rev 80 (3):1107–213

Xiao B, Heath R, Saiu P, Leiper FC, Leone P, Jing C et al. (2007) Structural basis for AMP binding to mammalian AMP-activated protein kinase. Nature 449(7161):496–500. doi:10.1038/nature06161

Xiao B, Sanders MJ, Underwood E, Heath R, Mayer FV, Carmena D et al. (2011) Structure of mammalian AMPK and its regulation by ADP. Nature 472(7342):230–3. doi:10.1038/nature09932

Xie M, Zhang D, Dyck JR, Li Y, Zhang H, Morishima M et al. (2006a) A pivotal role for endogenous TGF-beta-activated kinase-1 in the LKB1/AMP-activated protein kinase energy-sensor pathway. Proc Natl Acad Sci USA 103(46):17378–83. doi:10.1073/pnas.0604708103

Xie Z, Dong Y, Zhang M, Cui MZ, Cohen RA, Riek U et al. (2006b) Activation of protein kinase C zeta by peroxynitrite regulates LKB1-dependent AMP-activated protein kinase in cultured endothelial cells. J Biol Chem 281(10):6366–75. doi:10.1074/jbc.M511178200

Yamaguchi S, Katahira H, Ozawa S, Nakamichi Y, Tanaka T, Shimoyama T et al. (2005) Activators of AMP-activated protein kinase enhance GLUT4 translocation and its glucose transport activity in 3T3-L1 adipocytes. Am J Physiol Endocrinol Metab 289(4):E643–9. doi:10.1152/ajpendo.00456.2004

Yaniv Y, Juhaszova M, Wang S, Fishbein KW, Zorov DB, Sollott SJ (2011) Analysis of mitochondrial 3D-deformation in cardiomyocytes during active contraction reveals passive structural anisotropy of orthogonal short axes. PLoS One 6(7):e21985. doi:10.1371/journal.pone.0021985

Young LH (2008) AMP-activated protein kinase conducts the ischemic stress response orchestra. Circulation 117(6):832–40. doi:10.1161/CIRCULATIONAHA.107.713115

Zaha VG, Young LH (2012) AMP-activated protein kinase regulation and biological actions in the heart. Circ Res 111(6):800–14. doi:10.1161/CIRCRESAHA.111.255505

Zhang P, Hu X, Xu X, Fassett J, Zhu G, Viollet B et al. (2008) AMP activated protein kinase-alpha2 deficiency exacerbates pressure-overload-induced left ventricular hypertrophy and dysfunction in mice. Hypertension 52(5):918–24. doi:10.1161/HYPERTENSIONAHA.108.114702

Zhang BB, Zhou G, Li C (2009) AMPK: an emerging drug target for diabetes and the metabolic syndrome. Cell Metab 9(5):407–16. doi:10.1016/j.cmet.2009.03.012

Zhu L, Chen L, Zhou XM, Zhang YY, Zhang YJ, Zhao J et al. (2011) Structural insights into the architecture and allostery of full-length AMP-activated protein kinase. Structure 19(4):515–22. doi:10.1016/j.str.2011.01.018

Zmijewski JW, Banerjee S, Bae H, Friggeri A, Lazarowski ER, Abraham E (2010) Exposure to hydrogen peroxide induces oxidation and activation of AMP-activated protein kinase. J Biol Chem 285(43):33154–64. doi:10.1074/jbc.M110.143685

Zorov DB, Filburn CR, Klotz LO, Zweier JL, Sollott SJ (2000) Reactive oxygen species (ROS)-induced ROS release: a new phenomenon accompanying induction of the mitochondrial permeability transition in cardiac myocytes. J Exp Med 192(7):1001–14

Zou MH, Hou XY, Shi CM, Nagata D, Walsh K, Cohen RA (2002) Modulation by peroxynitrite of Akt- and AMP-activated kinase-dependent Ser1179 phosphorylation of endothelial nitric oxide synthase. J Biol Chem 277(36):32552–7. doi:10.1074/jbc.M204512200

Zou MH, Kirkpatrick SS, Davis BJ, Nelson JS, Wiles WG, Schlattner U et al. (2004) Activation of the AMP-activated protein kinase by the anti-diabetic drug metformin in vivo. Role of mitochondrial reactive nitrogen species. J Biol Chem 279(42):43940–51. doi:10.1074/jbc.M404421200

Part V
Systems Biology of Microorganisms

Chapter 12
Temporal Partitioning of the Yeast Cellular Network

Douglas B. Murray, Cornelia Amariei, Kalesh Sasidharan, Rainer Machné, Miguel A. Aon, and David Lloyd

Abstract A plethora of data is accumulating from high throughput methods on metabolites, coenzymes, proteins, lipids and nucleic acids and their interactions as well as the signalling and regulatory functions and pathways of the cellular network. The frozen moment viewed in a single discrete time sample requires frequent repetition and updating before any appreciation of the dynamics of component interaction becomes possible. Even then in a sample derived from a cell population, time-averaging of processes and events that occur in out-of-phase individuals blur the detailed complexity of single cell.

Continuously grown cultures of yeast spontaneously self-synchronise and provide resolution of detailed temporal structure. Continuous online monitoring (O_2 electrode and membrane-inlet mass spectrometry for O_2, CO_2 and H_2S; direct fluorimetry for NAD(P)H and flavins) gives dynamic information from timescales of minutes to hours. When these data are supplemented with mass spectrometry-based metabolomics and transcriptomics, the predominantly oscillatory behaviour of network components becomes evident, where respiration cycles between increased oxygen consumption (oxidative phase) and decreased oxygen consumption (reductive phase). This ultradian clock provides a coordinating function that links mitochondrial energetics and redox balance to transcriptional regulation, mitochondrial structure and organelle remodelling, DNA duplication and chroma-

D.B. Murray (✉) • C. Amariei • K. Sasidharan
Institute for Advanced Biosciences, Keio University, Tsuruoka, Japan
e-mail: dougie@ttck.keio.ac.jp

R. Machné
Institute for Theoretical Biochemistry, University of Vienna, Vienna, Austria

M.A. Aon
School of Medicine, Johns Hopkins University, Baltimore, MD, USA
e-mail: maon1@jhmi.edu

D. Lloyd
School of Biosciences, Cardiff University, Wales, UK

M.A. Aon et al. (eds.), *Systems Biology of Metabolic and Signaling Networks*, Springer Series in Biophysics 16, DOI 10.1007/978-3-642-38505-6_12,

tin dynamics. Ultimately, anabolism and catabolism become globally partitioned: mediation is by direct feedback loops between the energetic and redox state of the cell and chromatin architecture via enzymatic co-factors and co-enzymes.

Multi-oscillatory outputs were observed in dissolved gases with 12-h, 40-min and 4-min periods, and statistical self-similarity in Power Spectral and Relative Dispersional analyses: i.e. complex non-linear behaviour and a functional scale-free network operating simultaneously on several timescales. Fast sampling (at 10 or 1 Hz) of NAD(P)H fluorescence revealed subharmonic components of the 40-min signal at 20,10 and 3–5 min. The latter corresponds to oscillations directly observed and imaged by 2-photon microscopy in surface-attached cells. Signalling between time domains is suggested by studies with protonophore effectors of mitochondrial energetics. Multi-oscillatory states impinge on the complex reactome (where concentrations of most chemical species oscillate) and network functionality is made more comprehensible when in vivo time structure is taken into account.

12.1 Temporal Aspects of Integrative Yeast Cellular Function

Whereas our appreciation of the importance of spatial relationships between the myriad molecular components of living systems is well developed, temporal aspects of cell organisation are somewhat neglected. The intricate coordination of reactions, processes and events required during the life of the cell necessitates understanding across and between multiple timescales. Thus, it is not sufficient to consider only the slow process of growth and division (or budding in the case of *Saccharomyces cerevisiae*), the most evident as observed by simple light microscopy and taking a matter of hours (Lloyd et al. 1982b), but we must account for the integration of events from the membrane-associated biophysical domain (μs–ms), up through the metabolic timescale (seconds), the biosynthesis of macromolecules, transcription and translation (minutes), to assembly and remodelling of organelles (Aon and Cortassa 1997). Whereas spatio-temporal coincidence, convergence and simultaneity are often evident functional necessities, separation of incompatible processes must also be accommodated (Lloyd and Rossi 1992, 2008; Lloyd 2006a, 2008; Sasidharan et al. 2012).

Although as a matter of convenience and ease of study, the different time domains have been considered and researched as being separate; the resulting concept of hierarchical organisation is a misleading oversimplification. Interactions between faster and slower processes result in a heterarchical system (Yates 1992; Murray et al. 2007).

12.2 Continuous Culture of Yeast: An Ideal System for Study

Temporal compartmentalisation of the progress of energy generation, metabolic transformations, synthesis and assembly of membranes and organelles, as well as the organisation of chromosomal dynamics and the cell division cycle, requires studies either in single cells or in synchronous populations of cells or organisms. Whilst the former is even now restricted, the latter is usually problematic because of limitations associated with the preparation of material without perturbation. The observation that budding yeasts form stable oscillatory dynamics during continuous culture was first reported only a year after the DNA double helix model (Watson and Crick 1953; Finn and Wilson 1954). However it was not until the early 1990s, when high temporal resolution, computer acquisition became available that this spontaneous self-synchrony of dense (~5 $\times$ 10^8 organisms/ml) cultures of *S. cerevisiae* could readily be monitored (Fig. 12.1) to reveal short period and multi-timescale oscillatory dynamics (Satroutdinov et al. 1992; Keulers et al. 1994). The strain primarily employed (IFO-0233, IFO, Institute of Fermentation, Osaka, Japan) was an acid-tolerant diploid yeast (Naiki and Yamagata 1976).

During batch growth glucose is consumed and the culture produces biomass, CO_2, H_2S and ethanol as well as many other fermentation products. The pH is controlled at 3.4 and air flow rate is kept constant. The air flow rate is calculated for each reactor according to its specific oxygen transfer coefficient (Mueller et al. 2012). When ethanol is completely used up, depletion of trehalose and glycogen in the second stage of this diauxic growth process results in the initiation of oscillatory respiration, as indicated by 40-min cycles of dissolved O_2 and CO_2 (Murray 2004). A steady and continuous supply of growth medium at this stage provides material adequate both for long-term monitoring of the organisms through many thousands of generations over a period of many months, and discrete time samples for biochemical analyses. The rapidly responding probes immersed in the culture give either continuous outputs (e.g. for NAD(P)H fluorescence) or frequently sampled voltages at 0.1–10 Hz (Keulers et al. 1996a; Murray et al. 2007; Sasidharan et al. 2012). The most convenient readout, dissolved O_2, can also be monitored by membrane inlet mass spectrometry (MIMS) (Roussel and Lloyd 2007) as can CO_2, H_2S or ethanol. Off-gas measurements for CO_2, O_2 and H_2S can either be measured by MIMS (Keulers et al. 1996a) or by an array of dedicated sensors (Murray et al. 2011) to quantify transfer rates. Near-infrared spectroscopy can be used online for ethanol, glucose, NH_3, glutamine and biomass (Yeung et al. 1999).

The oscillatory state of this autonomously self-sustained system is observed when the cultures are supplied with glucose, ethanol and acetaldehyde as the main carbon source (Keulers et al. 1996b), implying mechanistic differences between respiratory and glycolytic oscillations. It exerts control on the production of hundreds of metabolites and many transcription factors (Murray et al. 2007). The genome-wide oscillation in transcription is pivotal (Klevecz et al. 2004; Li and

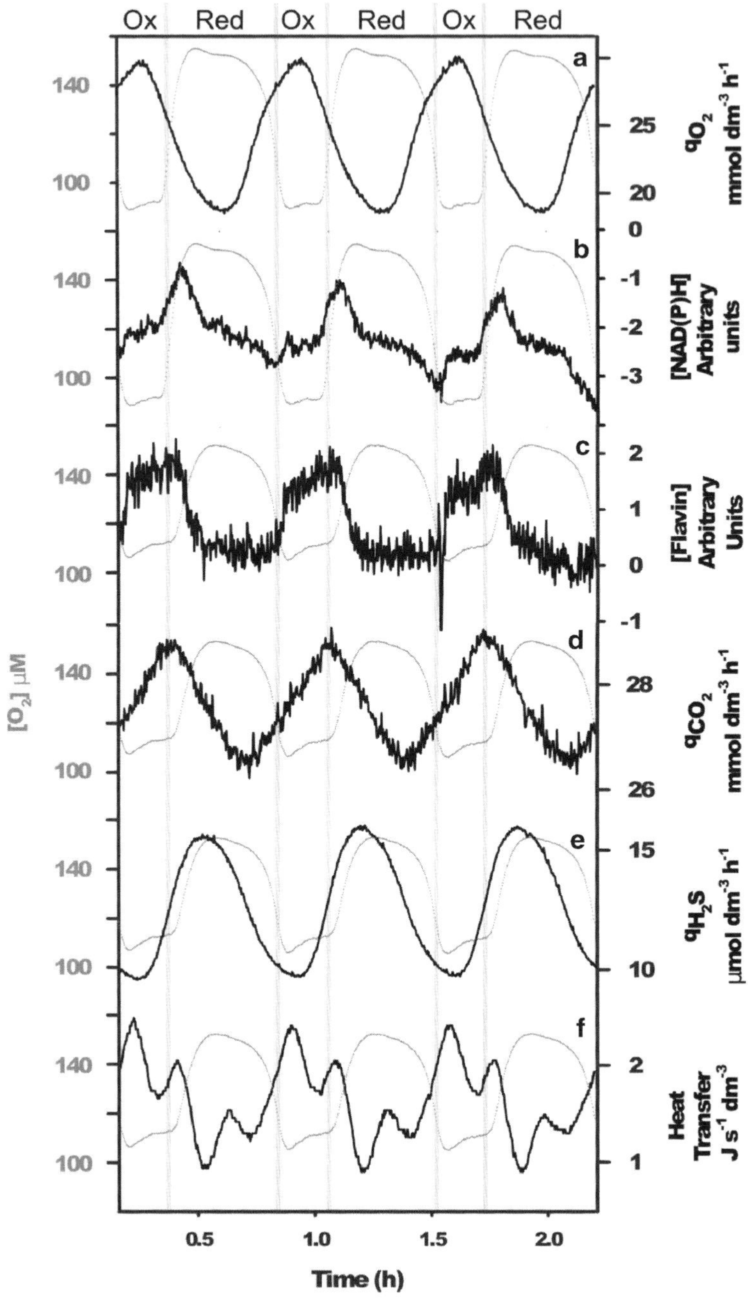

Fig. 12.1 The respiratory oscillation in yeast during continuous growth in fermentors where dissolved oxygen (measured by a dissolved oxygen electrode) is shown in all graphs as a *grey dotted line*. Oxygen uptake rate (qO_2; **a**) was calculated from the partial pressure of the O_2 in the fermentor off-gas using a zirconium oxygen sensor. NAD(P)H (**b**) and oxidised flavin (**c**) were

Klevecz 2006). A multidimensional clustering approach identified seven temporal clusters that formed a coherent growth programme in these data. When compared to a large compendium of independent datasets, it has been shown that the formation of these clusters is both transcription factor and chromatin state dependent. This leads to the conclusion that a simple feedback between energetics and chromatin state may be one of the primary regulatory loops involved in defining the transcriptional landscape (Machné and Murray 2012). Furthermore, the DNA synthesis is intimately timed during each respiratory cycle (Klevecz et al. 2004), although the exact link between the cell division cycle and the respiratory oscillation is unclear, as DNA synthesis is not limited to one phase of the oscillation. Moreover, respiratory oscillations occurred under conditions where changes in DNA synthesis were not detectable (Slavov et al. 2011).

Periods can range from 35 min to several hours depending on environmental conditions (Finn and Wilson 1954; von Meyenburg 1969; Sohn and Kuriyama 2001a; Murray and Lloyd 2006; Slavov and Botstein 2011; Machné and Murray 2012). However, the common modes of oscillation appear to be between 40 min and 5 h. The period is temperature compensated (Murray et al. 2001) and thus its timekeeping characteristics (Edwards and Lloyd 1978, 1980; Lloyd et al. 1982a; Marques et al. 1987) place it in the ultradian time domain (as defined by its characteristic of cycling many times during a day). That similar clock control can be demonstrated when ethanol medium is used further distinguishes these oscillations from glycolytic and cell-cycle associated oscillations, both of which are characterised as having highly temperature-dependent periods (Lloyd 2006a). Moreover, several other properties of the respiratory oscillation are shared with the circadian clock, e.g. period sensitivity to Li^+ and type-A monoamine oxidase inhibitors (Salgado et al. 2002).

12.3 Phase Definitions Guided by Real-Time Monitoring of Redox State

Here, we define the phase of the oscillation by the dissolved O_2 trace (Fig. 12.1), as this is one of the most stable measurements and responds rapidly to changes in culture concentrations (Murray et al. 1998; Murray 2006). We use the minimum first derivative of this trace to define the reference start point; the difference

Fig. 12.1 (continued) measured using an online flurorimeter. Carbon dioxide excretion rate (qCO_2; **d**) was calculated from the partial pressure of CO_2 in the fermentor off-gas using an IR sensor. Hydrogen sulphide excretion rate (qH_2S; **e**) was calculated from the partial pressure of H_2S in the fermentor off-gas using a silver nitrate sensor. Heat transfer (**f**) was calculated by Fourier's equation from the reactor temperature and the controlling bath temperature. The *grey line* represents the minimum and maximum first derivatives, i.e. the demarcation between the oxidative (*Ox*) and reductive (*Red*) phases (Sasidharan et al. 2012)

between this and the maximum first derivative is defined as the oxidative phase (where oxygen uptake and electron flux though the electron transport chain is at a maximum; Fig. 7.1a). The time between the maximum first derivative and the next minimum first derivative is defined as the reductive phase (conversely oxygen uptake rate is lowest).

Nicotinamide nucleotide (NAD(P)H) redox state (Fig. 12.1b), the most useful single indicator of intracellular redox state (Chance et al. 2005), was maximally reduced just before dissolved O_2 reached maximal values as the culture entered the reductive phase (Murray et al. 1999). Flavin fluorescence emission (oxidised FAD and FMN; Fig. 12.1c) indicates predominant coenzyme oxidation during the oxidative phase (Sasidharan et al. 2012). Waveforms of these cofactors are more complex than those for respiration, but show major phase relationships with dissolved O_2 on a 40-min cycle. Maximum CO_2 production rates occurred at the end of the oxidative phase (Fig. 12.1d), whereas highest H_2S production rates were seen as dissolved O_2 was attaining its maximum (Fig. 12.1e). Major peaks and troughs of heat transfer from the culture corresponded closely with maxima and minima in respiration (Fig. 12.1f) may indicate uncoupling of the electron transport chain from energy production and accompanying thermogenesis (Lloyd 2003; Jarmuszkiewicz et al. 2009; Murray et al. 2011). The two shoulders of heat production also separate the three active periods of transcriptional activity previously described (Klevecz et al. 2004; Li and Klevecz 2006) and thus may also correlate with chromosomal rearrangements during the remodelling of the transcriptional landscape (Machné and Murray 2012).

12.4 Carbon Metabolism

Glucose (Satroutdinov et al. 1992; Keulers et al. 1996a), ethanol (Keulers et al. 1996b; Murray et al. 1999) or acetaldehyde (Keulers and Kuriyama 1998) have been used as major carbon sources. Glucose or ethanol gave respiratory periods of about 40 min, and with 290 mM acetaldehyde as primary carbon source it was approximately 80 min. Acetaldehyde, acetate, ethanol, dissolved O_2 and CO_2 production all showed high signal-to-noise ratio oscillation, with acetaldehyde and acetate approximately in phase with O_2 uptake rate (i.e. these products were elevated during the stages of enhanced respiration). Potentially, inhibitory levels of acetaldehyde and acetate were never exceeded; ethanol production from acetaldehyde occurred periodically during the oscillation and was accompanied by decreased acetaldehyde conversion to acetate and acetyl-CoA. Diminished rates of acetaldehyde flux to the TCA cycle may indicate that the decreased respiratory activity was due to some inhibitory effect on the mitochondrial respiratory chain. The importance of acetaldehyde (rather than O_2 or ethanol) as a specific culture synchronising agent was suggested (Keulers and Kuriyama 1998), as high aeration rates leading to loss of CO_2 and the volatile aldehyde led to de-synchronisation. Phase shift experiments (Murray et al. 2003) demonstrated that acetaldehyde was indeed a synchronisation agent. However, further work is required to unravel how carbon metabolism is regulated during the oscillation; specifically, how intracellular carbon flux distributions develop dynamically through each cycle.

12.5 Sulphur Metabolism

H_2S, a potent respiratory inhibitor, reached a maximum (1.5 μM) just prior to minimal respiration rates (Sohn et al. 2000) and then decreased to 0.2 μM before the restoration of the high respiration state. Perturbants such as 50 μM glutathione, 50 μM $NaNO_2$ or 4.5 mM acetaldehyde transiently increased H_2S levels to more than 6 μM. Phase shifting of the oscillation by additions of 0.77 μM $(NH_3)_2S$ enabled a phase response curve to be obtained (Murray et al. 2003); thus the easily oxidised, rapidly diffusing gas, H_2S, acts as an intercellular messenger that amplifies the respiratory oscillation. The binuclear Cu-haem reaction centre of cytochrome c oxidase, the terminal electron transport component of the mitochondrial respiratory chain, is likely to be its target (Lloyd 2006b). However, phase-response curves indicate that it does not act by itself as a synchronising agent.

H_2S is evolved as an intermediary metabolite of the sulphur uptake pathway (Sohn and Kuriyama 2001a). Pulse injection of 100 μM cysteine or methionine altered the timing of H_2S production and perturbed the respiratory oscillation (Sohn and Kuriyama 2001b). Modelling of the feedback inhibition of sulphate uptake by cysteine suggests a major contribution to the respiratory oscillations (Wolf et al. 2001; Henson 2004). Transcript concentration of the high affinity sulphate permease (*SUL2*) and all the transcripts of the sulphate assimilation pathway showed especially high amplitude oscillation, and their peak preceded H_2S generation by 8–10 min (Murray et al. 2007). Hydrogen sulphide production from cysteine, catalysed by mitochondrial cystathione-γ-lyase in the mitochondria of mammalian cells, was recently demonstrated to be involved in the regulation of energy metabolism (Fu et al. 2012). There is a yeast homolog of the protein responsible (*CYS3*); however, it is unclear if Cys3p localises to the mitochondria and if this mechanism of H_2S production occurs in yeast.

The network of sulphate uptake and sulphur amino acid production is shown in Fig. 12.2. This network is intricately interwoven with the regulation of redox state. A direct product of this network is glutathione, and glutathione reductase is responsible for NADPH-dependent cycling of the GSH-GSSG system that buffers the redox state of the cell (Sohn et al. 2005a). Oscillatory dynamics of *GSH1* and *GLR1* transcript abundance, activity of glutathione reductase and the pool sizes of cysteine and glutathione indicated that redox buffering plays a critical role during the oscillation. Moreover, glutathione addition gave a phase-related perturbation of the respiratory oscillation (Murray et al. 1999; Sohn et al. 2000). The effects of pulse injection of thiol redox modifying agents (diethylmaleate, *N*-ethylmaleimide), of inhibitors of glutathione reductase (*DL*-butathionine [*S*,*R*]-sulphoxamine) or of glutathione synthesis (5-nitro-2-furaldehyde) further

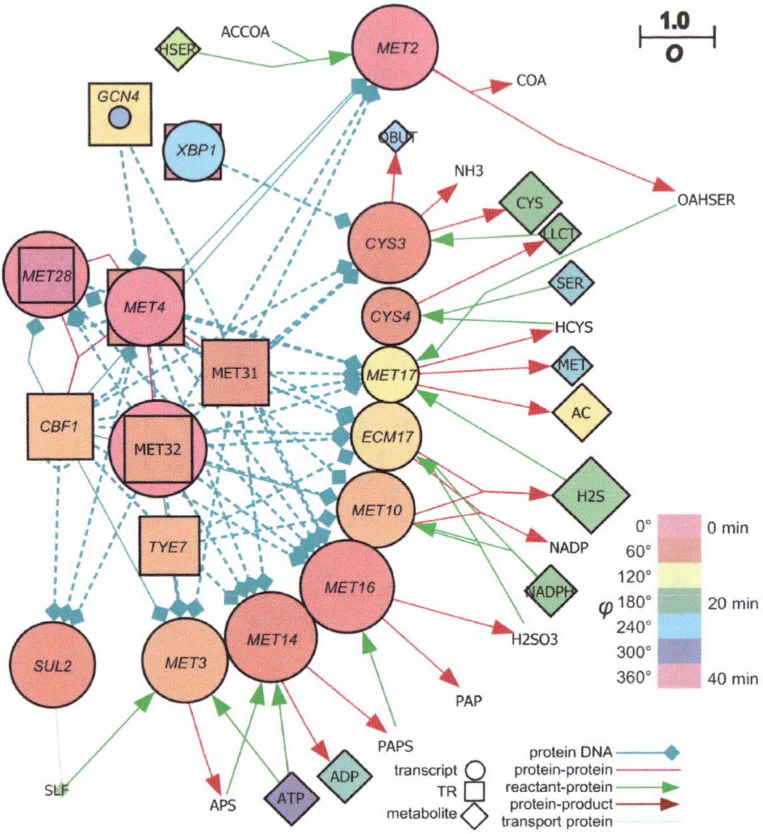

Fig. 12.2 Network derived for sulphur assimilation from the top oscillating ($O \geq 0.750$) transcripts and metabolites. The key provides a guide to the network. *Coloured circles* represent transcript abundance, *diamonds* metabolites and *squares* represent the transcription factor activity. If the transcription factor activity of a node was greater than its transcript concentration, the *square* was placed behind the *circle*, otherwise the *square* was placed in front of the *circle*. The nodes are *coloured* according to the phase angle (ϕ), and the oscillation strength (O) is indicated by the size of the node. *SLF* sulphate, *LLCT* cystathione, *HSER* homoserine, *OASER* *o*-acetylhomoserine, *OBUT* 2-oxobutanoate, *AC* acetate (Murray et al. 2007)

defined the tight coupling between redox state and the regulation of oscillatory dynamics. Cellular per-oxidative adducts, as measured by the levels of lipid peroxidation products, oscillates out of phase with levels of dissolved O_2 (Kwak et al. 2003). Pulse addition at minima of dissolved O_2 of 100 μM *N*-acetylcysteine (which scavenges H_2O_2 and hydroxyl radicals) perturbed the respiratory oscillation and attenuated H_2S production to 63 % of its normal amplitude in the next 40-min cycle. Then the respiratory oscillation damped out, only returning after 20 h. The non-toxic free radical scavenger, ascorbic acid as well as the inhibitor of catalase (3-aminotriazole) or superoxide dismutase (N,N-diethyldithiocarbamate) suggest that endogenously produced reactive oxygen species play a role in intracellular signalling during the oscillation. H_2O_2 (500 μM) added at a minimum of dissolved O_2 both perturbed the respiratory oscillation and elevated H_2S production in the subsequent cycle. Menadione causes the generation of superoxide and perturbed the oscillation in a similar manner as H_2O_2 when added at 500 μM.

Further work on the effects of glutathione perturbation on branched-chain and sulphur-containing amino acids tends to suggest that the observed effects are more closely related to amino acid metabolism and H_2S generation than with cellular redox state per se (Sohn et al. 2005b). This interpretation was supported by more in-depth analyses of the transcriptome and metabolome (Murray et al. 2007).

12.6 The Spatio-Temporal Self-Organisation of the Reactome

With the advent of high-throughput analyses, e.g. microarray-based multiple assays and mass spectrometric metabolomics, the yeast cellular network is the most fully characterised among all eukaryotes. During its life cycle and in its response to environmental challenges (e.g. on exposure to heat), between 20 and 50 % of the yeast protein coding transcriptome alters (Gasch and Werner-Washburne 2002). This has been confirmed by genome-wide and metabolome-wide elucidation of properties of the respiratory oscillations in yeast (Klevecz and Murray 2001; Klevecz et al. 2004; Li and Klevecz 2006; Murray et al. 2007; Slavov et al. 2011). The high-throughput methods required new computational approaches to curate and better understand the implications of vast array of new data. High-throughput chromatin immuno-precipitates hybridised to DNA microarray containing the probes for the upstream regulatory sequences or DNA tiling microarrays (ChIP-chip) revealed that the underlying system structures involved in global transcription can be profoundly altered in response to environmental stimuli (Harbison et al. 2004; MacIsaac et al. 2006). The topology of the protein–protein interaction network has been approached by series of immuno-precipitation, 2-hybrid and mass spectrometry analyses (Ito et al. 2001; Gavin et al. 2006; Krogan et al. 2006). More recently, the focus has turned to global modes of regulation initiated by distinct protein complexes that dynamically modify chromatin structure (Basehoar et al. 2004; Whitehouse et al. 2007; Tsankov et al. 2010). In conjunction with this, a concerted effort by the bioinformatics community has resulted in a series of advances that have revolutionised the manipulation and correlation of data obtained (e.g. KEGG, SGD or SwissPROT).

Although these advances have led to a deeper understanding of the structure of the cell network, they have done little to advance our understanding of cellular dynamics. Moreover, little has been done to combine the derived networks together, and even less has been done to model large-scale datasets to produce a coherent view of the formation of cellular phenotypes. With this in mind, we recently combined computational and statistical network approaches, with high quality transcriptional and metabolomic data (Fig. 12.3a), to analyse the global landscape

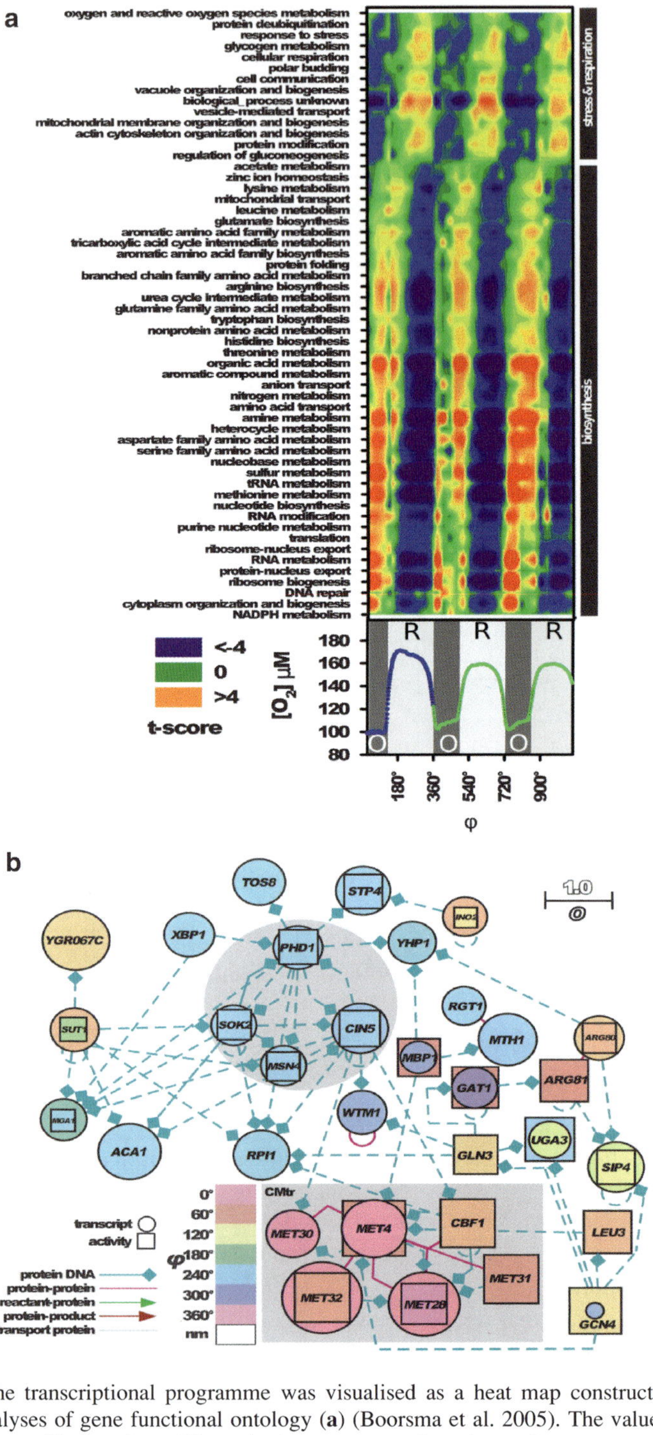

Fig. 12.3 The transcriptional programme was visualised as a heat map constructed from the statistical analyses of gene functional ontology (**a**) (Boorsma et al. 2005). The values were then plotted against oscillation phase. Biosynthetic processes and respiratory/stress events were clearly separated during the oscillation (*black boxes*), occurring (120–180°) out of phase with their

of the yeast oscillatory phenotype (Murray et al. 2007). This revealed that the entire biochemical network or reactome self-organises into two distinctive regions: oxidative and reductive. Transcription during the oxidative phase was almost exclusively focused on biosynthesis, and a clear temporal programme was initiated starting with nucleotide biosynthesis and ending with acetate metabolism. This programme spanned 10–12 min, and metabolical, physiological and morphological processes (e.g. sulphate uptake (Fig. 12.2), amino acid biosynthesis, S-phase, etc.) occurred 10–14 min after the peak in their transcriptional activity. Statistical analysis of transcription factor binding targets and the reconstruction of a yeast protein–protein interaction map (from multiple sources) implicated the temporal construction and destruction of a transcription factor complex (Fig. 12.3b), comprising Cbf1p (centromere binding factor), Met4p, Met28p, Met31p and Met32p (methionine regulation factors) and Gcn4p (general control protein involved in nitrogen catabolite repression); these processes orchestrated the majority of transcriptional changes of sulphate uptake and amino acid synthesis, and by targeting other transcription factors. The gene targets of the complex formed by these transcription factors produced the largest amplitude oscillation of the measured transcripts, and their metabolic products GSH, H_2S and *S*-adenosylmethionine also oscillated. Network analysis of the oscillatory transcription factor network (comprising some 33 transcriptional regulators) indicated that this works by temporally spacing gene transcription via the formation of multiple input feed-forward genetic circuits.

Out of phase with oxidative phase transcription, a much larger group of transcripts (~80 % of the most oscillatory transcripts) showed a peak production in the reductive phase. Many transcripts encode for proteins that encode mitochondrial assembly, respiration, carbohydrate catabolism and the stress response proteins. Moreover, many of these processes are shown to occur 14–18 min later, indicating that transcription actually precedes subsequent activity of the gene products. It was clear from the annotation and statistical analysis of network structure that this group of transcriptional regulatory proteins controlling the expression of these processes are the most highly interconnected nodes in the yeast network. These have complex regulatory patterns and are key components of the differentiation responses in yeast (for pseudohyphae and spore formation). However, the regulation of this system, whose targets show a strong oscillatory pattern and peak in the reductive phase, remained largely enigmatic.

Recently, new computational analyses aimed at elucidating a more global regulatory system that modulate these transcripts have implicated energetic state (ATP:ADP ratio) as a key factor (Machné and Murray 2012). When transcript data

Fig. 12.3 (continued) respective phenotype of biosynthetic. The heat maps were ordered according to the phase angle (ϕ) of the measurements' peak production. Activities of the transcriptional regulators were derived from the statistical analysis of their target transcripts and a *ball-and-stick network* representation of the transcriptional regulators (**b**). See Fig. 12.2 for a guide to the network. The *shaded box* indicates a transcription factor complex comprising of Cbf1p and the Met transcriptional regulators (Murray et al. 2007)

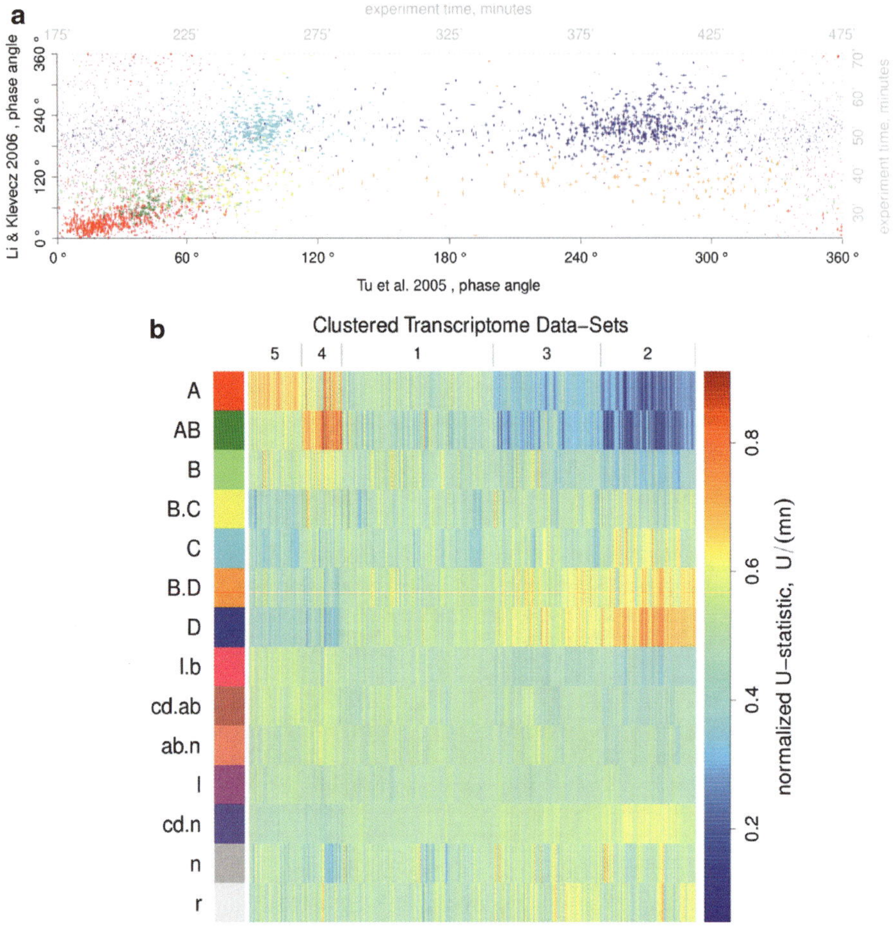

Fig. 12.4 *Clustered phase–phase plot* transcript time course profiles for the 0.7-h (Li and Klevecz 2006) and 5-h (Tu et al. 2005) period datasets (**a**). The *black axes* represent the phase angles of the samples, and the *grey axes* map these phase angles to the real time. The cluster names are given at the side of the lower panel (**b**). The transcripts in these clusters were then used to cross-cluster a collection of 1,327 transcriptome microarray hybridisations (McCord et al. 2007). The hue represents the normalised rank sum where *yellow to red* indicates an up-regulation, and *cyan to blue* represents a down-regulation of the any target hybridisation compared to the cluster. The *column numbers* indicate clusters of hybridisations derived using the SOTA algorithm and plotted in decreasing order (from *left* to *right*) using cluster A as a reference (Machné and Murray 2012)

from two oscillation periods (Tu et al. 2005; Li and Klevecz 2006) were compared by a Fourier transform model-based clustering approach (Machné and Murray 2012), we could define a temporal programme of five common co-expression clusters (Fig. 12.4a); there were also two clusters that showed differential expression between the two periods and two clusters that oscillated but with a lower signal-noise ratio than the main consensus clusters. For brevity we will concentrate on the five consensus clusters. The programme consisted of cell structural and metabolic changes and as it is cyclical we arbitrarily define the start point as the minimum first derivative of residual dissolved oxygen concentration (Fig. 12.1a). Three distinct clusters were observed in the oxidative phase, and two were observed in the reductive phase. Functionally, the first common oxidative phase cluster (A) consisted mainly of structural function, e.g. nucleolar synthesis, cytoplasmic ribosome biogenesis and RNA polymerase I and III. The next common cluster (AB) was enriched in cytoplasmic ribosomal proteins and translation. The final oxidative phase cluster comprised mostly of anabolic reactions (B), such as amino acid biosynthesis. The first reductive phase cluster (C) was enriched mitochondrial ribosomal transcripts, i.e. coding for a structural component. The final common reductive phase cluster (D) was primarily a metabolic cluster and comprised of many catabolic pathways and the genes involved in redox regulation and the general stress response. Therefore, both the oxidative and reductive phases start with a structural bias and finish with a metabolic bias. The period differences observed during the two oscillations (40 min and 300 min) are thought to be due to differences in growth conditions and strain, and they led an interesting context-dependent development of temporal gene expression, where the reductive phase clusters C and D were co-expressed in the 40-min oscillation and the oxidative phase clusters A, AB and B were co-expressed during the oxidative phase.

Strikingly, when these clusters were compared to a compendium of 1,327 microarray hybridisations (McCord et al. 2007), the oxidative and reductive phase clusters were differentially expressed in the majority of the array hybridisations (Fig. 12.4b). This analysis indicates that most of the experiments carried out in yeast effect a global hitherto unknown regulation that is epitomised by the respiratory oscillation. Further analysis supported a role for the general growth rate response where genes up-regulated during rapid growth corresponded with oxidative phase genes (A, AB and B) and those down-regulated corresponded with reductive phase genes (C and D) (Brauer et al. 2008; Slavov and Botstein 2011; Machné and Murray 2012). Moreover, the environmental stress response (Gasch and Werner-Washburne 2002) tended to correlate negatively with the growth rate response. These analyses support a much more basic mechanism for gene regulation that switches the system from the "growth" to "stress" responses. Indeed, it also makes it difficult to define what the "growth" or "stress" responses are as both these conditions alternate in the stable, unperturbed environment that is afforded by continuous culture. Implicit in this regulation is an oscillation in ATP: ADP ratio (Akiyama and Tsurugi 2003; Lloyd and Murray 2007; Machné and Murray 2012) as this is a major determinant of growth rate and is intricately interwoven with the redox state of the cell and the membrane status of the mitochondria. The appropriate chemical and membrane potentials must be maintained to generate ATP (Chance and Williams 1955; Murray et al. 2011). Maximum ATP availability occurs at the beginning of the oxidative phase and minimum ATP availability occurs at the beginning of the reductive phase. Therefore, we argue that the "growth" and "stress" responses are better described as a global switch between catabolic and anabolic processes. The oscillation in respiration serves as a mechanism to temporally partition these metabolic networks. The importance of mitochondrial respiratory control in the generation of temperature-compensated ultradian clock-driven (epigenetic) oscillatory metabolism in a range of lower eukaryotic organisms (three yeasts and five protists species) has been experimentally demonstrated (Edwards and Lloyd 1978; Lloyd and Rossi 1992; Murray et al. 2001).

Generally, there is a lack of annotation for many of the reductive phase transcripts, which in part, may be due to the focus of research activity by the yeast community on glucose repressed growth, where differentiation programmes and respiratory catabolism are repressed. Therefore, we analysed the promoter architectures, i.e. the regional sequence biases, in vitro and in vivo nucleosome positioning, to reveal significant differences between each cluster (Machné and Murray 2012). For example the promoters of the genes in cluster A (ribosomal) had large nucleosome free regions that were enriched with AT containing motifs (well known to have a lowered affinity for nucleosomes). Local nucleosome positions can be classified by their "fuzziness", i.e. as clearly defined positions or poorly defined. Nucleosome position is dictated by the DNA sequence and enzymatically, e.g. as the ATP-dependent remodelling complexes RSC and Isw2 complexes (Zhang et al. 2011). It appears that these act differentially, where the RSC activates (Lorch et al. 1998) and Isw2 represses transcription of genes (Vincent et al. 2008) based on their promoter configuration. In our analyses, we show the structural clusters A and C have very well-defined nucleosomes while clusters B and D have fuzzy nucleosome positions, presumably related to the development of structure to function between A to B and C to D. A caveat in in vivo nucleosome positioning data is that the analyses were almost exclusively carried out in rapidly growing exponential phase cultures, therefore the measured positioning may well be context dependent. This model is supported by the strict ATP dependence of in vitro promoter configurations (Zhang et al. 2011). Therefore, we proposed a simple dual negative feedback loop involved in the regulation of catabolic and anabolic processes, which would have a high potential to autonomously oscillate (Fig. 12.5). When ATP availability is high anabolism is initiated by ATP-dependent remodelling by the RSC on the anabolic and growth genes in cluster A, AB and B. The protein products of these genes then produce cellular growth leading to the down-regulation of these genes as ATP is consumed. In conjunction with this, increased ATP availability early in the oxidative phase leads to Isw2 remodelling of nucleosomes over the promoters of the catabolic genes. This leads to a repression of the catabolic genes during the oxidative phase, leading to a decrease in expression. Once ATP concentration declines late in the oxidative phase, the nucleosomes relax and catabolic genes are transcribed. It is apparent that anything which stresses the cell and causes a drop in ATP production (such as ROS damage of membranes) or mutations that cause the balance between catabolism and anabolism to be altered, will cause a rearrangement of chromatin structure, leading to stress and catabolic transcript production (without the need for specific transcription factors). Indeed, this model is supported in higher eukaryotes by previously reported in vivo and in vitro ultradian oscillations in nucleosome remodelling in glucocorticoid and

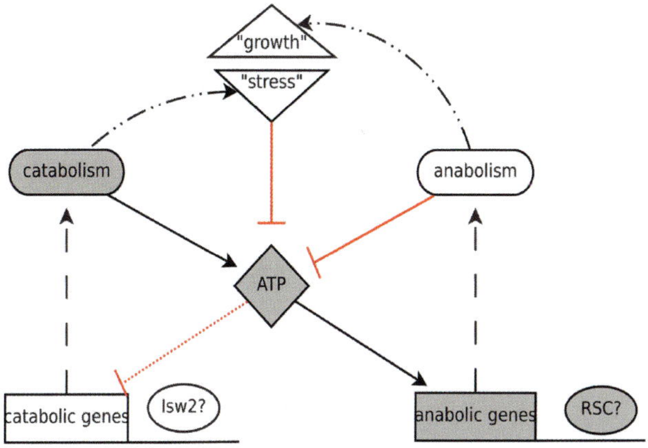

Fig. 12.5 The proposed dual negative feedback of energetics on gene expression via chromatin remodelling. During the oxidative phase (*grey*) catabolism increases ATP synthesis rate leading to a high ATP:ADP ratio thereby activating the transcription of anabolic genes. This leads to a "growth-like" transcriptional response. This is potentially mediated by ATP-dependent nucleosome remodelling (*RSC*), which simultaneously represses the transcription of catabolic genes. When respiratory activity switches (resulting in the onset of the reductive phase), the protein products of the anabolic genes, e.g. amino acid and protein synthesis enzymes, lower the ATP: ADP ratio. This results in a reconfiguration of the promoter structure where the production of anabolic transcripts decreases and the catabolic transcripts increase, i.e. typically that of a "stress-like" response. Supporting this concept, the initial response of many cellular stresses is a decrease in the ATP:ADP ratio caused by disruption of protein, membrane and organelle function

oestrogen systems, where pulses of stimulant or ATP caused a damped ultradian oscillation in nucleosome structure (Métivier et al. 2003; Nagaich et al. 2004).

12.7 Organelle Remodelling: Mitochondrial Changes During the Respiratory Cycle

Our data indicates that the reaction network is separated into catabolic and anabolic processes. However, it is currently unclear how biophysical structure and physical compartmentalisation integrate into this "programme". The spatio-temporal heterogeneity of cellular structure greatly influences every aspect of the reaction kinetics of the whole cellular network (Hiroi and Funahashi 2006). The reactome therefore has an intricate relationship with the organisation of DNA, RNA, protein, membrane and organelle structure both in space and time. The metabolites and transcripts involved in the core amphibolic carbon metabolism (glycolysis, gluconeogenesis and the tricarboxylic cycle) include some of the major oscillatory components of the cell. In a respiring system the mitochondria and the integrity of its structure are main determinants of the relative rates of the amphibolic pathways of carbon metabolism. Additionally, the dual negative feedback loop on transcription (Fig. 12.5) indicates mitochondrial activity must be tightly integrated with the oscillatory dynamics.

The high amplitude excursions of dissolved O_2 (80–170 μM) during the respiratory oscillations are accompanied by an in-phase modulation of NAD(P)H directly as continuously monitored in the fermenter by fluorimetry (Murray et al. 1999; Lloyd et al. 2002b). This key indicator of intracellular redox state (Chance et al. 2005) is pivotal at the core of the entire cellular network (Lloyd 2003; Lloyd et al. 2003; Lloyd and Murray 2005, 2006; Murray and Lloyd 2006). Electron micrographs of thin sections of yeasts rapidly fixed at different stages of the 40-min cycle showed marked ultra-structural changes in the mitochondria (Lloyd et al. 2002a, b; Sasidharan et al. 2012). The extremes of conformational state correspond to those originally described for liver mitochondria (Hackenbrock 1966, 1968) as "orthodox" or "condensed". In the former, a relatively large matrix volume with the inner membrane is closely opposed to the outer membrane: this state was identified at high levels of dissolved O_2 (when respiration rate was low). In the condensed form, the cristae became more clearly defined as the intermembrane compartment was larger. This corresponds to the energised state (Chance and Williams 1955). It has been understood for many years that massive and rapid changes in ion concentrations between the two mitochondrial compartments and the cytosol accompany or drive these changes and protonophores and ionophores that uncouple mitochondrial energy conservation from electron transport by collapse of inner membrane electrochemical membrane potential perturb mitochondrial structure (Hackenbrock 1968; Mitchell and Moyle 1969). The determination of total mitochondrial content of cytochromes b, c_1, c and aa_3 as measured in difference spectra at 77K showed that any changes were below the levels of detectability. However, the physiological redox states of cytochromes c and aa_3 in vivo indicated that high respiration was associated with elevated reduction of these two redox components. Effects of two protonophores (m-chlorocarbonylcyanide phenylhydrazone, CCCP, and 5-chloro-t-butyl-2′-chloro-4′-nitrosalicylanilide, S-13) (Fig. 12.6a, b) were dramatic and similar. The well-established acceleration of respiratory rates on addition of uncouplers rapidly drives down the dissolved O_2 in the fermenter. At increased concentrations, the subsequent respiratory cycle was prolonged, and more than five cycles were required for recovery to the normal cycle time. At higher concentrations both uncouplers produced very interesting effects that provided new insights (Lloyd 2003). For instance, at 10 μM CCCP the uncoupling effect was more evident and the dissolved O_2 remains low for more than 5 h, during which no oscillation was observed. Recovery to normal amplitudes required more than 20 h, although the respiratory oscillations were restored before this, albeit with greatly diminished amplitude. Another very significant observation was that uncoupler treatment gives a lasting complex waveform where the 40-min oscillation had an 8-h envelope waveform, most likely stemming from cell-division cycle dynamics (Fig. 12.6c, d). Thus, interference with mitochondrial energy generation can induce an alignment of cell division cycle controls with ultradian clock control (Fig. 12.7). Further

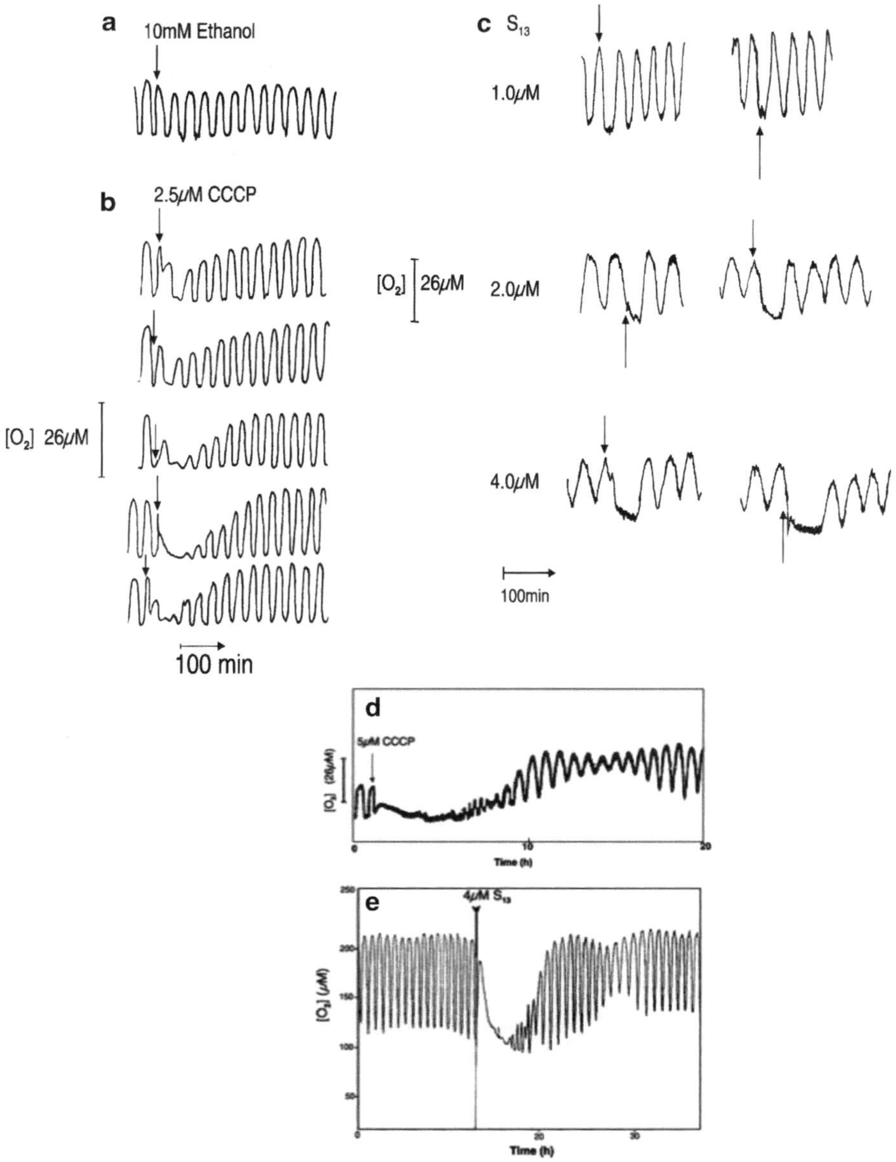

Fig. 12.6 The pulse addition of (**a**) ethanol (10 mM), and the protonophores (**b**) CCCP (2.5 μM) or S_{13} (1, 2 and 4 μM) during the respiratory oscillation causes phase and amplitude shifts (Lloyd 2003)

addition of uncoupler extends cell division time and causes desynchronisation of the population or causes the clock to stop. Further research is required to delineate clock interference from loss of cell–cell synchrony.

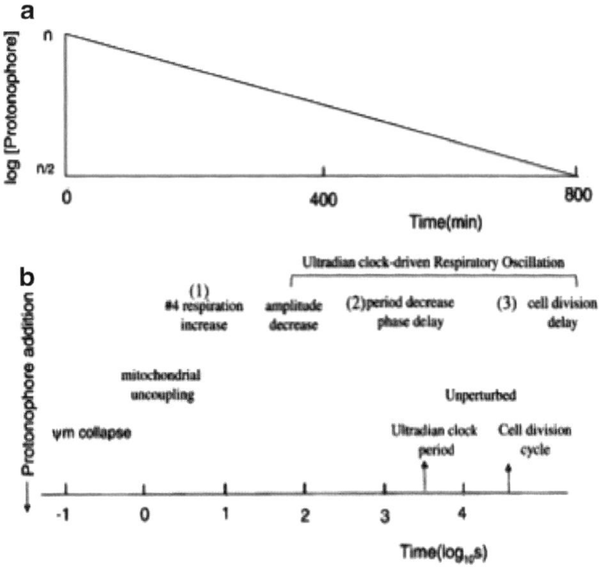

Fig. 12.7 (**a**) Protonophore concentration decreases during continuous dilution of *Saccharomyces cerevisiae* culture. (**b**) Escalating consequences over multiple scales (0.1–10,000 s)

In terms of the kinetics of mitochondrial processes, it is quite remarkable that these organelles are characteristically associated with an extremely dynamic potentiality (as determined by their rapid metabolic responses in vitro after extraction from organisms). However, in vivo they behave quite differently and their changing properties and functions during an oscillation cycle are evidently constrained (Lloyd and Edwards 1984; Marques et al. 1987). We have suggested that inside cells or organisms mitochondria not only display rapid kinetics but also have slower oscillatory modes that suggest that "they dance to a tune that is played by a piper performing elsewhere in the cell". In situ, the mitochondrion is enslaved to the slower beat of the nucleo-cytosolic system.

12.8 The Multi-Oscillatory State

The use of membrane inlet mass spectrometry to monitor dissolved gas levels (O_2, CO_2 and H_2S; Fig. 12.8) directly in the continuous culture at 12 s interval sampling time showed a predominant large-amplitude period of approximately 13.6 h, with superimposed 40 min, and an intermittent 4-min oscillation (Roussel and Lloyd 2007). A metabolic attractor was constructed using more than 36,000 points collected at 15 s intervals from a single experiment, lasting for 3 months. Computation of the leading Lyapounov exponent ($0.752 \pm 0.004\ h^{-1}$; 95 % confidence), which is clearly demarcated from 0, indicated the dynamics on this attractor was chaotic. This chaotic behaviour was further supported by the Poincaré plots of data taken from a time frame of weeks during the 40-min cycle, while amplitude

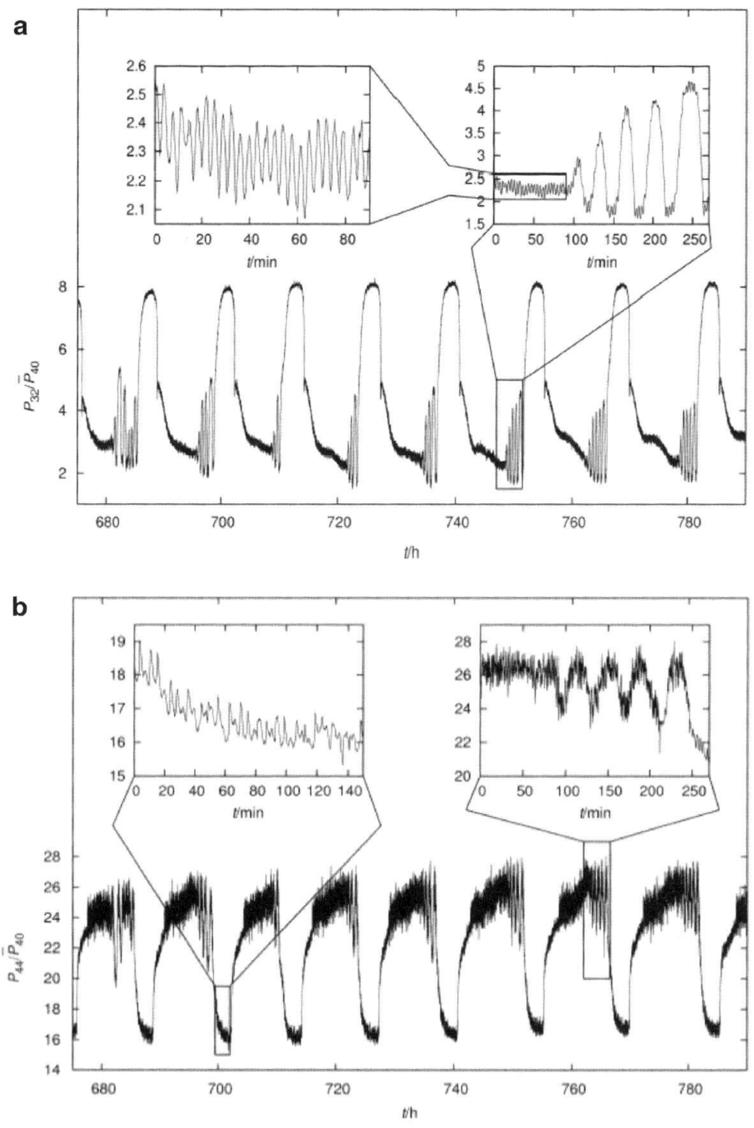

Fig. 12.8 The multi-oscillatory state of *S. cerevisiae* during synchronised continuous growth. Relative MIMS signals of the $m/z = 32$ and 44 components referred to Argon ($m/z = 40$) versus time. These mass components correspond, respectively, to O_2 (**a**) and to CO_2 (**b**). Time is given in hours after the start of fermentor continuous operation (Aon et al. 2008)

formed a linear relationship between successive cycles; period showed non-linear chaotic behaviour (Murray and Lloyd 2006). When the attractors were reconstructed from the dissolved oxygen data measured in cultures, where conditions were changed to near the limits that support rapid growth pH (2.8 or below) or temperature (25 °C and 36 °C), strange attractor-like behaviour with clear bifurcation points was observed.

In order to reveal the "scale-invariant" (fractal or self-similar) features of the multi-oscillator, well-established approaches previously applied to time series were employed (Aon et al. 2008). Thus, Relative Dispersional Analyses (RDA) of both the O_2 and CO_2 data indicated that the observed multi-oscillatory dynamics correspond to statistical fractals, as judged by the perfect correlation between oscillators in the 13-h, 40-min and 4-min time domains. Thus, double-log plots exhibit an inverse power relationship with a fractal dimension D_f (=1.0) implying that RD is constant with scale (i.e. the time series looks statistically self-similar at all scales). Long-term memory on timescales from minutes to hour is implicit in this oscillatory behaviour. Power Spectral Analysis (PSA) also indicated an inverse power law proportional to $1/f^{\beta}$, and the value of $\beta = 1.95$ is close to that characteristic of coloured noise, and again is as expected for the time series of a system that displays deterministic chaos. The fractal nature of a chaotic time series has been explained previously (Lloyd and Lloyd 1995). Moreover, and most significantly, these characteristics indicate a functional scale-free statistically self-similar network that operates simultaneously on several timescales (Aon et al. 2008) so as to provide coherence between time domains. A computational model indicated that both in yeast and cardiomyocytes, underlying mechanisms of scale-free behaviour are similar. This scale-free behaviour is supported by wavelet analyses of NAD(P)H fluorimetry data sampled at 100 Hz (Fig. 12.9) and revealed how the multioscillatory states correlated during the respiratory oscillation (Sasidharan et al. 2012). In analysis we revealed a 4-, 12- and 20-min signals, whose amplitude was modulated according to phase.

The 4-min period had not been observed previously in continuous cultures of yeast, but may be related to oscillatory autofluorescence emission from NAD(P)H in both mitochondrial and cytosolic compartments observed in contiguous single-layered films of cells perfused with buffer-containing glucose (Aon et al. 2007). Two-photon excitation of cells loaded with appropriate fluorescent probes indicated oscillations in inner mitochondrial membrane potential ($\Delta\Psi_m$) and superoxide radical anions ($O_2^{-\bullet}$) with definite phase relationships with the nicotinamide nucleotide redox states. The inner membrane anion channel has a peripheral benzodiazepine receptor, and an inhibitor, 4-Chlorodiazepam, attenuates the oscillation both of the redox state and of the $O_2^{-\bullet}$ as had previously been shown in cardiomyocytes (Aon et al. 2003; Brown et al. 2008). This out-of-phase relationship was lost after inhibitor treatment and washout. Superoxide dismutase addition was ineffective at blocking the oscillations indicating that the signalling function of this $O_2^{-\bullet}$ serves intracellular rather than an intercellular role (Aon et al. 2008). Thereby, yeast mitochondrial populations function as a network of coupled oscillators through metabolite-linked communication (Cortassa et al. 2011). Inhibiting mitochondrial respiration at the level of cytochrome oxidase with H_2S abates all oscillatory frequencies including the 40-min-period ultradian clock, therefore providing proof-of-principle that multi-scale timekeeping is an emergent property of the overall network involved in metabolism, growth and proliferation in yeast, as

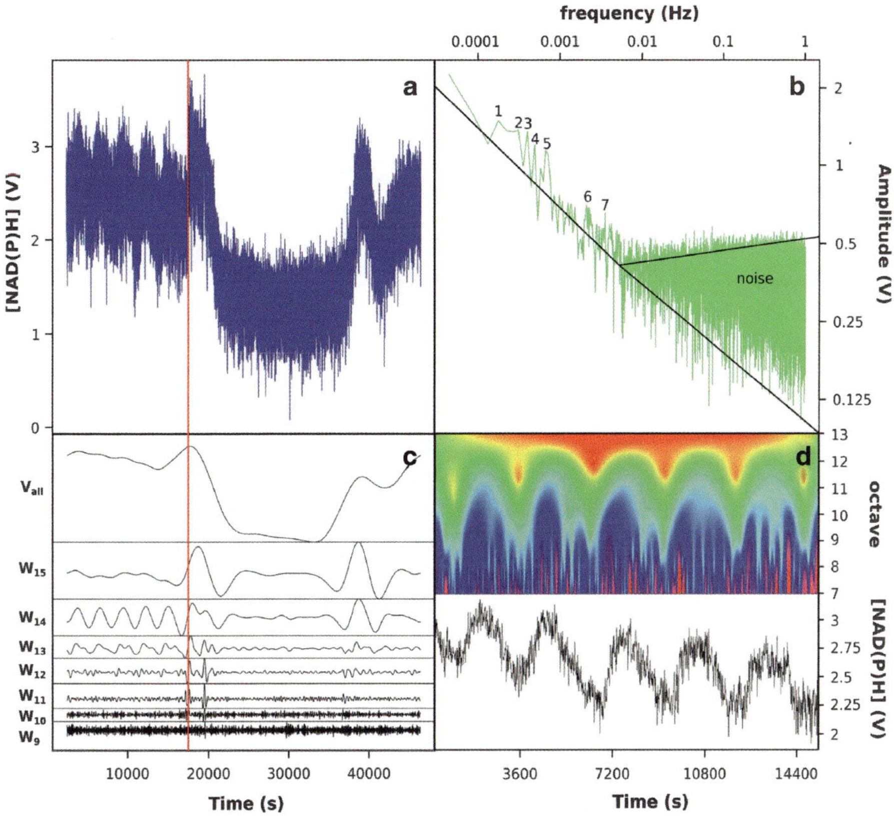

Fig. 12.9 Signal processing of the complex signal produced from continuous online measurement of NAD(P)H initially sampled at 10 Hz (**a**). Discrete Fourier Transformation (*DFT*) spectra reveal that the relation between the amplitude was linear until 0.05 Hz indicating scale-free dynamics in this region (**b**); below this we observed a region of *coloured noise*. Discrete Wavelet Transformation (*DWT*) using the Daubechies wavelet was then used to process the signal where windows (*W*) that had significant correlation were shown (**c**). The data was down-sampled to 1 Hz to reduce computation cost. Continuous wavelet transformation (*CWT*) using the derivative of Gaussian wavelet (*DOG*) of data down-sampled to 0.1 Hz reveals the finer grain temporal events of the signal (**d**). The heat map intensity indicates the correlation of the signal to the wavelet. The *vertical line* in (**a**) and (**c**) represents the time of $(NH_3)_2S$ addition. (Sasidharan et al. 2012)

the overall impact of the perturbation was shown over all temporal scales (Sasidharan et al. 2012). A similar effect is seen with 4′-Cl-diazepam, where addition to whole yeasts of this inhibitor of IMAC, control of $O_2^{-\bullet}$ efflux from mitochondria, attenuates the minute period oscillations of $\Delta\Psi_m$ and NADP(H) reduction states (Aon et al. 2007, 2008) as monitored optically. In a continuous culture this inhibitor decreased the amplitude of oscillations of dissolved O_2 (D.B. Murray, unpublished experiments). Furthermore, the period-lengthening effects of Li^+ and monoamine oxidase (type A) inhibitors on the 40-min yeast clock (Salgado et al. 2002) echo results seen with a variety of circadian oscillators (Engelman et al. 1976;

Delini-Stula et al. 1988). Together with the effects of protonophores (Sect. 12.7) through time domains from milliseconds to cell division times (Fig. 12.7), it becomes evident that signalling between timescales is mediated either directly or by the downstream effects of a number of small effector molecules (H^+, $O_2^{-\bullet}$, and H_2S).

Redox cycles lie close to core of global scale-free cellular dynamics (Rapkine 1931; Mano 1977; Lloyd and Murray 2005, 2006, 2007; Lloyd et al. 2012). The significance of scale-free temporal organisation for organelle, cell and organism timekeeping cannot be overstated as, potentially, what affects one timescale affects them all: a fundamental property of dynamic fractals (Aon et al. 2008).

12.9 Concluding Remarks

> *"What are called structures are slow processes of long duration, functions are quick processes of short duration"* (von Bertalanffy, Problems of Life 1952).

The respiratory oscillation percolates through almost every facet of the biochemistry and biology of cellular processes, and although transcription feed-forward protein feedback loops are important for the regulation and long-term stability of the oscillation, they do not act in isolation. Current dogma dictates that there is a distinctive hierarchy in cellular processes where DNA forms RNA to form proteins that biochemically alter other proteins or produce metabolites; however, biological networks cannot be so conveniently modularised. Therefore, we must conclude that real biological networks are heterarchical (where theoretically each sub-system is of equal importance) rather than hierarchical in their organisation (if a hierarchy does exist co-factors and co-enzymes would surely take the number one spot). This viewpoint does not preclude hierarchical structure in the sub-systems that comprise the heterarchy. Therefore, cell fate is dictated by the integrated output of the reactome rather than by specific structures in the system. Thus a highly interconnected reactome responds in a redox-phase-dependent manner, which can robustly adapt to most perturbative influences.

Our data and observations, although incomplete, provide a compelling picture of the holistic organisation of yeast. In an autodynamic system, where almost every facet of cell biology oscillates without external perturbation the concept of cause and effect is meaningless. Cell physiology is organised so that parallel interlocked events occur throughout all of the constituents of the system, i.e. macromolecules (DNA, RNA, proteins, lipids) and small molecules (metabolites, gases), as the organism adapts to the environmental context. At first glance this would appear to introduce a high degree of complexity to the system. However, our comparative study on the 1,327 independent array hybridisations illustrates (Fig. 12.4b) if a condition is changed the transcriptome (and presumably its output) locks into either anabolic or catabolic modes. Furthermore, ATP availability and utilisation is implicit in the formation of these transcriptional states. ATP synthesis in respiring cells is a function of the redox potential and the ability of the mitochondria to generate a membrane potential. These processes occur at very different frequencies and, thus, the physical structure and the chemical state of the cell must be integrated on multiple timescales. We show that by altering either state we can phase lock the system into either a catabolic or anabolic mode and that these perturbations influence the time structure of the entire cell system (Sasidharan et al. 2012).

In conclusion, rather than complexity, a heterarchical oscillatory system can provide simple rules for the global organisation of the cell, its response to the environment and the development of the dynamic architecture of the phenotype (Klevecz and Murray 2001; Chin et al. 2012).

Acknowledgements DL and DBM are grateful to the Royal Society and the Japan Society for the Promotion of Science for supporting this work. RM and DBM are grateful to the Vienna Science and Technology Fund (WWTF), for funding an exchange grant (MA07-30). KS, DBM and CA are supported in part by funds from Yamagata Prefectural Government and Tsuruoka-city. DBM is also supported by a Japan partnering award (Japan Science and Technology agency and the Biotechnology and Biological Sciences Research Council, UK) and a Japan Society for the Promotion of Science Grant-in-aid.

References

Akiyama S, Tsurugi K (2003) The *GTS1* gene product facilitates the self-organization of the energy metabolism oscillation in the continuous culture of the yeast *Saccharomyces cerevisiae*. FEMS Microbiol Lett 228:105–10

Aon MA, Cortassa S (1997) Dynamic biological organization: fundamentals as applied to cellular systems. Springer, New York

Aon MA, Cortassa S, Marbán E, O'Rourke B (2003) Synchronized whole cell oscillations in mitochondrial metabolism triggered by a local release of reactive oxygen species in cardiac myocytes. J Biol Chem 278:44735–44

Aon MA, Cortassa S, Lemar KM, Hayes AJ, Lloyd D (2007) Single and cell population respiratory oscillations in yeast: a 2-photon scanning laser microscopy study. FEBS Lett 581:8–14

Aon MA, Roussel MR, Cortassa S, O'Rourke B, Murray DB, Beckmann M, Lloyd D (2008) The scale free network organization of yeast and heart systems biology. PLoS One 3:e3624

Basehoar AD, Zanton SJ, Pugh BF (2004) Identification and distinct regulation of yeast TATA box-containing genes. Cell 116:699–709

Boorsma A, Foat BC, Vis D, Klis FM, Bussemaker HJ (2005) T-profiler: scoring the activity of predefined groups of genes using gene expression data. Nucleic Acids Res 33:W592–W595

Brauer MJ, Huttenhower C, Airoldi EM et al (2008) Coordination of growth rate, cell cycle, stress response, and metabolic activity in yeast. Mol Biol Cell 19:352–67

Brown DA, Aon MA, Akar FG, Liu T, Sorarrain N, O'Rourke B (2008) Effects of 4′-chlorodiazepam on cellular excitation-contraction coupling and ischaemia-reperfusion injury in rabbit heart. Cardiovasc Res 79:141–9

Chance B, Williams GR (1955) Respiratory enzymes in oxidative phosphorylation. III. The steady state. J Biol Chem 217:409–27

Chance B, Nioka S, Warren W, Yurtsever G (2005) Mitochondrial NADH as the bellwether of tissue O_2 delivery. Adv Exp Med Biol 566:231–42

Chin SL, Marcus IM, Klevecz RR, Li CM (2012) Dynamics of oscillatory phenotypes in *Saccharomyces cerevisiae* reveal a network of genome-wide transcriptional oscillators. FEBS J 279:119–30

Cortassa S, Aon MA, O'Rourke B, Winslow RL (2011) Metabolic control analysis applied to mitochondrial networks. Conf Proc IEEE Eng Med Biol Soc 2011:4673–6
Delini-Stula A, Radeke E, Waldmeier PC (1988) Basic and clinical aspects of the activity of the new monoamine oxidase inhibitors. Psychopharmacol Ser 5:147–58
Edwards SW, Lloyd D (1978) Oscillations of respiration and adenine nucleotides in synchronous cultures of *Acanthamoeba castellanii*: mitochondrial respiratory control in vivo. Microbiology 108:197–204
Edwards SW, Lloyd D (1980) Oscillations in protein and RNA content during synchronous growth of *Acanthamoeba castellanii*. Evidence for periodic turnover of macromolecules during the cell cycle. FEBS Lett 109:21–6
Engelman W, Bollig I, Hartmann R (1976) [The effects of lithium ions on circadian rhythms (author's transl.)]. Arzneimittelforschung 26(6):1085–6
Finn RK, Wilson RE (1954) Fermentation process control, population dynamics of a continuous propagator for microorganisms. J Agric Food Chem 2:66–69
Fu M, Zhang W, Wu L, Yang G, Li H, Wang R (2012) Hydrogen sulfide (H2S) metabolism in mitochondria and its regulatory role in energy production. Proc Natl Acad Sci USA 109:2943–8
Gasch AP, Werner-Washburne M (2002) The genomics of yeast responses to environmental stress and starvation. Funct Integr Genomics 2:181–92
Gavin A-C, Aloy P, Grandi P et al (2006) Proteome survey reveals modularity of the yeast cell machinery. Nature 440:631–6
Hackenbrock CR (1966) Ultrastructural bases for metabolically linked mechanical activity in mitochondria. I. Reversible ultrastructural changes with change in metabolic steady state in isolated liver mitochondria. J Cell Biol 30:269–97
Hackenbrock CR (1968) Ultrastructural bases for metabolically linked mechanical activity in mitochondria. II. Electron transport-linked ultrastructural transformations in mitochondria. J Cell Biol 37:345–69
Harbison CT, Gordon DB, Lee TI et al (2004) Transcriptional regulatory code of a eukaryotic genome. Nature 431:99–104
Henson MA (2004) Modeling the synchronization of yeast respiratory oscillations. J Theor Biol 231:443–58
Hiroi N, Funahashi A (2006) Kinetics of dimension restricted conditions. Humana, NJ, USA
Ito T, Chiba T, Ozawa R, Yoshida M, Hattori M, Sakaki Y (2001) A comprehensive two-hybrid analysis to explore the yeast protein interactome. Proc Natl Acad Sci U S A 98:4569–74
Jarmuszkiewicz W, Woyda-Ploszczyca A, Antos-Krzeminska N, Sluse FE (2009) Mitochondrial uncoupling proteins in unicellular eukaryotes. Biochim Biophys Acta 1797:792–9
Keulers M, Kuriyama H (1998) Extracellular signalling in an oscillatory yeast culture. In: Holcombe WML, Paton R, Holcombe M (eds) Information processing in cells and tissues. Plenum, New York, pp 85–94
Keulers M, Asaka T, Kuriyama H (1994) A versitile data acquisition system for physiological modelling of laboratory fermentation processes. Biotechnol Tech 8:879–84
Keulers M, Satroutdinov AD, Suzuki T, Kuriyama H (1996a) Synchronization affector of autonomous short-period-sustained oscillation of *Saccharomyces cerevisiae*. Yeast 12:673–82
Keulers M, Suzuki T, Satroutdinov AD, Kuriyama H (1996b) Autonomous metabolic oscillation in continuous culture of *Saccharomyces cerevisiae* grown on ethanol. FEMS Microbiol Lett 142:253–8
Klevecz RR, Murray DB (2001) Genome wide oscillations in expression. Wavelet analysis of time series data from yeast expression arrays uncovers the dynamic architecture of phenotype. Mol Biol Rep 28:73–82
Klevecz RR, Bolen J, Forrest G, Murray DB (2004) A genomewide oscillation in transcription gates DNA replication and cell cycle. Proc Natl Acad Sci U S A 101:1200–05
Krogan NJ, Cagney G, Yu H et al (2006) Global landscape of protein complexes in the yeast *Saccharomyces cerevisiae*. Nature 440:637–43

Kwak WJ, Kwon G-S, Jin I, Kuriyama H, Sohn H-Y (2003) Involvement of oxidative stress in the regulation of H2S production during ultradian metabolic oscillation of *Saccharomyces cerevisiae*. FEMS Microbiol Lett 219:99–104

Li CM, Klevecz RR (2006) A rapid genome-scale response of the transcriptional oscillator to perturbation reveals a period-doubling path to phenotypic change. Proc Natl Acad Sci U S A 103:16254–59

Lloyd D (2003) Effects of uncoupling of mitochondrial energy conservation on the ultradian clock-driven oscillations in *Saccharomyces cerevisiae* continuous culture. Mitochondrion 3:139–46

Lloyd D (2006a) The ultradian clock: not to be confused with the cell cycle. Nat Rev Mol Cell Biol. doi:10.1038/nrm1980-c1

Lloyd D (2006b) Hydrogen sulfide: clandestine microbial messenger? Trends Microbiol 14:456–62

Lloyd D (2008) Respiratory oscillations in yeasts. Adv Exp Med Biol 641:118–40

Lloyd D, Edwards SW (1984) Epigenetic oscillations during the cell cycles of lower eukaryotes are coupled to a clock: life's slow dance to the music of time. In: Edmunds L (ed) Cell cycle clocks. Marcel Dekker, New York, pp 27–46

Lloyd AL, Lloyd D (1995) Chaos: its significance and detection in biology. Biol Rhythm Res 26:233–52

Lloyd D, Murray DB (2005) Ultradian metronome: timekeeper for orchestration of cellular coherence. Trends Biochem Sci 30:373–7

Lloyd D, Murray DB (2006) The temporal architecture of eukaryotic growth. FEBS Lett 580:2830–35

Lloyd D, Murray DB (2007) Redox rhythmicity: clocks at the core of temporal coherence. Bioessays 29:465–73

Lloyd D, Rossi EL (1992) Ultradian rhythms in life processes: an inquiry into fundamental principles. Springer, London

Lloyd D, Rossi EL (2008) Ultradian rhythms from molecules to mind: a new vision of life. Springer, New York

Lloyd D, Edwards SW, Fry JC (1982a) Temperature-compensated oscillations in respiration and cellular protein content in synchronous cultures of *Acanthamoeba castellanii*. Proc Natl Acad Sci U S A 79:3785–8

Lloyd D, Poole RK, Edwards SW (1982b) The cell division cycle: temporal organization control of cellular growth and reproduction. Academic, London

Lloyd D, Salgado EJ, Murray DB, Eshantha L, Salgado J, Turner MP (2002a) Respiratory oscillations in yeast: clock-driven mitochondrial cycles of energization. FEBS Lett 519:41–4

Lloyd D, Salgado LEJ, Turner MP, Suller MTE, Murray DB (2002b) Cycles of mitochondrial energization driven by the ultradian clock in a continuous culture of *Saccharomyces cerevisiae*. Microbiology 148:3715–24

Lloyd D, Lemar KM, Salgado LE, Gould TM, Murray DB (2003) Respiratory oscillations in yeast: mitochondrial reactive oxygen species, apoptosis and time; a hypothesis. FEMS Yeast Res 3:333–9

Lloyd D, Cortassa S, O'Rourke B, Aon MA (2012) What yeast and cardiomyocytes share: ultradian oscillatory redox mechanisms of cellular coherence and survival. Integr Biol (Camb) 4(1):65–74

Lorch Y, Cairns BR, Zhang M, Kornberg RD (1998) Activated RSC-nucleosome complex and persistently altered form of the nucleosome. Cell 94:29–34

Machné R, Murray DB (2012) The yin and yang of yeast transcription: elements of a global feedback system between metabolism and chromatin. PLoS One 7:e37906

MacIsaac KD, Wang T, Gordon DB, Gifford DK, Stormo GD, Fraenkel E (2006) An improved map of conserved regulatory sites for *Saccharomyces cerevisiae*. BMC Bioinformatics 7:113

Mano Y (1977) Interaction between glutathione and the endoplasmic reticulum in cyclic protein synthesis in sea urchin embryos. Dev Biol 61:273–86

Marques N, Edwards SW, Fry JC, Halberg F, Lloyd D (1987) Temperature-compensated ultradian variation in cellular protein content of *Acanthamoeba castellanii* revisited. Prog Clin Biol Res 227A:105–19
McCord RP, Berger MF, Philippakis AA, Bulyk ML (2007) Inferring condition-specific transcription factor function from DNA binding and gene expression data. Mol Syst Biol 3:100
Métivier R, Penot G, Hübner MR et al (2003) Estrogen receptor-α directs ordered, cyclical, and combinatorial recruitment of cofactors on a natural target promoter. Cell 115:751–63
Mitchell P, Moyle J (1969) Estimation of membrane potential and pH difference across the cristae membrane of rat liver mitochondria. Eur J Biochem 7:471–84
Mueller S, Machné R, Murray DB (2012) A new dynamic model for highly efficient mass transfer in aerated bioreactors and consequences for kLa. Biotechnol Bioeng 109(12):2997–3006. doi:10.1002/bit.24594
Murray DB (2004) On the temporal self-organisation of *Saccharomyces cerevisiae*. Curr Genomics 5:665–7
Murray DB (2006) The respiratory oscillation in yeast phase definitions and periodicity. Nat Rev Mol Cell Biol. doi:10.1038/nrm1980-c2
Murray DB, Lloyd D (2006) A tuneable attractor underlies yeast respiratory dynamics. Biosystems 90:287–94
Murray DB, Engelen FA, Keulers M, Kuriyama H, Lloyd D (1998) NO+, but not NO., inhibits respiratory oscillations in ethanol-grown chemostat cultures of *Saccharomyces cerevisiae*. FEBS Lett 26:297–99
Murray DB, Engelen FA, Lloyd D, Kuriyama H (1999) Involvement of glutathione in the regulation of respiratory oscillation during a continuous culture of *Saccharomyces cerevisiae*. Microbiology 145:2739–45
Murray DB, Roller S, Kuriyama H, Lloyd D (2001) Clock control of ultradian respiratory oscillation found during yeast continuous culture. J Bacteriol 183:7253–59
Murray DB, Klevecz RR, Lloyd D (2003) Generation and maintenance of synchrony in *Saccharomyces cerevisiae* continuous culture. Exp Cell Res 287:10–15
Murray DB, Beckmann M, Kitano H (2007) Regulation of yeast oscillatory dynamics. Proc Natl Acad Sci U S A 104:2241–6
Murray DB, Haynes K, Tomita M (2011) Redox regulation in respiring *Saccharomyces cerevisiae*. Biochim Biophys Acta 1810:945–58
Nagaich AK, Walker DA, Wolford R, Hager GL (2004) Rapid periodic binding and displacement of the glucocorticoid receptor during chromatin remodeling. Mol Cell 14:163–74
Naiki N, Yamagata S (1976) Isolation and some properties of copper-binding proteins found in a copper-resistant strain of yeast. Plant Cell Physiol 17:1281–95
Rapkine L (1931) Chemical processes during cellular division. Ann Physiol Physicochem 7:382
Roussel MR, Lloyd D (2007) Observation of a chaotic multioscillatory metabolic attractor by real-time monitoring of a yeast continuous culture. FEBS J 274:1011–18
Salgado LE, Murray DB, Lloyd D (2002) Some antidepressant agents (Li^+, monoamine oxidase type A inhibitors) perturb the ultradian clock in *Saccharomyces cerevisiae*. Biol Rhythm Res 33:351–61
Sasidharan K, Tomita M, Aon MA, Lloyd D, Murray DB (2012) Time structure of the yeast metabolism in vivo. Adv Exp Med Biol 736:359–79
Satroutdinov AD, Kuriyama H, Kobayashi H (1992) Oscillatory metabolism of *Saccharomyces cerevisiae* in continuous culture. FEMS Microbiol Lett 77:261–7
Slavov N, Botstein D (2011) Coupling among growth rate response, metabolic cycle, and cell division cycle in yeast. Mol Biol Cell 22:1997–2009
Slavov N, Macinskas J, Caudy A, Botstein D (2011) Metabolic cycling without cell division cycling in respiring yeast. Proc Natl Acad Sci USA 108:19090–5
Sohn H-Y, Kuriyama H (2001a) Ultradian metabolic oscillation of *Saccharomyces cerevisiae* during aerobic continuous culture: hydrogen sulphide, a population synchronizer, is produced by sulphite reductase. Yeast 18:125–35

Sohn H-Y, Kuriyama H (2001b) The role of amino acids in the regulation of hydrogen sulfide production during ultradian respiratory oscillation of *Saccharomyces cerevisiae*. Arch Microbiol 176:69–78

Sohn H-Y, Murray DB, Kuriyama H (2000) Ultradian oscillation of *Saccharomyces cerevisiae* during aerobic continuous culture: hydrogen sulphide mediates population synchrony. Yeast 16:1185–90

Sohn H-Y, Kum E-J, Kwon G-S, Jin I, Adams CA, Kuriyama H (2005a) *GLR1* plays an essential role in the homeodynamics of glutathione and the regulation of H2S production during respiratory oscillation of *Saccharomyces cerevisiae*. Biosci Biotechnol Biochem 69:2450–4

Sohn H-Y, Kum E-J, Kwon G-S, Jin I, Kuriyama H (2005b) Regulation of branched-chain, and sulfur-containing amino acid metabolism by glutathione during ultradian metabolic oscillation of *Saccharomyces cerevisiae*. J Microbiol 43:375–80

Tsankov AM, Thompson DA, Socha A, Regev A, Rando OJ (2010) The role of nucleosome positioning in the evolution of gene regulation. PLoS Biol 8:e1000414

Tu BP, Kudlicki A, Rowicka M, McKnight SL (2005) Logic of the yeast metabolic cycle: temporal compartmentalization of cellular processes. Science 310:1152–8

Vincent JA, Kwong TJ, Tsukiyama T (2008) ATP-dependent chromatin remodeling shapes the DNA replication landscape. Nat Struct Mol Biol 15:477–84

von Bertalanffy L (1952) Problems of life: an evaluation of modern biological thought. C. A. Watts & Company, London

von Meyenburg HK (1969) Energetics of the budding cycle of *Saccharomyces cerevisiae* during glucose limited aerobic growth. Arch Microbiol 66:289–303

Watson JD, Crick FH (1953) Molecular structure of nucleic acids; a structure for deoxyribose nucleic acid. Nature 171:737–8

Whitehouse I, Rando OJ, Delrow J, Tsukiyama T (2007) Chromatin remodelling at promoters suppresses antisense transcription. Nature 450:1031–5

Wolf J, Sohn H-Y, Heinrich R, Kuriyama H (2001) Mathematical analysis of a mechanism for autonomous metabolic oscillations in continuous culture of *Saccharomyces cerevisiae*. FEBS Lett 499:230–4

Yates FE (1992) Fractal applications in biology: scaling time in biochemical networks. Methods Enzymol 210:636–75

Yeung KS, Hoare M, Thornhill NF, Williams T, Vaghjiani JD (1999) Near-infrared spectroscopy for bioprocess monitoring and control. Biotechnol Bioeng 63:684–93

Zhang Z, Wippo CJ, Wal M, Ward E, Korber P, Pugh BF (2011) A packing mechanism for nucleosome organization reconstituted across a eukaryotic genome. Science 332:977–80

Chapter 13
Systems Biology and Metabolic Engineering in Bacteria

Johannes Geiselmann

Abstract Complete metabolic maps are currently available for a number of important bacteria. Even when these maps are not experimentally confirmed, the topology of the metabolic network can be reconstructed from the genome sequence. Despite this extensive information, we still lack a good understanding of metabolic adaptations, the interactions of metabolism with gene regulation and tools for predicting the metabolic consequences of modifying the metabolic or genetic regulatory network of a bacterium. This chapter will briefly review current methods for analyzing bacterial metabolism from topological models and steady state techniques, such as flux balance analysis, to dynamical models using ordinary differential equations. Even though still incomplete, these models can predict the metabolic behavior of modified organisms. Using these tools, we can create novel metabolic pathways or optimize the yield of a desired metabolite. Focusing on *Escherichia coli*, we present examples of successful metabolic engineering using such systems-wide, rational approaches, integrating modeling and experiments. The conjunction of systems biology to metabolic engineering yields new insights into the fundamental functioning of the cell and opens the path to the biological production of a large variety of commodity chemicals.

13.1 Metabolism and Systems Biology

Microbes are relatively simple organisms, they grow in very diverse environments and their genomes are sequenced. For model organisms, such as *E. coli* or *Saccharomyces cerevisiae*, the metabolic network is very well mapped and many of the genetic and metabolic regulatory interactions have been characterized. These organisms are therefore an ideal study object for systems biology with the dual goal of

J. Geiselmann (✉)
Laboratory of Adaptation and Pathogenicity of Microorganisms, University Joseph Fourier, CNRS UMR5163, Grenoble, France
e-mail: hans.geiselmann@ujf-grenoble.fr

M.A. Aon et al. (eds.), *Systems Biology of Metabolic and Signaling Networks*, Springer Series in Biophysics 16, DOI 10.1007/978-3-642-38505-6_13,

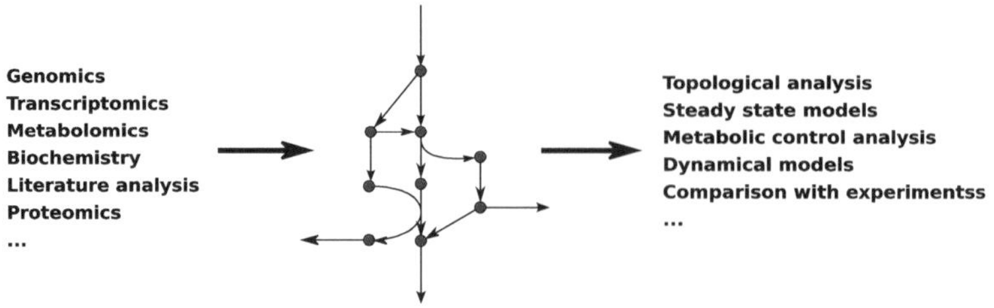

Fig. 13.1 The central role of the metabolic network. The metabolic network is central to a systems biology description of organism functioning. Numerous data sources, genetic, biochemical, and -omics data, are combined to construct a biochemical reaction network. This network describes all possible transformations between the metabolites (*blue circles*) of the organism. The transformations include inflow (e.g., glucose uptake) and outflow (e.g., acetate secretion) reactions symbolized by arrows without starting or ending metabolite. The reaction network forms the basis for all further analyses, from a topological description of the network to dynamical simulations and parameter estimation by comparison with experimental data

(1) understanding the functioning of the organism as a whole and (2) exploiting this knowledge for applications in biotechnology and synthetic biology.

The basis for understanding the functioning of microbial metabolism is a complete knowledge of the underlying metabolic network, i.e., the metabolites and the enzymes that interconvert these metabolites (Fig. 13.1). The next section argues that current -omics techniques have essentially achieved this goal. Based on this map, we can investigate the functional properties of the network at two levels: (1) during balanced growth, a hallmark of microbial physiology and (2) at transitions between different conditions, the "normal" lifestyle of most microorganisms. Finally, the metabolic network is, of course, not isolated in the cell, but embedded in a complex control structure involving metabolic, genetic, and posttranslational control mechanisms. Much progress has been made recently to incorporate all these interactions into integrated models, culminating in whole-cell models of an entire bacterium (Karr et al. 2012b). These models are not solely intellectual tools for understanding the functioning of an organism; they also form the basis of many applications in biotechnology (Feist and Palsson 2008).

13.2 Maps of Metabolic Networks in Bacteria

The first step in network reconstruction is the establishment of a reliable model of the network, i.e., identifying all components (metabolites) and their connections (enzymes and their reactions). The first metabolic models of the beginning of the 1990s were of limited scope, focusing, for example, on amino acid and nucleotide metabolism. The advent of the genomic era in the late 1990 has greatly contributed to the fast growth of the number and size of metabolic models (Kim et al. 2012). Numerous software tools have now been developed to help deriving a network connectivity and metabolic capacity of the network from genome annotation (Pitkänen et al. 2010). The preferred model organisms were *E. coli* and *S. cerevisiae*. The latest, integrated model of *E. coli* comprises almost 12,000 network components and over 13,000 reactions (Thiele et al. 2009). This model was created by the Palsson group, who has pioneered and is still at the forefront of the reconstruction of metabolic networks in *E. coli* and other organisms (Feist et al. 2009).

The second step after the reconstruction of network components and their connectivity is the comparison with available experimental data in order to validate the proposed structure. A prerequisite for the faithful reconstruction of a metabolic network is thus the availability of high-throughput, quantitative techniques for measuring the network components and their interactions (Yamada and Bork 2009). Even though the reconstruction of network connectivity is greatly facilitated by software tools, the automatic detection and repair network inconsistencies remain very difficult (Pitkänen et al. 2010). The best models still rely on manual curation.

Furthermore, even for very well studied organisms, such as *E. coli*, the experimental exploration of the metabolic network still yields surprises. Recently, Nakahigashi et al. (2009) have detected significant differences between the predictions derived from the very well established metabolic network of *E. coli* and the observed growth phenotypes of double knockout mutants. These additional reactions of the central carbon metabolism provide an alternative pathway for glucose breakdown. The remarkable fact about these new reactions is that their activation does not require any changes in gene expression. Such alternative pathways are certainly part of the features that convey robustness to metabolic networks. A purely bioinformatic analysis will almost certainly miss such reactions, reiterating the need for experimental validation of predicted network structures even for the best-studied organism. Despite certain shortcomings, these network reconstruction methods have been applied with great success to well-studied organisms (Kim et al. 2012), such as *E. coli* and *S. cerevisiae*, but also to less well studied organisms of particular fundamental or biotechnological interest, such as photosynthetic cyanobacteria (Montagud et al. 2010).

Even though the completeness of the metabolic network cannot be assured, the quality of current network reconstructions allows to pass on to the next step: calculating the phenotype produced by a metabolic network. This task consists essentially in predicting the metabolic fluxes through the network in different growth conditions. From these fluxes, we can calculate the phenotypes, such as growth rate, metabolic capacities, and yield of particular metabolite. There are two major modeling approaches for calculating the "behavior" of a metabolic network (Chen et al. 2012): (1) steady state, and (2) kinetic. The former can be applied to microorganisms during balanced growth, whereas the latter allows to assessing time-dependent responses of a microbe to changes in the environment.

13.3 Steady State Models

Steady state models attempt to predict the flux distribution during balanced growth, a condition reached in chemostats and at certain phases of a batch culture. Historically, most experiments have striven to reach balanced growth, since the constancy of the composition of the bacterial cell greatly facilitates the interpretation of the results. However, the constancy of composition is also the major experimental drawback of balanced growth. In order to obtain the large amount of data necessary to constrain a mathematical model, the experimental measurements have to be repeated under many different growth conditions.

A steady state model is a mathematical representation of the intracellular metabolic flows that can, or cannot, be observed directly. The basis on which all these models are built is the stoichiometry matrix (equivalent to the connectivity of the metabolic network, describing all possible reactions and their reaction stoichiometries) and a set of measured fluxes. These fluxes can be intracellular fluxes, which need tracer experiments, or uptake fluxes. In general, the combination of experimental datasets and topological models of the biological network leads to an under-determined system. Additional constraints have to be imposed in order to uniquely calculate the (steady state) behavior of the network. These constraints come from experimental monitoring of flux distribution, usually using 13C-labeling techniques, or predicting the flux distribution based on an objective function. The former method is called "metabolic flux analysis" (MFA), the latter "flux balance analysis" (FBA). Powerful tools for both approaches have been developed. Toya et al. (2011) have described the principles and available software tools for using MFA and FBA [see also (Dandekar et al. 2012)]. An excellent review of tools for constraint-based reconstruction and analysis methods can be found in Lewis et al. (2012).

Metabolic Flux Analysis (MFA) can be applied to metabolic networks of any complexity, e.g., cycles, parallel and reversible reactions. Labeling patterns provide very useful information, directly about the intracellular flux distribution. The procedure is normally initiated by adding the labeled compound (typically ^{13}C-glucose for central metabolism) to a steadily growing culture. The flux distribution is then derived by comparing the observed labeling pattern of metabolites with the predictions from an assumed flux distribution. The result is optimized in an iterative process of model adjustment and comparison to the experimental data. A limitation of this approach is it requires measuring fluxes under different experimental conditions. This constraint can be somewhat alleviated by parallel labeling experiments, where the same substrate carries multiple labels, or differently labeled substrates are added to the culture simultaneously (Crown and Antoniewicz 2012). With this approach specific fluxes can be measured precisely and efficiently to validate the proposed structure of a biochemical network. One disadvantage of multiple labeling is the higher cost of the experiment.

The other, commonly used technique for calculating metabolic fluxes is "flux balance analysis" (FBA). The principle is well described in Orth et al. (2010) and

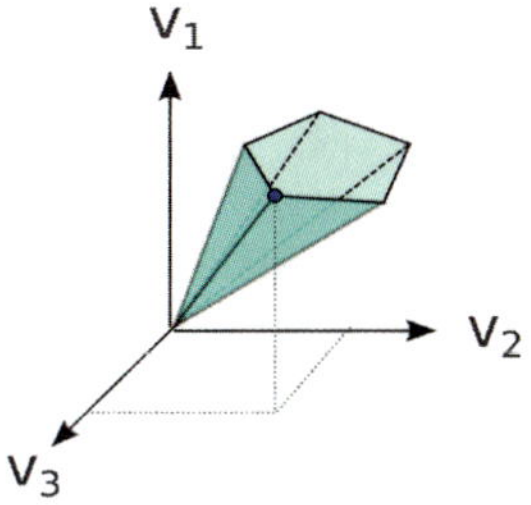

Fig. 13.2 Flux Balance Analysis. The *axes* represent metabolic fluxes. In this example, only three fluxes are considered. The possible fluxes can adopt any value of v_1, v_2, and v_3. Stoichiometry constraints and capacity constraints restrict the fluxes. The stoichiometry constraints are imposed by the reactions of the network, and capacity constraints derive from known limits of the reactions. For example, we may know that the flux v_1 has to lie between two known values: $a < v_1 < b$. Since these types of constraints are linear, possible fluxes now lie within the straight line boundaries of a polygonal cone starting at the origin. FBA then maximizes (or minimizes) an objective function (biomass, ATP production,. . .) that can be any linear combination of the fluxes. The optimal solution (*dark blue point*) lies on the edge of the allowable solution space

illustrated in Fig. 13.2. Like in MFA, the basis of FBA is the stoichiometry matrix, i.e., the complete topological description of the metabolic network. The stoichiometry matrix already imposes a first set of constraints on possible flux distributions. Additional constraints can be added, for example, in the form of maximal and minimal fluxes for a particular reaction. The second, and crucial, step of FBA consists on defining an "objective function" that will be maximized. A commonly used objective function is biomass, which amounts to adding a flux called "biomass" to the metabolic network. Different reactions can contribute to the biomass function, and the relative contributions can change according to growth conditions (Meadows et al. 2010). Mathematically, the constraints and the objective function form a system of linear equations that can be solved, using linear programming, to maximize the objective function. Similar to MFA, FBA does not require any knowledge of kinetic parameters and can rapidly be calculated even for large networks. Furthermore, prior knowledge of, e.g., reaction constants can be incorporated into the algorithm in the form of additional constraints.

The calculated optimal flux distribution depends on the choice of the objective function (Feist and Palsson 2010). Evolutionary arguments favor the biomass objective function for *E. coli* and other bacteria, at least for laboratory strains that have been grown for a long time on commonly used growth substrates. The basic operation for calculating biomass involves defining the macromolecular composition of the cell and thus the metabolites necessary for assembling these cellular constituents. This objective function can be further improved by considering the energy needed for macromolecular assembly, for example the number of ATP molecules needed for incorporating amino acids into proteins. Even more detailed formulations of the biomass include secondary metabolites such as vitamins and cofactors. Even though the biomass objective function can be calculated with high precision, different bacteria under different conditions may well be optimized for different objectives. For example, even for *E. coli*, at least five different objective functions, among which maximal biomass, ATP, or CO_2 production, are consistent with the observed flux distribution (Schuetz et al. 2012). Future research on the "real objective" of a bacterium subjected to a given growth condition will certainly prove fruitful for fundamental research as well as for biotechnological applications.

Several extensions of the basic FBA have been proposed. As early as 2002, the MOMA algorithm (minimization of metabolic adjustment) explicitly addressed the question of the objective function (Segrè et al. 2002). For wild type strains of *E. coli* it can reasonably be argued that evolution has optimized for growth (see above). However, in the case of genetically modified bacteria, where certain genes have been deleted, presumably not enough time has elapsed for metabolic network rewiring. In this context, searching for the model that minimally perturbs the fluxes of the wild type strain has proven more accurate than the pure biomass objective function. More recent work suggests that bacterial metabolism has been evolutionarily optimized for two competing goals: optimality under a given condition and minimum adjustment between conditions (Schuetz et al. 2012).

Several variants and extensions of MFA and FBA have been developed over the years. Metabolic pathway analysis adopts a pathway centered view and exploits thermodynamic and biochemical constraints to limit potential metabolic strategies (Bar-Even et al. 2012). The metabolic network is decomposed into individual pathways by elementary mode analysis (Trinh et al. 2009), thereby facilitating the analysis and prediction of cellular phenotypes, and the construction of metabolic networks. Following a related philosophy, resource balance analysis exploits the modularity of metabolic pathways and obtains good predictions for different growth phenotypes of *Bacillus subtilis* (Goelzer et al. 2011). Many of these variants are summarized in Kim et al. (2012).

MFA and FBA are descriptions of the metabolic system that predict the fluxes of metabolites across the system. They do not explicitly address the *control* of these fluxes. A powerful approach for analyzing the control structure of the metabolic system is called "metabolic control analysis" (MCA). The key concept is the "control coefficient" that describes the ratio of the relative change of a metabolic variable or function (such as a steady state flux or a metabolite concentration) and the relative change of a parameter of the system (Kremling et al. 2008; Lewis et al. 2012). The control coefficient can thus quantify, for example, the influence of enzyme concentrations on the flux of a metabolic pathway. More generally, MCA is a mathematical framework capable of relating local properties, such as enzyme activities, to global properties, such as the response of the entire system to an external perturbation. MCA not only provides a tool for predicting system behavior but also gives a deeper insight into system functioning by putting the accent on the underlying control logic. One drawback of MCA is the requirement of extensive experimental data for determining the system parameters, i.e., the control coefficients. Recent extensions to MCA, such as optimization-based MCA (OMCA) (Meadows et al. 2010), reduce the experimental burden by introducing simplifying assumptions. OMCA supposes that the metabolic network optimizes homeostasis. The fluxes of the network are thus correlated with metabolite concentration and enzyme activities. MCA and its extensions remain very useful tools for network understanding and they may gain in importance as more quantitative experimental data become available.

13.4 Dynamical Models

The natural habitat of most microorganisms is far from constant, forcing the microbe to adapt its metabolism to ever changing conditions. Even a typical batch culture in the laboratory puts the bacterium through cycles of feast and famine (Ferenci 2001). The same problem arises for production strains in industrial fermentors. In order to understand and optimize phenotypes under these conditions, we need to assess the system behavior not only during a particular steady state but also during transitions between different growth conditions. Otherwise stated, we need a dynamic description of the metabolic network.

Historically, such models derive from enzyme kinetics, a detailed description of an enzymatic reaction. The challenge for systems biology is to consider all enzymatic reactions of the cell simultaneously. Two approaches are possible for attaining this goal: (1) bottom-up, where we do precisely that: assemble the individual component reactions of a metabolic network, and (2) top-down, where we attempt to put dynamics into a genome-wide steady state model.

Generally, the first modeling approach is formulated as a system of differential equations, one equation for each enzymatic reaction (Fig. 13.3). This formalism has the advantage of being very general and leading to very precise predictions of system dynamics. However, this approach requires knowledge of (1) the topology of the metabolic and regulatory network, as well as (2) good estimates of all parameters of the kinetic equations. Such models are therefore limited to well-known organisms such as *E. coli*. Once all parameters and equations are assembled, the system behavior is calculated, "emerging" from the interactions of the individual components.

Because it is currently impossible to measure all the parameters of all the enzymatic reactions of an organism, such models focus on specific parts of the global network, for example, central metabolism. Kremling et al. (2008) have succeeded in compiling a complete model of glycolysis in *E. coli*. Such detailed models enable answering very specific questions. The Kremling model, for example, led to the discovery of specific metabolic regulatory mechanisms for system functioning. They could demonstrate the importance of a feed-forward loop linking the upper part of glycolysis to pyruvate kinase. Heinemann et al. have extended this approach and constructed a complete model of the central catabolism of *E. coli* (Kotte et al. 2010). One major advantage of such models is that metabolic and genetic regulatory mechanisms can be integrated seamlessly into a unified system description (also see below). The surprising result of Kotte and coworkers was that metabolic adaptations rely on distributed sensing of metabolic fluxes, where metabolites such as fructose-biphosphate or cAMP play key roles as flux sensors.

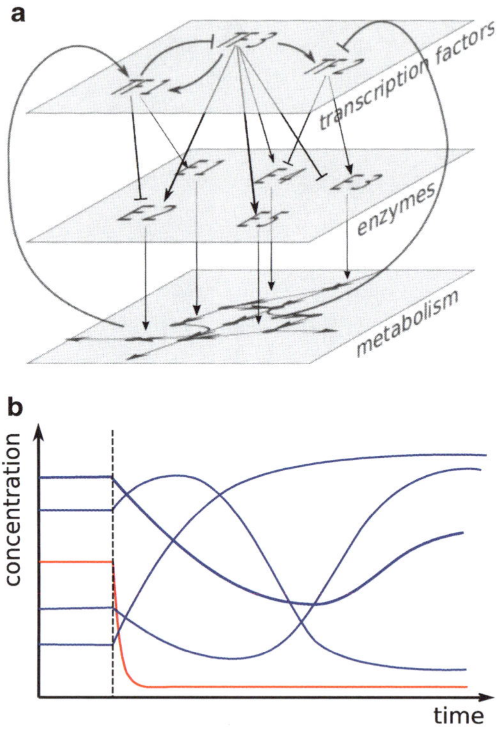

Fig. 13.3 Dynamical models of metabolism. While FBA (Fig. 13.2) can predict metabolite concentrations at steady state, i.e., the long-term behavior of a system in a constant environment, many important biological phenomena involve the transition between growth conditions. Only dynamical models can represent the system behavior in these situations. (**a**) In addition to the metabolic network (*bottom plane in the figure*), such models incorporate varying enzyme concentrations (*second plane*) and the regulation of gene expression by transcription factors or other mechanisms of signal transduction (*top plane*). The different levels of regulation have been separated here for better visualization, but all mechanisms are intertwined in the cell. For example, certain metabolites (such as cAMP) affect gene expression by modulating the activity of transcription factors (*arrows from the bottom plane to the top plane*). These models are formulated as a system of differential equations, one equation per system component. (**b**) The simulation of such a model predicts the time-varying concentration of each system component. Starting from a system at steady state (and therefore constant concentration of all components), an external parameter (e.g., the carbon source available in the growth medium) is varied at a specific time (indicated by the *dashed vertical line*). The rate of change of proteins (transcription factors and enzymes) is slow (represented by *blue lines* in the figure), whereas the rate of change of metabolite concentrations can be very rapid (*red line*)

This mode of "environmental detection" allows autonomous metabolic adaptation to changing environmental conditions, and has important implications for the construction and functioning of modified or artificial metabolic networks.

These large models of central catabolic pathways in *E. coli* represent probably the current limit of the pure differential equation-based approach to modeling the dynamics of metabolic networks, simply because quantitative data about enzymatic reactions are difficult to obtain. Furthermore, such data are usually measured in vitro and their relevance for the in vivo situation remains questionable. Attempts have therefore been made to extend the genomic-scale, steady state metabolic models by adding dynamics in different ways. As mentioned above, FBA, for example, strongly depends on the objective function and the constraints imposed on the system. Since these change during a typical growth experiment, making them time dependent leads to dynamic flux balancing. Meadows and colleagues have used this approach to successfully model *E. coli* fermentation in an industrial bioreactor, using FBA and time-dependent inputs, such as rate-dependent biomass composition (Meadows et al. 2010).

In principle, the data used for steady state network analysis could also be used to derive kinetic constants, and therefore a dynamic model of the network. Classic dynamical models are formulated as systems of nonlinear differential equations. Recovering all parameters of such a complicated system seems overambitious. However, in many situations, the nonlinear differential equation system is well approximated by lin-log, power-law, or S-system descriptions (Heijnen 2005). Using high-throughput data (Ishii et al. 2007), where substrate and reactant concentrations were quantified along with corresponding reaction fluxes in a series of steady state perturbation experiments, Berthoumieux et al. (2012) explored the possibility of identifying the parameters of the corresponding lin-log model. They concluded that even with such an extensive dataset and the linearizing model simplifications, only four out of 31 reactions, and 37 out of 100 parameters were identifiable.

Another approach was to take established steady state, genome-scale models and incorporate available kinetic information (Jamshidi and Palsson 2010). Enzymes are represented explicitly in this formulation. Different steady state measurements are used to estimate equilibrium constants, and subsequently kinetic constants of individual enzymatic reactions. These constants are then incorporated in the kinetic model using bilinear equations. This model faithfully describes system responses to external perturbations such as changes in the energy or redox state of the cell. The modeling algorithm makes extensive use of -omics data. Model construction thus profits from the vast amount of such data that are publicly available and avoids the "tedious" measurement of individual reaction constants necessary for the classical modeling approach. However, classical enzyme kinetic models, when available, are still a more faithful representation of system kinetics. Important biological phenomena, such as regulatory interactions via metabolites or second messengers, cannot be represented by the intrinsically linear approximation of FBA. Moreover, collective phenomena such as self-organization or synchronized oscillations are certainly beyond the realm of MFA or FBA.

13.4.1 Systems that Require Dynamical Models

These phenomena can only be adequately represented by nonlinear differential equations because the system behavior would simply be lost in a linear approximation. Models of such emergent phenomena are needed not only for understanding system properties but also for constructing novel systems by methods of synthetic biology. The first synthetic oscillatory circuit constructed and modeled was probably the "repressilator," composed of three regulators that repress each other in a cyclic arrangement (Elowitz and Leibler 1999). A more complex synthetic, oscillatory circuit, the "metabolator" combines genetic and metabolic regulations. The expression of the enzymes that interconvert acetyl coenzyme A and acetyl phosphate is regulated by the concentration of the second metabolite, acetyl phosphate (Fung et al. 2005). The nonlinear ODE model of the system correctly describes the observed oscillatory behavior, predicting properties such as the frequency of the oscillations as a function of the carbon flux through glycolysis. The synthetic constructs can be further extended to incorporate cell–cell communication (Song et al. 2008). An artificial predator–prey construct combines two strains of *E. coli*. Both strains produce a toxin molecule that kills the cell transcribing the corresponding gene. The predator strain sends a diffusible molecule to the prey strain, eliciting there the production of the toxin. The prey strain produces a different diffusible molecule that activates the production of an antidote in the predator strain. The predator thus depends on the prey for survival, but the prey will be killed in the presence of many predators. The observed population dynamics are oscillations of the number of predators and prey in anti-phase, as predicted by the dynamical model. The key to success of all these projects was parameter optimization of the system components based on a detailed dynamical model of the synthetic system. The number of system components was still relatively modest, which permitted the construction of a complete, nonlinear ODE model.

Intrinsically, nonlinear phenomena also govern major aspects of the "natural" physiology of the cell. Metabolic oscillations have been observed over 50 years ago and the first models describing these phenomena were developed soon thereafter (Song et al. 2008). Such oscillations concern many microorganisms and metabolites: for example, amidase activity in *Pseudomonas aeruginosa*, cAMP in *Dictyostelium discoideum*, lactose metabolism in *E. coli*, and glycolytic oscillations in yeast. This latter phenomenon has been studied in detail experimentally and theoretically. Continuous cultures of yeast spontaneously synchronize their cell physiology. During the 40 min of growth (corresponding to one mass doubling) in constant environmental conditions, the population traverses a cycle comprising a reductive and an oxidative phase (Klevecz et al. 2004). This metabolic oscillation affects all major cellular activities, ATP:ADP ratio, transcription, replication, and cell growth. A recent model proposes that the observed periodicity of cell physiology is controlled by ATP-dependent nucleosome remodeling (Thiele et al. 2009). The alternate anabolic and catabolic phases of the yeast metabolism could strongly affect the efficiency of biotechnological applications in this organism. However, these nonlinearities are not taken into account in current models of the yeast metabolism or in biotechnological applications aimed at maximizing product yield.

The collective behavior of cells has been exploited in synthetic biology to stabilize engineered circuits in a noisy cellular environment. Contrary to natural oscillators, which are extremely robust (Toya et al. 2011), synthetic circuits, such as the repressilator (Elowitz and Leibler 1999), are very sensitive to fluctuations of cellular components (noise). In the case of this oscillator, daughter cells may or may not inherit the phase and frequency of the mother cell's oscillation. Theory and modeling tell us that one way to remedy this shortcoming is to couple many unstable oscillators. The collection of cells should produce stable and precise oscillations. This prediction has recently been validated experimentally in *E. coli* (Prindle et al. 2012). These authors have constructed bacteria containing an oscillatory circuit responding to arsenite in the medium. Individual cells were coupled by the transmission of two diffusible molecules: a quorum-sensing molecule acyl-homoserine lactone (AHL) and the redox signaling molecule H_2O_2, produced by the periodic expression of NADH dehydrogenase. The bacteria were placed in a microfluidics device where AHL provided the intra-channel, short-range communication and H_2O_2, diffusing rapidly in the gas phase, was responsible for the coordination between channels. The device produced a very stable oscillation, coherent across a distance of 5 mm, the frequency of which revealed the concentration of arsenite in the medium. An explicit, partial differential equations model was used for the construction and optimization of the system. Explicit, dynamical models, taking into account nonlinear interactions in time and space, are necessary for understanding and engineering such systems. The increasing number of standard parts for synthetic biology along with the measurement of relevant parameters of these parts will allow modeling and construction of large, complex, nonlinear systems such as the one described above (Prindle et al. 2012).

13.5 Genetic Regulation of Metabolism

The integration of metabolism with gene regulation is important for metabolic engineering because changing metabolite concentrations affect gene expression, which, in turn, modulates enzyme activities (Keasling 2012). A complete metabolic model should therefore include the dynamics of metabolism and the connections to the genetic network and other regulatory mechanisms. This can be achieved in the differential equation formalism, for example, the Heinemann model mentioned above (Heinemann and Sauer 2010). A similar differential equation-based modeling approach of central nitrogen metabolism in *E. coli* predicts complex response patterns of the bacterium to diverse external and internal perturbations (Yuan et al. 2009). Modeling is here successfully used as a discovery tool of hitherto unknown regulatory mechanisms.

The improved FBA formalism of (Jamshidi and Palsson 2010) incorporates regulation into the steady state model, resulting in mass action stoichiometric simulation (MASS) models. Enzymes, and their changes in activity, are explicitly represented in the model. However, the model remains a linear model and the description of regulatory effects is limited to small perturbations of the steady state. Such a description will never be capable of describing transient behavior or oscillations. Another promising formalism, termed IOMA (integrated omics metabolic analysis) (Yizhak et al. 2010), uses the stoichiometry matrix as in FBA, but complements the description by Michaelis–Menten type kinetic rate equations. The model predictions are compared to proteomic and metabolomic data, and the optimal solution is obtained by quadratic programming. This methodology compares very well with other available algorithms for standard data sets in *E. coli*.

These success stories are not limited to *E. coli*. Szappanos et al. (2011) built an integrated model of yeast metabolism, including regulatory interactions, by quantitatively measuring interactions between more than 180,000 gene pairs encoding metabolic enzymes. They combined the regulatory model with the established metabolic network and developed a machine learning algorithm for comparing experimental data with model predictions. See Gerosa and Sauer (2011) and Reaves and Rabinowitz (2011) for recent reviews on integration of metabolism with different kinds of cellular regulatory mechanisms.

Enormous progress has been made with integrated models for bacteria with small genomes, in particular *Mycoplasma* species. These bacteria have a greatly reduced genome, containing only between 500 and 700 genes. Because they live in a relatively constant environment, their metabolism is simpler than that of larger bacteria such as *E. coli* or *Bacillus subtilis*. The integration of experimental data with a metabolic model of *Mycoplasma pneumoniae* has shown remarkable predictive power (Yus et al. 2009). The whole genome, integrated model of *Mycoplasma genitalium*, a human urogenital parasite containing only 525 genes, has recently been completed (Karr et al. 2012b). The same group has also developed a generic tool for assembling such models (Karr et al. 2012a). Even though the task is simplified by the size of the genome, eventually, models of comparable detail and predictive power will certainly become available for larger bacteria. The conceptual tools are largely in place.

13.6 Applications: Modifying Existing Networks and De Novo Design of Metabolic Pathways

As shown already by several examples above, modeling metabolic networks of microorganisms, and integrating these networks with regulatory interactions in the cell has led to new, fundamental functional mechanistic insights in these organisms. In addition, the knowledge can be used to rationally modify existing networks, or design networks de novo, for biotechnological applications (Oberhardt et al. 2009). As early as 2003, an algorithm called OptKnock for determining optimal gene knockouts to improve specific metabolic functions was developed and successfully applied to maximize respiration rates in *Geobacter sulfurreducens* (Burgard et al. 2003).

The construction of synthetic pathways is another class of applications that arise as we build upon knowledge gained from understanding the function of metabolic networks (Xu et al. 2012). The preferred chassis organisms for these synthetic biology projects remain *E. coli* and *S. cerevisiae* (Na et al. 2010). Utilizing databases of metabolic reactions similar to the ones employed for the reconstruction of metabolic networks from genome annotations, Parkya and collaborators proposed the OptStrain algorithm for pathway optimization by eliminating superfluous reactions or constructing novel pathways in *E. coli* (Pharkya et al. 2004). The additional dimension offered by synthetic biology is the possibility to explore many variants of a particular metabolic pathway. A particularly clever method has been pioneered by the Church group (Wang et al. 2009), according to which several billion variants of a given pathway can be explored in parallel within a couple of days. This strategy of combining synthetic biology with accelerated evolution has considerably improved the efficiency of the lycopen production pathway in *E. coli*. The question of which strategy—rational design of a pathway and fine-tuning of intermediate reaction steps or combinatorial exploration of a large number of variants of a particular pathway—is more efficient for the production of new chemicals remains open (Yadav and Stephanopoulos 2010). A combination of both strategies may prove the most promising (see following paragraph).

No matter how sophisticated the rational design of a genetic-metabolic network may be, there will always be "bugs" when the circuit is constructed in the host cell. A first remedy would be to devise a method for easily detecting the problems. For example, the imbalance of metabolic pathways often induces stress responses. The signature of these responses could be used in future diagnostic tools for strain optimization (Keasling 2012). Even though nonlinear models of complex systems are essential for designing gene-metabolic systems, experimental strategies will be needed for the final optimization of the construct. Recent experimental advances allow a combinatorial exploration of diverse expression levels of the constituent enzymes. A high throughput screen is used to select the "best" strain. This strategy has been used to maximize xylose and cellobiose utilization in yeast (Du et al. 2012). Xylose is an abundant pentose, but is inefficiently utilized by ethanol producing yeast strains. Optimal xylose assimilation relies on the balancing of enzymatic activities and cofactor usage. The best expression levels of the three key enzymes constituting the xylose assimilation pathway was obtained by screening strains (based on colony size), each one containing different combinations of about ten promoters of different strengths placed upstream of each of the three genes comprising the pathway. The optimized strain improved ethanol yield by more than 60 % with respect to the reference strain. A decisive advantage of the experimental approach to optimization over a purely computational approach is the possibility to adapt to varying behaviors of different strains. Indeed, the transcriptional profile of two different strains optimized for the same pathway is different (Du et al. 2012). Such fine adjustments are difficult, if not impossible, to predict from modeling alone. A combination of modeling and combinatorial exploration may therefore prove to be the most effective strategy to strain optimization (Yadav and Stephanopoulos 2010).

In addition to fine-tuning the expression of individual genes, modifying the global gene expression machinery can prove even more efficient in optimizing product yield. In a proof of concept study, Alper and Stephanopoulos showed that mutations in the major sigma factor of *E. coli*, σ^{70}, outperform traditional metabolic engineering approaches for improving ethanol tolerance or lycopene production (Alper and Stephanopoulos 2007). Our recent results in *E. coli* confirm the importance of global versus gene-specific regulation (Berthoumieux et al. 2013). We have measured the relative importance of global versus gene-specific factors for the regulation of promoter activities at growth transitions. Contrary to the commonly accepted paradigm that attributes a major importance to gene-specific regulations, we find that even for global regulators these interactions serve "only" to fine-tune the transcription of the target gene. These results emphasize the need for developing global, integrated models of gene expression and metabolism.

13.7 Conclusions and a Prospective

The pace of research in systems biology of microbial metabolism has tremendously accelerated since the availability of genome sequences and the deluge of -omics data (metabol-, prote-, transcript-omics). The different modeling approaches, top-down and bottom-up, are progressing rapidly and both ends will meet in the near future. Whole-cell models of an organism, including metabolism and all levels of regulation, have already become a reality. The advances made with small bacteria, such as *M. genitalium*, will have to be transposed to larger and more experimentally accessible bacteria, such as *E. coli*. These conceptual and experimental advances have already led the way to new fundamental discoveries about the functioning of microorganisms and the concepts and techniques are being exploited in biotechnological and industrial applications. The last decade of research has clearly demonstrated that a true understanding of a biological system necessarily involves mathematical modeling. Many modeling and experimental tools are available and are continuously improved. We can now ask many important questions about the functioning of an organism and obtain the answers.

The promise of combining synthetic biology and metabolic engineering is the design and construction of microorganisms that transform a starting chemical into a desired product: microorganisms will serve as "chemical factories." The tools for analyzing metabolic networks and calculating potential flux distributions are available (FBA, etc.). The database of metabolic networks of microorganisms is rapidly expanding. However, predictive, integrated models of metabolism and genetic regulation are still scarce. The major challenge for the future consists in developing such nonlinear models for systems of biotechnological interest. However, reliable models require good parameter estimates, and obtaining parameters for all cellular reactions remains a tantamount endeavor. The task is simplified by the development of reusable modules, in line with the biobricks of iGEM. The basic modules can be well characterized and then assembled in different ways to obtain the desired circuit. The development of computer-aided design (CAD) tools will be necessary to allow assembly and in silico testing of the new circuit. These CAD tools should not only help in vector and chromosome construction (as is generally the case today), but also include functions for predicting fluxes and help in designing the optimal regulatory interactions. Once assembled in the organism, high throughput screening methods will have to be used to fine-tune and debug the de novo designed system. No fundamental obstacles separate us from this future.

References

Alper H, Stephanopoulos G (2007) Global transcription machinery engineering: a new approach for improving cellular phenotype. Metab Eng 9(3):258–67

Bar-Even A, Flamholz A, Noor E, Milo R (2012) Rethinking glycolysis: on the biochemical logic of metabolic pathways. Nat Chem Biol 8(6):509–17

Berthoumieux S, Brilli M, Kahn D, Jong H, Cinquemani E (2012) On the identifiability of metabolic network models. J Math Biol. doi:10.1007/s00285-012-0614-x

Berthoumieux S, de Jong H, Baptist G, Pinel C, Ranquet C, Ropers D, Geiselmann J (2013) Shared control of gene expression in bacteria by transcription factors and global physiology of the cell. Mol Syst Biol 9:634

Burgard AP, Pharkya P, Maranas CD (2003) Optknock: a bilevel programming framework for identifying gene knockout strategies for microbial strain optimization. Biotechnol Bioeng 84 (6):647–57

Chen N, del Val IJ, Kyriakopoulos S, Polizzi KM, Kontoravdi C (2012) Metabolic network reconstruction: advances in in silico interpretation of analytical information. Curr Opin Biotechnol 23(1):77–82

Crown SB, Antoniewicz MR (2012) Parallel labeling experiments and metabolic flux analysis: past, present and future methodologies. Metab Eng 16:21–32

Dandekar T, Fieselmann A, Majeed S, Ahmed Z (2012) Software applications toward quantitative metabolic flux analysis and modeling. Brief Bioinform. doi:10.1093/bib/bbs065

Du J, Yuan Y, Si T, Lian J, Zhao H (2012) Customized optimization of metabolic pathways by combinatorial transcriptional engineering. Nucleic Acids Res 40(18):e142

Elowitz MB, Leibler S (1999) A synthetic oscillatory network of transcriptional regulators. Nature 403:335–8

Feist AM, Palsson BØ (2008) The growing scope of applications of genome-scale metabolic reconstructions using *Escherichia coli*. Nat Biotechnol 26(6):659–67

Feist AM, Palsson BO (2010) The biomass objective function. Curr Opin Microbiol 13(3):344–9

Feist AM, rd Herrgå MJ, Thiele I, Reed JL, Palsson BØ (2009) Reconstruction of biochemical networks in microorganisms. Nat Rev Microbiol 7(2):129–43

Ferenci T (2001) Hungry bacteria: definition and properties of a nutritional state. Environ Microbiol 3(10):605–11

Fung E, Wong WW, Suen JK, Bulter T, Lee SG, Liao JC (2005) A synthetic gene-metabolic oscillator. Nature 435(7038):118–22

Gerosa L, Sauer U (2011) Regulation and control of metabolic fluxes in microbes. Curr Opin Biotechnol 22(4):566–75

Goelzer A, Fromion V, Scorletti G (2011) Cell design in bacteria as a convex optimization problem. Automatica 47(6):1210–18

Heijnen JJ (2005) Approximative kinetic formats used in metabolic network modeling. Biotechnol Bioeng 91(5):534–45
Heinemann M, Sauer U (2010) Systems biology of microbial metabolism. Curr Opin Microbiol 13:1–7
Ishii N, Nakahigashi K, Baba T, Robert M, Soga T, Kanai A, Hirasawa T, Naba M, Hirai K, Hoque A, Ho PY, Kakazu Y, Sugawara K, Igarashi S, Harada S, Masuda T, Sugiyama N, Togashi T, Hasegawa M, Takai Y, Yugi K, Arakawa K, Iwata N, Toya Y, Nakayama Y, Nishioka T, Shimizu K, Mori H, Tomita M (2007) Multiple high-throughput analyses monitor the response of *E. coli* to perturbations. Science 316(5824):593–7
Jamshidi N, Palsson BØ (2010) Mass action stoichiometric simulation models: incorporating kinetics and regulation into stoichiometric models. Biophys J 98(2):175–85
Karr JR, Sanghvi JC, Macklin DN, Arora A, Covert MW (2012a) WholeCellKB: model organism databases for comprehensive whole-cell models. Nucleic Acids Res 41:787–92
Karr JR, Sanghvi JC, Macklin DN, Gutschow MV, Jacobs JM, Bolival B, Assad-Garcia N, Glass JI, Covert MW (2012b) A whole-cell computational model predicts phenotype from genotype. Cell 150(2):389–401
Keasling JD (2012) Synthetic biology and the development of tools for metabolic engineering. Metab Eng 14(3):189–95
Kim TY, Sohn SB, Kim YB, Kim WJ, Lee SY (2012) Recent advances in reconstruction and applications of genome-scale metabolic models. Curr Opin Biotechnol 23(4):617–23
Klevecz RR, Bolen J, Forrest G, Murray DB (2004) A genomewide oscillation in transcription gates DNA replication and cell cycle. Proc Natl Acad Sci USA 101(5):1200–05
Kotte O, Zaugg JB, Heinemann M (2010) Bacterial adaptation through distributed sensing of metabolic fluxes. Mol Syst Biol 6:355
Kremling A, Bettenbrock K, Gilles ED (2008) A feed-forward loop guarantees robust behavior in *Escherichia coli* carbohydrate uptake. Bioinformatics 24(5):704–10
Lewis NE, Nagarajan H, Palsson BO (2012) Constraining the metabolic genotype-phenotype relationship using a phylogeny of in silico methods. Nat Rev Microbiol 10(4):291–305
Meadows AL, Karnik R, Lam H, Forestell S, Snedecor B (2010) Application of dynamic flux balance analysis to an industrial *Escherichia coli* fermentation. Metab Eng 12(2):150–60
Montagud A, Navarro E, Navarro E, Fernández de Córdoba P, Urchueguía JF, Patil KR (2010) Reconstruction and analysis of genome-scale metabolic model of a photosynthetic bacterium. BMC Syst Biol 4(1):156
Na D, Kim TY, Lee SY (2010) Construction and optimization of synthetic pathways in metabolic engineering. Curr Opin Microbiol 13(3):363–70
Nakahigashi K, Toya Y, Ishii N, Soga T, Hasegawa M, Watanabe H, Takai Y, Honma M, Mori H, Tomita M (2009) Systematic phenome analysis of *Escherichia coli* multiple-knockout mutants reveals hidden reactions in central carbon metabolism. Mol Syst Biol 5:306
Oberhardt MA, Palsson BØ, Papin JA (2009) Applications of genome-scale metabolic reconstructions. Mol Syst Biol 5(320):320
Orth JD, Thiele I, Palsson BØ (2010) What is flux balance analysis? Nat Biotechnol 28(3):245–48
Pharkya P, Burgard AP, Maranas CD (2004) OptStrain: a computational framework for redesign of microbial production systems. Genome Res 14(11):2367–76
Pitkänen E, Rousu J, Ukkonen E (2010) Computational methods for metabolic reconstruction. Curr Opin Biotechnol 21(1):70–77
Prindle A, Samayoa P, Razinkov I, Danino T, Tsimring LS, Hasty J (2012) A sensing array of radically coupled genetic 'biopixels'. Nature 481(7379):39–44
Reaves ML, Rabinowitz JD (2011) Metabolomics in systems microbiology. Curr Opin Biotechnol 22(1):17–25
Schuetz R, Zamboni N, Zampieri M, Heinemann M, Sauer U (2012) Multidimensional optimality of microbial metabolism. Science 336(6081):601–04
Segrè D, Vitkup D, Church GM (2002) Analysis of optimality in natural and perturbed metabolic networks. Proc Natl Acad Sci USA 99(23):15112–17

Song H, Ozaki J, Collins CH, Barnet M, Arnold FH, Quake SR, Balagaddé FK, You L (2008) A synthetic *Escherichia coli* predator–prey ecosystem. Mol Syst Biol 4(187):187

Szappanos B, Kovács K, Szamecz B, Honti F, Costanzo M, Baryshnikova A, Gelius-Dietrich G, Lercher MJ, Jelasity M, Myers CL, Andrews BJ, Boone C, Oliver SG, Pál C, Papp B (2011) An integrated approach to characterize genetic interaction networks in yeast metabolism. Nat Genet 43(7):656–62

Thiele I, Jamshidi N, Fleming RMT, Palsson BØ (2009) Genome-scale reconstruction of *Escherichia coli*'s transcriptional and translational machinery: a knowledge base, its mathematical formulation, and its functional characterization. PLoS Comput Biol 5(3):e1000312

Toya Y, Kono N, Arakawa K, Tomita M (2011) Metabolic flux analysis and visualization. J Proteome Res 10(8):3313–23

Trinh CT, Wlaschin A, Srienc F (2009) Elementary mode analysis: a useful metabolic pathway analysis tool for characterizing cellular metabolism. Appl Microbiol Biotechnol 81(5):813–26

Wang HH, Isaacs FJ, Carr PA, Sun ZZ, Xu G, Forest CR, Church GM (2009) Programming cells by multiplex genome engineering and accelerated evolution. Nature 460(7257):894–8

Xu P, Vansiri A, Bhan N, Koffas MAG (2012) ePathBrick: a synthetic biology platform for engineering metabolic pathways in *E. coli*. ACS Synth Biol 1(7):256–66

Yadav VG, Stephanopoulos G (2010) Reevaluating synthesis by biology. Curr Opin Microbiol 13 (3):371–6

Yamada T, Bork P (2009) Evolution of biomolecular networks: lessons from metabolic and protein interactions. Nat Rev Mol Cell Biol 10(11):791–803

Yizhak K, Benyamini T, Liebermeister W, Ruppin E, Shlomi T (2010) Integrating quantitative proteomics and metabolomics with a genome-scale metabolic network model. Bioinformatics 26(12):i255–i60

Yuan J, Doucette CD, Fowler WU, Feng XJ, Piazza M, Rabitz HA, Wingreen NS, Rabinowitz JD (2009) Metabolomics-driven quantitative analysis of ammonia assimilation in *E. coli*. Mol Syst Biol 5(302):302

Yus E, Maier T, Michalodimitrakis K, van Noort V, Yamada T, Chen W-H, Wodke JAH, Wodke JAH, Güell M, Martínez S, Bourgeois R, Kühner S, Raineri E, Letunic I, Kalinina OV, Rode M, Herrmann R, Gutiérrez-Gallego R, Russell RB, Gavin A-C, Bork P, Serrano L (2009) Impact of genome reduction on bacterial metabolism and its regulation. Science 326(5957):1263–8

Index

M.A. Aon et al. (eds.), *Systems Biology of Metabolic and Signaling Networks*,
Springer Series in Biophysics 16, DOI 10.1007/978-3-642-38505-6,

D

Printed by Printforce, the Netherlands